LONDON MATHEMATICAL SOCIETY LECTURE NOTE SERIES

Managing Editor: Professor N.J. Hitchin, Mathematics Institute,
University of Oxford, 24–29 St Giles, Oxford OX1 3TG, United Kingdom

The titles below are available from booksellers, or, in case of difficulty, from Cambridge University Press.

46 *p*-adic Analysis: a short course on recent work, N. KOBLITZ
59 Applicable differential geometry, M. CRAMPIN & F.A.E. PIRANI
66 Several complex variables and complex manifolds II, M.J. FIELD
86 Topological topics, I.M. JAMES (ed)
87 Surveys in set theory, A.R.D. MATHIAS (ed)
88 FPF ring theory, C. FAITH & S. PAGE
89 An F-space sampler, N.J. KALTON, N.T. PECK & J.W. ROBERTS
90 Polytopes and symmetry, S.A. ROBERTSON
92 Representations of rings over skew fields, A.H. SCHOFIELD
93 Aspects of topology, I.M. JAMES & E.H. KRONHEIMER (eds)
96 Diophantine equations over function fields, R.C. MASON
97 Varieties of constructive mathematics, D.S. BRIDGES & F. RICHMAN
98 Localization in Noetherian rings, A.V. JATEGAONKAR
99 Methods of differential geometry in algebraic topology, M. KAROUBI & C. LERUSTE
100 Stopping time techniques for analysts and probabilists, L. EGGHE
104 Elliptic structures on 3-manifolds, C.B. THOMAS
105 A local spectral theory for closed operators, I. ERDELYI & WANG SHENGWANG
107 Compactification of Siegel moduli schemes, C.-L. CHAI
109 Diophantine analysis, J. LOXTON & A. VAN DER POORTEN (eds)
113 Lectures on the asymptotic theory of ideals, D. REES
114 Lectures on Bochner-Riesz means, K.M. DAVIS & Y.-C. CHANG
116 Representations of algebras, P.J. WEBB (ed)
119 Triangulated categories in the representation theory of finite-dimensional algebras, D. HAPPEL
121 Proceedings of *Groups - St Andrews 1985*, E. ROBERTSON & C. CAMPBELL (eds)
128 Descriptive set theory and the structure of sets of uniqueness, A.S. KECHRIS & A. LOUVEAU
130 Model theory and modules, M. PREST
131 Algebraic, extremal & metric combinatorics, M.-M. DEZA, P. FRANKL & I.G. ROSENBERG (eds)
132 Whitehead groups of finite groups, ROBERT OLIVER
133 Linear algebraic monoids, MOHAN S. PUTCHA
134 Number theory and dynamical systems, M. DODSON & J. VICKERS (eds)
137 Analysis at Urbana, I, E. BERKSON, T. PECK, & J. UHL (eds)
138 Analysis at Urbana, II, E. BERKSON, T. PECK, & J. UHL (eds)
139 Advances in homotopy theory, S. SALAMON, B. STEER & W. SUTHERLAND (eds)
140 Geometric aspects of Banach spaces, E.M. PEINADOR & A. RODES (eds)
141 Surveys in combinatorics 1989, J. SIEMONS (ed)
144 Introduction to uniform spaces, I.M. JAMES
146 Cohen-Macaulay modules over Cohen-Macaulay rings, Y. YOSHINO
148 Helices and vector bundles, A.N. RUDAKOV *et al*
149 Solitons, nonlinear evolution equations and inverse scattering, M. ABLOWITZ & P. CLARKSON
150 Geometry of low-dimensional manifolds 1, S. DONALDSON & C.B. THOMAS (eds)
151 Geometry of low-dimensional manifolds 2, S. DONALDSON & C.B. THOMAS (eds)
152 Oligomorphic permutation groups, P. CAMERON
153 L-functions and arithmetic, J. COATES & M.J. TAYLOR (eds)
155 Classification theories of polarized varieties, TAKAO FUJITA
156 Twistors in mathematics and physics, T.N. BAILEY & R.J. BASTON (eds)
158 Geometry of Banach spaces, P.F.X. MÜLLER & W. SCHACHERMAYER (eds)
159 Groups St Andrews 1989 volume 1, C.M. CAMPBELL & E.F. ROBERTSON (eds)
160 Groups St Andrews 1989 volume 2, C.M. CAMPBELL & E.F. ROBERTSON (eds)
161 Lectures on block theory, BURKHARD KÜLSHAMMER
162 Harmonic analysis and representation theory, A. FIGA-TALAMANCA & C. NEBBIA
163 Topics in varieties of group representations, S.M. VOVSI
164 Quasi-symmetric designs, M.S. SHRIKANDE & S.S. SANE
166 Surveys in combinatorics, 1991, A.D. KEEDWELL (ed)
168 Representations of algebras, H. TACHIKAWA & S. BRENNER (eds)
169 Boolean function complexity, M.S. PATERSON (ed)
170 Manifolds with singularities and the Adams-Novikov spectral sequence, B. BOTVINNIK
171 Squares, A.R. RAJWADE
172 Algebraic varieties, GEORGE R. KEMPF
173 Discrete groups and geometry, W.J. HARVEY & C. MACLACHLAN (eds)
174 Lectures on mechanics, J.E. MARSDEN
175 Adams memorial symposium on algebraic topology 1, N. RAY & G. WALKER (eds)
176 Adams memorial symposium on algebraic topology 2, N. RAY & G. WALKER (eds)
177 Applications of categories in computer science, M. FOURMAN, P. JOHNSTONE & A. PITTS (eds)
178 Lower K- and L-theory, A. RANICKI
179 Complex projective geometry, G. ELLINGSRUD *et al*
180 Lectures on ergodic theory and Pesin theory on compact manifolds, M. POLLICOTT
181 Geometric group theory I, G.A. NIBLO & M.A. ROLLER (eds)
182 Geometric group theory II, G.A. NIBLO & M.A. ROLLER (eds)

183 Shintani zeta functions, A. YUKIE
184 Arithmetical functions, W. SCHWARZ & J. SPILKER
185 Representations of solvable groups, O. MANZ & T.R. WOLF
186 Complexity: knots, colourings and counting, D.J.A. WELSH
187 Surveys in combinatorics, 1993, K. WALKER (ed)
188 Local analysis for the odd order theorem, H. BENDER & G. GLAUBERMAN
189 Locally presentable and accessible categories, J. ADAMEK & J. ROSICKY
190 Polynomial invariants of finite groups, D.J. BENSON
191 Finite geometry and combinatorics, F. DE CLERCK *et al*
192 Symplectic geometry, D. SALAMON (ed)
194 Independent random variables and rearrangement invariant spaces, M. BRAVERMAN
195 Arithmetic of blowup algebras, WOLMER VASCONCELOS
196 Microlocal analysis for differential operators, A. GRIGIS & J. SJÖSTRAND
197 Two-dimensional homotopy and combinatorial group theory, C. HOG-ANGELONI *et al*
198 The algebraic characterization of geometric 4-manifolds, J.A. HILLMAN
199 Invariant potential theory in the unit ball of $\mathbf{C}^n$, MANFRED STOLL
200 The Grothendieck theory of dessins d'enfant, L. SCHNEPS (ed)
201 Singularities, JEAN-PAUL BRASSELET (ed)
202 The technique of pseudodifferential operators, H.O. CORDES
203 Hochschild cohomology of von Neumann algebras, A. SINCLAIR & R. SMITH
204 Combinatorial and geometric group theory, A.J. DUNCAN, N.D. GILBERT & J. HOWIE (eds)
205 Ergodic theory and its connections with harmonic analysis, K. PETERSEN & I. SALAMA (eds)
207 Groups of Lie type and their geometries, W.M. KANTOR & L. DI MARTINO (eds)
208 Vector bundles in algebraic geometry, N.J. HITCHIN, P. NEWSTEAD & W.M. OXBURY (eds)
209 Arithmetic of diagonal hypersurfaces over finite fields, F.Q. GOUVÊA & N. YUI
210 Hilbert C*-modules, E.C. LANCE
211 Groups 93 Galway / St Andrews I, C.M. CAMPBELL *et al* (eds)
212 Groups 93 Galway / St Andrews II, C.M. CAMPBELL *et al* (eds)
214 Generalised Euler-Jacobi inversion formula and asymptotics beyond all orders, V. KOWALENKO *et al*
215 Number theory 1992–93, S. DAVID (ed)
216 Stochastic partial differential equations, A. ETHERIDGE (ed)
217 Quadratic forms with applications to algebraic geometry and topology, A. PFISTER
218 Surveys in combinatorics, 1995, PETER ROWLINSON (ed)
220 Algebraic set theory, A. JOYAL & I. MOERDIJK
221 Harmonic approximation, S.J. GARDINER
222 Advances in linear logic, J.-Y. GIRARD, Y. LAFONT & L. REGNIER (eds)
223 Analytic semigroups and semilinear initial boundary value problems, KAZUAKI TAIRA
224 Computability, enumerability, unsolvability, S.B. COOPER, T.A. SLAMAN & S.S. WAINER (eds)
225 A mathematical introduction to string theory, S. ALBEVERIO, J. JOST, S. PAYCHA, S. SCARLATTI
226 Novikov conjectures, index theorems and rigidity I, S. FERRY, A. RANICKI & J. ROSENBERG (eds)
227 Novikov conjectures, index theorems and rigidity II, S. FERRY, A. RANICKI & J. ROSENBERG (eds)
228 Ergodic theory of $\mathbf{Z}^d$ actions, M. POLLICOTT & K. SCHMIDT (eds)
229 Ergodicity for infinite dimensional systems, G. DA PRATO & J. ZABCZYK
230 Prolegomena to a middlebrow arithmetic of curves of genus 2, J.W.S. CASSELS & E.V. FLYNN
231 Semigroup theory and its applications, K.H. HOFMANN & M.W. MISLOVE (eds)
232 The descriptive set theory of Polish group actions, H. BECKER & A.S. KECHRIS
233 Finite fields and applications, S. COHEN & H. NIEDERREITER (eds)
234 Introduction to subfactors, V. JONES & V.S. SUNDER
235 Number theory 1993–94, S. DAVID (ed)
236 The James forest, H. FETTER & B. GAMBOA DE BUEN
237 Sieve methods, exponential sums, and their applications in number theory, G.R.H. GREAVES *et al*
238 Representation theory and algebraic geometry, A. MARTSINKOVSKY & G. TODOROV (eds)
239 Clifford algebras and spinors, P. LOUNESTO
240 Stable groups, FRANK O. WAGNER
241 Surveys in combinatorics, 1997, R.A. BAILEY (ed)
242 Geometric Galois actions I, L. SCHNEPS & P. LOCHAK (eds)
243 Geometric Galois actions II, L. SCHNEPS & P. LOCHAK (eds)
244 Model theory of groups and automorphism groups, D. EVANS (ed)
245 Geometry, combinatorial designs and related structures, J.W.P. HIRSCHFELD *et al*
246 p-Automorphisms of finite p-groups, E.I. KHUKHRO
247 Analytic number theory, Y. MOTOHASHI (ed)
248 Tame topology and o-minimal structures, LOU VAN DEN DRIES
249 The atlas of finite groups: ten years on, ROBERT CURTIS & ROBERT WILSON (eds)
250 Characters and blocks of finite groups, G. NAVARRO
251 Gröbner bases and applications, B. BUCHBERGER & F. WINKLER (eds)
252 Geometry and cohomology in group theory, P. KROPHOLLER, G. NIBLO, R. STÖHR (eds)
253 The q-Schur algebra, S. DONKIN
254 Galois representations in arithmetic algebraic geometry, A.J. SCHOLL & R.L. TAYLOR (eds)
255 Symmetries and integrability of difference equations, P.A. CLARKSON & F.W. NIJHOFF (eds)
257 An introduction to noncommutative differential geometry and its physical applications, 2ed, J. MADORE
258 Sets and Proofs, S.B. COOPER & J.K. TRUSS (eds)
259 Models and Computability, S.B. COOPER & J.K. TRUSS (eds)
260 Groups St Andrews 1997 in Bath, I, C.M. CAMPBELL *et al* (eds)
261 Groups St Andrews 1997 in Bath, II, C.M. CAMPBELL *et al* (eds)
264 Recent trends in algebraic geometry, K. HULEK, F. CATANESE, C. PETERS & M. REID (eds)

London Mathematical Society Lecture Note Series. 259

Models and Computability

Invited papers from Logic Colloquium '97 - European Meeting of the Association for Symbolic Logic, Leeds, July 1997

Edited by

S. Barry Cooper
University of Leeds

John K. Truss
University of Leeds

CAMBRIDGE UNIVERSITY PRESS
Cambridge, New York, Melbourne, Madrid, Cape Town, Singapore,
São Paulo, Delhi, Dubai, Tokyo, Mexico City

Cambridge University Press
The Edinburgh Building, Cambridge CB2 8RU, UK

Published in the United States of America by Cambridge University Press, New York

www.cambridge.org
Information on this title: www.cambridge.org/9780521635509

First published 1999

A catalogue record for this publication is available from the British Library

ISBN 978-0-521-63550-9 Paperback

Contents

Preface

Basic science, and within that pure mathematics, has a unique ability to surprise and change our view of the world we live in. But more often than not, its fundamental 'relevance' has emerged in ways impossible to have anticipated. As has often been remarked, that of the best basic science (say of non-Euclidean geometry, or of Hilbert spaces, or of the universal Turing machine) is independent of limited views of potential applicability.

Logic Colloquium '97, held in Leeds, England, 6th – 13th July, 1997, set out to reflect all that was best in contemporary logic, and **Models and Computability** and **Sets and Proofs** comprise two volumes of refereed articles, mainly based on the invited talks given at that meeting. Thanks to the programme committee (its other members being George Boolos, Sam Buss, Wilfrid Hodges, Martin Hyland, Alistair Lachlan, Alain Louveau, Yiannis Moschovakis, Leszeck Pacholski, Helmut Schwichtenberg, Ted Slaman and Hugh Woodin) and the special sessions organisers (Klaus Ambos-Spies, Sy Friedman, Wilfrid Hodges, Gerhard Jaeger, Steffen Lempp, Anand Pillay and Helmut Schwichtenberg), the editors have been able to call on a rich and distinguished array of authors. It is of great regret that one of our programme committee members was not able to see the success to which he had substantially contributed, and the **British Logic Colloquium Lecture**, given by Paul Benacerraf, took the form of a tribute to his memory. It would be difficult for us to improve on the introduction to Professor Benacerraf's article (on p.27 of **Sets and Proofs**) provided by the following extract from the comments received from the referee (necessarily anonymous):

> 'Ever since Paul Benacerraf published "What numbers could not be" (1965) and "Mathematical truth" (1973), his views have been seminal in the development of philosophy of mathematics, and for this reason one can expect that any paper by him that revisits the issues he first discussed in those papers (see footnote 1) will be of immediate interest for the subject. Also, the present paper is written as a very personal tribute to George Boolos, a student of Paul Benacerraf, whose early death, at the age of 55, deprived philosophy of mathematics of one of its leading – and one may also say – best loved contributors, and the Association of Symbolic Logic of its serving President, so it is very particularly fitting that this paper should be published in the proceedings of a meeting of the ASL.'

Logic Colloquium '97 was also the first such conference in Britain since the death of Robin Gandy (the first president of the British Logic Colloquium) on 20

November, 1995. As observed by Andrew Hodges[1]:

> 'Robin Gandy's death on 20 November 1995 has ended the strongest living link with Alan Turing, with whom he was all of intimate friend, student and colleague. He inherited all Turing's mathematical books and papers; and thereafter also carried forward part of Turing's intellectual tradition; more precisely he took on the subject that Turing lost interest in, by becoming a pre-eminent British figure in the revival and renewal of mathematical logic.'

Appropriately, Gerald Sacks, Robin's long-time friend and occasional research collaborator, gave the **Robin Gandy Lecture** (eloquently introduced by Joe Shoenfield). The original invitation to Professor Sacks had suggested a theme of "Computability Theory – The First Sixty Years", perhaps with Robin's 1988 article on "The Confluence of Ideas in 1936" in mind. But Gerald, undoubtedly grand but never grandiose, responded in his own, very personal, way, and the paper based on his lecture (on pages 367–376 of this volume) provides, among other things, fascinating background to the development of such landmarks of contemporary computability theory as the Sacks Density Theorem.

An element of arbitrariness in the allocation of topics and particular papers between the two volumes has been unavoidable. The prominence of effective model theory as a conference topic was one determining factor in the particular distribution adopted. Other decisions, such as the inclusion of the two papers from the 'Philosophy of Proof' part of the conference programme in **Models and Computability**, were more practical in origin. We hope that the overall benefits of convenience to the reader will be sufficient compensation.

Together, we hope that this volume and its companion **Sets and Proofs** will provide readers with with a comprehensive guide to the current state of mathematical logic, and while not pretending to the definitiveness of a handbook, perhaps communicating more of the excitement of a subject in flight. All the authors are leaders in their fields, some articles pushing forward the technical boundaries of the subject, others providing readable and authoritative overviews of particular important topics. (All the contributors have been encouraged to include a good introduction, putting their work in context.) A number of papers can be expected to become classics, essential to any good library (individual or institutional).

In any project of this magnitude, it is impossible to thank all those who have helped. Special thanks are due to all the authors (and to the small number who tried, and failed to deliver!), and to the host of referees who coped with tight deadlines without complaint. On the technical side, we would like to thank Margaret Williams, Audrey Landford, Tim Hainsworth, Frank Drake, David Knapp, Zarina Akhtar, Eric Cole, Kevin McEvoy, and Ben Salzberg and Benjamin Thoma at Blue

[1] At http://www.turing.org.uk/turing/scrapbook/robin.html.

Sky Research. Finally, thanks to Rebecca Mikulin and Roger Astley at Cambridge University Press for their advice and inexhaustible (it seemed) patience.

All royalties accruing from the sale of **Models and Computability** and **Sets and Proofs** go directly to the British Logic Colloquium.

We dedicate this volume to the memory of Robin Gandy.

S. Barry Cooper
John K. Truss
Leeds, November 1998

Continuous Functionals of Dependent and Transfinite Types

Ulrich Berger[1]
Mathematisches Institut der Universität München

1 Introduction

In this paper we study some extensions of the Kleene-Kreisel continuous functionals [7, 8] and show that most of the constructions and results, in particular the crucial density theorem, carry over from finite to dependent and transfinite types. Following an approach of Ershov we define the continuous functionals as the total elements in a hierarchy of Ershov-Scott-domains of partial continuous functionals. In this setting the density theorem says that the total functionals are topologically dense in the partial ones, i.e. every finite (compact) functional has a total extension. We will extend this theorem from function spaces to dependent products and sums and universes. The key to the proof is the introduction of a suitable notion of density and associated with it a notion of co-density for dependent domains with totality. We show that the universe obtained by closing a given family of basic domains with totality under some quantifiers has a dense and co-dense totality provided the totalities on the basic domains are dense and co-dense and the quantifiers preserve density and co-density. In particular we can show that the quantifiers Π and Σ have this preservation property and hence, for example, the closure of the integers and the booleans (which are dense and co-dense) under Π and Σ has a dense and co-dense totality. We also discuss extensions of the density theorem to iterated universes, i.e. universes closed under universe operators.

From our results we derive a dependent continuous choice principle and a simple order-theoretic characterization of extensional equality for total objects. Finally we survey two further applications of density: Waagbø's extension of the Kreisel-Lacombe-Shoenfield-Theorem showing the coincidence of

[1]I wish to thank Dag Normann for continuous and fruitful collaboration. A first version of this work was written when visiting the University of Uppsala in 1995. My sincere thanks go to Viggo Stoltenberg-Hansen and his colleagues for their hospitality and their help. I am grateful to the Swedish Government agency TFR and the Deutsche Forschungsgemeinschaft for supporting this research.

the hereditarily effectively continuous hierarchy of transfinite types with Beeson's realizability model, and Normann's representation theorem for S1-S9 computability in ${}^3E$.

Proofs will be given only for the main theorems and some crucial lemmas. More details can be found in the author's Habilitationsschrift [4].

2 Dependent domains with totality

We introduce Ershov-Scott domains and continuous families thereof, sketch how universes are constructed, and discuss the notion of totality in general.

2.1 Domains

A *Ershov-Scott domain* [6, 21], called *domain* for short, is a directed-complete partial order (dcpo) which is algebraic, countably based, and bounded-complete. If $(D, \sqsubseteq)$ is a domain, we let D_0 denote the set of compact elements of D. We will mainly use notations and results from [22] concerning the basics of domain theory. We let $[D \to E]$ denote the continuous function space, $D \times E$ the cartesian product, and $D + E$ the separated sum. The latter is defined by $D + E := \{0\} \times D \cup \{1\} \times E \cup \{\bot\}$ with least element $\bot$, and $(i, x) \sqsubseteq (j, y)$ iff $i = j$ and $x \sqsubseteq y$.

An *embedding* is a continuous function $\eta: D \to E$ such that there exists a continuous function $\eta^-: E \to D$ satisfying $\eta^- \circ \eta = \mathrm{id}_D$ and $\eta \circ \eta^- \sqsubseteq \mathrm{id}_E$. The function η^-, which is uniquely determined by η, is called the *projection* associated with η. $\eta: D \to E$ is called *good*, if η^- preserves arbitrary suprema, i.e. $\eta^-(\bigsqcup B) = \bigsqcup \eta^-[B]$ for every bounded set $B \subseteq E$.

We let DOM denote the category of Ershov-Scott domains with good embeddings as morphisms. The reason for considering good instead of arbitrary embeddings will become clear in section 3.1.

The operations $\to$, $\times$ and $+$ define continuous bifunctors on DOM in a canonical way. We explicate this only for the function space: If $\eta: D \to D'$ and $\xi: E \to E'$ are morphisms in DOM then $(\eta \to \xi): [D \to E] \to [D' \to E']$ is defined by $(\eta \to \xi)(f) = \xi \circ f \circ \eta^-$.

We call a category $\mathcal{C}$ *d-complete*, if every direct system in $\mathcal{C}$, i.e. every functor $F: I \to \mathcal{C}$ where I is a directed partial order, has a co-limit in $\mathcal{C}$

Proposition. The category DOM is d-complete.

2.2 Parametrizations

If $F: D \to \mathrm{DOM}$ is a continuous functor from a domain D (considered as a category) into a category DOM, then the pair (D, F) is called a *parametriza-*

tion.

The notion of a parametrization as well as the following notations are taken from [18]. If $x \sqsubseteq y \in D$, then $F[x,y]: F(x) \to F(y)$ denotes the morphism given by the functoriality of F. Continuity of F means that for every directed set $A \subseteq D$ we have $(F[x, \bigsqcup A])_{x \in A} = \lim(F[x,y])_{x \sqsubseteq y \in A}$. If $x \sqsubseteq y$ we define for $u \in F(x)$ and $v \in F(y)$, $u^{(y)} := F[x,y](u)$ and $v_{(x)} := F[x,y]^{-}(v)$ respectively.

We let PAR denote the category of parametrizations. A morphism in PAR is given by $(\eta, \tau): (D,F) \to (E,G)$, where $\eta: D \to E$ is a good embedding (i.e. a morphism in DOM), and $\tau: F \circ \eta^{-} \to G$ is a natural transformation. Composition is defined by

$$(\eta_1, \tau_1) \circ (\eta, \tau) := (\eta_1 \circ \eta, \lambda y_1 . \tau_1(y_1) \circ \tau(\eta_1^{-}(y_1))).$$

Proposition. The category PAR is d-complete.

Note that we may pack any finite list of domains $D^0, \ldots, D^{k-1}$ into a parametrization $\langle D^0, \ldots, D^{k-1} \rangle := (\{0, \ldots, k-1\}_{\perp}, F)$, where $F(\perp) := \{\perp\}$ and $F(i) := D^i$ for $i < k$.

Let $(D,F) \in$ PAR. We define

$$\Sigma(D,F) := \{(x,u) \mid x \in D,\ u \in F(x)\},\ (x,u) \sqsubseteq (y,v) :\Leftrightarrow x \sqsubseteq y \text{ and } u^{(y)} \sqsubseteq v,$$

$$\Pi(D,F) := \{f \in \Pi_{x \in D} F(x) \mid f \text{ p-continuous}\},\ f \sqsubseteq g :\Leftrightarrow \forall x \in D\,(f(x) \sqsubseteq g(x)),$$

where f is *p-continuous* if $f(x)^{(y)} \sqsubseteq f(y)$ whenever $x \sqsubseteq y \in D$ (section property), and $f(\bigsqcup A) = \bigsqcup\{ f(x)^{(\bigsqcup A)} \mid x \in A\}$ for all directed sets $A \subseteq D$ (continuity). In [18] it is shown that $\Sigma(D,F)$ and $\Pi(D,F)$ are domains.

For PAR-morphisms $(\eta,\tau): (D,F) \to (E,G)$ we define $\Sigma(\eta,\tau): \Sigma(D,F) \to \Sigma(E,G)$ and $\Pi(\eta,\tau): \Pi(D,F) \to \Pi(E,G)$ by

$$\begin{aligned} \Sigma(\eta,\tau)(x,u) &= (\eta(x), \tau(\eta(x))(u)), \\ \Pi(\eta,\tau)(f) &= \lambda y \in E\,.\,\tau(y)(f(\eta^{-}(y))). \end{aligned}$$

It is easy to verify that these are good embeddings.

Proposition. Σ: PAR $\to$ DOM and Π: PAR $\to$ DOM are continuous functors.

If (D,F) and $(\Sigma(D,F), G)$ are parametrizations, then G may as well be viewed as a continuous functor $G: D \to$ PAR by setting $G(x) := (F(x), G(x))$, where $G(x)(u) := G(x,u)$ for $u \in F(x)$ (large currying). Now if Φ is a *quantifier*, i.e. a continuous functor Φ: PAR $\to$ DOM (examples are Π and Σ), we may compose Φ and G and obtain another quantifier

$$\Phi(F,G) := (D, \lambda x \Phi(F(x), G(x))).$$

This construction will be frequently used in the sequel.

2.3 Universes

The idea of considering a universe of domains as a model of dependent types is due to Normann. Using a category theoretic fixed point technique we generalize Normann's construction.

Given a parametrization (A, B) (considered as a family of base types $B(a)$ indexed by $a \in A$) and quantifiers $\vec{\Phi} = \Phi^1, \ldots, \Phi^k$ we may construct the *universe over* (A, B) *closed under* $\vec{\Phi}$ as the parametrization $(S, I) \in \mathrm{PAR}$ defined recursively by

$$S = A + \mathrm{par}(S, I) + \ldots + \mathrm{par}(S, I)$$

where $\mathrm{par}(S, I) := (\Sigma s \in S)[I(s) \to S]$ and, denoting by $\beta: A \to S$, and $\varphi^i: \mathrm{par}(S, I) \to S$ $(i = 1, \ldots, k)$ the respective injections into the sum (constructors),

$$I(\beta(a)) = B(a),$$

$$I(\varphi^i(s, f)) = \Phi^i(I(s), I \circ f).$$

More precisely (S, I) is the initial fixed point of some continuous functor from PAR to PAR which exists by d-completeness of PAR and, strictly speaking, instead of the equalities above we have isomorphisms natural in the respective parameters.

In [12] and [24], concrete constructions of (S, I) for $\vec{\Phi} = \Pi, \Sigma$ and $B = \langle \mathbb{B}_\perp, \mathbb{N}_\perp \rangle$ are given. 'Concrete' means that the elements of S_0 and $I(s_0)_0$ and their order relation are constructed directly by an inductive definition. In [16] similar constructions are studied in the framework of information systems.

2.4 Totality

Our general idea is to model a type by a subset of 'total' elements of a domain. Since the notion of totality is not absolute, but relative to the construction of the domain, we have to investigate how a quantifier propagates totality. This gives rises to different notions of totality which we are going to clarify now.

A *totality on a domain* $D \in \mathrm{DOM}$ is a subset D_* of D.

A *totality on a parametrization* $(D, F) \in \mathrm{PAR}$ is a pair (D_*, F_*) where $D_* \subseteq D$ and $F_*(x) \subseteq F(x)$ for all $x \in D_*$.

A *totality on a quantifier* $\Phi: \mathrm{PAR} \to \mathrm{DOM}$ is a mapping Φ_* assigning to every totality (D_*, F_*) on a parametrization (D, F) a subset $\Phi_*(D_*, F_*)$ of the domain $\Phi(D, F)$.

This specifies possible candidates for totality. Now we fix some *standard totalities.*

The standard totality on a domain of the form $M_\perp$ is M.

The standard totalities on the quantifiers Π, Σ: PAR $\to$ DOM are given as follows. Let (D_*, F_*) be a totality on $(D, F) \in$ PAR.

$$\begin{aligned} \Pi_*(D_*, F_*) &:= \{ f \in \Pi(D, F) \mid \forall x \in D_* . f(x) \in F_*(x) \}, \\ \Sigma_*(D_*, F_*) &:= \{ (x, u) \in \Sigma(D, F) \mid x \in D_* \wedge u \in F_*(x) \}. \end{aligned}$$

In order to define a standard totality on a universe (S, I) over (A, B) closed under $\vec{\Phi}$, we need a totality (A_*, B_*) on the parametrization (A, B), and totalities Φ^i_* on the quantifiers Φ^i. We define inductively a set $S_{\mathrm{wf}} \subseteq S$ of *well-founded types*, and simultaneously for every $s \in S_{\mathrm{wf}}$ a set $I_{\mathrm{tot}}(s) \subseteq I(s)$ of *total objects* of type s.

If $a \in A_*$, then $\beta(a) \in S_{\mathrm{wf}}$, and $I_{\mathrm{tot}}(\beta(a)) := B_*(a)$.

If $(s, f) \in \mathrm{par}(S, I)$ such that $s \in S_{\mathrm{wf}}$ and $f(x) \in S_{\mathrm{wf}}$ for all $x \in I_{\mathrm{tot}}(s)$, then $\varphi^i(s, f) \in S_{\mathrm{wf}}$, and $I_{\mathrm{tot}}(\varphi^i(s, f)) := \Phi^i_*(I_{\mathrm{tot}}(s), I_{\mathrm{tot}} \circ f)$.

For $\vec{\Phi} = \Pi, \Sigma$ and $(A, B) = \langle \mathbb{N}_\perp, \mathbb{B}_\perp \rangle$ this definition is due to Normann.

3 Density theorems

It is the main concern of this paper to prove that in certain domains with totality the set of total elements is dense w.r.t. the Scott topology. Order theoretically a subset D_* of a domain D is dense if for every compact element $x_0 \in D_0$ there is some total element $x \in D_*$ such that $x_0 \sqsubseteq x$. It is clear that, since we are concerned with dependent domains, we also have to define density for a totality (D_*, F_*) on a parametrization (D, F). As one might expect, it will not be enough to simply require that for each $x \in D_*$ the set $F_*(x)$ is dense in $F(x)$. Rather some kind of 'uniform density' is needed. We will also be concerned with a companion of density called 'co-density', which will allow for smooth proofs of our main theorem saying that the standard totalities on the quantifiers Π and Σ preserve density and co-density.

3.1 Transporters

The notions of density and co-density which we will define in the next section use the fact that in a certain parametrization (D, F) it is possible to transport an element $u \in F(x)$ to any $F(y)$ 'without losing too much information'. Recall that if $x \sqsubseteq y$ or $x \sqsupseteq y$, then we may transport u to $F(y)$ using $F[x, y]$ or $F[y, x]^-$. The point is that we need to transport u to $F(y)$ also if x and y are *not* related.

Let (D, F) be a parametrization. For $x \sqsupseteq y \in D$ we set $F[x, y] := F[y, x]^-$. Hence, now the function $F[x, y]: F(x) \to F(y)$ is defined whenever x and y are comparable, i.e. $x \sqsubseteq y$ or $x \sqsupseteq y$.

We define $\mathcal{T}(D, F) := (\Pi(x, y) \in D \times D)[F(x) \to F(y)]$. We call $t \in \mathcal{T}(D, F)$ a *transporter* (for (D, F)) if for all $x, y, z \in D$

($\mathcal{T}1$) if x, y are comparable then $F[x, y] \sqsubseteq t(x, y)$,

($\mathcal{T}2$) $t(x, y)$ is strict,

($\mathcal{T}3$) $t(y, z) \circ t(x, y) \sqsubseteq t(x, z)$.

A *path* in a domain D is a nonempty finite sequence $\langle x^0, \ldots, x^k \rangle$ of elements in D such that x^i and x^{i+1} are comparable for all $i < k$. k is the *length* of α. For $x, y \in D$ we set

$$\mathrm{path}(x, y) := \{ \langle x^0, \ldots, x^k \rangle \mid \langle x^0, \ldots, x^k \rangle \text{ path},\ x^0 = x,\ x^k = y \}.$$

Let $(D, F) \in \mathrm{PAR}$. For $x, y \in D$ and $\alpha = \langle x^0, \ldots, x^k \rangle \in \mathrm{path}(x, y)$ we set

$$[\alpha] := F[x^{k-1}, x^k] \circ \ldots \circ F[x^0, x^1] \in [F(x) \to F(y)].$$

(D, F) is *consistent* if for all $x, y \in D$ the set $\{ [\alpha] \mid \alpha \in \mathrm{path}(x, y) \}$ is consistent, i.e. bounded in $[F(x) \to F(y)]$. In this case we set

$$t_F(x, y) := \bigsqcup \{ [\alpha] \mid \alpha \in \mathrm{path}(x, y) \} \in [F(x) \to F(y)].$$

For $x, y \in D$ and $u \in F(x)$ we set $u[y] := t_F(x, y)(u)$. Now ($\mathcal{T}3$) reads

$$\forall u \in F(x)\, u[y][z] \sqsubseteq u[z].$$

Proposition. A parametrization $(D, F) \in \mathrm{PAR}$ is consistent iff it has a transporter. In that case t_F is the least transporter for F.

Proof. We do not give the full proof, but only show where we use the fact that morphisms between domains are good embeddings. Assume that (D, F) is consistent. We show that t_F satisfies ($\mathcal{T}3$).

$$\begin{aligned} u[y][z] &= \bigsqcup \{ [\beta] \mid \beta \in \mathrm{path}(y, z) \} (\bigsqcup \{ [\alpha] \mid \alpha \in \mathrm{path}(x, y) \} (u)) \\ &= \bigsqcup \{ [\beta] (\bigsqcup \{ [\alpha](u) \mid \alpha \in \mathrm{path}(x, y) \}) \mid \beta \in \mathrm{path}(y, z) \} \\ &= \bigsqcup \{ [\beta]([\alpha](u)) \mid \alpha \in \mathrm{path}(x, y),\ \beta \in \mathrm{path}(y, z) \} \\ &\sqsubseteq u[z] \end{aligned}$$

The last equality holds, since $[\beta]$ is a composition of good embeddings and associated projections and hence preserves arbitrary suprema. □

Proposition. The category of consistent parametrizations considered as a full sub-category of the category PAR is d-complete.

A quantifier Φ: PAR $\to$ DOM is called consistent, if whenever (D, F) and $(\Sigma(D, F), G)$ are consistent, then $(D, \Phi(F, G))$ is consistent, too.

Proposition. All quantifiers and operators considered so far are consistent. Moreover, consistent continuous quantifiers and operators are closed under direct co-limits and composition. For Π and Σ we have the following estimates: Let (D, F) and $(\Sigma(D, F), G)$ be consistent parametrizations, and let $x, y \in D$.

$$\begin{aligned} f[y] &\sqsubseteq \lambda v \in F(y) \,.\, f(v[x])[(y, v)] \quad (f \in \Pi(F(x), G(x))) \\ (u, a)[y] &\sqsubseteq (u[y], a[(y, u[y])]) \quad ((u, w) \in \Sigma(F(x), G(x))). \end{aligned}$$

Lemma. If $(D, F) \in$ PAR is consistent and $g \in [E \to D]$ then $(E, F \circ g) \in$ PAR is consistent. For $y, y' \in E$ we have $t_{F \circ g}(y, y') \sqsubseteq t_F(g(y), g(y'))$, i.e. $u[y'] \sqsubseteq u[g(y')]$ for $u \in F(g(y))$. If in addition there is a function $h \in [D \to E]$ such that $g \circ h = \mathrm{id}_D$, and y, y' are in the range of h then $t_{F \circ g}(y, y') = t_F(g(y), g(y'))$, i.e. $u[y'] = u[g(y')]$.

3.2 Density and co-density

In this and the following section we tacitly assume that all parametrizations in question are consistent.

Let (D, F) be a parametrization and $x_0 \in D_0$ compact. We say that $t^0, \ldots, t^k \in [\Sigma(D, F) \to \mathbb{B}_\perp]$ *separate* $u_0, \ldots, u_k \in F(x_0)_0$ if

(s1) $\forall i \leq k \forall (x, u) \in \Sigma(D, F) \,.\, u_i \sqsubseteq u[x_0] \implies t^i(x, u) = \#\mathrm{t}$, and

(s2) $\bigcap_{i=0}^k (t^i)^{-1}[\#\mathrm{t}] = \emptyset$.

Let (D_*, F_*) be a totality on (D, F) and $x_0 \in D_0$.

(D_*, F_*) is *dense at* x_0 if

$$\forall u_0 \in F(x_0)_0 \exists d \in \Pi(D_*, F_*) \forall x \in D \,.\, u_0[x] \sqsubseteq d(x).$$

(D_*, F_*) is *co-dense at* x_0 if

$$\forall \vec{u} \in F(x_0)_0 \text{ inconsistent } \exists \vec{t} \in [\Sigma_*(D_*, F_*) \to_* \mathbb{B}] \,.\, \vec{t} \text{ separate } \vec{u}.$$

A totality (D_*, F_*) on (D, F) is *dense* respectively *co-dense*, if it is dense respectively co-dense at all $x_0 \in D_0$.

A totality $(\Sigma_*(D_*, F_*), G_*)$ on $(\Sigma(D, F), G)$ is is *dense* respectively *co-dense* at $x_0 \in D_0$, if it is dense respectively co-dense at (x_0, u_0) for all $u_0 \in F(x_0)_0$.

It is easy to see that if (D_*, F_*) is a dense totality on (D, F), then $F_*(x)$ is a dense subset of $F(x)$ w.r.t. the Scott-topology, for every $x \in D_*$.

Let $x_0 \in D_0$. We say that $t^0, \ldots, t^k \in [\Sigma(D, F) \to \mathbb{B}_\perp]$ *simultaneously separate inconsistent subsets of* $u_0, \ldots, u_k \in F(x_0)_0$ if

(s1) $\forall i \leq k \forall (x, u) \in \Sigma(D, F)(u_i \sqsubseteq u[x_0] \implies t^i(x, u) = \#t)$ and

(s$\tilde{2}$) $\forall J \subseteq \{0, \ldots, k\}(\{u_i \mid i \in J\}$ inconsistent $\implies \bigcap_{i \in J}(t^i)^{-1}[\#t] = \emptyset)$.

Lemma. Let the totality (D_*, F_*) on (D, F) be co-dense at x_0. Then for each $u_0, \ldots, u_k \in F(x)_0$ there are $t^0, \ldots, t^k \in [\Sigma(D_*, F_*) \to \mathbb{B}]$ simultaneously separating $u_0, \ldots, u_k$.

Lemma. Let (D_*, F_*) be a totality on (D, F), $E_* \subseteq E$ and $g \in [E_* \to_* D_*]$. $y_0 \in E_0$ such that $g(y_0) \in D_0$. Then: If (D_*, F_*) is dense (co-dense) at $g(y_0)$ then $(E_*, F_* \circ g)$ is dense (co-dense) at y_0.

3.3 Density for Π and Σ

Theorem. Let (D_*, F_*) and $(\Sigma_*(D_*, F_*), G_*)$ be totalities on (D, F) and $(\Sigma(D, F), G)$ respectively. Let $x_0 \in D_0$.

(1) If (D_*, F_*) is co-dense at x_0 and $(\Sigma_*(D_*, F_*), G_*)$ is dense at x_0, then $(D_*, \Pi_*(F_*, G_*))$ is dense at x_0.

(2) If (D_*, F_*) is dense at x_0 and $(\Sigma_*(D_*, F_*), G_*)$ is co-dense at x_0, then $(D_*, \Pi_*(F_*, G_*))$ is co-dense at x_0.

(3) If (D_*, F_*) is dense at x_0 and $(\Sigma_*(D_*, F_*), G_*)$ is dense at x_0, then $(D_*, \Sigma_*(F_*, G_*))$ is dense at x_0.

(4) If (D_*, F_*) is co-dense at x_0 and $(\Sigma_*(D_*, F_*), G_*)$ is co-dense at x_0, then $(D_*, \Sigma_*(F_*, G_*))$ is co-dense at x_0.

Proof. (1) Let $f_0 \in \Pi(F, G)(x_0)_0$ $(= \Pi(F(x_0), G(x_0))_0)$. Then there are $u_1, \ldots, u_k \in F(x_0)_0$ such that for all $u \in F(x_0)$

$$f_0(u) = f_0(y_0)^{(u)}, \quad \text{where } y_0 := \bigsqcup \{ u_i \mid i \in \{1, \ldots, k\}, u_i \sqsubseteq u \}.$$

By section 3.2 there are tests $t^1, \dots, t^k \in [\Sigma_*(D_*, F_*) \to_* \mathbb{B}]$ simultaneously separating inconsistent subsets of the set $\{u_1, \dots, u_k\}$. Set

$$\mathcal{J} := \{J \subseteq \{1, \dots, k\} \mid \{u_i \mid i \in J\} \text{ is consistent in } F(x_0)\}.$$

For $J \in \mathcal{J}$ define

$$u_J := \bigsqcup_{i \in J} u_i \in F(x_0)_0.$$

Hence for all $u \in F(x_0)$, $f_0(u) = f_0(u_J)^{(u)} = f_0(u_J)^{(x_0,u)}$, where $J := \{i \in \{1, \dots, k\} \mid u_i \sqsubseteq u\}$.

Since $(\Sigma_*(D_*, F_*), G_*)$ is dense at x_0, for each $J \in \mathcal{J}$ there is some $d_J \in \Pi_*(\Sigma_*(D_*, F_*), G_*)$ such that

$$(+) \qquad \forall (x,u) \in \Sigma(D,F)\ f_0(u_J)[(x,u)] \sqsubseteq d_J(x,u).$$

For $(x,u) \in \Sigma(D,F)$ we set $J(x,u) := \{i \in \{1, \dots, k\} \mid t^i(x,u) = \#\mathrm{t}\}$. By the choice of the t^i, $J(x,u) \in \mathcal{J}$, and hence $d_{J(x,u)}$ is defined. Set furthermore

$$U := \bigcap_{i=1}^{k} (t^i)^{-1}[\mathbb{B}]$$

which is an open subset of $\Sigma(D,F)$. Now for $x \in D$ and $u \in F(x)$ we define

$$d(x)(u) := \begin{cases} d_{J(x,u)}(x,u) & \text{if } (x,u) \in U \\ f_0[x](u) & \text{otherwise} \end{cases}$$

In order to prove that this works we first show that

$$(*) \qquad f_0[x](u) \sqsubseteq d_{J(x,u)}(x,u)$$

for all $(x,u) \in \Sigma(D,F)$.

Proof of $(*)$: Let $(x,u) \in \Sigma(D,F)$ and set $J := \{i \in \{1, \dots, k\} \mid u_i \sqsubseteq u[x_0]\}$. Then $f_0(u[x_0]) = f_0(u_J)^{(x_0, u[x_0])}$. Furthermore, by the choice of the t^i we have $J \subseteq J(x,u)$. Hence

$$\begin{aligned} f_0[x](u) &\sqsubseteq f_0(u[x_0])[(x,u)] \\ &= f_0(u_J)^{(x_0,u[x_0])}[(x,u)] \\ &\sqsubseteq f_0(u_J)[(x,u)] && \text{(by } (\mathcal{T}3)) \\ &\sqsubseteq f_0(u_{J(x,u)})[(x,u)] && \text{(since } J \subseteq J(x,u)) \\ &\sqsubseteq d_{J(x,u)}(x,u) && \text{(by } (+)). \end{aligned}$$

This proves $(*)$. From $(*)$ and the fact that U is open it follows immediately that d is continuous, i.e., $d \in \Pi(D, \Pi(F,G))$ (note that if $(x,u), (\tilde{x}, \tilde{y}) \in U$

and $(x,u) \sqsubseteq (\tilde{x},\tilde{y})$ then $J(x,u) = J(\tilde{x},\tilde{y})$). Furthermore, by $(*)$, we have $f_0[x] \sqsubseteq d(x)$ for all $x \in D$, and clearly $d \in \Pi_*(D_*, \Pi_*(F_*, G_*))$.

(2) Let $f_1, \ldots, f_k \in \Pi(F,G)(x_0)_0$ be inconsistent. Clearly there exists some $u_0 \in F(a_0)_0$ such that $\{f_1(u_0), \ldots, f_k(u_0)\} \subseteq G(x_0, u_0)_0$ is inconsistent. Since the totality (D_*, F_*) is dense at x_0 there is some $d \in \Pi_*(D_*, F_*)$ such that

$$\forall x \in D \; u_0[x] \sqsubseteq d(x).$$

Since the totality $(\Sigma_*(D_*, F_*), G_*)$ is co-dense at (x_0, u_0), there are tests $t^1, \ldots, t^k \in [\Sigma(\Sigma_*(D_*, F_*), G_*) \to_* \mathbb{B}]$ separating $f_1(u_0), \ldots, f_k(u_0)$. We define the tests $\tilde{t}^1, \ldots, \tilde{t}^k \in [\Sigma(D, \Pi(F,G)) \to \mathbb{B}_\perp]$ by

$$\tilde{t}^i(x,f) := t^i((x, d(x)), f(d(x))).$$

Clearly $\tilde{t}^i \in [\Sigma_*(D_*, \Pi_* \circ (F_*, G_*)) \to_* \mathbb{B}]$. In order to verify that $\tilde{t}^1, \ldots, \tilde{t}^k$ separate $f_1, \ldots, f_k$ assume first $f_i \sqsubseteq f[x_0]$. To prove $\tilde{t}^i(x,f) = \#\mathrm{t}$ it clearly suffices to show that $f_i(u_0) \sqsubseteq f(d(x))[(x_0, u_0)]$. We have

$$\begin{array}{rcll} f_i(u_0) & \sqsubseteq & f[x_0](u_0) & (\text{since} f_i \sqsubseteq f[x_0]) \\ & \sqsubseteq & f(u_0[x])[(x_0, u_0)] & \\ & \sqsubseteq & f(d(x))[(x_0, u_0)] & (\text{since } u_0[x] \sqsubseteq d(x)). \end{array}$$

Clearly $\bigcap_{i=1}^k (\tilde{t^i})^{-1}[\#\mathrm{t}] = \emptyset$.

(3) Let $(u_0, w_0) \in \Sigma(F,G)(x_0)_0$, i.e., $u_0 \in F(x_0)_0$ and $w_0 \in G(x_0, u_0)_0$. Since (D_*, F_*) and $(\Sigma_*(D_*, F_*), G_*)$ are dense at x_0, there are total choice functions $d_1 \in \Pi_*(D_*, F_*)$, and $d_2 \in \Pi(\Sigma_*(D_*, F_*), G_*)$ such that

$$\forall x \in D \; u_0[x] \sqsubseteq d_1(x) \quad \text{and} \quad \forall (x,u) \in \Sigma(D,F) \; w_0[(x,u)] \sqsubseteq d_2(x,u).$$

Define $d \in \Pi(D, \Sigma(F,G))$ by

$$d(x) := (d_1(x), d_2(x, d_1(x))).$$

Clearly $d \in \Pi_*(D_*, \Sigma_* \circ (F_*, G_*))$. Furthermore

$$(u_0, w_0)[x] \sqsubseteq (u_0[x], w_0[(x, u_0[x])]) \sqsubseteq (d_1(x), d_2(x, d_1(x))) = d(x).$$

(4) Let $\{(u_1, w_1), \ldots, (u_k, w_k)\} \subseteq \Sigma(F,G)(x_0)_0$ be inconsistent. There are two cases.

Case 1: The set $\{u_1, \ldots, u_k\} \subseteq F(x_0)_0$ is inconsistent.
Let the tests $t^1, \ldots, t^k \in [\Sigma_*(D_*, F_*) \to_* \mathbb{B}]$ separate $u_1, \ldots, u_k$.
Define $\tilde{t}^1, \ldots, \tilde{t}^k \in [\Sigma(D, \Sigma(F,G)) \to \mathbb{B}_\perp]$ by

$$\tilde{t}^i(x, (u,r)) := t^i(x,u).$$

Clearly $\tilde{t}^i \in [\Sigma_*(D_*, \Sigma_*(F_*, G_*)) \to_* \mathbb{B}]$. It is easy to see that $\tilde{t}^1, \ldots, \tilde{t}^k$ separate $(u_1, w_1), \ldots, (u_k, w_k)$.

Case 2: $u_0 := \bigsqcup_{i=1}^k u_i$ exists. Then $\{w_1^{(x_0,u_0)}, \ldots, w_k^{(x_0,u_0)}\} \subseteq G(x_0, u_0)_0$ is inconsistent. Let $t^1, \ldots, t^k \in [\Sigma(\Sigma_*(D_*, F_*), G_*) \to_* \mathbb{B}]$ separate the compact elements $w_1^{(x_0,u_0)}, \ldots, w_k^{(x_0,u_0)}$. Define $\tilde{t}^1, \ldots, \tilde{t}^k \in [\Sigma(D, \Sigma(F, G)) \to \mathbb{B}_\perp]$ by

$$\tilde{t}^i(x, (u, w)) := t^i((x, u), w).$$

Clearly $\tilde{t}^i \in [\Sigma_*(D_*, \Sigma_*(F_*, G_*)) \to_* \mathbb{B}]$. In order to prove that $\tilde{t}^1, \ldots, \tilde{t}^k$ separate $(u_1, w_1), \ldots, (u_k, w_k)$ assume $(u_i, w_i) \sqsubseteq (u, w)[x_0]$. Then $u_i \sqsubseteq u_i[x_0]$ and $w_i \sqsubseteq w[(x_0, u_i)]$. Therefore $\tilde{t}^i(x, (u, w)) = t^i((x, u), w) = \#\text{t}$. Clearly $\tilde{t}^i$ is total. □

This theorem immediately implies that for finite types s, the set $I_{\text{tot}}(s)$ of total elements of type s is dense in $I(s)$. This was proved first by Kleene [7] and Kreisel [8] (see also [23], 2.6.19 and [6]). In [3] the notion of co-density of a subset of a domain (called "totality" there) was introduced and a corresponding theorem with $\to$ and $\times$ instead of Π and Σ was proved.

3.4 Density for universes

Now we are going to show that the wellfounded universe $(S_{\text{wf}}, I_{\text{tot}})$ is dense and co-dense provided the totality on the base types is dense and co-dense and the totalities on the quantifiers preserve density and co-density and are furthermore 'nonempty' in a sense made precise below.

Let $\Phi\colon \text{PAR} \to \text{DOM}$ be a quantifier with totality Φ_*.

Φ_* is *dense-and-co-dense* if whenever $D \in \text{DOM}$, $x_0 \in D_0$, and parametrizations (D, F), $(\Sigma(D, F), G)$ with totalities (D_*, F_*), $(\Sigma_*(D_*, F_*), G_*)$ which are dense and co-dense at x_0 are given, then the totality $(D_*, \Phi_*(F_*, G_*))$ on the parametrization $(D, \Phi(F, G))$ is dense and co-dense at x_0.

Φ_* is *nonempty* if there is an operation sel_Φ selecting continuously (in an obvious sense) for every $(D, F) \in \text{PAR}$ a continuous function $\text{sel}_\Phi(D, F)\colon D \times \Pi(D, F) \to \Phi(D, F)$ which is 'total', i.e. if (D_*, F_*) is a totality on (D, F), $x \in D_*$ and $f \in \Pi_*(D_*, F_*)$, then $\text{sel}_\Phi(D, F)(x, f) \in \Phi_*(D_*, F_*)$.

By the density theorem in section 3.3 we know that the standard totalities on Π and Σ are both dense-and-co-dense. They are also nonempty, with the following obvious witnesses:

$$\begin{aligned}
\text{sel}_\Pi(D, F)(x, f) &:= f, \\
\text{sel}_\Sigma(D, F)(x, f) &:= (x, f(x)).
\end{aligned}$$

Theorem. Let $\vec{\Phi}$ be quantifiers with dense-and-co-dense and nonempty totalities $\vec{\Phi}_*$, and let (A, B) be a parametrization with a dense and co-dense totality (A_*, B_*) such that in addition $\Pi_*(A_*, B_*)$ is nonempty. Let (S, I) be the partial universe over (A, B) closed under $\vec{\Phi}$, and let $(S_{\mathrm{wf}}, I_{\mathrm{tot}})$ be the well-founded totality on (S, I). Then $(S_{\mathrm{wf}}, I_{\mathrm{tot}})$ is dense and co-dense.

Proof. We use the notation introduced in section 2.4. We first define a total choice function $h \in \Pi_*(S_{\mathrm{wf}}, I_{\mathrm{tot}})$. Let $g \in \Pi_*(A_*, B_*)$. We define h recursively:

$$h(s) = \begin{cases} g(a) & \text{if } s = \beta(a), \\ \mathrm{sel}_{\Phi_i}(I(t), I \circ f)(h(t), h \circ f) & \text{if } s = \varphi^i(t, f), \\ \bot & \text{if } s = \bot. \end{cases}$$

By induction on the definition of $s \in S_{\mathrm{wf}}$ one easily shows that $h(s) \in I_{\mathrm{tot}}(s)$.

Before proceeding further with the proof we introduce some auxiliaries. We define a continuous function $p \colon S \to \mathrm{par}(S, I)$ by $p(s) := (t, f)$ if $s = \varphi^i(t, f)$, $p(s) = \bot$ otherwise. Let $p^0 \in [S \to S]$ and $p^1 \in (\Pi s \in S)[I(p(s)) \to S]$ its components, i.e. $p(s) = (p^0(s), p^1(s))$. Let furthermore $'\beta, '\Phi^1, \ldots, '\Phi^k$ be different symbols and define a continuous function $\tau \colon S \to \{'\beta, '\Phi^1, \ldots, '\Phi^k\}_\bot$ as follows: $\tau(s) = \bot$ iff $s = \bot$; $\tau(s) = '\beta$ iff $s = \beta(a)$ for some $a \in A$; $\tau(s) = '\Phi^i$ iff $s = \varphi^i(t, f)$ for some $(t, f) \in \mathrm{par}(S, I)$. Hence if $s \in S_{\mathrm{wf}}$ and $\tau(s) = '\Phi^i$, then

$$I_{\mathrm{tot}}(s) = \Phi^i_*(I_{\mathrm{tot}}(p(s)), I_{\mathrm{tot}} \circ p^1(s)).$$

Since the universe was generated as the initial fixed point of a continuous functor it is possible to associate with every compact $s_0 \in S_0$ a natural number $\mathrm{rk}(s_0)$, the *rank* of s_0. Intuitively this is the stage in the inductive definition of S where s_0 was 'generated' (for details see [4]). The rank function has the property that if $s_0 = \varphi^i(t_0, f_0) \in S_0$, then $\mathrm{rk}(t_0) < \mathrm{rk}(s_0)$ and $\mathrm{rk}(f_0(x)) < \mathrm{rk}(s_0)$ for all $x \in I(t_0)$ (clearly $t_0, f_0(x) \in S_0$).

Now let us show that $(S_{\mathrm{wf}}, I_{\mathrm{tot}})$ is dense and co-dense at every compact $s_0 \in S_0$ using induction on the *rank* of s_0 We only treat the most interesting case, namely $s_0 = \varphi^i(t_0, f_0) \in S_0$.

Set $L := \{s \in S_* \mid \tau(s) = '\Phi_i\}$. Clearly $p^0 \in [L \to_* S_{\mathrm{wf}}]$ and $p^1 \in (\Pi_* s \in L)[I_{\mathrm{tot}}(p^0(s)) \to_* S_{\mathrm{wf}}]$. Since $\mathrm{rk}(p^0(s_0)) < \mathrm{rk}(s_0)$, by induction hypothesis, the totality $(S_{\mathrm{wf}}, I_{\mathrm{tot}})$ on (S, I) is dense and co-dense at $p^0(s_0)$. Hence, by the last lemma in section 2.4, the totality $(L, I_{\mathrm{tot}} \circ p^0)$ on $(S, I \circ p^0)$ is dense and co-dense at s_0. Furthermore $\mathrm{rk}(p^1(s_0)(x_0)) < \mathrm{rk}(s_0)$ for all $x_0 \in I(p^0(s_0))_0$. Hence, by induction hypothesis, $(S_{\mathrm{wf}}, I_{\mathrm{tot}})$ is dense and co-dense at $p^1(s_0)(x_0)$, and, by the lemma, the totality $(\Sigma_*(L, I_{\mathrm{tot}}), \lambda(s, x) I_{\mathrm{tot}}(p^1(s)(x)))$ is dense and co-dense at (s_0, x_0) for all $x_0 \in I(p^0(s_0))_0$. Hence, since Φ^i_* is dense-and-co-dense, the totality $(L, \lambda s \Phi^i_*(I_{\mathrm{tot}}(p^0(s)), I_{\mathrm{tot}} \circ p^1(s)))$ is dense and co-dense at s_0.

In order to show that $(S_{\mathrm{wf}}, I_{\mathrm{tot}})$ is dense at s_0 we let $x_0 \in I(s_0)_0$. Since

$(L, \lambda s \Phi^i_*(I_{\mathrm{tot}}(p^0(s)), I_{\mathrm{tot}} \circ p^1(s)))$ is dense at s_0, there is some

$$\tilde{d} \in \Pi_*(L, \lambda s \Phi^i_*(I_{\mathrm{tot}}(p^0(s), I_{\mathrm{tot}} \circ p^1(s))))$$

such that $x_0[s] \sqsubseteq \tilde{d}(s)$ for all $s \in S$. Define $d \in \Pi(S, I)$ by

$$d(s) := \begin{cases} \tilde{d}(s) & \text{if } \tau(s) = \text{'}\Phi^i, \\ h(s) & \text{if } \bot \neq \tau(s) \neq \text{'}\Phi^i, \\ \bot_{I(s)} & \text{if } \tau(s) = \bot. \end{cases}$$

This is well-defined, since if $\tau(s) = \text{'}\Phi^i$, then

$$I(s) = \Phi^i(I(p^0(s)), I \circ (p^1(s))).$$

Clearly $d \in \Pi_*(S_{\mathrm{wf}}, I_{\mathrm{tot}})$. We have to show that $x_0[s] \sqsubseteq d(s)$ for all $s \in S$. If $\tau(s) \neq \text{'}\Phi^i$ this holds, since then clearly $x_0[s] = \bot_{I(s)}$.

If $s = \varphi^i(t, f)$ it suffices to show that $x_0[s] \sqsubseteq \tilde{d}(s)$, i.e.

$$t_I(s_0, s)(x_0) \sqsubseteq \tilde{d}(s).$$

We will get this with the help of the second part of the last lemma in section 3.1. Define $D := \mathrm{par}(S, I)$, and $F: D \to DOM$, $F(s, f) := \Phi_i(I(s), I \circ f)$. Clearly $p \circ \varphi^i = \mathrm{id}_D$, and if $s = \varphi^i(t, f)$ then $\varphi^i(p(s)) = s$. Furthermore $F(p(s)) = \Phi^i(I(p^0(s)), I \circ p^1(s))$ for all $s \in S$. Now assume $\tau(s) = \text{'}\Phi^i$. Then

$$t_I(s_0, s)(x_0) \sqsubseteq t_F(p(s_0), p(s))(x_0) = t_{F \circ p}(s_0, s) \sqsubseteq \tilde{d}(s).$$

The proof that the totality $(S_{\mathrm{wf}}, I_{\mathrm{tot}})$ on (S, I) is co-dense at s_0 is not difficult but somewhat technical. Due to lack of space we omit it. □

Corollary. Let (S, I) be the partial universe over the base types $\mathbb{N}_\bot$ and $\mathbb{B}_\bot$ closed under Π and Σ and let $(S_{\mathrm{wf}}, I_{\mathrm{tot}})$ denote the corresponding total universe. Then $(S_{\mathrm{wf}}, I_{\mathrm{tot}})$ is dense and co-dense in (S, I). In particular for every well-founded type $s \in S_{\mathrm{wf}}$ the set of total elements $I_{\mathrm{tot}}(s)$ is a dense subset of $I(s)$.

Proof. Clearly the base family $\langle \mathbb{N}, \mathbb{B} \rangle$ is dense and co-dense. By the density theorem for Π and Σ in section 3.3 the quantifiers Π and Σ preserve density and co-density and, as we saw at the beginning of the present section, are nonempty. Hence by the theorem above the result follows. □

Normann proved density for the total universe over the integers closed under Π (see e.g. [14]).

3.5 Extension to iterated universe operators

Let quantifiers $\vec{\Phi} = \Phi^1, \ldots, \Phi^k$ be given. It follows from general category-theoretic considerations that the operation mapping a family of base types (A, B) to the universe (S, I) over (A, B) closed under $\vec{\Phi}$ defines a continuous functor

$$\mathcal{U}[\vec{\Phi}]: \mathrm{PAR} \to \mathrm{PAR}$$

which we call the *universe operator* associated with the quantifiers $\vec{\Phi}$. Composing $\mathcal{U}[\vec{\Phi}]$ with the obvious functor $\delta: \mathrm{PAR} \to \mathrm{DOM}$, $\delta(D, F) = D$, we obtain a continuous functor

$$\delta \circ \mathcal{U}[\vec{\Phi}]: \mathrm{PAR} \to \mathrm{DOM}$$

i.e. a parametrization, called the *universe quantifier* associated with $\vec{\Phi}$. Hence for (S, I) as above $(\delta \circ \mathcal{U}[\vec{\Phi}])(A, B) = S$. Now we may construct the next universe operator $\mathcal{U}[\vec{\Phi}, \delta \circ \mathcal{U}[\vec{\Phi}]]: \mathrm{PAR} \to \mathrm{PAR}$ which performs in addition closure under $\delta \circ \mathcal{U}[\vec{\Phi}]$. In this way we obtain iterated universe operators $\mathcal{U}^n: \mathrm{PAR} \to \mathrm{PAR}$ by setting

$$\mathcal{U}^n := \mathcal{U}[\Pi, \Sigma, \delta \circ \mathcal{U}^1, \ldots, \delta \circ \mathcal{U}^{(n-1)}].$$

Instead of just $\delta \circ \mathcal{U}^i$ we could as well have plugged in $\mathcal{U}^i$ (then we had to define what it means to perform closure under a continuous functor $\Psi: \mathrm{PAR} \to \mathrm{PAR}$), but this would lead to essentially the same hierarchy.

Now let totalities $\vec{\Phi}_*$ for the quantifiers $\vec{\Phi}$ be given. Using the definition of wellfounded totality for universes in section 2.4 we see that the quantifier $\delta \circ \mathcal{U}[\vec{\Phi}]$ bears a standard wellfounded totality $\delta \circ \mathcal{U}_{\mathrm{wf}}[\vec{\Phi}_*]$ mapping a totality (A_*, B_*) on (A, B) to S_{wf} where $S = (\delta \circ \mathcal{U}[\vec{\Phi}])(A, B)$.

The theorem in 3.4 may be now restated as follows.

Theorem. Let $\vec{\Phi}$ be quantifiers with dense-and-co-dense and nonempty totalities $\vec{\Phi}_*$, then the wellfounded totality $\delta \circ \mathcal{U}_{\mathrm{wf}}[\vec{\Phi}_*]$ on the universe quantifier $\delta \circ \mathcal{U}[\vec{\Phi}]$ is dense-and-co-dense and nonempty.

Using the iterated universe operators $\mathcal{U}^n$ we obtain iterated universes over the booleans and the integers

$$(S^n, I^n) := \mathcal{U}^n(\langle \mathbb{B}_\perp, \mathbb{N}_\perp \rangle).$$

the corresponding totalities are

$$(S^n_{\mathrm{wf}}, I^n_{\mathrm{tot}}) := \mathcal{U}^n_{\mathrm{wf}}(\langle \mathbb{B}, \mathbb{N} \rangle),$$

where

$$\mathcal{U}^n_{\mathrm{wf}} := \mathcal{U}_{\mathrm{wf}}[\Pi_*, \Sigma_*, \delta \circ \mathcal{U}^1_{\mathrm{wf}}, \ldots, \delta \circ \mathcal{U}^{(n-1)}_{\mathrm{wf}}].$$

Corollary. The wellfounded totalities $(S^n_{\text{wf}}, I^n_{\text{tot}})$ on the iterated universes (S^n, I^n) are dense-and-co-dense and nonempty.

The reader may have observed that we did not close our universes under the well-ordering type constructor which could be defined as a quantifier $W\colon \text{PAR} \to \text{DOM}$ by $W(D, F) := E$, where E is defined by the recursive domain equation $E = (\Sigma x \in D)[F(x) \to E]$. The reason is that W is in some obvious sense included in the quantifier $\delta \circ \mathcal{U}^1$. Hence the universes (S^n, I^n) are closed under W, for $n \geq 2$.

4 Applications

We survey some applications of our density theorems, which are, despite the last one, generalizations of the corresponding facts for finite types.

In the sequel we denote by (S, I) and $(S_{\text{wf}}, I_{\text{tot}})$ any universe (S^n, I^n) with its wellfounded totality $(S^n_{\text{wf}}, I^n_{\text{tot}})$.

4.1 Continuous choice

Proposition. Let (D_*, F_*) be a dense totality on (D, F), and let $g \in [\Sigma_*(D_*, F_*) \to \mathbb{B}]$ be such that

$$\forall x \in D_* \exists u \in F_*(x) \,.\, g(x, u) = \#\text{t}.$$

Then there is $h \in \Pi_*(D_*, F_*)$ such that

$$\forall x \in D_* \,.\, g(x, h(x)) = \#\text{t}.$$

Proof. Let $(x_n, u_n)_{n \in \mathbb{N}}$ be an enumeration of the countable set $\Sigma(D, F)_0$. Choose for every $n \in \mathbb{N}$ some $d(n) \in \Pi_*(D_*, F_*)$ such that $u_n[x] \sqsubseteq d_n(x)$ for all $x \in D$. Define

$$h(x) := \begin{cases} d_n(x) & \text{if } g(x, d_n(x)) = \#\text{t and } \forall k < n \; g(x, d_k(x)) = \#\text{f}, \\ \bot_{F(x)} & \text{if no such } n \text{ exists.} \end{cases}$$

Clearly h is continuous. Let $x \in D_*$. By assumption there exists some $u \in F_*(x)$ such that $g(x, u) = \#\text{t}$. Since g is continuous there is some $n \in \mathbb{N}$ such that $x_n \sqsubseteq x$, $u_n^{(x)} \sqsubseteq u$, and $g(x_n, u_n^{(x)}) = \#\text{t}$. Since $u_n^{(x)} \sqsubseteq u_n[x] \sqsubseteq d_n(x)$, we have $g(x, d_n(x)) = \#\text{t}$, by monotonicity of g. Let $n \in \mathbb{N}$ be minimal with $g(x, d_n(x)) = \#\text{t}$. Then $g(x, d_k(x)) = \#\text{f}$ for all $k < n$, since g and all d_k are total. Therefore $h(x) = d_n(x)$. □

Theorem. Let $\sigma(s, f) \in S_{\text{wf}}$ and $g \in I_{\text{tot}}(\sigma(s, f) \to \text{boole})$ such that

$$\forall x \in I_{\text{tot}}(s) \exists u \in I_{\text{tot}}(f(x)) \,.\, g(x, u) = \#\text{t}.$$

Then there is $h \in I_{\text{tot}}(\pi(s,f))$ such that

$$\forall x \in I_{\text{tot}}(s) \,.\, g(x, h(x)) = \#\text{t}.$$

Proof. By the corollary to the density theorem for universes, $(S_{\text{wf}}, I_{\text{tot}})$ is dense. Therefore $(I_{\text{tot}}(s), I_{\text{tot}} \circ f)$ is dense as well. Hence we may apply the proposition above with $(D_*, F_*) := (I_{\text{tot}}(s), I_{\text{tot}} \circ f)$. □

This theorem shows that the hierarchy $(S_{\text{wf}}, I_{\text{tot}})$ satisfies a quantifier-free axiom of choice. For finite type this has been established already by Kreisel [8] (see also [23], 2.6.20).

4.2 Extensionality

In our next application we show that in the total universe $(S_{\text{wf}}, I_{\text{tot}})$ extensional equality coincides with consistency in the domain ordering, and – as a consequence – that all well-founded types and their total objects are extensional, i.e. respect extensional equality.

Two elements x, y of a domain D are called *consistent*, written $x \uparrow y$, if the set $\{x, y\}$ is bounded, i.e. $x \sqcup y$ exists in D.

We call a totality (D_*, F_*) on (D, F) *natural* if D_* is upwards closed in D (i.e. if $x \in D_*$ and $x \sqsubseteq y \in D$, then $y \in D_*$), $F_*(x)$ is upwards closed in $F(x)$ for every $x \in D_*$, and for $x, y \in D_*$, if $x \sqsubseteq y$ then for every $v \in F(y)$, $v \in F_*(y)$ iff $v_{(x)} \in F_*(x)$. This corresponds to the notion of a total parametrization as introduced in [13]. Our terminology comes from the fact that a natural totality can be explained in categorical terms as a natural transformation (see [4]).

Proposition. All totalities on parametrizations considered so far, in particular the well-founded totality $(S_{\text{wf}}, I_{\text{tot}})$ on (S, I), are natural.

Lemma. Let (D_*, F_*) be natural and dense on (D, F), and $(\Sigma_*(D_*, F_*), G_*)$ co-dense on $(\Sigma(D, F), G)$. Then for $(x, f), (y, g) \in \Sigma_*(D_*, \Pi_*(F_*, G_*))$

$$(x,f) \uparrow (y,g) \iff x \uparrow y \wedge \forall u \in F_*(x), v \in F_*(y) \,. \\ (x,u) \uparrow (y,v) \rightarrow ((x,u), f(u)) \uparrow ((y,v), g(v)).$$

Proof. If $(x, f) \uparrow (y, g)$ then clearly $x \uparrow y$ and $((x,u), f(u)) \uparrow ((y,v), g(v))$ whenever $(x,u) \uparrow (y,v)$. If $(x,f) \not\uparrow (y,g)$ and $x \uparrow y$, then, letting $z := x \sqcup y \in D_*$, $f^{(z)} \not\uparrow g^{(z)}$. Hence there is $w_0 \in F(z)_0$ such that $f^{(z)}(w_0) \not\uparrow g^{(z)}(w_0)$. Since $F_*(z)$ is dense in $F(z)$ there is $w \in F_*(z)$ with $w_0 \sqsubseteq w$. Set $u := w_{(x)}$ and $v := w_{(y)}$. Since (D_*, F_*) is natural, we have $u \in F_*(x)$ and $v \in F_*(y)$. Clearly $(x,u) \uparrow (y,v)$. But $((x,u), f(u)) \not\uparrow ((y,v), g(v))$, since

$$((x,u), f(u)) \uparrow ((z,w), f^{(z)}(w)) \not\uparrow ((z,w), g^{(z)}(w)) \uparrow ((y,v), g(v))$$

and, by co-density, $\uparrow$ is an equivalence relation on $\Sigma_*(\Sigma_*(D_*, F_*), G_*)$. $\square$

We define inductively a binary relation $\sim$ on S_{wf} (extensional equality for well-founded types), and with every pair $(s, \tilde{s}) \in \sim$ explicitly a relation $\approx_{(s,\tilde{s})} \subseteq I_{\mathrm{tot}}(s) \times I_{\mathrm{tot}}(\tilde{s})$ (extensional equality of total objects of extensionally equal types). Let $s, \tilde{s} \in S_{\mathrm{wf}}$.

(i) If $s = \tilde{s} = \mathrm{nat}$ or $s = \tilde{s} = \mathrm{boole}$, then $s \sim \tilde{s}$ and $\approx_{(s,\tilde{s})}$ is the equality relation on $I_{\mathrm{tot}}(s)$.

(ii) If $s = \square(t, f)$ and $\tilde{s} = \square(\tilde{t}, \tilde{f})$, where $\square \in \{\pi, \sigma\}$ such that $t \sim_n \tilde{t}$ and $(\forall x, \tilde{x} \,.\, x \approx_{(t,\tilde{t})} \tilde{x} \rightarrow f(x) \sim \tilde{f}(\tilde{x}))$, then $s \sim \tilde{s}$ and
if $\square = \pi$ then $g \approx_{(s,\tilde{s})} \tilde{g} :\iff \forall x, \tilde{x} \,.\, x \approx_{(t,\tilde{t})} \tilde{x} \rightarrow g(x) \approx_{(f(x),\tilde{f}(\tilde{x}))} \tilde{g}(\tilde{x}))$,
if $\square = \sigma$ then $(x, u) \approx_{(s,\tilde{s})} (\tilde{x}, \tilde{u}) :\iff x \approx_{(t,\tilde{t})} \tilde{x} \wedge u \approx_{(f(x),\tilde{f}(\tilde{x}))} \tilde{u}$.

Theorem. For all $s, \tilde{s} \in S_{\mathrm{wf}}$, $s \sim \tilde{s}$ iff $s \uparrow \tilde{s}$. If $s \sim \tilde{s}$, then for all $x \in I_{\mathrm{tot}}(s)$ and $\tilde{x} \in I_{\mathrm{tot}}(\tilde{s})$, $x \approx_{(s,\tilde{s})} \tilde{x}$ iff $(s, x) \uparrow (\tilde{s}, \tilde{x})$ in $\Sigma(S, I)$.

Proof. Induction on the definition of S_{wf} using the lemma above and the fact that $(S_{\mathrm{wf}}, I_{\mathrm{tot}})$ is dense and co-dense. $\square$

Corollary. The well-founded types and their total objects are extensional, i.e. $s \sim s$ and $x \approx_{(s,s)} x$ holds for all $s \in S_{\mathrm{wf}}$ and $x \in I_{\mathrm{tot}}(s)$. In particular for $g \in I_{\mathrm{tot}}(\pi(s, f))$ this means that $x \approx_{(s,s)} \tilde{x}$ implies $g(x) \approx_{(f(x),f(\tilde{x}))} g(\tilde{x})$.

This corollary can also be obtained without density by proving that e.g. two well-founded types are extensionally equal iff their infimum is well-founded (this argument is due to Longo and Moggi [11]) which immediately implies extensionality. But note that the test, whether the infimum is well-founded, is as complex as well-foundedness itself, whereas consistency is a Π^0_1-predicate.

4.3 Effective operations

So far we have not mentioned effectivity explicitly. However it should be clear that all domains, parametrizations, and continuous functors considered are in some fairly obvious sense effective. This could be made precise by either introducing numberings for the compact elements of the domains [3, 5, 6], or by representing domains by information systems [9, 16, 20]. In [3] and [20] effective versions of the density theorem and the continuous choice principle for functionals of finite types are proved.

In the following we assume that all domains and parametrizations considered are effective. For a domain D we let D^{eff} denote the set of elements $x \in D$ such that the set of compact approximations, $\{x_0 \in D_0 \mid x_0 \sqsubseteq x\}$, is recursively enumerable (w.r.t. a suitable numbering of the compacts). We let $(S^{\mathrm{eff}}_{\mathrm{wf}}, I^{\mathrm{eff}}_{\mathrm{tot}})$ denote the hereditarily effective analogue of the wellfounded hierarchy $(S_{\mathrm{wf}}, I_{\mathrm{tot}})$ introduced in section 2.4, with base types $\mathbb{N}$ and $\mathbb{B}$. More precisely

- nat, boole $\in S^{\text{eff}}_{\text{wf}}$, and $I^{\text{eff}}_{\text{tot}}(\text{nat}) := \mathbb{N}$, $I^{\text{eff}}_{\text{tot}}(\text{boole}) := \mathbb{B}$.

- If $(s,f) \in \Sigma(S,I)^{\text{eff}}$ such that $f(x) \in S^{\text{eff}}_{\text{wf}}$ for all $x \in I^{\text{eff}}_{\text{tot}}(s)$, then $\pi(s,f) \in S^{\text{eff}}_{\text{wf}}$ and $\sigma(s,f) \in S^{\text{eff}}_{\text{wf}}$, and $I^{\text{eff}}_{\text{tot}}(\pi(s,f)) := \Pi_*(I^{\text{eff}}_{\text{tot}}(s), I^{\text{eff}}_{\text{tot}} \circ f)^{\text{eff}}$, $I^{\text{eff}}_{\text{tot}}(\sigma(s,f)) := \Sigma_*(I^{\text{eff}}_{\text{tot}}(s), I^{\text{eff}}_{\text{tot}} \circ f)^{\text{eff}}$.

This must not be confused with the hierarchy $(S_{\text{wf}} \cap S^{\text{eff}}, \lambda s.I_{\text{tot}}(s) \cap I(s)^{\text{eff}})$ which is incomparable with $(S^{\text{eff}}_{\text{wf}}, I^{\text{eff}}_{\text{tot}})$, even when restricted to finite types. For example, letting $\text{nat}^k \in S_{\text{wf}}$ be such that $I_{\text{tot}}(\text{nat}^0) = \mathbb{N}$, $I_{\text{tot}}(\text{nat}^{k+1}) = [I_{\text{tot}}(\text{nat}^k) \to \mathbb{N}$, a non wellfounded recursive tree without infinite recursive branches defines an element in $I^{\text{eff}}_{\text{tot}}(\text{nat}^2) \setminus I_{\text{tot}}(\text{nat}^2)$; the Fan-functional is an element of $I_{\text{tot}}(\text{nat}^3) \cap I(\text{nat}^3)^{\text{eff}} \setminus I^{\text{eff}}_{\text{tot}}(\text{nat}^3)$ (cf. [23]).

All results about $(S_{\text{wf}}, I_{\text{tot}})$ carry over to $(S^{\text{eff}}_{\text{wf}}, I^{\text{eff}}_{\text{tot}})$. In particular consistency characterizes extensional equality (4.2). Following [24] we denote the hierarchy $(S^{\text{eff}}_{\text{wf}}, I^{\text{eff}}_{\text{tot}})$ modulo extensional equality by ECFT (effectively continuous functionals of transfinite type). In [24] it is shown that ECFT is isomorphic to a corresponding hierarchy of functionals, called HEOT (hereditarily effective operations of transfinite types). HEOT is based on recursive function application and generalizes the structure HEO of hereditarily effective operations of finite types. In modern terms it is a subcategory of the category PER of partial equivalence relations on the integers. In [2] Beeson introduced a slightly richer hierarchy and used it to interpret Martin-Löf Type Theory.

Theorem (Normann, Waagbø). ECFT and HEOT are effectively isomorphic.

The proof uses the recursion theorem as well as a generalization of the Kreisel-Lacombe-Shoenfield-Theorem. The latter hinges on density and co-density [5, 3]. We cannot go into further details here. For finite types the corresponding result is due to Kreisel and others (cf. [23], 2.6.21).

One should also mention a recent result of Loader [10] who has analyzed the equational theory of a lambda calculus with inductive types in a PER model. He proves that this model is extensional and fully abstract and its equational theory is maximally consistent. As an intermediate step he introduces another model based on so-called set domains (a variant of Girard's qualitative domains) with totality, proves the corresponding results for the total domain model using a density theorem, and then shows that the total domain model and the PER model validate the same equations. Loader also introduces a separation property of the total objects which is closely related to our notion of co-density.

4.4 Representing computations in ^3E

One of Normann's main motivations for studying a transfinite hierarchy of types was to compare its complexity with Kleene's S1-S9 computations relative to certain (noncontinuous and non-effective) higher type functionals. As an example we consider the functional $^3E \in \mathrm{Tp}(3)$ representing quantification over number theoretic functions. Here $(\mathrm{Tp}(n))_{n\in\mathbb{N}}$ denotes the 'full' hierarchy, $\mathrm{Tp}(0) := \mathbb{N}$, $\mathrm{Tp}(n+1) :=$ the set of all total functions from $\mathrm{Tp}(n)$ to $\mathbb{N}$, and

$$^3E(F) := \begin{cases} 0 & \text{if } F(f) = 0 \text{ for all } f \in \mathrm{Tp}(2)), \\ 1 & \text{otherwise.} \end{cases}$$

Below we set $(S, I) := \mathcal{U}[\Pi](\langle \mathbb{N}_\perp \rangle)$.

Theorem (Normann). The closure ordinal of the inductive definition of the well-founded hierarchy $(S_{\mathrm{wf}}, I_{\mathrm{tot}})$ is the least ordinal not S1-S9 computable in ^3E and any $f \in \mathrm{Tp}(1)$.

An analogous result holds for recursion in ^2E and the hereditarily effective well-founded hierarchy $(S^{\mathrm{eff}}_{\mathrm{wf}}, I^{\mathrm{eff}}_{\mathrm{tot}})$. In [14] Normann proves the difficult part of the theorem, namely that the least ordinal not recursive in ^3E and any $f \in \mathrm{Tp}(1)$ is bounded by the closure ordinal of $(S_{\mathrm{wf}}, I_{\mathrm{tot}})$, by simulating S1-S9 computations in ^3E and some $\vec{f} \in \mathbb{N} \cup \mathrm{Tp}(1)$ in the hierarchy $(S_{\mathrm{wf}}, I_{\mathrm{tot}})$ in the following way. To every tuple $(e, \vec{f}, n)$ a type $S(e, \vec{f}, n) \in S$ and an object $\alpha(e, \vec{f}, n) \in I(e, \vec{f}, n)$ are assigned, such that whenever $\{e\}(^3E, \vec{f})$ is defined then for all $n \in \mathbb{N}$

(a) $C(e, \vec{f}, n) \in S_{\mathrm{wf}}$,

(b) $\alpha(e, \vec{f}, n) \in I_{\mathrm{tot}}(e, \vec{f}, n)$ iff $\{e\}(\vec{f}) = n$.

Density is used here to define for every $(s, x) \in \Sigma_*(S_{\mathrm{wf}}, I_{\mathrm{tot}})$ a total function $h(s, x) \colon \mathbb{N} \to \mathbb{N}$ encoding the behaviour of x on a dense subset of $I_{\mathrm{tot}}(s)$. Using the functions $h(s, x)$, which play the role of Kleene's associates, it is possible to define $C(e, \vec{f}, n) \in S_{\mathrm{wf}}$ and $\alpha(e, \vec{f}, n)$ with the desired properties.

5 Concluding remarks

In this paper we have studied models of dependent types and universes, and hence, implicitly, interpretations of constructive type theory. These kinds of interpretations, which have been worked out in detail e.g. in [2], [18], [16], [24], have been criticized by type theorists, since they are carried out in a framework of classical set theory. Note however that we did not use weird set theoretic constructions, but principles, like inductive-recursive definitions, which are discussed in current approaches to constructive set theory [1]. Moreover such interpretations make intuitionistic type theories accessible to classical mathematicians and they provide additional information about the theory, e.g. justification or validity of continuity- choice-, and extensionality-principles as studied in [8],[23] (see also the applications of density in 4.3, 4.1 and 4.2). Similar aspects have gained increasing interest in computer science where type theory is also considered as a (very rich) functional programming language One can use a denotational semantics, for example a domain-theoretic one as in this paper, to analyze the behaviour of the programming language. Examples are full abstraction and adequacy results [19], [18], and (in)completeness results regarding expressive power [19].

Let us return to type theory as a logical system. It is natural to interpret a type as the total elements of a domain. In particular a type corresponding to a false proposition should be interpreted as a domain without total elements. But this is in conflict with our principle that the total elements are always dense, in particular nonempty! One can give an interpretation without density [24], but loses then a lot of nice structural properties. Another way out of this problem could be a modified realizability interpretation in the spirit of [8]. In such an interpretation the total elements (which are always dense) are to be considered as 'potential realizers' whereas the 'actual realizers' form a proper subset (which may be empty). Details of this approach have to be worked out.

It is not difficult to iterate the universe operators transfinitely by introducing a universe of operators closed under the operation $\vec{\Phi}_* \mapsto \mathcal{U}_{\mathrm{wf}}[\vec{\Phi}_*]$. This could be pushed even further by introducing higher type universes of operators. A formal approach to similar ideas has been developed by Palmgren in [17] who suggested an extension of intuitionistic type theory by 'higher universe operators'. Using the machinery developed so far, density theorems for these hierarchies should not offer too big difficulties. However, this would not take us very far. The next point of real interest seems to be a 'Mahlo-universe' enjoying closure properties similar to a recursively Mahlo ordinal. Recently Normann [15] has defined such a universe and has shown that it can be used to represent Kleene recursion in a type three functional known as the superjump. The representation works without density, but with density a much smoother representation is to be expected.

References

[1] P. Aczel. The type theoretic interpretation of constructive set theory: inductive definitions. In R. Barcan Marcus et al., editors, *Logic, Methodology and Philosophy of Science VII*, North-Holland, pp. 17-49, 1986.

[2] M. Beeson. Recursive models for constructive set theories. *Annals of Mathematical Logic*, 23:127-178, 1982.

[3] U. Berger. Total sets and objects in domain theory. *Annals of Pure and Applied Logic*, 60:91-117, 1993.

[4] U. Berger. Continuous Functionals of Dependent and Transfinite Types. Habilitationsschrift. München, 1997.

[5] Y. L. Ershov. Hereditarily effective operations. *Algebra i Logika*, 15(6):642-654, 1976.

[6] Y. L. Ershov. Model C of partial continuous functionals. *Logic Colloquium 76*, 455-467, R. Gandy and M. Hyland, editors, North-Holland, 1977.

[7] S. C. Kleene. Countable functionals. In A. Heyting, editor, *Constructivity in Mathematics*, 81-100, North-Holland, 1959.

[8] G. Kreisel. Interpretation of analysis by means of constructive functionals of finite types. In A. Heyting, editor, *Constructivity in Mathematics*, 101-128, North-Holland, 1959.

[9] K. G. Larsen, G. Winskel. Using Information Systems to Solve Recursive Domain Equations Effectively. Proceedings of the Conference on Abstract Datatypes, Sophia-Antipolis, France, *Lecture Notes in Computer Science*, 173:109-129, Springer, 1984.

[10] R. Loader. Equational theories for inductive types. *Annals of Pure and Applied Logic*, 84:175-217, 1997.

[11] G. Longo, E. Moggi. The hereditarily partial effective functionals and recursion theory in higher types. *Journal of Symbolic Logic*, 49:1319-1332, 1984.

[12] L. Kristiansen, D. Normann. Semantics for some constructors of type theory. In Behara, Fritsch, Lintz, editors, *Symposia Gaussiana*, 201-224, de Gruyter, 1995.

[13] D. Normann. Categories of domains with totality. *Preprint Series, Inst. Math. Univ. Oslo*, 4, 1997.

[14] D. Normann. Closing the gap between the continuous functionals and recursion in ${}^3E$. *Proceedings of the Sacks conference MIT 1993*, *Archive for Mathematical Logic*, 36(6):405-436, 1997.

[15] D. Normann. A Mahlo-universe of effective domains with totality. This volume.

[16] E. Palmgren. An Information System Interpretation of Martin-Löf's Partial Type Theory with Universes. *Information and Computation*, 106(1):26-60, 1993.

[17] E. Palmgren. On Universes in Type Theory. Draft, University of Uppsala, 1996.

[18] E. Palmgren, V. Stoltenberg-Hansen. Domain interpretations of Martin-Löf's partial type theory. *Annals of Pure and Applied Logic*, 48:135-196, 1990.

[19] G. Plotkin. LCF considered as a programming language. *Theoretical Computer Science*, 5:223-255, 1977.

[20] H. Schwichtenberg. Density and Choice for Total Continuous Functionals. In *Kreiseliana. About and Around Georg Kreisel*, P. Odifreddi, editor, A. Peters, Wellesley, 335-362, 1996.

[21] D. Scott. Domains for denotational semantics. In *Automata, Languages and Programming, Lecture Notes in Computer Science*, 140:577-613, Springer, 1982.

[22] V. Stoltenberg-Hansen, I. Lindström, E. Griffor. Mathematical theory of domains. Cambridge University Press, 1993.

[23] A. S. Troelstra. Metamathematical Investigations of Intuitionistic Arithmetic and Analysis. *Lecture Notes in Mathematics*, 344, Springer, 1973.

[24] G. Waagbø. Domains-with-Totality Semantics for Intuitionistic Type Theory. PhD-thesis, University of Oslo, 1997.

Degree-Theoretic Aspects of Computably Enumerable Reals*

Cristian S. Calude, Richard Coles, Peter H. Hertling
and Bakhadyr Khoussainov

Department of Computer Science
University of Auckland
Private Bag 92019, Auckland
New Zealand

{cristian,coles,hertling,bmk}@cs.auckland.ac.nz

Abstract

A real α is computable if its left cut, $L(\alpha)$, is computable. If $(q_i)_i$ is a computable sequence of rationals computably converging to α, then $\{q_i\}$, the corresponding set, is always computable. A computably enumerable (c.e.) real α is a real which is the limit of an increasing computable sequence of rationals. It has a left cut which is c.e. We study the Turing degrees of representations of c.e. reals, that is the degrees of increasing computable sequences converging to α. For example, every representation A of α is Turing reducible to $L(\alpha)$. Every noncomputable c.e. real has both a computable representation and a noncomputable representation of degree $\deg_T L(\alpha)$. In fact, the representations of any noncomputable c.e. real are downwards dense, and yet not every c.e. Turing degree below $\deg_T L(\alpha)$ necessarily contains a representation of α.

1 Introduction

Computability theory essentially studies the relative computability of sets of natural numbers. Since Gödel introduced a method for coding structures

*The first and fourth authors were partially supported by AURC A18/XXXXX/62090/F3414056, 1996. The second author was supported by a UARC Post-Doctoral Fellowship, and the third author was supported by DFG Research Grant No. HE 2489/2-1.

using natural numbers, computability has been applied to many areas of mathematics, for example, to the theory of linear orders, to group theory and to real analysis. In this paper we will consider an application of computability theory to the real numbers.

The real numbers $\mathbb{R}$ may be defined in several different ways. In classical analysis, the reals are those entities which are the limit of a Cauchy sequence. In mathematical logic the real numbers are defined as Dedekind cuts of sets. Robinson [14] considered effective versions of Cauchy sequences and Dedekind cuts. In this paper the effective version of both Dedekind cuts and Cauchy sequences play a role.

We first consider effective converging sequences of rationals. A computable real number will be one that can be expressed as the limit of a computable sequence of rationals that converges computably. Alternatively, and equivalently, a computable real can be defined as a real with a computable left (or right) Dedekind cut. All relevant notions will be made formal in later sections.

It follows that every rational is a computable real, and further, that many well-known irrationals are computable, for example π and e. Furthermore, the set of all computable reals is a real-algebraically closed field.

Specker [17] gave the first example that computable sequences of rationals may converge in a noneffective way. He coded the Halting Problem into a computable sequence of rationals and then showed that if this sequence converged computably then there would be an algorithm for computing the Halting Problem.

For a similar reason, it is not possible to decide whether a computable sequence of rationals that converges computably, converges to 0 say. Hence the equality between two computable reals is also undecidable. (See Rice [13] or Calude [2] for example.)

We now consider effective Dedekind cuts. A natural set associated with a real α is $L(\alpha) = \{q \in \mathbb{Q} \mid q < \alpha\}$; it corresponds to the left Dedekind cut of the real α. Soare [15] studied computability theoretic properties of Dedekind cuts. With a subset A of natural numbers we associate the real $\alpha = 0.A(0)A(1)A(2)\ldots$ where $A(i) = 1$ if $i \in A$ and $A(i) = 0$ if $i \notin A$. We also write $\alpha = 0.\chi_A$. By interpreting this as a binary string we observe that we are dealing with reals in the interval $[0, 1]$. [1]

Soare showed, for example, that $A \leqslant_{tt} L(\alpha)$ but $L(\alpha)$ is not necessarily truth-table reducible to A, although $L(\alpha) \leqslant_T A$. Furthermore, $L(\alpha)$ is always

[1]Our choice of which real to associate with a set of natural numbers differs slightly from that in Soare [15]. With a subset A of natural numbers Soare associated a real number in the interval $[0, 2]$, namely $\Phi(A) = \Sigma_{n\in A}2^{-n}$, and $\Phi(\emptyset) = 0$. However from the point of view of Turing reducibility there is no difference because if $\alpha = 0.\chi_A$ then $\Phi(A) = 2\alpha$, hence $L(\alpha) \equiv_{tt} L(\Phi(A))$ and $A \equiv_T L(\alpha) \equiv_T L(\Phi(A))$.

a semirecursive set. (See Jockusch [7] for much more on semirecursive sets.)

It is not difficult to see that A is a computable set if and only if $L(\alpha)$ is a computable set. As Soare points out, if we replace computable by computably enumerable, then this equivalence no longer holds. If A is a computably enumerable set then $L(\alpha)$ is also a computably enumerable set. However the converse fails. In this paper we will define the property of a set being strongly ω-c.e. and see that if $L(\alpha)$ is computably enumerable then A is a strongly ω-c.e. set. This observation is made in Calude et al [4, Theorem 4.1]. Also note that it is a classical result of computable analysis that for $\alpha \in [0,1]$, $L(\alpha)$ is c.e. if and only if $[\alpha, 1]$ is recursively closed (see Ko [8]), or equivalently that $[\alpha, 1]$ is a Π_1^0-class (see Cenzer and Remmel [5]).

In a lot of work in computable analysis, emphasis has been put on computability and the question whether certain objects (real numbers, functions, etc.) are computable or not. Less is known about noncomputable real numbers. See Pour-El and Richards [12] or Weihrauch [19] for a development of computable analysis. We also cite Martin-Löf [11] and Bridges and Richman [1].

A computably enumerable real is defined to be the limit of an increasing computable sequence of rationals. In this paper we focus on the Turing degrees of increasing computable sequences of rationals that converge to some real α. The layout of the paper is the following.

We begin with some notation in Section 2. In Section 3, first we consider Turing degrees of computable sequences of rationals which converge computably. The only degree we get this way is $\mathbf{0}$. Then we consider Turing degrees of arbitrary computable sequences of rationals. It turns out that we can fix an arbitrary real number which can be obtained as the limit of such a sequence and restrict ourselves to sequences with this limit, and still get all c.e. degrees. Then we introduce computably enumerable reals.

In Section 4 we take a closer look at the Turing degrees of increasing computable sequences converging to c.e. reals. It turns out that the set of degrees obtained by looking at increasing computable sequences of rationals converging to a fixed c.e. real has a lot to do with splitting. Any such degree is below $L(\alpha)$. We show that for every noncomputable c.e. real α, there are infinitely many c.e. Turing degrees that contain sets of rationals which are increasing computable sequences converging to α.

The construction of increasing computable sequences converging to a real α is a dynamic process, and so many computability strategies that depend on waiting for some situation to occur in a construction, for example waiting for a partial computable function to halt on a particular argument, or waiting for some number to be enumerated into a set, are in conflict with such a construction. Because of this, in Section 5 we are able to show that there is a computably enumerable real α and a computably enumerable Turing degree below $L(\alpha)$ that does not contain an increasing computable sequence that converges to α. We conclude the paper with Section 6 by stating some open questions suggested by the above results.

For more background on computability theory see Soare [16] for example.

2 Preliminaries

When dealing with computability we think in terms of computations on natural numbers. In this paper we are concerned with constructing sequences of rationals having certain computability theoretic properties, and so we fix a standard computable bijection $\theta : \mathbb{Q} \mapsto \mathbb{N}$. When we work with sets of rationals we identify them with the subsets of $\mathbb{N}$ that are their images under θ, and similarly we move from sets of natural numbers to sets of rationals via θ^{-1}. If $A \subseteq \mathbb{Q}$ then we write $\theta(A)$ for the set $\{\theta(q) \mid q \in A\}$, and similarly for $B \subseteq \mathbb{N}$ and $\theta^{-1}(B)$.

For a set A we write $|A|$ to denote the cardinality of A. We let $A \setminus B$ denote $\{x \mid x \in A \,\&\, x \notin B\}$. For a set A, we define $A(i) = 1$ if $i \in A$ and $A(i) = 0$ if $i \notin A$. We write $A\restriction_x$ for the set $\{n \mid n \in A \,\&\, n < x\}$. If A is a set of natural numbers then we write $\alpha = 0.\chi_A$ to denote the real $\alpha = 0.A(0)A(1)A(2)\ldots$, in binary notation.

Let $\Phi_0, \Phi_1, \ldots$ be a standard listing of all Turing functionals. We write $\Phi_e(A)$ to denote the e-th functional acting on oracle A. We often write Φ_e in place of $\Phi_e(\emptyset)$. If $\Phi(A)(x)$ is a halting computation then we write $\Phi(A)(x)\downarrow$, and similarly, if $\Phi(A)(x)$ is not a halting computation then we write $\Phi(A)(x)\uparrow$. We append $[s]$ to parameters to denote their value after s steps, for example $\Phi(A)(x)\uparrow[s]$ means that the computation $\Phi(A)(x)$ has not halted after s steps.

Let W_e denote the e-th computably enumerable set, that is $W_e = \operatorname{dom} \Phi_e$. We let K denote the set $\{x \mid \Phi_x(x)\downarrow\}$. We relativise K to subsets $A \subseteq \mathbb{N}$, writing K^A for the set $\{x \mid \Phi(A)(x)\downarrow\}$.

If $A = \Phi(B)$ for some Turing functional Φ then we say that A is Turing reducible to B and write $A \leqslant_T B$. Further, we let $A \equiv_T B$ denote $A \leqslant_T B$ and $B \leqslant_T A$. We form equivalence classes of subsets of $\mathbb{N}$ via the equivalence relation $\equiv_T$ and write $\boldsymbol{a}, \boldsymbol{b}$, and so on to denote the equivalence classes. We define $\boldsymbol{a} \leqslant \boldsymbol{b}$ if there is some $A \in \boldsymbol{a}$ and $B \in \boldsymbol{b}$ such that $A \leqslant_T B$. The Turing degrees then form a partial order with respect to the ordering above which we denote by $\mathcal{D}(\leqslant)$.

The Turing jump operator is defined on subsets of $\mathbb{N}$ as $A' = K^A$. Hence $K \equiv_T \emptyset'$. We say a set A is low if $A' \equiv_T K$. Recall that a set A is Δ^0_2 if and only if $A \leqslant_T K$.

Define $A \oplus B = \{2n \mid n \in A\} \cup \{2n+1 \mid n \in B\}$. It is not difficult to see that $\deg_T(A \oplus B)$ is the least upper bound of $\deg_T(A)$ and $\deg_T(B)$, and so $\mathcal{D}(\leqslant)$ forms an upper semi-lattice. Whenever we simply write degree it is understood to be Turing degree.

We define the degree of a real α, $\deg_T(\alpha)$, to be the degree of A, where $0.\chi_A$ is the fractional part of α. Note that there is either a unique such set A or there are two, one finite and one cofinite.

With a finite set $X = \{x_1, x_2, \ldots x_k\}$ we can associate the canonical index $y = 2^{x_1} + 2^{x_2} + \ldots + 2^{x_k}$. Let D_y denote the finite set with canonical index y and D_0 denotes $\emptyset$. A computable approximation to a Δ^0_2 set A is a sequence $(D_{f(i)})_i$ of sets $D_{f(i)}$ for $i \in \mathbb{N}$ for some computable function f such that $A(x) = \lim_i D_{f(i)}(x)$, for all x.

We denote sequences of rationals by $(q_i)_i$ and by increasing sequence we shall mean strictly increasing. For $q \in \mathbb{Q}$ we define $q(x) = i$ if the xth bit of the binary representation containing infinitely many ones of the fractional part of q is i.

3 Computable sequences of rationals

We consider the Turing degrees of two classes of computable sequences of rationals and introduce computably enumerable reals.

Definition 1. A sequence $(q_i)_i$ of rationals is called *computable* if the total function $g : \mathbb{N} \mapsto \mathbb{N}$ defined by $\theta^{-1} \circ g(i) = q_i$ for all i is computable.

So a sequence is computable if we can (uniformly) effectively decide the ith member. However, we may be unable to decide the rationals that do not occur in the sequence. If $(q_i)_i$ is a sequence of rationals, we denote the set $\{q \in \mathbb{Q} \mid \exists i \in \mathbb{N}\,(q = q_i)\}$ by $\{q_i\}$. For computable sequences of rationals it is obvious that $\{q_i\}$ is a computably enumerable set. We will be interested in the Turing degrees of such sets.

First we look at sequences which define the "simplest" kind of real number, in terms of computability.

Definition 2. A sequence $(r_i)_i$ of reals *converges computably* to a real α if there is a total computable function $g : \mathbb{N} \mapsto \mathbb{N}$ such that for each n, $|r_i - \alpha| \leqslant 2^{-n}$ for all $i \geqslant g(n)$. We call g a *modulus of convergence* function for $(r_i)_i$.

Definition 3. A real α is *computable* if there is a computable sequence of rationals that converges computably to α.

As noted in the introduction, we could equivalently define computable reals to be those reals α such that $L(\alpha)$ is a computable set.

Theorem 4. *If a sequence $(q_i)_i$ of rationals is computable and converges computably, then the set $\{q_i\}$ is computable.*

Proof. Let $(q_i)_i$ be a computable sequence of rationals converging computably to α. Then there is a total computable function g such that for each n, $|q_i - \alpha| \leqslant 2^{-n}$ for all $i \geqslant g(n)$. We give a procedure for deciding if $p \in \{q_i\}$ for an arbitrary rational p. We distinguish three cases.

(1) α is irrational.

To decide $p \in \{q_i\}$ perform the following procedure:

Enumerate intervals $(q_k - 3 \cdot 2^{-n}, q_k + 3 \cdot 2^{-n})$ with $k \geqslant g(n)$ until finding the first such interval with $p \notin (q_k - 3 \cdot 2^{-n}, q_k + 3 \cdot 2^{-n})$. Such an interval will be found because $p \neq \alpha$ and $(q_i)_i$ converges to α.

Then $q_l \in (q_k - 3 \cdot 2^{-n}, q_k + 3 \cdot 2^{-n})$ for all $l \geqslant k$. Hence $p \in \{q_i\}$ if and only if $p \in \{q_0, \ldots, q_{k-1}\}$.

(2) α is rational and $\alpha \in \{q_i\}$.

To decide $p \in \{q_i\}$ perform the following procedure:

Check whether $p = \alpha$. If yes, conclude $p \in \{q_i\}$. If $p \neq \alpha$ then carry out the procedure in (1). The same argument as in (1) applies.

(3) α is rational and $\alpha \notin \{q_i\}$.

To decide $p \in \{q_i\}$ perform the following procedure:

Check whether $p = \alpha$. If yes, conclude $p \notin \{q_i\}$. If $p \neq \alpha$ then carry out the procedure in (1). The same argument as in (1) applies.

□

Remark 5. The procedure in the last proof is not uniform in the sequence $(q_i)_i$ and a modulus of convergence g. Indeed, a uniform procedure does not exist as one sees by considering the following list of sequences: $(r_i^{(j)})_i$ for $j = 0, 1, 2, \ldots$, where $r_i^{(j)} = 0$ for all $i \in \mathbb{N}$, if $j \in K$, and $r_i^{(j)} = \frac{1}{i+1}$ for all $i \in \mathbb{N}$, if $j \notin K$.

The next situation to consider is what happens if we relax the condition that the sequence should converge computably. Which reals do we get as limits of computable sequences $(q_i)_i$ and which degrees do we get as degrees of $\{q_i\}$?

Proposition 6. *For a real $\alpha \in [0, 1]$ the following two conditions are equivalent:*

1. *There exists a computable sequence $(q_i)_i$ of rationals converging to α.*

2. $\alpha = 0.\chi_A$ *for some* Δ_2^0 *set* A.

Proof. ($\rightarrow$) We can assume that all rationals q_i lie in the unit interval $[0,1]$. Define $x \in A[s]$ if and only if $x < s$ and $q_s(x) = 1$. Then $A = \lim_s A[s]$ is a Δ_2^0 set and $\alpha = 0.\chi_A$.

($\leftarrow$) Suppose $\alpha = 0.\chi_A$ where A is a Δ_2^0 set and $\{A[s]\}_{s\in\mathbb{N}}$ is a computable approximation to A. Let $q_i = 0.\chi_{A[s]}$. Then clearly $(q_i)_i$ is a computable sequence converging to α. □

Furthermore, if $\alpha = 0.\chi_A$ for A a Δ_2^0 set, then we can code every c.e. Turing degree into a computable sequence of rationals converging to α.

Theorem 7. *Suppose* $\alpha = 0.\chi_A$ *for a* Δ_2^0 *set* A. *Then for every c.e. degree* $\boldsymbol{b}$ *there exists a computable sequence* $(q_i)_i$ *with limit* α *such that* $\{q_i\}$ *has degree* $\boldsymbol{b}$.

Proof. Let $(p_i)_i$ be a computable sequence converging to α such that $\{p_i\}$ is infinite. We can construct a computable subsequence $(r_j)_j$ of $(p_i)_i$ such that $\theta(r_j)$ is strictly increasing. Let B be an arbitrary infinite c.e. set of natural numbers and $b_0, b_1, b_2, \ldots$ be an effective injective enumeration of B. Then the sequence $(q_i)_i = (r_{b_i})_i$ is a computable sequence of rationals, it converges to α, and we claim that $\{q_i\} \equiv_T B$. Indeed, a natural number m is in B iff r_m is in $\{q_i\}$. Conversely, for an arbitrary rational number s we can decide $s \in \{q_i\}$ by first asking whether $s \in \{r_i\}$. This is decidable because $\theta(r_i)$ is strictly increasing. If the answer is positive we compute the unique number b with $r_b = s$, and ask whether $b \in B$. □

So far we have considered arbitrary computable sequences of rationals that converge. As we remarked in the introduction, if the left cut $L(\alpha)$ is c.e. and $\alpha = 0.\chi_A$ then A may not be a c.e. set. By restricting our attention to increasing computable sequences we make the following definition of computably enumerable real.

Definition 8. A real α is *computably enumerable* if there is an increasing computable sequence of rationals converging to α.

Any computable real is also computably enumerable. Furthermore, if α is a computable real, then every increasing computable sequence of rationals converging to α converges computably, see Calude and Hertling [3]. There one can also find information about the speed of convergence of increasing computable sequences of rationals with a noncomputable modulus of convergence. In Theorem 10 below we give a characterization of the c.e. reals. and classify those sets A for which $\alpha = 0.\chi_A$ has $L(\alpha)$ c.e. We first motivate a definition that is useful for us.

Suppose $(q_i)_i$ is an increasing computable sequence of rationals in $[0,1]$ converging to $\alpha = 0.\chi_A$. There is a natural computable approximation to A derived from $(q_i)_i$, namely $A = \lim_s A[s]$ where $A[s] = \{n \mid n < s \ \& \ q_s(n) = 1\}$. Of course each $A[s]$ is a finite set. Furthermore, if $x \in A[s]$ but $x \notin A[s+1]$ then because $(q_i)_i$ is an increasing sequence there must be some $y < x$ such that $y \in A[s+1] \setminus A[s]$.

Definition 9. Let A be a Δ_2^0 set. We say that A is *strongly ω-c.e* if there is a computable approximation $(A[s])_s$ to A such that

1. $A[0] = \emptyset$,
2. $x \in A[s] \setminus A[s+1] \implies \exists y < x(y \in A[s+1] \setminus A[s])$.

Then the following theorem characterises the c.e. reals.

Theorem 10 (Calude, Hertling, Khoussainov and Wang [4]). *The following conditions are equivalent for a real $\alpha \in [0,1]$:*

1. *α is a c.e. real.*
2. *$L(\alpha)$ is c.e.*
3. *There is a strongly ω-c.e. set A such that $\alpha = 0.\chi_A$.*

Proof. (2) $\to$ (1). Suppose $L(\alpha)$ is c.e. Then it is easy to generate an increasing computable sequence of rationals from an effective enumeration of $L(\alpha)$.

(1) $\to$ (3). The assertion holds for $\alpha = 0$. Suppose $\alpha > 0$ and that $(q_i)_i$ is an increasing computable sequence of rationals in $[0,1]$ converging to α. Define $A = \lim_s A[s]$ where $A[s] = \{x \mid x < s \ \& \ q_s(x) = 1\}$. Then of course $\alpha = 0.\chi_A$. It is also clear that A is strongly ω-c.e.

(3) $\to$ (2). Let $\alpha = 0.\chi_A$ for some strongly ω-c.e. set A. Let $q_s = 0.\chi_{A[s]}$ where $\{A[s]\}_{s\in\mathbb{N}}$ is a computable approximation to A satisfying Definition 9. Then $L(\alpha)$ can be enumerated from an enumeration of $\{q_s \mid s \in \mathbb{N}\}$. □

Corollary 11. *If A is a strongly ω-c.e. set then A is of c.e. degree.*

Proof. We have already mentioned that $L(0.\chi_A) \equiv_T A$ for $A \subseteq \mathbb{N}$. This together with Theorem 10 gives the assertion. □

4 Representations of computably enumerable reals

For the remainder of the paper we only consider c.e. reals.

Definition 12. A set $B \subseteq \mathbb{Q}$ of rationals is called a *representation* of α if there is an increasing computable sequence $(q_i)_i$ of rationals with limit α and $\{q_i\} = B$. We identify B with $\theta(B)$, its image under the bijection $\theta : \mathbb{Q} \mapsto \mathbb{N}$ and also call $\theta(B)$ a representation of α.

We look at the Turing degrees of representations of c.e. reals. As we noted in the introduction, $\deg_T(\alpha) = \deg_T(L(\alpha))$.

Lemma 13. *Every c.e. degree is the degree of $L(\alpha)$ for some c.e. real α.*

Proof. Let A be a c.e. set of degree $\boldsymbol{a}$ and let α be the c.e. real equal to $0.\chi_A$. Then it is clear that $L(\alpha) \equiv_T A$. □

We direct the reader to Soare [15] for related work on the relative computability of cuts of arbitrary reals.

It turns out that the question about the Turing degrees of c.e. reals is intimately connected with the splitting properties of the c.e. Turing degrees. We first recall some notions from computability theory. (See Downey and Stob [6] for a survey of splitting theorems.)

Definition 14. A *splitting* of a c.e. set A is a pair of disjoint c.e. sets A_1 and A_2 such that $A_1 \cup A_2 = A$. Then we say that A_1 and A_2 *form a splitting* of A and that each of the sets A_1 and A_2 is a *half of a splitting of* A.

It is easy to see that if A_1 and A_2 form a splitting of a c.e. set A, then $A \equiv_T A_1 \oplus A_2$. The following two lemmata show the connection between representations of c.e. reals and splitting.

Lemma 15. *If B is a representation of a c.e. real α, then B is an infinite half of a splitting of $L(\alpha)$.*

Proof. It is clear that any representation B of a c.e. real α is an infinite c.e. subset of $L(\alpha)$. We have to show that also $L(\alpha) \setminus B$ is c.e. Let $(q_i)_i$ be the increasing (computable!) sequence of rationals with $B = \{q_i\}$. The set $L(\alpha)$ is c.e. We can for each element $p \in L(\alpha)$ wait until we find a q_j with $p \leqslant q_j$ (as rationals), and choose p if and only if $p \notin \{q_0, \dots, q_j\}$. Hence, we can enumerate $L(\alpha) \setminus B$. □

Lemma 16. *Let B be a representation of a c.e. real α. For a subset $C \subseteq B$ the following two conditions are equivalent:*

1. *C is a representation of α.*

2. *C is an infinite half of a splitting of B.*

Proof. ($\rightarrow$) As the proof of Lemma 15.

($\leftarrow$) Let $(q_i)_i$ be the increasing computable sequence of rationals with $B = \{q_i\}$, let C be an infinite half of a splitting of B, and let D be the other half of this splitting. We construct an increasing rational sequence $(p_i)_i$ with limit α and $C = \{p_i\}$ by going through the list $(q_i)_i$, by waiting for each element q_i until it is enumerated either in C or in D, and by choosing it if and only if it is enumerated in C. □

It follows from Lemma 15 that $L(\alpha)$ is an upper bound for the degrees of representations of α.

Corollary 17. *If B is a representation of a c.e. real α, then $B \leqslant_T L(\alpha)$.*

For the special case of computable reals we then get the following:

Corollary 18. *If α is a computable real, then every representation of α is computable.*

We consider the following partial orders.

Definition 19. For a c.e. real α, let $\mathcal{L}(\alpha)$ be the partial order (with respect to Turing reducibility) of those c.e. Turing degrees below $\deg_T(L(\alpha))$ that contain a representation of α.

Proposition 20. *For every c.e. real α, $\mathcal{L}(\alpha)$ is an upper-semi lattice.*

Proof. Let α be a c.e. real. Then $\mathcal{L}(\alpha)$ is closed under the usual join operation on Turing degrees. Indeed suppose $\boldsymbol{a}, \boldsymbol{b} \in \mathcal{L}(\alpha)$ with A and B being representations of α in $\boldsymbol{a}$ and $\boldsymbol{b}$, respectively. Let $C = A \cup B$. Then C is the representation of α formed by effectively enumerating the two sequences of A and B in increasing order as rationals. We claim that $\deg_T(C) = \boldsymbol{a} \cup \boldsymbol{b}$, that is $\theta(C) \equiv_T \theta(A) \oplus \theta(B)$. It is obvious that $\theta(C) \leqslant_T \theta(A) \oplus \theta(B)$. For the converse we observe that by Lemma 16 ($\rightarrow$) the set A is a half of a splitting of C, hence $\theta(A) \leqslant_T \theta(C)$, the same for B. □

We now study this upper-semi lattice further. We first prove that $\boldsymbol{0}$ and $\deg_T(L(\alpha))$ are in $\mathcal{L}(\alpha)$.

Proposition 21. *For any c.e. real α there is a computable representation of α.*

Proof. Take an arbitrary representation of α. The classical result that every infinite c.e. set contains an infinite computable subset yields the assertion. □

Furthermore we can construct a noncomputable representation as follows.

Theorem 22. *Every noncomputable c.e. real α has a noncomputable representation.*

Proof. Fix an increasing computable sequence $(q_i)_i$ converging to α such that $\{q_i\}$ is computable. We construct a noncomputable representation B such that $(p_i)_i$ is a subsequence of B and B is not the complement of any c.e. set.

At stage $s = 0$ let $b_0 = q_0$.

At stage $s + 1$ we have already constructed $B[s] = \{b_0, \ldots, b_{k_s}\}$ where $b_0 < \ldots < b_{k_s}$ (as rationals) and $b_{k_s} = q_s$.

If there is a least $e < s + 1$ such that $W_e[s] \cap B[s] = \emptyset$ and an $x \in W_e[s]$ with $q_s < x \leqslant q_{s+1}$ then let $b_{k_s+1} = x$, $b_{k_s+2} = q_{s+1}$ and $k_{s+1} = k_s + 2$.

If there is no such e then let $b_{k_s+1} = q_{s+1}$ and $k_{s+1} = k_s + 1$.

We complete the construction by letting $B = \bigcup_s B[s]$.

Clearly $(b_i)_i$ is an increasing computable sequence of rationals converging to α. It remains to show that B is not computable.

Suppose B is a computable set. Then let e be the least index such that $B = \overline{W_e}$. Let s_0 be a stage such that for all $i < e$ and all $s \geqslant s_0$ we have $W_i[s] \cap B[s] \neq \emptyset$ or there is no $x \in W_i[s]$ with $q_s < x \leqslant q_{s+1}$. We will show that for all $p > q_{s_0}$ (as rationals), $p \in L(\alpha)$ is decidable, contradicting the hypothesis of the theorem. To compute $p \in L(\alpha)$, enumerate B and W_e until p occurs in one of them. If $p \in B$ then $p \in L(\alpha)$. Otherwise $p \in W_e$ and we claim that $p \notin L(\alpha)$. For suppose $p \in L(\alpha)$, then at some least stage $t > s_0$, $q_t < p \leqslant q_{t+1}$, and the construction enumerates some $p' \in B$ for $q_t < p' \leqslant q_{t+1}$ and $p' \in W_e$. This contradicts $B \cap W_e = \emptyset$ and hence B is not a computable set. □

We now show that $\deg_T(L(\alpha))$ is in the lattice.

Theorem 23. *Let α be a c.e. real. Then α has a representation of degree $L(\alpha)$. Furthermore, every representation of α can be extended to a representation of degree $L(\alpha)$.*

Proof. Let $(p_i)_i$ be an increasing computable sequence of rationals converging to α. We shall construct a new computable sequence $(q_i)_i$ of rationals such that $\{q_i\}$ is a representation of α with $\{q_i\} \equiv_T L(\alpha)$. Additionally we define $l_i = \max\{\theta(p_j) \mid j \leqslant i\}$ for all i, and we will define a sequence $(j_i)_i$ of natural numbers with $q_{j_i} = p_i$ for all i. We start with $j_0 = 0$ and $q_0 = p_0$. Given j_i with $q_{j_i} = p_i$, we define $j_{i+1} > j_i$ such that

$$j_{i+1} - j_i = |\{q \in \mathbb{Q} \mid p_i < q \leqslant p_{i+1} \;\&\; \theta(q) \leqslant l_{i+1}\}|$$

and for $m = 1, \ldots, j_{i+1} - j_i$ we define the numbers q_{j_i+m} as the rational numbers in this set in increasing order.

It is obvious that $(q_i)_i$ is an increasing computable sequence of rationals converging to α, and $q_{j_i} = p_i$ for all i. From Corollary 17 we know $\{q_i\} \leqslant_T L(\alpha)$. We still have to prove $L(\alpha) \leqslant_T \{q_i\}$. Let $p \in \mathbb{Q}$. In order to decide $p \in L(\alpha)$ we compute the minimal k with $l_k \geqslant \theta(p)$. Then we check whether $p \leqslant q_{j_k}$. If $p \leqslant q_{j_k}$, then clearly $p \in L(\alpha)$. If $p > q_{j_k}$, then $p \in L(\alpha)$ if and only if $p \in \{q_i\}$.

We give an alternative proof for the first assertion of the theorem by a slightly different construction. It shows that we can obtain a representation of α of degree $L(\alpha)$ consisting only of dyadic rational numbers.

Fix an increasing computable sequence $(p_i)_i$ of dyadic rationals with limit α with increasing denominator:

$$p_i = \frac{2n_i + 1}{2^{k_i}}$$

for a computable sequence $(n_i)_i$ of integers and a computable, increasing sequence $(k_i)_i$ of natural numbers. We shall construct a new computable sequence $(q_i)_i$ of rationals such that $\{q_i\}$ is a representation of α having Turing degree $\deg_T(L(\alpha))$. We will define a sequence $(j_i)_i$ of natural numbers. We will have $q_{j_i} = p_i$ for all i.

We start with $j_0 = 0$ and $q_0 = p_0$. Given j_i with $q_{j_i} = p_i$, we set

$$q_{j_i+m} = q_{j_i} + \frac{m}{2^{k_{i+1}}}$$

for $m = 1, \ldots, (p_{i+1} - p_i) \cdot 2^{k_{i+1}}$ and

$$j_{i+1} = j_i + (p_{i+1} - p_i) \cdot 2^{k_{i+1}}.$$

Of course, $(q_i)_i$ is an increasing computable sequence of rationals converging to α since $q_{j_i} = p_i$ for all i.

We have to show $L(\alpha) \leqslant_T \{q_i\}$. If α is a rational then $L(\alpha)$ is computable and hence $\leqslant_T \{q_i\}$. So we assume that α is irrational.

The important property of the set $\{q_i\}$ is that if it contains a dyadic number $\frac{2n+1}{2^k}$, then it contains all dyadic numbers in the interval $(\frac{2n+1}{2^k}, \alpha)$ whose denominator is at most 2^k. But $\{q_i\}$ does not contain any number greater than α. Furthermore, the denominator of the dyadic number q_{j_i} is at least $2^{k_i} \geqslant 2^i$. Hence, given $\{q_i\}$ as an oracle, for an arbitrary natural number l we can compute a dyadic rational $\frac{2n+1}{2^k}$ with $k \geqslant l$ and such that the interval $(\frac{2n+1}{2^k}, \frac{2n+3}{2^k})$ contains α. Using $\{q_i\}$, for a given rational number r, we can decide whether $r < \alpha$ by computing such an interval which contains α but not r (any sufficiently small interval containing the irrational number α will not contain r) and checking whether r lies to the left or to the right of this interval. □

Corollary 24. *Every c.e. degree contains a representation of a c.e. real.*

Proof. By Lemma 13 and Theorem 23. □

By Lemma 15 every representation of a c.e. real α is a half of a splitting of $L(\alpha)$. The following result shows that there is a representation of α of the same degree as the other half. It is also a strengthening of Theorem 23: take B to be a computable representation in order to obtain the first part of Theorem 23.

Theorem 25. *Suppose B is a representation of a c.e real α. Then there is a representation C of α such that $C \equiv_T L(\alpha) \setminus B$.*

Proof. Let (b_i) be the increasing computable sequence such that $B = \{b_i\}$. Let $(p_i)_i$ be a representation of α such that $\{p_i\}$ is computable and $\{p_i\} \cap \{b_i\} = \emptyset$. We construct a new increasing computable sequence of rationals $(c_i)_i$ such that $\{c_i\} \equiv_T L(\alpha) \setminus B$. Define $l_i = \max\{\theta(p_j) \mid j \leqslant i\}$ for all i, and we will define a sequence $(j_i)_i$ of natural numbers with $c_{j_i} = p_i$ for all i. We start with $j_0 = 0$ and $c_0 = p_0$. Let b_{p_i} denote the least rational in B which is greater than p_i. Then given j_i with $c_{j_i} = p_i$, define $j_{i+1} > j_i$ such that

$$j_{i+1} - j_i = |\{q \in \mathbb{Q} \mid p_i < q \leqslant p_{i+1} \,\&\, \theta(q) \leqslant l_{i+1} \,\&\, q \notin \{b_0, \dots, b_{p_{i+1}}\}\}|,$$

and for $m = 1, \dots, j_{i+1} - j_i$ we define c_{j_i+m} to be those rational numbers in this set in increasing order. Let $C = \{c_i\}$.

It is clear that $(c_i)_i$ is an increasing computable sequence of rationals converging to α, since $c_{j_i} = p_i$ for all i. We now show that $C \equiv_T L(\alpha) \setminus B$.

First, $C \leqslant_T L(\alpha) \setminus B$ as follows. Let $p \in \mathbb{Q}$. If $p \notin L(\alpha) \setminus B$ then $p \notin C$. Otherwise, if $p \in L(\alpha) \setminus B$, enumerate C until reaching a least c_i such that $c_i \geqslant p$. Then $p \in C$ if and only if $p \in \{c_0, \dots, c_i\}$.

Secondly, $L(\alpha) \setminus B \leqslant_T C$ as follows. Let $p \in \mathbb{Q}$. Compute the least k such that $l_k \geqslant \theta(p)$ and then check whether $p \leqslant c_{j_k}$. If $p \leqslant c_{j_k}$ then enumerate B until reaching a least b_i such that $p \leqslant b_i$, and conclude $p \in L(\alpha) \setminus B$ if and only if $p \notin \{b_0, \dots, b_i\}$. Otherwise, $p > c_{j_k}$ and we can conclude that $p \in L(\alpha) \setminus B$ if and only if $p \in C$. □

So we have established that for noncomputable c.e. reals α, $|\mathcal{L}(\alpha)| \geqslant 2$. Are there intermediate representations? That is, for every noncomputable c.e. real α, is there a representation B such that $\emptyset <_T B <_T L(\alpha)$? We call upon a classical result from computability theory; see Soare [16, Chapter VII.3] for more details.

Theorem 26 (Sacks Splitting Theorem). *Let A and D be given noncomputable c.e. sets. Then there are low c.e. sets B and C such that $A = B \cup C$, $B \cap C = \emptyset$ and $D \nleqslant_T B, C$.*

By Theorem 23 every noncomputable c.e. real has a noncomputable representation.

Corollary 27. *Let α be a noncomputable c.e. real, let A be a noncomputable representation of α, and let D be a noncomputable c.e. set. Then there is a noncomputable representation B of α such that $B \leqslant_T A$, $D \not\leqslant_T B$ and B is low.*

Proof. Apply Sacks Splitting Theorem to A and D. At least one of the obtained sets B and C is noncomputable, hence also infinite and by Lemma 16 ($\leftarrow$) a representation of α. □

Remark 28. It is possible to construct directly a low noncomputable representation B avoiding the upper cone of a c.e. D. This is done via a finite priority argument combining the construction of Theorem 22 with the usual techniques of Sacks restraint and lowness requirements. See Soare [16, Chapter VII] for these techniques.

So how many representations do noncomputable c.e. reals have?

Corollary 29. *Let α be a noncomputable c.e. real and A a noncomputable representation of α. Then there is a low representation B of α such that $\emptyset <_T B <_T A$.*

Proof. Apply Corollary 27 with $D = A$. □

Repeated application of this corollary shows that every noncomputable c.e. real has infinitely many representations of different degree, and in fact the representations of noncomputable c.e. reals are downwards dense. That is, if A is a noncomputable representation of α, then there is a noncomputable representation B of α such that $B <_T B$.

5 The cone below $L(\alpha)$

In the light of the above results it is natural to ask whether every c.e. degree in the cone below $L(\alpha)$ contains a representation of α. We say that α realises the cone if for all c.e. sets $A \leqslant_T L(\alpha)$ there is a computable increasing sequence of rationals converging to α of degree A.

Once more, splitting properties of the c.e. Turing degrees play a role. Let

$$S(A) = \{\mathbf{c} \mid \exists A_1(A_1 \text{ is a half of a splitting of } A \text{ and } \deg_T(A_1) = \mathbf{c})\}.$$

Definition 30. A computably enumerable set A has the *Universal Splitting Property* (USP) if $S(A) = \{\mathbf{b} \mid \mathbf{b} \leqslant \deg_T(A)\}$. A is non-USP otherwise.

Lerman and Remmel were motivated to study this property from investigations by Remmel in effective algebra.

Theorem 31 (Lerman and Remmel [9, 10]). *There is a computably enumerable degree $\boldsymbol{a}$ such that every c.e. set of degree $\boldsymbol{a}$ is non-USP.*

Such a degree is called completely non-USP.

Theorem 32. *There is a noncomputable c.e. real α that does not realise the cone.*

Proof. By Theorem 31 take a c.e. degree $\boldsymbol{a}$ which is completely non-USP. Let α be a c.e. real such that $L(\alpha) \in \boldsymbol{a}$. Since $\boldsymbol{a}$ is completely non-USP, there is a c.e. degree $\boldsymbol{b}$ such that $\boldsymbol{0} < \boldsymbol{b} < \boldsymbol{a}$ and $\boldsymbol{b}$ contains no half of a splitting of $L(\alpha)$. Now suppose $\{q_i\}$ is a representation of α. By Lemma 15 $\{q_i\}$ is a half of a splitting of $L(\alpha)$, and hence cannot have Turing degree $\boldsymbol{b}$. □

Remark 33. It is possible to construct directly a noncomputable c.e. real that does not realise the cone. The construction is a finite injury priority argument with strategies that resemble those needed to construct sets without the USP together with technology to deal with computable sequences of rationals.

6 Open Questions

We conclude with several open questions concerning the Turing degrees of increasing computable sequences of rationals converging to reals. Again there is a connection with the splitting properties of c.e. sets and to the structure of $S(A)$ for c.e. sets A.

1. Is $\mathcal{L}(\alpha)$ dense for all noncomputable c.e. reals α?
2. Is there a noncomputable c.e. real α that realises the cone? If there is a completely USP c.e. degree, then the answer is yes.
3. Are there c.e. reals α and β such that $\mathcal{L}(\alpha) \ncong \mathcal{L}(\beta)$?
4. Characterize the c.e. reals β such that $\mathcal{L}(\alpha) \cong \mathcal{L}(\beta)$ for a given α.

References

[1] Bridges, D. S. and Richman, F., *Varieties of Constructive Mathematics*, Cambridge University Press, Cambridge, 1987.

[2] Calude. C., *Theories of Computational Complexity*, North-Holland, Amsterdam, 1988.

[3] Calude, C. S. and Hertling, P., Computable approximations of reals: An information-theoretic analysis, *Fundamenta Informaticae,* 33 (1998), 105–120.

[4] Calude, C. S., Hertling, P., Khoussainov, B., Wang, Y., Recursively enumerable reals and Chaitin Ω numbers, in M. Morvan, C. Meinel, D. Krob (eds.). *STACS'98, Proceedings of the 15th Annual Symposium on Theoretical Aspects of Computer Science, Paris,1998*, Lectures Notes in Computer Science 1373, Springer-Verlag, Berlin, 1998, 596–606.

[5] Cenzer, D. and Remmel, J. B., Π_1^0 classes in mathematics, in *Handbook of Recursive Mathematics,* eds. Y. Ershov, S. Goncharov, A. Nerode and J. Remmel, Elsevier, to appear.

[6] Downey, R. and Stob, M., Splitting theorems in recursion theory, *Annals of Pure and Applied Logic,* 65 (1993). 1–106.

[7] Jockusch, C. G., Semirecursive sets and positive reducibility, *Trans. Amer. Math. Soc.* 131 (1968), 420–436.

[8] Ko, Ker-I, *Complexity Theory of Real Functions,* Birkhäuser, Boston 1991.

[9] Lerman, M. and Remmel, J. B., The universal splitting property, I, in D. van Dalen, D. Lascar and T. J. Smiley, eds., *Logic Colloquium '80* (North-Holland, Amsterdam), 1982, 181–208.

[10] Lerman, M. and Remmel, J. B., The universal splitting property, II, *J. Symbolic Logic* 49 (1984), 137–150.

[11] Martin-Löf, P., *Notes on Constructive Mathematics*, Almqvist & Wiksell, Stockholm, 1970.

[12] Pour-El, M. B. and Richards, J. I., *Computability in Analysis and Physics,* Springer-Verlag, Berlin, 1989.

[13] Rice, H. G., Recursive real numbers, *Proc. Amer. Math. Soc.* 5 (1954), 784–791.

[14] Robinson, R. M., Review of "Peter, R., 'Rekursive Funktionen', Akad. Kiado, Budapest 1951", *J. Symbolic Logic* 16 (1951), 282.

[15] Soare, R. I., Recursion theory and Dedekind cuts, *Trans. Amer. Math. Soc.* 140 (1969), 271–294.

[16] Soare, R. I., *Recursively Enumerable Sets and Degrees,* Springer-Verlag, New York, 1987.

[17] Specker, E., Nicht konstruktiv beweisbare Sätze der Analysis, *J. Symbolic Logic,* 14 (1949), 145–158.

[18] Turing, A. M., On computable numbers, with an application to the Entscheidungsproblem, *Proc. Amer. Math. Soc.* 42 (1936), 230–265, corrections ibid. 43 (1937) 544–546.

[19] Weihrauch, K., *Computability,* Springer- Verlag, Berlin, 1987.

SIMPLICITY AND INDEPENDENCE FOR PSEUDO-ALGEBRAICALLY CLOSED FIELDS

ZOÉ CHATZIDAKIS

CNRS - Paris 7

Simple theories were introduced by Shelah in 1980 in [S]. Recently, Kim and Pillay proved several important results on simple theories, which revived interest in them. Several people are now actively studying these theories, and finding analogs of classical results from the stable context.

A famous conjecture states that all stable fields are separably closed. Thus, examples of fields with a simple theory are of particular interest. The first examples of fields with a simple theory were obtained by Hrushovski in [H], see also [HP]: the perfect bounded pseudo-algebraically closed fields (a field is bounded if for each integer n it only has finitely many Galois extensions of degree n). In [CP], the assumption of perfection on the field was dropped.

The model theory of pseudo-algebraically closed fields (henceforth called PAC) was extensively studied in the seventies and eighties, starting with the work of Ax on pseudofinite fields [A], up to a complete description of the elementary invariants given by Cherlin, van den Dries and Macintyre [CDM], see also the papers by Ershov [E1,E2]. Besides the classical fields (real closed, algebraically closed, separably closed, p-adically closed), the PAC fields form maybe the best known class of fields, in any case the only class with explicit invariants.

In this paper, we show (Theorem 3.9): a PAC field whose theory is simple, is bounded. Note that if the assumption of pseudo-algebraic closure could be dropped, then the conjecture on stable fields would be proved: Poizat ([Po2], Thm. 5.10) has shown that a stable field is either separably closed or has infinitely many extensions of degree n.

The paper starts with two chapters of review, and in chapter 3 we prove the main result. In chapter 4 we study forking in the case of ω-free PAC fields, and introduce the notions of weak independence and strong independence. The first one is implied by non-forking, but has awkward properties. The second notion is more congenial but much stronger than non-forking.

Typeset by $\mathcal{A}_{\mathcal{M}}\mathcal{S}$-TEX

It puts in prominence a notion of "generic extension" of a type, which might be of interest.

1. Review on fields. We review briefly well-known results in field theory. The notions and results specific to positive characteristic (p-bases, p-independence, etc...) can be found in Bourbaki [B]. The other unreferenced results come from Chapter III of Lang's book [L]. We assume a good knowledge of Galois theory.

(1.1) Notation. Let A be a field. Then A^{alg} denotes the (field-theoretic) algebraic closure of A, and A^s the separable closure of A (i.e., the elements of A^{alg} which are separably algebraic over A). We denote by $G(A)$ the absolute Galois group of A, i.e., $G(A) = \mathcal{G}al(A^s/A)$. We often identify $G(A)$ with $Aut(A^{alg}/A)$.

If A and B are subfields of some larger field Ω, we denote by $A[B]$ or $B[A]$ the subring of Ω generated by A and B, and by AB the quotient field of $A[B]$.

If the characteristic of A is $p > 0$, then the map $x \mapsto x^p$ defines a monomorphism $A \to A$; the image of A under this homomorphism is denoted by A^p. We also define $A^{1/p} = \{a \in A^{alg} \mid a^p \in A\}$ and $A^{1/p^\infty} = \{a \in A^{alg} \mid a^{p^n} \in A \text{ for some } n \in \mathbb{N}\}$.

In what follows we will work within models of a complete theory T of fields, and if A is a subfield of a model F of T, then $acl(A)$ will denote the (model-theoretic) algebraic closure of A in F and $dcl(A)$ the (model-theoretic) definable closure of A in F.

(1.2) p-bases, p-independence. Let F be a field of characteristic $p > 0$. Then F^p is a subfield of F, isomorphic to F via the map $x \mapsto x^p$. Thus F is naturally an F^p-vector space. We say that $B \subseteq F$ is *p-independent in F* if the set M of all monomials in B of the form $b_1^{i(1)} \cdots b_n^{i(n)}$ with $b_1, \dots, b_n \in B$ and $0 \le i(1), \dots, i(n) \le p-1$, is independent in the F^p-vector space F. If furthermore M is a basis of the F^p-vector space F, then we call B a *p-basis of F*. Note that B is a p-basis of F if and only if B is a maximal p-independent subset of F (and then $F = F^p[B]$).

Any p-independent subset of F extends to a p-basis of F, and any two p-bases of F have the same cardinality. The size of a p-basis of F is called the *degree of imperfection* of F. The following are easy consequences of the definition:

(1) Let $B \subset F$. Then B is p-independent in F if and only if for every $b \in B$, $b \notin F^p[B \setminus \{b\}]$.

(2) Let $B \subset F$ be p-independent in F. Then B is a p-basis of F if and only if $F^p[B] = F$.

If E is a subfield of F, we say that B is a *p-basis of F over E* if the set M of all monomials in B of the form $b_1^{i(1)} \cdots b_n^{i(n)}$ with $b_1, \ldots, b_n \in B$ and $0 \leq i(1), \ldots, i(n) \leq p-1$, is a basis of the EF^p-vector space F. Then $F = EF^p[B]$.

The size of a p-basis of F over E is called the *degree of imperfection of F over E*. For properties of p-bases, see (1.13) below.

(1.3) The λ-functions. Let F be a field of characteristic $p > 0$. For each n fix an enumeration $m_{i,n}(\bar{x})$ of the monomials $x_1^{i(1)} \cdots x_n^{i(n)}$ with $0 \leq i(1), \ldots, i(n) \leq p-1$. Define the $(n+1)$-ary functions $\lambda_{i,n} : F^n \times F \to F$ as follows:

If the n-tuple $\bar{a}$ is not p-independent, or if the $(n+1)$-tuple $(\bar{a}, b)$ is p-independent, then $\lambda_{i,n}(\bar{a}, b) = 0$. Otherwise, the $\lambda_{i,n}(\bar{a}, b)$ satisfy

$$b = \sum_{i=0}^{p^n-1} \lambda_{i,n}(\bar{a}, b)^p m_{i,n}(\bar{a}).$$

Note that these functions depend on the field F, and that the above properties define them uniquely. They are first-order definable in the field F. We will refer to them as: the "*λ-functions defined on F*".

(1.4) Linear disjointness. We refer to Chapter III of Lang [L] for the proofs and details. Unless otherwise stated, we work inside some large algebraically closed field Ω, and K, L, M, E are subfields of Ω.

Assume that $K \subseteq L, M$. We say that L is *linearly disjoint from M over K* if every finite set of elements of L that is linearly independent over K remains linearly independent over M in the field composite LM.

Even though the definition is asymmetric, the property is symmetric: L is linearly disjoint from M over K if and only if M is linearly disjoint from L over K. Thus we will also say: L and M are *linearly disjoint over K*.

The following are equivalent:

(1) L and M are linearly disjoint over K.
(2) The canonical map $L \otimes_K M \to L[M]$ is an isomorphism.
(3) If $B \subset L$ is a basis of the K-vector space L, then B is a basis of the M-vector space $L[M]$.

(1.5). Let $K \subseteq L, M$, and assume that L is algebraic over K. Then L and M are linearly disjoint over K if and only if $[K(a) : K] = [M(a) : M]$ for every finite tuple a from L. Assume that L is a finite Galois extension of K, linearly disjoint from M over K. Then $[L : K] = [ML : M]$ implies that $\mathcal{G}al(LM/M)$ is canonically isomorphic to $\mathcal{G}al(L/K)$ via the restriction map.

This has the following consequences:

(1) If L is a Galois extension of K, then L and M are linearly disjoint over K if and only if $L \cap M = K$, if and only if the restriction map : $\mathcal{G}al(LM/M) \to \mathcal{G}al(L/K)$ is an isomorphism (of profinite groups).

(2) If L and M are linearly disjoint over K and are Galois extensions of K, then $\mathcal{G}al(LM/K)$ is canonically isomorphic to $\mathcal{G}al(L/K) \times \mathcal{G}al(M/K)$.

(3) Note that (1) can fail when L is not Galois: consider e.g., $a \neq b \in \mathbf{Q}^{alg}$ such that $a^3 = b^3 = 2$; then $\mathbf{Q}(a) \cap \mathbf{Q}(b) = \mathbf{Q}$, and $[\mathbf{Q}(a,b) : \mathbf{Q}(a)] = 2 < [\mathbf{Q}(b) : \mathbf{Q}]$, which shows that $\mathbf{Q}(a)$ and $\mathbf{Q}(b)$ are not linearly disjoint over $\mathbf{Q}$.

(1.6). Let $K \subseteq L, M$, and assume that E is a subfield of L containing K. Then L and M are linearly disjoint over K if and only if E and M are linearly disjoint over K and L and EM are linearly disjoint over E.

(1.7) Let $K \subseteq L$, and let $u_1, \dots, u_n$ be a tuple of elements of Ω which are algebraically independent over L. Then $K(u_1, \dots, u_n)$ and L are linearly disjoint over K.

(1.8) Let $\{L_i \mid i \in I\}$ be a family of extensions of K contained in Ω. We say that $\{L_i \mid i \in I\}$ is linearly disjoint over K, if for every $i \in I$, the field L_i and the field composite of $\{L_j \mid j \in I,\ j \neq i\}$ are linearly disjoint over K. Note that, by (1.6), if $<$ is a linear ordering on I, this is equivalent to: for every i, the field L_i and the field composite of $\{L_j \mid j < i\}$ are linearly disjoint over K.

(1.9) Algebraic independence or freeness. Let $K \subseteq L, M$ be fields. We say that L and M are *free over* K, or that L and M are *algebraically independent over* K, if every finite set of elements of L which is algebraically independent over K remains algebraically independent over M. Again, this notion is symmetric.

If L and M are linearly disjoint over K, then L and M are algebraically independent over K. The converse is not true: if L is algebraic over K, then L is algebraically independent over K from any extension of K.

(1.10) Separable extensions. Let $K \subseteq L$ be fields of characteristic $p > 0$. We say that L is a *separable extension* of K if the fields L and $K^{1/p}$ are linearly disjoint over K. If L is algebraic over K, this is equivalent to $L \subseteq K^s$, and in that case we have $KL^p = L$.

The following conditions are equivalent:

(1) L is a separable extension of K.
(2) L and K^{1/p^∞} are linearly disjoint over K.
(3) Whenever u is a finite tuple from L, then $K(u)$ has a transcendence basis $\{t_1, \dots, t_m\}$ over K, such that $K(u)$ is separably algebraic over $K(t_1, \dots, t_m)$.
(4) L^p and K are linearly disjoint over K^p.

(5) Any p-basis of K remains p-independent in L.
(5') Any p-basis of K is contained in a p-basis of L.
(6) Some p-basis of K remains p-independent in L.
(6') Some p-basis of K is contained in a p-basis of L.
(7) K is closed under the λ-functions of L.

Remark. One extends the definition to the characteristic 0 case as follows: if $char(K) = 0$, then *any* field extension of K is *separable over* K.

(1.11) Let $K \subseteq L \subseteq M$. If M is separable over L and L is separable over K then M is separable over K.

If M is separable over K, then L is separable over K, but M is not necessarily separable over L (e.g.: $K \subset K(t) \subset K(t^{1/p})$).

(1.12) Assume that L is a separable extension of K and that M is an extension of K algebraically independent from L over K. Then LM is a separable extension of M.

Assume that L and M are separable extensions of K, which are algebraically independent over K. Then LM is a separable extension of L and of M (and also of K).

(1.13) Properties of p-bases. Let K be a field of characteristic $p > 0$ and L a separable extension of K. The following facts are easy consequences of the definitions:

(1) If B is a p-basis of L over K, then the elements of B are algebraically independent over K.
(2) Assume that L is finitely generated over K. Then the degree of imperfection of L over K equals the transcendence degree of L over K. Any p-basis of L over K is a (separating) transcendence basis of L over K.
(3) A subset B of L is p-independent over K if and only if for every element $b \in B$, $b \notin KL^p[B \setminus \{b\}]$.
(4) A subset B of L which is p-independent over K is a p-basis of L over K if and only if $L = KL^p[B]$.
(5) Assume that B_0 is a p-basis of K and B_1 is a p-basis of L over K. Then $B_0 \cup B_1$ is a p-basis of L.

(1.14) p-independent extensions. Assume that L and M are separable extensions of K and $char(K) = p > 0$. We say that L and M are *p-independent over* K, if any subset of L which is p-independent over K remains p-independent over M (in the field LM).

Choose a p-basis B_0 of K, and extend it to p-bases B_1 of L and B_2 of M. The following are equivalent:

(1) L and M are p-independent over K.
(2) $B_1 \cup B_2$ is a p-basis of LM.
(3) $(B_1 \setminus B_0) \cup (B_2 \setminus B_0)$ is p-independent over K in the field LM.

Remark. Note that if L and M are linearly disjoint over K, then they are p-independent over K: assume that $B = B_1 \setminus B_0$ is not p-independent over M in LM. Then some non-trivial M-linear combination of elements of $L^p[B]$ is 0. By linear disjointness, some non-trivial K-linear combination of these elements equals 0, which contradicts our assumption on B.

(1.15) Regular extensions. Let $K \subseteq L$. We say that L is a *regular extension of* K if the fields L and K^{alg} are linearly disjoint over K. In that case, the restriction map : $G(L) \to G(K)$ is onto.

The following conditions are equivalent:

(1) L is a regular extension of K.
(2) L is a separable extension of K, and L and K^s are linearly disjoint over K.
(3) L is a separable extension of K, and $L \cap K^s = K$.
(4) L is a separable extension of K, and the restriction map : $G(L) \to G(K)$ is onto.

(1.16) Properties of regular extensions. Let $K \subseteq L, M$.

(1) If L is a regular extension of K and M is a regular extension of L, then M is a regular extension of K.
(2) If M is regular over K and $L \subseteq M$, then L is regular over K.
(3) Assume that L and M are linearly disjoint over K. Then L is regular over K if and only if LM is regular over M.
(4) If L and M are algebraically independent over K and L is a regular extension of K, then L and M are linearly disjoint over K. Thus LM is a regular extension of M.
(5) If L and M are regular extensions of K, and algebraically independent over K, then LM is a regular extension of L, M and K.

(1.17) A particular case: algebraic closure of a set within a model. Let $A \subseteq K$, and consider the algebraic closure $acl(A)$ of A in the model K. Then $acl(A)$ is a field and K is a regular extension of $acl(A)$.

Proof. Any element of K which is separably algebraic over $acl(A)$ is clearly in $acl(A)$, which implies $K \cap acl(A)^s = acl(A)$. By (1.15) it suffices to show that K is a separable extension of $acl(A)$. If $char(K) = 0$ then K is a separable extension of any subfield, and we are done. Otherwise, the λ-functions of K are definable in K by (1.3). Thus $acl(A)$ is closed under the λ-functions of K, which implies that K is a separable extension of $acl(A)$ by (1.10).

(1.18) Algebraic sets, varieties and regular extensions. Details can be found in Chapter III of [L]. Let K be a field, n an integer. A subset V of Ω^n is called *an algebraic set*, or a *Zariski closed set*, if $V = \{a \in \Omega^n \mid f_1(a) = \dots = f_m(a) = 0\}$ for some polynomials $f_i(X) \in \Omega[X]$, $X = (X_1, \dots, X_n)$. We denote by $V(K)$ the set $V \cap K^n$.

If the polynomials $f_1(X), \dots, f_m(X) \in K[X]$, then we say that V is *definable over* K, or that V is a K*-closed set.* The set V is K*-irreducible* if it is not the proper union of two proper K-closed sets. The set V is called *irreducible* (or *absolutely irreducible*, or a *variety*) if it is Ω-irreducible.

If V is defined over K, then V is irreducible if and only if it is K^s-irreducible. Every algebraic set V decomposes into a finite union of irreducible closed sets, say $V_1, \dots, V_m$, and this decomposition is unique up to a permutation, if one assumes that $V_i \subseteq V_j$ implies that $i = j$. The sets V_i are called the *irreducible components* of V.

To an algebraic set V we associate the ideal $I(V) = \{f(X) \in \Omega[X] \mid f(a) = 0 \text{ for all } a \in V\}$.

We say that V is *defined over* K if and only if $I(V)$ is generated by $I(V) \cap K[X]$. For every algebraic set V, there is a unique smallest field over which V is defined, called the *field of definition of* V.

Note: if V is defined over K, then it is definable over K. The converse is not true in general, but we have: if V is definable over K then V is defined over K^{1/p^∞}

Assume that V is definable over K. Then V is K-irreducible if and only if $I(V) \cap K[X]$ is a prime ideal. Assume that V is K-irreducible. Then the irreducible components of V are defined over K^{alg}, and are permuted transitively by $Aut(K^{alg}/K)$.

(1.19). Assume that V is K-irreducible and defined over K. Then V is irreducible if and only if the field of quotients of $K[V] =_{\text{def}} K[X]/I(V) \cap K[X]$ is a regular extension of K.

(1.20) Pseudo-algebraically closed fields. A field F is said to be *pseudo-algebraically closed* (PAC) if every (absolutely irreducible) variety V defined over F has an F-rational point, i.e., a point with all its coordinates in F.

(1.21) Properties of PAC fields. Let E and F be fields, with F PAC.

(1) (10.7 in [FJ]) An algebraic extension of a PAC field is also PAC.

(2) Let E be a regular field extension of F. Then F has an elementary extension F^* containing E. If the degree of imperfection of F is infinite and B is a p-independent subset of E containing a p-basis of F, then F^* can be chosen so that B is a p-basis of F^*. This is an immediate consequence of the definition of PAC and of (1.19).

(3) ((4.5) in [CP]) If $E \subseteq F$, then $acl(E)$ is obtained by closing E under the λ-functions of F and taking the relative (field-theoretic) algebraic closure in F.

(1.22) Definition. We say that the field F is *bounded* if F has finitely many separably algebraic extensions of degree n for all $n > 1$.

Lemma. Assume that F is bounded, and let F^* be an elementary extension of F (in a language containing the language of fields). Then the restriction map : $G(F^*) \to G(F)$ is an isomorphism.

Proof. Equivalently, we need to show that the separable closure of F^* is F^*F^s. Fix $n > 1$, and let $L_1, \ldots, L_N$ be the separable extensions of F of degree n. Since $F \prec F^*$, the extensions L_iF^* have degree n over F^* and are distinct. On the other hand, the phrase: "F has exactly N separable extensions of degree n" is expressible by a first-order sentence of the language of fields, and is therefore satisfied by F^*. Hence $L_1F^*, \ldots, L_NF^*$ are precisely the extensions of F^* of degree n. Thus, for every $n \in \mathbb{N}$, any separable extension of F^* of degree n over F^* is contained in F^*F^s. This proves our assertion.

2. Reviews on forking and simplicity.

(2.1) Definitions and notation. We work in a large model M of a complete theory T, which we assume in a countable language. Let $A \subseteq B$, x a tuple of variables, and let $p(x)$ be a complete type over B, $q(x) = p(x)|_A$.

(1) If C is the image of B under some automorphism of M, we denote by $p_C(x)$ the corresponding image of $p(x)$.

(2) We say that $p(x)$ *forks over* A if there is some A-indiscernible sequence $(B_i : i < \omega)$ of realizations of $tp(B/A)$ such that $\bigcup\{p_{B_i}(x) : i < \omega\}$ is inconsistent.

(3) The theory T is *simple* if for any complete type $p(x)$ over a set B, there is some subset A of B, of cardinality at most that of T, such that $p(x)$ does not fork over A.

(4) Let $\varphi(x, y)$ be a formula. We say that $\varphi(x, y)$ is *represented* in $p(x)$ if there is a tuple b in B such that $\varphi(x, b) \in p(x)$.

(5) Assume that $A \prec M$. We say that p is an *heir* of $q = p|_A$ if every formula represented in $p(x)$ is also represented in $p(x)|_A$.

(5) Assume that $A \prec M$. We say that p is a *coheir* of $q = p|_A$ if whenever a realises p, then $tp(B/A \cup \{a\})$ is an heir of $tp(B/A)$. Equivalently, if for every formula $\varphi(x, b) \in p(x)$, there is $c \in A$ such that $\varphi(c, b)$ holds.

(2.2). The following are easy consequences of the definition of forking.

(1) (Weak right-transitivity) Let $A \subseteq B \subseteq C$. If $tp(a/C)$ does not fork over A, then $tp(a/C)$ does not fork over B, and $tp(a/B)$ does not fork over A.

(2) (Weak left-transitivity) Let $A \subseteq B$, a, b tuples from M. If $tp(ab/B)$ does not fork over Ab and $tp(b/B)$ does not fork over A, then $tp(ab/B)$ does not fork over A.

(2.3) Properties and remarks. We keep the assumptions and notation of (2.1), and also assume that $A \prec M$.

(1) ([Po] 11.01, or [P] 1.15) q extends to a complete type over B which is an heir of q.
(2) ([Po] 12.10 or [P] 1.16) q extends to a complete type over B which is a coheir of q.
(3) Assume that p is a coheir of q. Then p does not fork over A.

Proof of (3). Let $(B_i)_{i\in\omega}$ be an A-indiscernible sequence with $B_0 = B$, and consider a formula $\varphi(x, b) \in p(x)$. By assumption, there is a tuple $a \in A$ such that $M \models \varphi(a, b)$. Since the B_i's realise $tp(B/A)$, $M \models \varphi(a, b_i)$, where b_i is the tuple from B_i corresponding to b. This shows that $\bigcup_i p_{B_i}$ is consistent.

Remarks. (1) Hence, any type over a model A has a non-forking extension to B.

(2) In general it is not true that if p is an heir of its restriction to A, then p does not fork over A. This is because of the possible failure of the symmetry of forking.

(2.4) Properties of simple theories (Kim [K]). Assume that the theory T is simple. Let $A \subseteq B$ and a a tuple. Then
(1) (Symmetry of forking) $tp(a/B)$ forks over A if and only if there is a finite tuple b of elements of B such that $tp(b/A \cup a)$ forks over A.
(2) (Transitivity). Let $B \subseteq C$. Then $tp(a/C)$ does not fork over A if and only if $tp(a/C)$ does not fork over B and $tp(a/B)$ does not fork over A.

3. PAC fields with a simple theory are bounded.

Let F be a field, F^* an elementary extension of F, and a, b two tuples in F^*. In this chapter we first isolate conditions that must be satisfied by a and b if $tp(a/Fb)$ does not fork over F (Theorem 3.5). To show this, we first need to show that our conditions are consistent with $tp(a/F) \cup tp(b/F)$, i.e., that we may move b over F so that a and b satisfy these conditions over F; this is done in Proposition 3.2. We are then able to show the main result (Theorem 3.9): a PAC field F with a simple theory must be bounded.

(3.1) Lemma. Let L and M be regular extensions of K, contained in some large algebraically closed field Ω and linearly disjoint over K. Then
(1) L^s and M^s are linearly disjoint over K^s.
(2) Assume that L_1, M_1 and K_1 are Galois extensions of L, M and K respectively, and that L_1 and M_1 are regular extensions of K_1. Then L_1 and M_1 are linearly disjoint over K_1, and

$$\mathcal{G}al(L_1M_1/LM) \simeq \mathcal{G}al(L_1/L) \times_{\mathcal{G}al(K_1/K)} \mathcal{G}al(M_1/M)$$
$$=_{\rm def} \{(g, h) \in \mathcal{G}al(L_1/L) \times \mathcal{G}al(M_1/M) \mid g|_{K_1} = h|_{K_1}\}.$$

In particular, $[L_1M_1 : LM] = [L_1 : K_1][M_1 : K_1][K_1 : K]$.

(3) Let N be a finite separable extension of LM. There are finite Galois extensions L_1, M_1 and K_1 of L, M, K respectively, such that L_1 and M_1 are regular extension of K_1 and $N \subseteq L_1M_1$.

Proof. (1) By (1.16)(4), since L^s is a regular extension of K^s, and L and M are algebraically independent over K.

(2) By (1.16)(4), L_1 and M_1 are linearly disjoint over K_1. Note that $LM \cap K^s = K$ because LM is a regular extension of K by (1.16)(5), and therefore $\mathcal{G}al(LMK_1/LM) \simeq \mathcal{G}al(K_1/K)$. Hence, to show the next assertion, it suffices to show that L_1M and M_1L are linearly disjoint over LMK_1, and that the restriction maps $\mathcal{G}al(L_1M/LMK_1) \to \mathcal{G}al(L_1/LK_1)$ and $\mathcal{G}al(LM_1/LMK_1) \to \mathcal{G}al(M_1/MK_1)$ are isomorphisms.

Applying (1.6), we obtain first that $(MK_1)L_1 = L_1M$ and M_1 are linearly disjoint over MK_1, and then that $(LMK_1)M_1 = LM_1$ and L_1M are linearly disjoint over LMK_1.

L_1 is a Galois extension of LK_1, linearly disjoint from MK_1 over LK_1. But by (1.5) this implies that the restriction map: $\mathcal{G}al(L_1M/LMK_1) \to \mathcal{G}al(L_1/LK_1)$ is an isomorphism. And similarly, the restriction map $\mathcal{G}al(LM_1/LMK_1) \to \mathcal{G}al(M_1/MK_1)$ is an isomorphism.

The last assertion is obvious.

(3) Choose L_2 and M_2 finite Galois over L and M respectively, such that $N \subseteq L_2M_2$. Define $K_1 = (L_2M_2) \cap K^s$, and $L_1 = L_2K_1$, $M_1 = M_2K_1$. Then L_2 is a regular extension of $L_2 \cap K^s$, and by (1.16), $L_1 = L_2K_1$ is a regular extension of K_1. Similarly M_1 is a regular extension of K_1.

(3.2) Proposition. Let $F \prec F^*$ be fields, and a, b tuples from F^*. Assume that $tp(a/F(b))$ is an heir of $tp(a/F)$. Then the fields $A = acl(Fa)$ and $B = acl(Fb)$ satisfy the following conditions:

(A) A and B are linearly disjoint over F.

(B) F^* is a separable extension of AB.

(C) $acl(AB) \cap A^sB^s = AB$.

Proof. By (1.17), F^* is a regular extension of A, of B, and of $acl(AB)$. Observe also that since the type over $F \cup \{a\}$ of any finite tuple in A is isolated and the type over $F \cup \{b\}$ of any finite tuple in B is algebraic, our assumption implies that $tp(A/B)$ is an heir of $tp(A/F)$.

(A) Choose $a_1, \dots, a_n \in A$ and $b_1, \dots, b_n \in B$, and assume that $a_1b_1 + \dots + a_nb_n = 0$. Then the formula $\varphi(\bar{x}, \bar{y}) : x_1y_1 + \dots + x_ny_n = 0$ is represented in $tp(a_1, \dots, a_n/B)$, which implies that there are $c_1, \dots, c_n \in F$ such that $a_1c_1 + \dots + a_nc_n = 0$, i.e., that $a_1, \dots, a_n$ are not linearly independent over F.

(B) Assume that $char(F) = p > 0$. By (A), we know that A and B are linearly disjoint over F. This implies that if B_1 is a p-basis of A over F and B_2 is a p-basis of B over F, then $B_1 \cup B_2$ is a p-basis of AB over F

(Remark (1.14)).

We need to show that $B_1 \cup B_2$ remains p-independent over F in F^*. Note that each of the sets B_1 and B_2 remains p-independent over F in F^*.

Let $\bar{a}$ be an n-tuple of distinct elements of B_1, $\bar{b}$ an m-tuple of distinct elements of B_2, and assume that $\bar{a} \cup \bar{b}$ is not p-independent over F in F^*. Then there is a polynomial $f(\bar{X}, \bar{Y}, \bar{Z})$ such that $F^* \models \exists \bar{z}\ f(\bar{a}, \bar{b}, \bar{z}^p) = 0$, and $f \in F[\bar{X}, \bar{Y}, \bar{Z}]$ is of degree $\leq p-1$ in each of the variables of $\bar{X} \cup \bar{Y}$. By assumption, this formula is represented in $tp(\bar{a}/F)$, i.e., for some $\bar{c} \in F^m$, we have: $F^* \models \exists \bar{z}\ f(\bar{a}, \bar{c}, \bar{z}^p) = 0$, which contradicts our assumption on $\bar{a}$.

(C) Since F^* is a regular extension of both A and B, $acl(AB) \cap A^s = A$ and $acl(AB) \cap B^s = B$. By (A) and (3.1), A^s and B^s are linearly disjoint over F^s.

Assume that $acl(AB) \cap A^s B^s \neq AB$, and choose a finite algebraic extension N of AB witnessing that fact. By (3.1)(3), there are finite Galois extensions E of F, E_1 of A and E_2 of B such that E_1 and E_2 are regular extensions of E and $N \subseteq E_1 E_2$. Then

$$[E_1 E_2 : acl(AB) \cap E_1 E_2] < [E_1 E_2 : AB] = [E_1 : EA][E_2 : EB][E : F].$$

Let α, β, γ be such that $E = F(\alpha)$, $E_1 = A(\beta)$ and $E_2 = B(\gamma)$. Choose tuples $\bar{a}$ in A, $\bar{b}$ in B, such that the minimal monic polynomial of β over EA is of the form $f(Y, \bar{a}, \alpha)$ for some $f \in F[Y, \bar{U}, X]$, and the minimal monic polynomial of γ over EB is of the form $g(Z, \bar{b}, \alpha)$ for some $g \in F[Z, \bar{V}, X]$. Let $h(X) \in F[X]$ be the minimal polynomial of α over F. Then the ideal generated by $h(X), f(Y, \bar{a}, X), g(Z, \bar{b}, X)$ is a prime ideal of $AB[X, Y, Z]$ (it is precisely the ideal of polynomials in $AB[X, Y, Z]$ vanishing at (α, β, γ)). Let $\varphi(\bar{y})$ and $\psi(\bar{x}, \bar{y})$ be the $\mathcal{L}(F)$-formulas expressing that the ideals generated respectively by $h(X), g(Z, \bar{y}, X)$ in $F[X, Z]$ and by $h(X), f(Y, \bar{x}, X), g(Z, \bar{y}, X)$ in $F[X, Y, Z]$ are prime (such formulas exist, see e.g. [DS]), and assume that $F^* \models \neg\psi(\bar{a}, \bar{b})$. Then some tuple $\bar{c}$ from F satisfies $\varphi(\bar{c}) \wedge \neg\psi(\bar{a}, \bar{c})$. This implies that the ideal of $F^*[X, Y, Z]$ generated by $h(X), f(Y, \bar{a}, X), g(Z, \bar{c}, X)$ is not prime, and therefore that the polynomial $g(Z, \bar{c}, \alpha)$ is not irreducible over $F^*(\alpha, \beta) = F^* E_1$, as $F^*(\alpha, \beta)$ is canonically isomorphic to $F^*[X, Y]/(h(X), f(Y, \bar{a}, X))$.

The polynomial $g(Z, \bar{c}, \alpha)$ is irreducible over $F(\alpha)$ because $F^* \models \varphi(\bar{c})$, and its roots are in F^s since $\alpha \in E^s$ and $\bar{c} \in F$. Hence, the fact that $g(Z, \bar{c}, \alpha)$ does not remain irreducible over $F^* E_1$ implies that $F^* E_1$ is not a regular extension of E. Since F^* is a regular extension of A, $F^* E_1$ is a regular extension of E_1 by (1.6); by definition, E_1 is a regular extension of E, and therefore $F^* E_1$ is a regular extension of E, which gives us a contradiction.

Thus $F^* \models \psi(\bar{a}, \bar{b})$, and $E_1 E_2 \cap acl(AB) = AB$.

(3.3) Remarks. Conditions (A) and (B) are not surprising. Keeping the notation of (3.2), and working in the complete theory SCF of the separable closure F^s of F, we have: $tp_{SCF}(a/Fb)$ is the heir of $tp(a/F)$ if and only if the λ-closures of $F(a)$ and $F(b)$ in F^* (or equivalently in $(F^*)^s$) satisfy conditions (A) and (B).

Condition (C) is new. Note that if F is bounded, condition (C) is vacuous: indeed, we then have $(F^*)^s = F^*F^s$ (Lemma 1.22), which implies that $A^s = AF^s$, $B^s = BF^s$ and $acl(AB)^s = acl(AB)F^s$; hence $acl(AB) \cap A^sB^s = acl(AB) \cap ABF^s = AB$ because $acl(AB)$ is a regular extension of F.

(3.4) An example. Notation as in (3.3). Assume that F has infinitely many Galois extensions of degree 2 and has characteristic $\neq 2$. Then the equivalence relation $\sim$ on $F^\times$ (the non-zero elements of F) defined by $x \sim y$ if and only if xy^{-1} is a square in F, has infinitely many classes.

Assume now that the $\sim$-equivalence classes $[a]_\sim$ and $[b]_\sim$ are not in $F^\times/\sim$. Proposition 3.2 shows that there is b_1 realising $tp(b/F)$ such that $b \not\sim b_1$. Thus if $x \sim b \in tp(a/Fb)$ then $tp(a/Fb)$ forks over F: by the definition of forking and because $(x \sim b) \wedge (x \sim b_1)$ is not satisfiable.

(3.5) Theorem. Let $F \prec F^*$ be fields, and a, b tuples from F^*. Assume that $tp(a/F(b))$ does not fork over F. Then the fields $A = acl(Fa)$ and $B = acl(Fb)$ satisfy the following conditions:
(A) A and B are linearly disjoint over F.
(B) F^* is a separable extension of AB.
(C) $acl(AB) \cap A^sB^s = AB$.

Proof. By assumption, $tp(A/B)$ does not fork over F, and we may choose a sequence $(B_i)_{i\in\mathbf{N}}$ of realisations of $tp(B/A)$, indiscernible over F and such that $tp(B_{i+1}/acl(B_0\cdots B_i))$ is an heir of $tp(B/F)$ for any i. By Proposition 3.2, the fields $acl(B_0\cdots B_i)$ and B_{i+1} satisfy conditions (A), (B) and (C) over F.

(A) Assume by way of contradiction that A and B are not linearly disjoint over F, and let $b_1, \ldots, b_n \in B$ be linearly independent over F and $a_1, \ldots, a_n \in A$, $(a_1, \ldots, a_n) \neq (0, \ldots, 0)$, such that $a_1b_1 + \cdots + a_nb_n = 0$. For $i \in \mathbf{N}$, let $(b_{1,i}, \ldots, b_{n,i})$ be the n-tuple of B_i corresponding to $(b_1, \ldots, b_n)$. Consider the system $\Sigma(x_1, \ldots, x_n)$:

$$x_1b_{1,i} + x_2b_{2,i} + \cdots + x_nb_{n,i} = 0, \qquad \text{for all } i \in \mathbf{N}.$$

Since the $b_{j,i}$'s are linearly independent over F, the determinant of any n of these equations is non-zero. This implies that $(0, 0, \ldots, 0)$ is the only solution of Σ, and therefore that $tp(a_1, \ldots, a_n/F) \cup \Sigma(x_1, \ldots, x_n)$ is inconsistent. Hence $tp(a_1, \ldots, a_n/B)$ forks over F, a contradiction.

(B) We may therefore assume that (A) holds. By (1.6) this implies that A and the field composite of $\{B_i \mid i \in \mathbf{N}\}$, are linearly disjoint over F. If (B) does not hold, then $p = char(F)$ is positive and there are tuples $a_1, \dots, a_n$ in A and $b_1, \dots, b_m$ in B, which are p-independent over F in A and B respectively, but do not remain p-independent over F in F^*. Take such tuples with m minimal. This implies that $b_m \in F(F^*)^p[a_1, \dots, a_n, b_1, \dots, b_{m-1}]$.

For $0 \leq i \leq n$, let $b_{i,1}, \dots, b_{i,m} \in B_i$ be the elements corresponding to $b_1, \dots, b_m$, and let $C = F(b_{i,j} \mid 0 \leq i \leq n, 1 \leq j \leq m)$. Then $[CF^{*p} : FF^{*p}] = p^{m(n+1)}$.

Since $(b_{i,1}, \dots, b_{i,m})$ realises $tp(b_1, \dots, b_m/A)$ for $i \leq n$ we have:

$$[CF^{*p}(a_1, \dots, a_n) : FF^{*p}] \leq [FF^{*p}(a_1, \dots, a_n) : FF^{*p}] \times \prod_{i=0}^{n} [FF^{*p}(a_1, \dots, a_n, b_{i,1}, \dots, b_{i,m}) : FF^{*p}(a_1, \dots, a_n)]$$

$$= p^n p^{(m-1)(n+1)} = p^{m(n+1)-1},$$

which gives a contradiction. Hence (B) holds.

For (C): Assume that $acl(AB) \cap A^s B^s$ strictly contains AB. As in the proof of (3.2), there are finite Galois extensions E of F, E_1 of A and E_2 of B, such that E_1 and E_2 are regular extensions of E, and $[E_1E_2 : acl(AB) \cap E_1E_2] < [E_1E_2 : AB] = [E_1 : A][E_2 : EB]$. Since $acl(AB)$ is a regular extension of A, E_1 and $acl(AB)$ are linearly disjoint over A. By (1.6), this implies that $E_1B \cap acl(AB) = AB$. Hence $E_1(acl(AB) \cap E_1E_2)$ is a proper Galois extension of E_1B, contained in E_1E_2. This and the regularity of $acl(AB)$ over A imply:

$$[E_1E_2 : E_1(acl(AB) \cap E_1E_2)] < [E_1E_2 : E_1B] = [E_2 : EB],$$
$$[E_1 acl(AB) : acl(AB)] = [E_1 : A].$$

For $i \in \mathbf{N}$ let K_i be the Galois extension of B_i corresponding to E_2, and let $L_i = acl(AB_i) \cap E_1K_i$. Since the B_i's are linearly disjoint over F, and K_i is a regular extension of $E = E_2 \cap F^s$, Lemma 3.1 (2) gives

$$[K_0 \cdots K_n : B_0 \cdots B_n] = \Big(\prod_{i=0}^{n} [K_i : EB_i]\Big)[E : F]$$
$$= [E_2 : EB]^{n+1}[E : F].$$

Also, $[E_1K_i : E_1L_i] = [E_1E_2 : E_1(acl(AB) \cap E_1E_2)] \leq [E_2 : EB]/2$ for $i \in \mathbf{N}$, and $[E_1L_i : L_i] = [E_1 : A]$. Hence

$$[E_1K_0 \cdots K_n : L_0 \cdots L_n] \leq [E_1 : A] \prod_{i=0}^{n} [E_1K_i : E_1L_i]$$
$$\leq [E_1 : A]([E_2 : EB]/2)^{n+1}.$$

On the other hand, from $L_0 \cdots L_n \subseteq acl(AB_0 \cdots B_n)$ we get

$$[E_1K_0 \cdots K_n : L_0 \cdots L_n] \geq [E_1K_0 \cdots K_n : acl(AB_0 \cdots B_n) \cap E_1K_0 \cdots K_n].$$

Since $acl(AB_0 \ldots B_n)$ is a regular extension of $acl(B_0 \ldots B_n)$ (see (1.17)), $acl(AB_0 \cdots B_n) \cap K_0 \cdots K_n = acl(B_0 \cdots B_n) \cap K_0 \cdots K_n$, and therefore

$$[E_1K_0 \cdots K_n : acl(AB_0 \cdots B_n) \cap E_1K_0 \cdots K_n] \geq [K_0 \cdots K_n : acl(B_0 \cdots B_n) \cap K_0 \cdots K_n].$$

Since the B_i's satisfy condition (C), $acl(B_0 \cdots B_n) \cap K_0 \cdots K_n = B_0 \cdots B_n$. Hence, we obtain

$$[E_1K_0 \cdots K_n : L_0 \cdots L_n] \geq [K_0 \cdots K_n : B_0 \cdots B_n].$$

Thus for all n, we have: $[E : F][E_2 : EB]^{n+1} \leq [E_1 : A]([E_2 : EB]/2)^{n+1}$, which is absurd.

(3.6) Remark. With the notation above, the non-forking of $tp(A/B)$ over F implies in fact the stronger property:
(D) For every subfield E of B, $acl(AB) \cap acl(EA)^s B^s = acl(EA)B$.
Indeed, if $tp(A/B)$ does not fork over F then it does not fork over $acl(FE)$.

(3.7) The example revisited. Assume that F is a PAC field of characteristic $\neq 2$, with infinitely many Galois extensions of degree 2. We will assume that F has size $\aleph_1$. The proof that $\mathrm{Th}(F)$ is not simple proceeds as follows:

Step 1: Show that the type $\Sigma(x) = \{(x+c) \not\sim (x+d) \mid c, d \in F, c \neq d\}$ is consistent. (This produces $\aleph_1$ new $\sim$-equivalence classes, which are equi-algebraic over F).

Step 2: Find an elementary extension K of F, containing elements c_α, $\alpha < \aleph_1$, which are algebraically independent over F and in distinct $\sim$-equivalence classes in K.

Step 3: Let d_α, $\alpha < \aleph_1$, be distinct elements of F. Show that $\Phi(x) = \Sigma(x) \cup \{(x + d_\alpha) \sim c_\alpha\}$ is consistent.

Clearly, if a realises $\Phi(x)$ and $tp(a/K)$ does not fork over a subset C of K, then $tp(a/K)$ does not fork over $acl(FC)$. Hence $acl(FC)$ must contain all the c_α's, which implies that C is uncountable.

(3.8) Lemma. Let F be an uncountable PAC field, t an element transcendental over F, and fix $n \geq 5$. Then $F(t)$ has a Galois extension M, regular over F, with $\mathcal{G}al(M/F(t))$ isomorphic to the direct product of $|F|$ copies of A_n (the group of even permutations on n elements).

Proof. Let k be a countable elementary substructure of F, $\kappa = |F|$, and $b_\alpha, \alpha < \kappa$, a transcendence basis of F over k. By a result of F. Pop (Theorem A in [Pp], see also Theorem 6.4 of [HJ] for a more elementary proof), there is a polynomial $f(u, X) \in k[u, X]$ such that the splitting field L of $f(u, X)$ over $k(u)$ is regular over k and $\mathcal{G}al(L/k(u)) \simeq A_n$.

For each α, consider the polynomial $f(b_\alpha t, X)$, and its splitting field K_α over $k(b_\alpha t)$. Because b_α and t are algebraically independent over k, each of them is also transcendental over $k(b_\alpha t)$. This implies that $L_\alpha = K_\alpha(t)$ is a regular extension of both $k(t)$ and $k(b_\alpha)$, and that $\mathcal{G}al(L_\alpha/k(b_\alpha, t)) \simeq A_n$.

L_α is a regular extension of $k(b_\alpha)$, and is algebraically independent from F over $k(b_\alpha)$. By (1.16)(3), $M_\alpha =_{\text{def}} FL_\alpha$ is a regular extension of F. We also have: $\mathcal{G}al(M_\alpha/F(t)) = A_n$.

We claim that the Galois extensions M_α, $\alpha < \kappa$, are linearly disjoint over $F(t)$. Indeed, assume that for some α, M_α is not linearly disjoint from the field composite N of $\{M_\beta \mid \beta \neq \alpha\}$. Then, $M_\alpha \cap N$ is a proper Galois extension of $F(t)$ (by (1.5)(1)). Since A_n is simple, this implies that $M_\alpha \subseteq N$, and therefore also that $L_\alpha \subseteq N$. The sets $\{b_\alpha\}$, $\{t\}$ and $\{b_\beta \mid \beta \neq \alpha, \beta < \kappa\}$ are independent over k. Hence, by Remark (1.9)(2) of [CH],

$$(*) \qquad k(t, b_\alpha)^{alg} \cap (k(t, b_\beta \mid \beta \neq \alpha)^{alg} F^{alg}) = k(t)^{alg} k(b_\alpha)^{alg}.$$

By the definition of L_α (and the properties of $f(u, X)$), we know that $\mathcal{G}al(k(t)^s L_\alpha/k(t)^s(b_\alpha)) = A_n$ and $\mathcal{G}al(k(b_\alpha)^s L_\alpha/k(b_\alpha)^s(t)) = A_n$. Since A_n is simple non-abelian and $k(t)^s(b_\alpha) \cap k(b_\alpha)^s(t) = k^s(b_\alpha, t)$, this implies that L_α cannot be contained in $k(t)^s k(b_\alpha)^s$. From $N \subseteq k(t, b_\beta \mid \beta \neq \alpha)^{alg} F^{alg}$ and $(*)$, we deduce that L_α cannot be contained in N. This gives us a contradiction, and shows that the M_α's are linearly disjoint over $F(t)$.

Let M be the field composite of $\{M_\alpha \mid \alpha < \kappa\}$. Then $\mathcal{G}al(M/F(t))$ is isomorphic to the product of κ copies of A_n. Assume that M is not regular over F. Then $F^s \cap M$ strictly contains F. Since A_n is simple non-abelian, this implies that some M_α is contained in $F^s(t)$, but this contradicts the regularity of M_α over F.

(3.9) Theorem. Let F be a PAC field, and assume that Th(F) is simple. Then F is bounded.

Proof. We will assume that F is unbounded and Th(F) is simple, and reach a contradiction. Let n be such that F has infinitely many separable extensions of degree n. Then F has infinitely many Galois extensions of degree $\leq n!$. Choose m minimal such that F has infinitely many Galois extensions of degree m. Then F has a finite Galois extension L such that infinitely many of these extensions contain L and are linearly disjoint over L. If the theory of F is simple, then so is the theory of L; by (1.21)(1),

L is also PAC. We may therefore assume that $F = L$, i.e., that there are infinitely many algebraic extensions of F, which are linearly disjoint over F and with Galois group over F isomorphic to some fixed simple group G. The strategy of the proof is the one given in (3.7).

We may assume that F has size $\aleph_1$, and that F^* is an elementary extension of F which is sufficiently saturated. Using compactness and our assumption on $G(F)$, we know that F has an elementary extension F', with a Galois extension E such that $\mathcal{G}al(E/F') \simeq G \simeq \mathcal{G}al(EF^s/F'F^s)$ (i.e., $G(F')$ has a quotient isomorphic to G, which does not come from a Galois extension of F). Use this remark to build an increasing sequence F_α, $\alpha < \aleph_1$, of elementary substructures of F^*, such that $F_0 = F$, each F_α has a Galois extension E_α, which is linearly disjoint from F^s_β over F_β for every $\beta < \alpha$, and with $\mathcal{G}al(E_\alpha/F_\alpha) \simeq G$.

Let $K = \bigcup_{\alpha<\aleph_1} F_\alpha$, and let $L_\alpha = KE_\alpha$. Then the extensions L_α are Galois over K with Galois group isomorphic to G, and they are linearly disjoint over K. By assumption, E_α is a regular extension of $\bigcup_{\beta<\alpha} F_\beta$; this implies that the field composite of $\{E_\beta \mid \beta \leq \alpha\}$ is a regular extension of the field composite of $\{E_\beta \mid \beta < \alpha\}$. Let L be the field composite of $\{L_\alpha \mid \alpha < \kappa\}$; then L is also the field composite of $\{E_\alpha \mid \alpha < \kappa\}$, and the above, and transitivity of regularity, show that L is a regular extension of K.

Choose an element a transcendental over K, and let n be such that G embeds into A_n (e.g., $n = 2|G|$). By (3.8), we can find a family $\{M_\alpha \mid \alpha < \aleph_1\}$, of Galois extensions of $F(a)$, which are linearly disjoint over $F(a)$, with Galois group over $F(a)$ isomorphic to A_n, and such that their composite M is a regular extension of F. For each $\alpha < \kappa$ fix an embedding i_α : $\mathcal{G}al(L_\alpha/K) \to \mathcal{G}al(M_\alpha/F(a))$, and consider the subgroup $H_\alpha = \{(g, i_\alpha(g) \mid g \in \mathcal{G}al(L_\alpha/K)\}$ of $\mathcal{G}al(L_\alpha M_\alpha/K(a)) \simeq \mathcal{G}al(L_\alpha/K) \times \mathcal{G}al(M_\alpha/F(a))$. Let N_α be the subfield of $L_\alpha M_\alpha$ fixed by H_α. From the definition of H_α we conclude that $L_\alpha M_\alpha = L_\alpha N_\alpha = M_\alpha N_\alpha$.

Let N be the field composite of $\{N_\alpha \mid \alpha < \kappa\}$. We want to show that N is a regular extension of K. We know that L is a regular extension of F, Galois over K, M is a regular extension of F, Galois over $F(a)$, and L and M are algebraically independent over F. By (3.1) this implies that $\mathcal{G}al(LM/K(a)) \simeq \mathcal{G}al(L/K) \times \mathcal{G}al(M/F(a))$, and therefore that $\mathcal{G}al(LM/N)$ is canonically isomorphic to $\prod_{\alpha<\kappa} H_\alpha$, and in particular projects onto $\mathcal{G}al(L/K)$. Our assumptions on L, M also imply that LM is a regular extension of L; hence $N \cap K^s \subseteq N \cap (LM \cap K^s) = N \cap L = K$, which shows that N is a regular extension of K.

Consider now the extension N' defined as follows: if *char*$(F) = 0$ then $N' = N$; if *char*$(F) = p > 0$ then N' is obtained by adjoining to N all

p^n-th roots of a. Then $N' \cap K^s = K$. If $char(F) = p > 0$, then N' is a separable extension of $K(a^{1/p^n} \mid n \in \mathbf{N})$, and $K(a^{1/p^n} \mid n \in \mathbf{N})$ is separable over K. Hence N' is a regular extension of K, with the same p-basis as K if the characteristic is positive.

By (1.21)(2), we may assume that $N' \subseteq F^*$. Consider now $tp(a/K)$. Since Th(F) is simple, there is a countable subfield E of K such that $tp(a/K)$ does not fork over E. We will show that this gives a contradiction. Indeed, choose $\alpha < \aleph_1$ such that $E \subseteq F_\alpha$. Then, $tp(a/K)$ does not fork over F_α. Thus, by Theorem 3.5

$$(**) \qquad acl(K(a)) \cap acl(F_\alpha(a))^s K^s = acl(F_\alpha(a))K.$$

Observe that since F^* contains all p^n-th roots of a, then $acl(Ka) = K(a)^{alg} \cap F^*$, and $acl(F_\alpha(a)) = F_\alpha(a)^{alg} \cap F^*$. Consider $N_{\alpha+1}$. It is contained in F^* and is algebraic over $K(a)$, and therefore is contained in $acl(K(a))$. It is also contained in $L_{\alpha+1}M_{\alpha+1} \subseteq K^s F(a)^s$. Hence $N_{\alpha+1} \subseteq acl(K(a)) \cap acl(F_\alpha(a))^s K^s$, and therefore $N_{\alpha+1} \subseteq acl(F_\alpha(a))K$ by (**). From $E_{\alpha+1} \subseteq L_{\alpha+1} \subseteq M_{\alpha+1}N_{\alpha+1}$ we deduce that

$$E_{\alpha+1} \subseteq acl(F_\alpha(a))KM_{\alpha+1} \subseteq F_\alpha(a)^{alg}K.$$

Since $E_{\alpha+1} \subseteq F^s_{\alpha+1} \subseteq K^s$ and a is transcendental over K, we get $E_{\alpha+1} \subseteq F^s_\alpha K$, and then $E_{\alpha+1} \subseteq F^s_\alpha F_{\alpha+1}$. But this contradicts the choice of $E_{\alpha+1}$.

Remark. The above construction can be easily modified to give any order type of forking.

4. A strong notion of independence for ω-free PAC fields.

Throughout this chapter F will be an ω-free PAC field. We start with the definition and some properties of ω-free PAC fields.

(4.1) ω-free PAC fields. Recall that a field F is ω-free if some countable elementary substructure F_0 of F has absolute Galois group $G(F_0)$ isomorphic to $\hat{F}_\omega$, the free profinite group on $\aleph_0$ generators. An ω-free field is certainly unbounded: $\mathbf{Z}/p\mathbf{Z}^{\aleph_0}$ is a quotient of $\hat{F}_\omega$. Their elementary theory is however very pleasant, for a description see [FJ]. If F is ω-free PAC and κ-saturated, then:

(1) Let $E = acl(E)$ be a subfield of F of size $< \kappa$, and a, b two tuples from F. Then $tp(a/E) = tp(b/E)$ if and only if there is an E-isomorphism f between $acl(Ea)$ and $acl(Eb)$ which sends the tuple a to the tuple b.

(2) Let $E = acl(E)$ be a subfield of F, and L a regular extension of E of size $< \kappa$. Then there is an E-embedding $f : L \to F$ such that $F \cap f(L)^s = f(L)$. If $[L : L^p] \leq [F : F^p]$ then f can be chosen so that F is a regular extension of $f(L)$.

Proof. These results are well-known, but I was unable to find a direct reference. For (1), use Proposition 18.9 of [FJ] and the following two observations: F is separable over $acl(Ea)$ and $acl(Eb)$. Extend f to an isomorphism with domain F. Then $f(F)$ is also ω-free PAC, and $tp(a/E) = tp(b/E)$ if and only if $f(F) \equiv_{acl(E)} F$. For (2), use Lemma 24.32 of [FJ] and the saturation of F.

(4.2) I believe that properties (A), (B) and (D) characterise non-forking for ω-free PAC fields. However, while the properties (A) and (B) are well-behaved, property (D) is not. To avoid repetitions, we will introduce a definition: we say that the tuple a is *weakly independent from the tuple* b *over* E, if $acl(Ea)$ and $acl(Eb)$ satisfy conditions (A), (B) and (D) over $acl(E)$. Below and in (4.4) we give some examples of this bad behaviour.

(4.3) Proposition. Let F be an ω-free PAC field. We can find elements a, b, c (in some elementary extension of F), such that a is weakly independent from (b, c) over $F(b)$, and from b over F, but a is not weakly independent from (b, c) over F.

Proof. We will assume that F is of characteristic 0. The proof is almost identical in positive characteristic, (using cube roots instead of square roots in characteristic 2), but one has to be a little careful with the perfect closure.

Let a, b, c be algebraically independent transcendental elements over F, and consider the field $L = F(a, b, c, \sqrt{c(a + b + c)})$. By (4.1)(2), we may assume that L is contained in an elementary extension F^* of F, and that F^* is a regular extension of L. Then $acl(F(a, b)) = F(a, b)$, $acl(F(a, c)) = F(a, c)$ and $acl(F(b, c)) = F(b, c)$. Conditions (A) and (B) are immediate. Then

$$\begin{aligned} F^* \cap acl(F(a,b))^s acl(F(b,c))^s &= L \cap (F(a,b)^s F(b,c)^s) \\ &= F(a,b)F(a,c). \end{aligned}$$

Since $tr.deg(F(b, c)/F(b)) = 1$, this shows condition (D), and therefore that a is weakly independent from (b, c) over $F(b)$. Similarly, $F^* \cap acl(F(a))^s acl(F(b))^s = F(a, b)$, and a is weakly independent from b over F. However, a is not weakly independent from (b, c) over F: looking at the subfield $F(b + c)$ of $F(b, c)$, one gets:

$$F^* \cap acl(F(a, b + c))^s acl(F(b, c))^s = L$$

because $\sqrt{a + b + c} \in (F(a, b + c))^s$ and $\sqrt{c} \in F(b, c)^s$; on the other hand, $L \cap F(a, b + c)^s = F(a, b + c)$ and $L \cap F(b, c)^s = F(b, c)$.

(4.4) Proposition. Let F be an ω-free PAC field. Then forking is not fully left-transitive, i.e., we can find a, b, c such that $tp(a, b/F(c))$ does not fork over F, but $tp(a/F(b, c))$ forks over $F(c)$.

Proof. Again, for simplicity, we will assume that the characteristic is 0. The proof is similar in the positive characteristic case.

Let a, b, c be algebraically independent transcendental elements over F, and consider the field $L = F(a, b, c, \sqrt{a(b+c)})$. As in (4.3), we may assume that $L \subseteq F^*$, where F^* is a regular extension of L and an elementary extension of F.

Let c_i, $i \in \mathbf{N}$, be a sequence of indiscernibles over F, with $c_0 = c$. Then, either they are all equal, or they are algebraically independent over F. In the first case there is nothing to prove. Assume we are in the second case; we may assume that (a, b) are algebraically independent over $K = acl(F, c_i \mid i \in \mathbf{N})$. Then the extension $M = K(a, b, \sqrt{a(b+c_i)} \mid i \in \mathbf{N})$ is a regular extension of K and of $F(a, b)$. For each i, we have $M \cap F(a, b, c_i)^s = F(a, b, c_i, \sqrt{a(b+c_i)})$. By (4.1)(2), we may assume that F^* is a regular extension of M. By (4.1)(1), all c_i's realise $tp(c/F(a, b))$. This shows that $tp(a, b/F(c))$ does not fork over F.

On the other hand: $L \subseteq F(a)^s F(b, c)^s$, and $acl(F(a, b)) = L \cap F(a, b)^s = F(a, b)$, $acl(F(b, c)) = L \cap F(b, c)^s = F(b, c)$. This shows that a is not weakly independent from (b, c) over $F(b)$, and therefore that $tp(a/F(b, c))$ forks over $F(b)$ by Theorem 3.5.

(4.5) Definition. Let $E = acl(E)$ be a subfield of F, and a, b two tuples from F. We say that a and b are *strongly independent over* F if they satisfy conditions (A) and (B), and moreover: $F \cap acl(E(a, b))^s = acl(E(a))acl(E(b))$.

This notion is clearly symmetric and transitive.

(4.6) Lemma. Let $E = acl(E)$ be a subfield of F, and a, b, c_1, c_2 tuples from F. Assume that a and c_1 are strongly independent over E, and b and c_2 are strongly independent over E, and that c_1 and c_2 realise the same type over E. Assume moreover, if the degree of imperfection of F is finite, that E contains a p-basis for F.

Then there is c realising $tp(c_1/acl(Ea)) \cup tp(c_2/acl(Eb))$, which is strongly independent from (a, b) over E.

Proof. Let $A = acl(Ea)$, $B = acl(Eb)$, and $C_1 = acl(Ec_1)$, $C_2 = acl(Ec_2)$. Let C be an E-isomorphic copy of C_1 such that C and $acl(AB)$ satisfy the conditions (A) and (B) over E. Consider now $L = Cacl(AB)$. It is a regular extension of C and of $acl(AB)$. By (4.1)(2), we may therefore assume that F is a regular extension of L. By (4.1)(1) and our strong independence assumptions, $tp(C/A) = tp(C_1/A)$ and $tp(C/B) = tp(C_2/B)$. Also clearly C and $acl(AB)$ are strongly independent over E.

(4.7) Corollary. If a and b are strongly independent over E, then $tp(a/Eb)$ does not fork over E.

Proof. Immediate by the definition of forking and (4.6).

(4.8) Remarks. Note that, given a and b, $tp(a/E)$ has a unique extension to $E(b)$ such that a and b are strongly independent over E. It is clearly a sort of "generic extension".

Lemma 4.6 exemplifies this uniqueness: it says that if two types over two sets of parameters containing E but not necessarily independent over E, are "strong" non-forking extensions of $tp(c/E)$, then they have a common extension to a "strong" non-forking extension.

REFERENCES

[A] J. Ax, The elementary theory of finite fields, Annals of Math. 88 (1968), 239–271.

[B] N. Bourbaki, XI, Algèbre Chapître 5, Corps commutatifs, Hermann, Paris 1959.

[CH] Z. Chatzidakis, E. Hrushovski, The model theory of difference fields, to appear in Trans. A.M.S.

[CP] Z. Chatzidakis, A. Pillay, Generic structures and simple theories, to appear in Ann. P. Appl. Logic.

[CDM] G. Cherlin, L. van den Dries, A. Macintyre, Decidability and Undecidability Theorems for PAC-Fields, Bull. AMS 4 (1981), 101-104.

[DS] L. van den Dries, K. Schmidt, Bounds in the theory of polynomials rings over fields. A non-standard approach. Invent. Math. 76 (1984), 77–91.

[E1] Ju. L. Ershov, Regularly closed fields, Soviet Math. Doklady 31 (1980), 510–512.

[E2] Ju. L. Ershov, Undecidability of regularly closed fields, Alg. and Log. 20 (1981), 257–260.

[FJ] M. Fried, M. Jarden, Field Arithmetic, Ergebnisse 11, Springer Berlin-Heidelberg 1986.

[HJ] D. Haran, M. Jarden, Regular split embedding problems over complete local fields, Forum Mathematicum 10 (1998) no 3, 329 – 351.

[H] E. Hrushovski, Pseudo-finite fields and related structures, manuscript 1991.

[HP] E. Hrushovski, A. Pillay, Groups definable in local fields and pseudo-finite fields, Israel J. of Math. 85 (1994), 203 – 262.

[K] B. Kim, Forking in simple unstable theories, J. London Math. Soc. 57 (1998), 257–267.

[KP] B. Kim, A. Pillay, Simple theories, Ann. P. Appl. Logic 88 Nr 2-3 (1997), 149–164.

[L1] S. Lang, Introduction to algebraic geometry, Addison-Wesley Pub. Co., Menlo Park 1973.

[P] A. Pillay, An introduction to stability theory, Oxford Logic Guide 8, Clarendon Press, Oxford, 1983.

[Po1] B. Poizat, Cours de Théorie des Modèles, Nur Al-Mantiq Wal-Ma'rifah, Paris 1985.

[Po2] B. Poizat, Groupes stables, Nur Al-Mantiq Wal-Ma'rifah, Paris 1987.

[Pp] F. Pop, Embedding problems over large fields, Ann. of Math. (2) 144 (1996), no. 1, 1–34.

[S] S. Shelah, Simple unstable theories, Ann. P. Appl. Logic 19 (1980), 177–203.

Clockwork or Turing U/universe? – Remarks on Causal Determinism and Computability

S. Barry Cooper [†]

University of Leeds
Leeds LS2 9JT
England

ABSTRACT. The relevance of the Turing universe as a model for complex physical situations (that is, those showing both computable and incomputable aspects) is discussed. Some well-known arguments concerning the nature of scientific reality are related to this theoretical context.

The close relationship between computability, mechanism and causal determinacy[1] is basic to post-Newtonian science, and underpins the familiar notion of a Laplacian 'clockwork' Universe, which provided a clear mathematical model of physical reality for over a century and a half — the discovery in the 1930s of the possibility of an explicit mathematical description of the model providing a key ingredient in its eventual demise. Acceptance of such a model has never been total of course, even amongst scientists. But in recent times the need for a mathematical alternative to that of Laplace, subsuming and extending according to the changing theoretical and empirical environment, has become increasingly overdue. The purpose of this note is to argue that the genesis of such a precise and intuitively natural model lies in Alan Turing's response, in a more limited mathematical context, to the newly discovered incomputabilities of the decade 1927–1936.

[†] We would like to thank P. Odifreddi for a number of helpful comments on the first draft of this paper. Supported by E.C. Human Capital and Mobility network 'Complexity, Logic and Recursion Theory'.

1991 *Mathematics Subject Classification.* Primary 03D25, 03D30; Secondary 03D35.

[1] Or, with David Hume [1739], [1748] in mind, *apparent* causal determinacy.

Typeset by $\mathcal{A}\mathcal{M}\mathcal{S}$-TEX

1. Laplacian determinism

A perceived algorithmic content for reality is peculiar to our time and culture. Its origins are commonly traced back to the ancient Greeks, as is the development of the notion of *proof*, providing a useful infrastructure for mathematical and scientific truth, having been rediscovered (largely via Arab texts) in the late Middle Ages and developed during and after the Renaissance.

The extent to which scientific activity, at least since the time of Newton, has been directed towards the inductive identification of the algorithmic content of nature is illustrated by the emphasis on *prediction* in the formulation of satisfactory theoretical explanations (see for example Casti [1990]). According to Einstein [1950], p. 54:

> When we say that we understand a group of natural phenomena, we mean that we have found a constructive theory which embraces them.

The key element in the revolution in conceptual framework which Newton was responsible for was the new mathematics. This secured the theoretical basis for the mechanistic Universe implicit in Laplace's [1819] description of his predictive[2] 'demon':

> Given for one instant an intelligence which could comprehend all the forces by which nature is animated and the respective situations of the beings who compose it — an intelligence sufficiently vast to submit these data to analysis — it would embrace in the same formula the movements of the greatest bodies and those of the lightest atom; for it, nothing would be uncertain and the future, as the past, would be present to its eyes.

Newton's embracing of the idea of gravity at a distance, independent of any transmitting medium, involved a replacement of Descartes' extreme mechanistic framework by one in which algorithmic content became an accepted replacement for proximate interaction. This revolutionary change in how causality was viewed has been related (see Betty Jo Dobbs [1991] or Richard Westfall [1984]) to Newton's alchemical interests, his awareness of the possible significance of a more global causality presaging current dissatisfaction with existing models of determinism. Such a broad view of mechanism became increasingly necessary, as during the nineteenth and early twentieth century new fields, electrical and magnetic, with no clear

[2]In focusing on the ontological content of determinism rather than the epistemological, the predictive element of Laplace's formulation, which is open to differing and confusing interpretations, will be ignored in favour of the widespread association of Laplacian determinism with mechanism.

explanation in terms of the interactions of physical bodies, assumed the scientific centre-stage.

Over time, a closer inspection of the Universe's mechanical credentials revealed a number of problems. While twentieth century questioning of Newtonian (in the sense of mechanistic) rationalism was paralleled (cf. Pope, Blake, Berkeley and Coleridge) by that of the eighteenth and early nineteenth century.

David Hume was the first to refer to fundamental difficulties in actually pinning down the basic workings of a putative mechanical Universe, in that the intuitive sense of there being a *connection* between a cause and effect appeared to lack formal content. The working scientist, from before Newton even, has been able to ignore the underlying, and very real, conceptual deficiency (which we return to in section 5), which makes it difficult to even formulate what we mean by determinism.

Moreover, the apparent fragmentation of science into heterogeneous bodies of knowledge, each with its own individual methodological and technical frameworks (see, for example, Dupré [1993]), belies the expected reductionism characteristic of a machine. This further relates to the ostensible gaps in known computability of material phenomena, and (see sections 2 and 5) to the more formal attempts to extend the picture of the role of mechanism as it relates to complex material systems.

Finally, changes to the dominant scientific theories have weakened the intuitive basis for mechanism. It is possible to argue (Earman [1986]) that even:

> Newtonian space-time, whose structure is rich enough to support the possibility of Laplacian determinism, nevertheless proves to be a none too friendly environment.

And of current theories, only special relativity comes with convincing deterministic credentials. Associating mechanism with standard formulations of computability (as in the following section) promises precise criteria by which to decide the significance of such theories for Laplacian determinism. In doing this one must follow Kreisel [1974] in distinguishing between *phenomena* and *theories* — a theory is mechanistic if " every sequence of natural numbers or every real number which is well defined (observable) *according to theory* is recursive or, more generally, recursive in the data (which, according to the theory, determine the observations considered)". Shipman [1998] notes that even very successful theories such as QED (the theory of quantum electro-dynamics — see Lawrie [1990], pp. 201–212), described by Feynman [1985] as "the jewel of physics ... our proudest possession", do "not have a fully satisfactory mathematical and computational foundation". Geroch and Hartle [1986] point to the possible incomputability of physically measurable numbers predicted by Hartle's version of quantum gravity, noting that the implementation of the algorithm suggested by the theory needs one to detect effectively homeomorphism affecting pairs of simplicial 4-manifolds (not possible in general — see Haken [1973]). But in the absence of relevant techniques for *excluding* computable models of these theories, their uncertain status tells us little about the underlying reality.

As already mentioned, an association between computability (in the practical sense) and mechanism goes back to the time of Newton and before, but the *identification* between material structures and those governing information is a comparatively recent development. Different insights arise according to the particular perspective — for Turing [1936], [1950] it was how real phenomena can be described in terms of mathematical models of computability, while for Shannon [1948] (representing the other main theme) it was the recognition of informational structure based on physical laws. Accordingly, one can describe the Laplacian model of the informational structure of the Universe in terms of a (schematically) well-understood shifting around of information content according to the second law of thermodynamics (which one can roughly paraphrase as saying that any nontrivial restructuring of information content corresponding to a change in an isolated physical system cannot be perfectly reversed, due to a loss in *available* information). That is, there is no *creation* of enhanced information content, although there can be (algorithmically translatable) transmutation. There may be local concentrations of high information content (e.g., in biological organisms), but a concomitant entropic loss of a proportion of the antecedent information content to disorderly, non-retrievable manifestations. Incomputability, if it does arise, can only originate with initial conditions and can never be explained in terms of what happens in the observable Universe. Or so the argument goes.

In recent times, a version of Laplacian determinism has been given more precise form via a detailed analysis, and corresponding restatement, of Church's thesis. See Odifreddi [1989] for a guide through the more arcane (but still important) subtleties of terminology[3], and Odifreddi [1996] for a useful introduction to some very relevant contributions of Georg Kreisel which we will need to return to later.

[3]For us, the term 'computable' will be firmly based on the notion of Turing computability and (see section 2) mathematical models of the physical universe will be non-discrete. Although the discussion will certainly point to the probability of an *analogue* computability which transcends Turing computability — depending on the controlled reproducibility of certain (Turing) incomputable natural phenomena coming out of a posited non-mechanistic determinism — the use of terminology will be familiar and, as far as possible, standard, a choice to be justified by the eventual placing of such an extended notion of computability within the framework of classical computability theory.

2. Occurrent incomputability in Nature

To find a single body of *empirical* evidence which is clearly inconsistent with a narrowly mechanistic Laplacian determinism, one must first look to the quantum level. However, much that appears strange there does not in itself require a theoretical interpretation taking us beyond the classical framework of Turing computability.[4]

The familiar overview of the determinism provided by quantum theory is that the quantum-mechanical state of a system of particles, specified as precisely as it can be by its *wave function*, and subject to Schrödinger's equation, contains insufficient information to give more than a probabilistic computation of the precise locations or momenta of the particles at a later time. The lack of determination of the quantum-mechanical parameters, according to Heisenberg's Uncertainty Principle, is a feature of reality. But the quantum states themselves, in the absence of measurements, can be calculated over time, which means that a consistent interpretation of the reality (or unreality) underlying the quantum superpositions is sufficient for a deterministic picture. (Popper's 'propensities', for example, give a formalisation of the intimate connection between the quantum-mechanical probabilities attached to particular states and the contextual contingencies involved — see Popper [1983], p. 351.) What *does* radically challenge determinism — as is expressed in the famous example of Shrödinger's cat — is what happens at the margins of the quantum and classical domains, for instance where a measurement is made, when a superposition is turned into an actual outcome via a so-called 'collapse of the wave function'. This leads to the so-called *measurement problem*, asking for an explanation of exactly why this collapse, with its associated probabilities, takes the particular form it does. But it is the evidence, both theoretical and empirical, of nonlocal causality which is most damaging to the clockwork model. The associated incomputabilities, undeniable but apparently beyond explanation in terms of classical computability/incomputability theory, are discussed in more detail in the next section.

Incomputability also emerges, but less assuredly, in physical situations involving mathematical non-linearity. Kreisel [1967] distinguishes between classical systems and *cooperative phenomena* (not known to have Turing computable behaviour), and proposes [1970] (p. 143, Note 2) a collision problem related to the 3-body problem as a possible source of incomputability, suggesting that this might result in "an analog computation of a non-recursive function (by repeating collision experiments sufficiently often)"

[4]For instance 'quantum computation' (originating with Benioff [1982] and Feynman [1982], with more specific proposals from David Deutsch [1985] and Feynman [1986]) appears to hold few surprises for the classical recursion theorist.

(see also Kreisel [1974]). The role of non-linearity (for instance in relation to chaotic situations) is suggestive of that of primitive recursion in the derivation of recursive functions capable of enumerating noncomputable sets.

An ingredient to the history of chaos theory has been the characterisation, due to Claude Shannon [1948] of physical phenomena — or more precisely, the orderliness or otherwise, of such phenomena — as *information*. From the slightly adjusted perspective of computability theory, one still obtains a multitude of different examples of the generation of informational complexity via very simple rules, and of the emergence of new regularities. The existence of so-called 'strange attractors' (see for example the two classic papers of Robert Shaw [1984], [1981]) provide a small-scale illustration of the macroscopic emergence (see section 5) of new forms from a causal context involving many diverse histories.

A good source of examples of the synthetical approach to finding precise mathematical models of functional complexity in nature, is provided by particular physicalist approximations to mental processes. For instance, more recent connectionist theories potentially transcend the Turing computability arising from the classical McCulloch and Pitts [1943] artificial neuron formalism, as observed by Smolensky [1988], p. 3:

> There is a reasonable chance that connectionist models will lead to the development of new somewhat-general-purpose self-programming, massively parallel analog computers, and a new theory of analog parallel computation: they may possibly even challenge the strong construal of Church's Thesis as the claim that the class of well-defined computations is exhausted by those of Turing machines.

But (and this is relevant to what follows) the analog computations cannot, it seems, be achieved within the discrete computational environment successfully analysed by Turing [1936].

The role of particular *presentations* of mathematical structures (not just in Minkowskian geometry) is basic to any extrinsic discussion of computability or otherwise in the material universe. Although science depends on the fact that observation of the world is independent of any particular description of it (in terms of notational reals, say), the particular presentation must be appropriate to the formal development of the underlying reality. (For instance, it is formally possible that a (Turing) computable relation on a countable set of reals may not be computably presentable as a relation on the natural numbers.) Ignoring such considerations[5], physical atomism and an overview of scientific practice suggest a reduction to discrete sys-

---

[5]cf. Feynman's [1982] suggested resolution of the uneasy relationship between reality and its discrete representations: "It is really true, somehow, that the physical world is representable in a discretized way, and ... we are going to have to change the laws of physics."

tems — all of which seem to point to the computability, or mechanism, of nature (see for example Church [1957], Kolmogorov and Uspenskii [1958], Kreisel [1965], Greenspan [1973], [1980], [1982], Gandy [1980], La Budde [1980], Vichniac [1984] and Toffoli [1984]). But a more basic problem is that discrete representations suppress many of the asymptotic features of the Universe as a relational system, not only undermining natural ingredients of standard theoretical analysis, but severing the link with higher levels of logical structure. The latter may not appear to have much practical significance, since in dealing with relatively small systems those relevant aspects which might otherwise emerge from the system's logical structure are determined mechanistically within a larger causal context. In general one is not unduly concerned with the extent of the material environment involved (the underlying logic is intrinsic to even finite immanently developing structures). But in practice, particular discrete representations cannot be selected without bypassing the essentially dynamic nature of the information content represented. One should be clear that one is not talking about the physical realisation of asymptotes, but about algorithmic actualisation which, according to their role in the determination of the global properties of the system, define reals essential to their description. In any case, the observational evidence of incomputability in nature directs ones attention to mathematical models which offer at least a reasonable chance of providing a theoretical explanation. Any discrete model *consistent with current physical theory* will, in as much as it can provide a recursive simulation (see Odifreddi [1989], pp. 109–113), be a necessarily probabilistic one describing *possible* behaviour (expressed as a sequence of states with non-zero probability). As we shall see in the next section, this and the absence of a generally accepted realist interpretation of current physical theory effectively entails a crudely built in incomputability, and an inability to convincingly present the classical and quantum levels of the material universe within a coherent logical framework. For the moment it will be assumed (cf. Pour-El and Richards [1989]) that the most natural approach is via computability in analysis.

However, one cannot precisely describe the exact nature of a global presentation of essential *information content* (and one must even allow the possibility of objects of higher type than reals being necessary) without a deeper understanding of how information is recorded in an immanently developing material environment. What are the mathematical and physical characteristics of the most basic unit of information? How does atomism, and its specific forms, relate to presentations of such information? For instance, what is the role of the various approaches to quantum gravity, and attempts to directly quantise general relativity (see Smolin [1991], [1993], Rovelli and Smolin [1990], Ashtekar, Lewandowski, Marolf, Mourão and Thiemann [1995]) or string theory (a seminal reference being Green, Shwarz and Witten [1987])? (Of course, given countability of a structure one may present the state of a particle as a real in terms of suitably presented relationships to others.) How does global context impact as locally available information content? How can an absolute structure of space-time be derived from the local? What is the relationship with the current working descriptions used in physics? Such questions may be hard to answer, but fortunately one can say a lot without a completely precise notion of local information content for the material universe. In section 5 below alternatives of optimal plausibility will be identified, sufficiently constrained for the development of a useful theoretical picture.

Confronted with incomputability of empirical origin, one is driven to look for natural parallels between the ways in which incomputability arises in the mathematical context and the relevant physical scenarios. One can of course accept high information content as a given in nature, but success of the scientific project has always been related to the search for mathematical structures through which observed complexity can be reduced to more fundamental features of the natural environment.

The way in which incomputability arises theoretically, by merely taking an overview of a sufficiently advanced mechanical process, is at first sight extremely simple, providing an obvious basis for any extrinsic approach to the problem. By analogy with Penrose's Mandelbrot question (see below), one can view the well-known failure to find any orderly pattern in the decimal expansion of π (cf. Gandy [1988], p. 66) as a manifestation of incomputability of the (computably enumerable) set of finite configurations contained therein. Davis, Matijasevič, Putnam and Robinson's proof (see Matijasevič [1970]) of the Diophantine nature of *all* computably enumerable sets shows how simple are the *mathematical* levers underlying incomputability. The main obstacle to a straightforward transfer into nature is the difficulty in identifying the physical link corresponding to the shift from the local mechanics to the more global manifestation of incomputability, and a resulting jump in information content with observable consequences of a similar level of logical complexity as that of the originating environment. Despite his apparent non-materialism in relation to the mind (see Hao Wang [1974]), one can detect in Gödel [1972], p. 306, an attempt to grapple with such problems, in particular in his observation (in relation to Turing's [1936] argument that "a machine can reproduce all steps that a human computer can perform"):

> ... that *mind, in its use, is not static, but constantly developing*, i.e., that we understand abstract terms more and more precisely as we go on using them, and that more and more abstract terms enter the sphere of our understanding. There may exist systematic methods of actualizing this development, which could form part of the

> procedure. Therefore, although at each stage the number and precision of the abstract terms at our disposal may be *finite*, both (and, therefore, also Turing's number of *distinguishable states of mind*) may *converge toward infinity* in the course of the application of the procedure.[6]

In the closing pages of Chaitin [1987], one finds parallel speculation concerning contingent incomputability of biological origin. Letting Ω be the halting probability for a suitably chosen universal computer U (p. 164 of the revised edition, 1992):

> We have seen that Ω is about as random, patternless, unpredictable and incomprehensible as possible; the pattern of its bit sequence defies understanding. However with computations in the limit, which is equivalent to having an oracle for the halting problem, Ω seems quite understandable: it becomes a computable sequence. Biological evolution is the nearest thing to an infinite computation in the limit that we will ever see: it is a computation with molecular components that has proceeded for 10^9 years in parallel over the entire surface of the earth. That amount of computing could easily produce a good approximation to Ω, except that that is not the goal of biological evolution. The goal of evolution is survival, for example, keeping viruses such as those that cause AIDS from subverting one's molecular mechanisms for their own purposes.
>
> This suggests to me a very crude evolutionary model based on the game of matching pennies, in which players use computable strategies for predicting their opponent's next play from the previous ones. I don't think it would be too difficult to formulate this more precisely and to show that prediction strategies will tend to increase in program-size complexity with time.
>
> *Perhaps biological structures are simple and easy to understand only if one has an oracle for the halting problem.* (italics added)

In order to overcome this obstacle to convergence of incomputability in nature and in theory, one needs to be more specific about what one means by a 'computable' set of reals or of a 'computable' function on the reals. For this, one must refer to the early work of Lacombe [1955a], [1955b] and Grzegorczyk [1955], [1957] framing the computability of a real-valued function f in terms of the recursiveness of the corresponding functional (see Pour-El and Richards [1989], Chapter 0, for further discussion). One notes that, modulo certain specifics connected with the role of the metric in analysis, the relevant definitions can be formulated naturally in terms of Turing

[6] Further, Kreisel [1972] refers to problems connected with the way in which new mental states are mechanically included in the computer program.

relative computability of reals. This framework, giving, for instance, a natural definition of 'computably enumerable' set of reals, is the familiar one in which reals and sets of reals are accessed piece-wise by a computing machine. However computability of information content in nature, unlike that of an observed mechanical process, is not a primary property and involves problematic features of the role of the reals as a presentation, as pointed out by Penrose [1989] in relation to the question of the computability of certain simply generated mathematical objects, such as the Mandelbrot and Julia sets (see Mandelbrot [1982]).

Penrose (p. 124) points to the apparent unpredictability of structure in computer generated approximations to the Mandelbrot set (the latter readily obtainable via the set's computably enumerable complement) as indications of an underlying incomputability:

> Now we witnessed ... a certain extraordinarily complicated looking set, namely the Mandelbrot set. Although the rules which provide its definition are surprisingly simple, the set itself exhibits an endless variety of highly elaborate structures.

And goes on to observe (essentially) that the incomputability of the identity on the reals leads one, via particular presentations, to a counterintuitive notion of a noncomputable geometric object in two dimensions, even. In this context, Blum and Smale [1993] argue convincingly for a suitably relaxed version of the classical definition of computable set of reals (their *decidable* sets) according to which the Mandelbrot set and most Julia sets still turn out to be incomputable, and in a more basic sense even than that suggested by Penrose's observations on structure. The persuasiveness of Blum and Smale (see also Blum, Cucker, Shub and Smale [1998]) arises from optimal positioning within the classical framework of Turing computability, and the close correspondence with intuition — either in relation to nature (which does not care about different presentations), or to everyday practice (where any limiting process of approximation can be satisfactorily terminated at some finite level).

So the Blum-Smale notion of decidable set of reals provides a formal framework for presenting the evolution of incomputability in nature, in particular, suggesting a scenario in which nature has no problem in producing the basis for sets of reals which can be presented as the halting set of some (reasonably formulated, cf. Blum and Smale) machine. The resulting incomputability will then provide a context at particular moments of time for further physical interaction. If this environment is fairly diffuse or limited in extent, the interactions can be captured classically (that is, mechanically). But if there is sufficient contiguity of causal influences — that is, the contingencies converge to produce systemic rather than classically local effects (see Bohm [1957] for a particularly graphic analysis of such processes)

— then the resulting state of an individual constituent particle can only be computed by taking into account overall the mathematical and logical structure encompassing the incomputability of information content present. This can be envisaged as the iteration of simple operations producing an incomputability of *texture* of time-space, this texture becoming apparent not just as time passes, but physically (in approximation) in finite time (cf. Pour-El and Richards [1983]). The process by which local absorption of this augmented incomputability of information content derived from such a context takes place depends on the presence of adequate form and contiguity to by-pass any entropic factors at work. As we shall see later, the process can be described in terms of the Turing jump, although to fully capture it theoretically one would expect that the notion of E-recursion (based on Kleene's [1959], [1963] formulation of recursiveness of objects of finite type), and particularly the notion of the 1-section in relation to the E-recursively enumerable degrees (see Sacks [1990], Part D), is needed. The Mandelbrot example, presenting digitally observable evidence of undecidability, points to a formal process of algorithmic re-presentation of the associated incomputability, with the global manifest locally. In nature, one encounters this at the onset of turbulence, or of complexity in weather systems, or, for that matter, the human brain (a climate in miniature),[7] with incomputability apparently feeding on itself as the threshold is passed at which nonlocal causality comes to dominate. While the empirical accessibility of incomputability is essential, irreducible higher type features, both set theoretical and recursion theoretic, may well be relevant to a full characterisation of the information content. But thinking in terms of nature collating the results of computably enumerated events while simultaneously computing relative to the total context presented as a Δ^0_2 characteristic function, one can envisage a reduction of this picture to the classical framework. One notes that while the key to the incomputability derivable classically from a computable function is a severing of the link between image and argument, the physical root is the independence of effect from scale of causal origin, characteristic of chaotic situations.

However, having relied on current knowledge of quantum events for circumstantial evidence of incomputability in nature, it is clear that in this case no such reduction is appropriate. The problem is that the incomputability indicated, although unavoidable, does not come with an explanation in terms of local causality, as will be discussed in more detail in the next section. In fact at first sight it is not hierarchically developing incomputability but that of basic laws of nature which is suggested, the underlying reason

[7] Of course, as Penrose [1994], p. 153, observes: "Once it is conceded that *some* physical action might be non-computational, the possibility is laid open for non-computational actions also in the physical brain, ...".

for this only becoming apparent in section 5 below. For the moment one notes that in mathematics one does not need to take incomputability as a given, and that it seems unlikely that the simple mathematical structures that give rise to incomputable phenomena are not reproducible in nature. And while understanding of creation itself may be out of reach (despite some cosmologists' legitimisations of creational scenarios in terms of what is perceived as being mysterious at the quantum level), it being impossible to qualify or quantify the information content of the Universe's 'boundary conditions', there is no real evidence that the absolute origins of the Universe are not governed by very basic mathematical structures. One cannot but agree with Alan Guth [1997], pp. 251–252, who (reminding one of the evolutionary cosmology of Charles Sanders Peirce) imagines in relation to the improbability of the big bang as a "singular act of creation", a biologist who "discovered a bacterium that belonged to no known species":

> Although she believes firmly that life on earth originated from non-living materials, the possibility that this particular cell is the result of such an improbable occurrence would be too preposterous to even consider.

But this is a topic we shall return to in section 5.

On the other hand, if one chooses to approach incomputability in nature via the everyday mathematics used to describe physical phenomena, evidence consistent with the previous schematic picture is provided by Pour-El and Richards [1983], [1989].

What is lacking, of course, is a mathematical model of the Universe which captures enough of its specific infrastructure to closely simulate the *development* of the observed hierarchical structure of information content. Such a framework should be capable of representing in an organic way the mutable balance between *entropy* (described by Hawking [1977] "as a measure of the disorder of a system or, equivalently, as a lack of knowledge of its precise state"), conventionally ruled by the second law of thermodynamics, and the (less familiar) hierarchically organised, nascent incomputabilities of nature. We will argue below that while the Turing universe tells us little about the *dynamic* relationship between computable and incomputable, it provides a particularly illuminating model of the fine structure of *actually existing* information content.

3. Nonlocality

The argument so far is for a 'quasi-mechanical' Universe in which systemic input to the causal relationships take one beyond the computability over time expected of a (Turing) machine-like universe. But the most radical challenge to Laplacian determinism and the associated computability comes at the quantum level where the very nature of causality comes into question, along with the Turing model of what mechanism remains (which, of course, gives the appearance of being considerable). However it is not the observed ambiguities of quantum phenomena — the associated probabilistic analysis is consistent with a deterministic interpretation of the underlying reality — which is so unusual. What is strange is what happens in the collapse of the wave function — concerning the reasons for which, according to Richard Feynman, "we have no idea" (Feynman, Leighton and Sands [1965]). The fact that the collapse appears to be global in origin, involving nonlocal communication of a seemingly causal nature, indicates firstly that some causality involving higher logical structure is involved. This is most clearly illustrated in the EPR thought experiment of Albert Einstein, Boris Podolsky and Nathan Rosen [1935], of which Bell's [1964] inequality provides an empirically testable (and tested, see Aspect, Dalibard and Roger [1982], Aspect, Grangier and Roger [1982]) version. Even though QED provides a mathematical formulation which seems to successfully transcend some of the conceptual and descriptive difficulties inherited from the classical context (e.g., wave/particle ambiguity), it does not remove the essential dichotomy between theoretical descriptions of quantum/classical reality or explain the apparent systemic nature of the transition between the two.

The so-called 'EPR paradox' is most usefully considered as an indicator of the incompleteness, as a description of physical reality, of quantum mechanics (and this seems to have been an important aspect of the thinking of Einstein himself). In outline (and the details of the original experiment, its recasting by Bell [1964], and its subsequent empirical scrutiny, can be found in many places — see, for example, Omnès [1994], chap. 9), Einstein and his two colleagues considered (in the more famous of their examples) the behaviour of two particles whose initial interaction means that their subsequent descriptions are derived from a single Schrödinger wave equation, although subsequently physical communication between the two, according to special relativity, may involve a significant delay. If one imagines the simultaneous measurement of the momentum of particle one and of the position of particle two, one can arrive at a complete description of the system. But according to the uncertainty principle, one cannot simultaneously quantify the position and momentum observables of a particle. The standard Copenhagen interpretation of quantum theory requires that the measurement relative to particle one should instantaneously make the measurement relative to particle two ill-defined (a process commonly described in terms of a collapse of the wave function).

One could try to explain this in various ways. Clearly it constituted a major challenge to any existing causal explanation. Kreisel [1971], p. 177, reminds us (in a mathematical context) that the appearance of a process is not necessarily a good guide to the computability of its results. The immediate question was whether there existed undiscovered, but qualitatively familiar, causal aspects of the material universe in terms of which this apparent inconsistency could be explained. This was what EPR expected:

> While we have thus shown that the wave function does not provide a complete description of the physical reality, we left open the question of whether or not such a description exists. We believe, however, that such a theory is possible.

Despite Einstein's later negative comments concerning the particular proposal of Bohm [1952], this has been taken (see Bell [1976]) as a tacit endorsement of a 'hidden variables' approach.

Some early objections to the hidden variables programme (most well-known being that of von Neumann[1932]) are disposed of in Bell [1966]. However, Bell [1964] proposed a testable version of the EPR experiment (recast, following a suggestion of Bohm, in terms of spin rather than position and momentum). This led to the experiments described in Aspect, Dalibard and Roger [1982] and Aspect, Grangier and Roger [1982], and a general acceptance[8] of the predictions of quantum theory as a description (but not a *complete* description) of what actually happens. This means that although there are quite viable *non-local* theories of hidden variables (presaged by de Broglie [1927] and his non-local 'pilot wave'), the complete description of quantum theory in terms of a hidden reality involving *only* local parameters is not possible.

So one can reasonably deduce that there is no mechanical Universe. One could, presumably, have computability emerging from a grossly non-mechanistic causal context, but there seems no reason for allowing such a possibility in this case. Experience tells us that computability of natural phenomena must derive via natural translation from material mechanism. That is, that if the material universe is to *behave* like a machine, then it essentially achieves this in a causal context via mechanism.

It is important to notice that it is not just to our intuitive sense of a mechanistic Universe which suffers, but also *computability*. One can no longer rely on establishing computability by predicting local phenomena via a local analysis. A computable causality depends not only on being able to computably characterise the nature of the causal factors at work, but on being able to effectively identify the particular causal context relevant to a given phenomenon. And any extended notion of 'locality' which is not physically based (for example based on Bohm's [1980] 'enfolded' reality), may retrieve determinism, but is not likely to deliver computability.

So one is left with the question of what is the appropriate mathematical framework, radical enough to clarify questions raised about the logical

[8] But see Franson [1985].

status of contemporary physics — in particular, its *completeness* (in regard to explanations of nonlocality and the measurement problem), and *consistency* (in particular concerning the contrasting descriptions of the classical and quantum worlds). There are well-known examples of structures in which the global properties appear strikingly undetermined by the local ones (cf. the M. C. Escher lithographs 'Waterfall' and 'Ascending and Descending', based on mathematical examples of Penrose and Penrose [1958]). But what is needed is an appropriate mathematical formulation of local causality, with a corresponding, mathematically nonrigid, global structure, capable of confirming the widespread intuition that nonlocality has important consequences for the nature of of the real world and our knowledge of it.

4. The Turing model

We have seen that at a number of levels the standard mathematical model of determinism in nature has been found lacking. Although at first sight this may not appear to be relevant to the working scientist, beyond a background confusion regarding the role of 'truth' in science and an accompanying pragmatism, it may well be that the resulting conceptual vacuum underlies a number of fundamental scientific and philosophical problems, and seriously undermines the hegemonic role of science in society generally. But the very persistence of the model suggests that any alternative, while achieving improved stability in the presence of high information content, will have to retain at least some of the Laplacian strengths of simplicity and aptness.

As we saw in section 2, there is plenty of evidence that structures derived from the Turing model are capable of enriching their information content via the iteration of familiar algorithmic processes. The rejection of determinism itself — for example, Dupré [1993]:

> There are two very powerful reasons for rejecting the doctrine of determinism. The first is that it seems almost entirely, or perhaps entirely, devoid of empirical support. The second is that our most successful scientific theories describe a probabilistic rather than a deterministic world.

— still entails the admission of incomputability, without denying the essential role of computability as a universal force for understanding and structural cohesion. So we are left with a Universe in which incomputability, even if not formally verified, is both theoretically likely and empirically forced on us, but which we have no means of making any sense of in any precise sense except via those computable relationships we can identify. Following Turing [1939], who attempted to use the notion of oracle computability give an explanatory context to Gödel's incompleteness phenomenon, the Turing model can be thought of as the minimal extension of the Laplacian one consistent with the observed incomputability of nature. Our choice of Turing reducibility (or one of its subreducibilities) is based on the empirical evidence of histories being reducible to the analysis of correlations between descriptions of essentially finitary circumstances, and the aesthetically based preference for models based on ourselves, as understood, as reflections of the functionality of the larger Universe.[9] Particular reductions will be intended to capture computable aspects of allowable historical processes, such as those derived from the actions of the basic forces of nature. Fundamental to the model will be transitivity, and a conventional 'arrow of time' identifying events which are permitted historical consequences of each other. From such basic considerations eventually arise closer and more striking correlations between specific physical phenomena and particular features of the Turing model. None of the definitions of the *strong reducibilities* (those properly contained in Turing reducibility) seem to fully capture the range of basic physical processes, nor do what is known of their theories (see Odifreddi [ta]) tie in with these deeper properties of the physical universe (see below).

For details of standard notation and terminology for the Turing degrees, see for example Soare [1987] or Odifreddi [1989].[10] A fuller description of Turing machines and their properties can be found in Davis [1958].

For instance, corresponding to the ith Turing machine, Φ_i denotes the ith partial computable (p.c.) functional $2^\omega \longrightarrow 2^\omega$. A set (or binary real) A is *Turing computable from*, or *Turing reducible to*, a set B ($A \leq_T B$) if and only if $A = \Phi_i^B$ for some $i \in \omega$, and A, B are *Turing equivalent* ($A \equiv_T B$) if and only if $A \leq_T B$ and $B \leq_T A$. Since $\equiv_T$ turns out to be an equivalence relation, one can (following Post [1948]) define the *degree of unsolvability* or *Turing degree* of A by

$$\deg(A) = \{X \in 2^\omega \mid A \equiv_T X\}.$$

So the Turing degrees are sets of objects, formally described in terms of reals, the essential information content of their members being algorithmically indistinguishable. We write $\leq$ for the partial ordering induced by $\leq_T$ on the set $\mathcal{D}$ of all degrees, $\mathbf{0}$ for the least degree, consisting of all computable sets of numbers (or, equivalently, of computable reals), and $\boldsymbol{\mathcal{D}}$ for the structure $\langle \mathcal{D}, \leq \rangle$.

Let $W_i^A = \operatorname{dom} \Phi_i^A$ denote the ith *computably enumerable in* A (A-*c.e.*) set ($W_i = W_i^\phi$ being the ith c.e. set). The *c.e. degrees* (collectively

[9]However see, for example, Post's [1965] speculations on the possible relevance of extensions of the familiar Turing model to an understanding of thought processes.

[10]But each must be read in the light of Soare [1996].

denoted by $\mathcal{E}$) are those containing c.e. sets. Feferman [1957] showed the c.e. degrees to be exactly those degrees containing (coded) recursively axiomatisable first-order theories, and there are many other examples of classes of natural, mechanically generated, objects from which the totality of c.e. degrees arise.

Kleene and Post [1954] defined the notion of *jump operator* on sets and degrees. The *jump* ($n+1$ *th jump*) of a set A is defined by $A' = A^{(1)} = \{x \mid x \in W_x^A\}$ ($A^{(n+1)} = (A^{(n)})'$). This induces a *jump operator* on degrees defined by $\mathbf{a}' = \deg(A')$, $A \in \mathbf{a}$, with the special properties that $\mathbf{a} < \mathbf{a}'$, and $\mathbf{a}'$ is the largest of the degrees of sets c.e. in $A \in \mathbf{a}$. Post's Theorem [1948] that $X \in \Delta_{n+1}^A \Leftrightarrow X \leq_T A^{(n)}$ attaches special importance to the ascending sequence $\mathbf{a}, \mathbf{a}', \ldots, \mathbf{a}^{(n)}, \ldots$. The most important noncomputable degree $\mathbf{0}'$ contains, for instance the (coded) undecidable axiomatic theories of Gödel [1934], as well as many other natural mathematical objects. We define the standard ω*-jump* of $\mathbf{a}$ by $\mathbf{a}^{(\omega)} = \deg(\oplus_{n\in\omega} A^{(n)})$, $A \in \mathbf{a}$.

Events originating with computable physical processes operating on given initial conditions lead to a notion related to that of A-c.e.: a set B is said to be ***computably enumerable in, and above*** a set A (or A-CEA) if $A <_T B$ and B is A-c.e. The CEA sets are those X-CEA for *some* X. There are corresponding degree theoretic notions. This is fortunate, in that it is easy to theoretically provide for a (possibly very small) entropic element in any reduction to an event CEA, due to the (relativised) Sacks splitting and density theorems (Sacks [1963], [1964], respectively).

The structure of $\mathcal{D}$ (and of local structures such as $\mathcal{E}$) turns out to be very rich, and to lead to great technical complexity. In fact, Simpson [1977] was able to characterise the first order theory of $\mathcal{D}$ as being recursively isomorphic to the second order theory of arithmetic.

It is clear that most of the evidence supporting the Laplacian model, making it so intuitively persuasive over such a long period, also supports the algorithmic aspects of the Turing one. While those incomputabilities so damaging to the Laplacian picture either provide the basic information content of the Turing model, or as we shall see (in the case of quantum phenomena arising from nonlocality) are accounted for by it.

Although it is convenient in framing the above definitions to distinguish between information and computable process, there are (for our purposes equivalent) formulations of computability theory, for instance based on Church's [1933] notion of λ-computability, which make no theoretical distinction between function and argument. This is not just mathematically attractive (with correspondingly elegant models, cf. Scott [1975a], [1975b]), but also seems in line with the primacy of process in nature, and the observed flexibility of relationship between matter and energy at the most basic (that is, quantum) level.

Anyway, given a set of reals with corresponding information content, empirical considerations have lead us to look to material expressions with mechanical basic relationships, and this has in turn suggested the Turing model. It is then the full range of such computable reductions, containing as it must a blueprint for any future development of a consistent infrastructure, and allowing for any eventual complexity of information content, which determines the overall logical coherence of the system and, specifically, via which the system must achieve a level of definition of its fundamental causal structure.

We saw above how problems with the logical coherence of the material universe surface via quantum considerations, such as from the EPR paradox, and how current theoretical solutions involve discussion of such logical characteristics of the Universe as its *consistency*. Gödel's incompleteness theorems have also been imported from logic to provide useful analogies relevant to some of the more mysterious aspects of the observed universe. However, in dealing with the real world, the axiomatic theories to which consistency and incompleteness relate seem incapable of providing *more* than illuminating analogies. What is needed is a precise explanation of what Leibniz [1714] describes as the "pre-established harmony" of a Universe exhibiting a high degree of systemic unity. One must look to the neglected and little understood notion of *definability* (and the closely related notion of *invariance*), within a mathematical structure which captures the algorithmic and underlying information content of the material universe, for a more precise analysis of how nature achieves its consistency. This will enable one to model a process whereby basic natural laws both determine and are determined according to hierarchical principles, and which thereby guarantees consistency and coherence. One notes that even though a system may be, according to certain reductionistically arrived at criteria, strictly finite, the incomputabilities and logical forms derived from the algorithmic content are intrinsically present and collectively comprise a logical blueprint for structural development.

Epistemologically, there is a close link between definability in terms of basic, perhaps physical derived, concepts and human (or at least scientific) understanding. Even in a mathematical context, for instance, Gödel [1946], p. 152, remarks that although the formal notion of ordinal definability might not completely capture the informal one of "comprehensibility by our mind", he believed it to provide "an adequate formulation in an absolute sense ... of [a set's] 'being formed according to a law' ". So there is the half-suggestion of 'definable in terms of something basic' as providing at least an approximation to what one can comprehend. And although Gödel [1964], p. 268, talks about the "remoteness from sense experience" of sets, which might still be perceptible via mathematical intuition — so there is no acceptance of an explicit link between sensory input and mathematical intuitions — in saying (Gödel [1995], p. 383) that "the certainty of mathematics is to be secured not by ... the manipulation of physical symbols, but rather by cultivating (deepening) knowledge of the abstract concepts themselves" one can detect an implicit assumption of an analogy between the process of verifying the role of formal definitions and the empirical process, in that they must be based on an epistemologically irreducible level of input. While (Gödel [1964], p. 268) although:

> ... mathematical intuition need not be conceived of as a faculty giving an *immediate* knowledge of the objects concerned ... as in the case of physical experience, we *form* our ideas of [mathematical] objects on the basis of something else which *is* immediately given ... [and which] may represent an aspect of objective reality ...

The notion of definability relevant to causal structures involving computable basic laws is that of *Turing definability*. (For a recursion theoretic overview of mathematical definability, see Slaman [1998].) Formally, a relation on reals is *(absolutely) Turing definable* if its degree theoretic counterpart can be described, using only the standard first-order language, in terms of the ordering relation $\leq$ on $\mathcal{D}$. Rogers [1967a], looking for a language independent notion, defined such a relation to be (Turing) *invariant* if and only if it was left degree theoretically unchanged under all automorphisms of $\mathcal{D}$. Of course, in the real world one often achieves a level of understanding via the establishment of relative definabilities, not just of a general kind, but involving specific objects, which may not by themselves be definable. This leads to the important formal notion of *definability relative to parameters* (see Slaman and Woodin [1986]). There is of course a corresponding notion of *relative invariance* of relations on the Turing universe, although little is known so far concerning the structures arising (see Cooper [1997]).

As one would expect in comparison with a Universe with many features and general characteristics accessible to human understanding, many of the naturally arising relations on $\mathcal{D}$ turn out to be Turing definable. Particular examples include the definability (Cooper [1994]) of the relation of computably enumerable in — and hence of $\mathcal{E}$, of CEA, of the Turing jump — giving that of $\mathbf{0}'$ and of each $\mathbf{0}^{(n)}$ for $n > 0$, of every level of the arithmetical (that is, Kleene-Post) hierarchy, of (Jockusch and Shore [1984]) the ω-jump and $\mathbf{0}^{\omega}$ (the degree of the theory of $\mathcal{D}$), and (see Nies, Shore and Slaman [1996], [ta]) of every *atomic double-jump class* — where one says that degrees $\mathbf{a}, \mathbf{b}$ with identical n^{th} jumps are n^{th}*-jump equivalent*, and the nth-atomic jump classes are the corresponding equivalence classes. One has to look to the local level (that is below $\mathbf{0}'$) to find specific non-invariant (and hence undefinable) relations. One defines the *high-low hierarchy* (usually relative to $\mathcal{E}$), to be comprised of the following *jump classes*:

$$\mathbf{a} \in \mathbf{High}_n \Leftrightarrow \mathbf{a}^{(n)} = \mathbf{0}^{(n+1)}, \qquad \mathbf{a} \in \mathbf{Low}_n \Leftrightarrow \mathbf{a}^{(n)} = \mathbf{0}^{(n)},$$

Then (Cooper [1997]) the class of *low* degrees (= $\mathbf{Low}_1$ is not definable in $\mathcal{E}$ (whereas by Nies, Slaman and Shore [1996], [ta], all the other jump classes are so definable).

It is of course a trivial observation that there is a universal ceiling on that level of definability of a real achievable degree theoretically. Accordingly, just as observational data provide the foundation for all scientific knowledge, at the theoretical level the nature of the observed world itself relies on the *available* definability of the system. This corresponds to the fact (previously noted) that it is in the nature of different presentations of a structure to particularise the essential informational content of a constituent part in ways which have no relevance to the underlying reality.

Before 1927, one might reasonably conjecture that this obvious limitation on definability is all there is, and that, correspondingly, the causal structure of the universe uniquely defines all its characteristics. One might take William James' graphic description (from an 1884 lecture to the Harvard Divinity School, [1897], p. 150, 1956 reprinted edn.) of an 'iron block' Universe as a metaphor for this conjecture:

> What does determinism profess? It professes that those parts of the universe already laid down absolutely appoint and decree what the other parts shall be. The future has no ambiguous possibilities hidden in its womb: the part we call the present is compatible with only one totality. Any other future complement than the one fixed from eternity is impossible. The whole is in each part, and welds it with the rest into an absolute unity, an iron block, in which there can be no equivocation or shadow of turning.

Formally it is subsumed in the so-called Bi interpretability Conjecture. The programme of research stimulated by this productive and mathematically attractive proposal eventually showed there to be deeper theoretical reasons for a locally lower ceiling on the level of definability or invariance achievable. Bi interpretability is fully discussed elsewhere (see for example Slaman [1991] or Nies, Shore and Slaman [1996]), but we will sketch in some of the essentials.

The notion of bi interpretability in computability theory seems to have been imported from model theory (see Ahlbrandt and Ziegler [1986]) by Harrington, although much of the subsequent development and application is associated with Slaman and Woodin [1986].

As described in Cooper [1997], bi interpretability extends some of the benefits of isomorphism (between known and less well-known structures) to apparently dissimilar pairs of structures. Choosing a representative X for a member $\mathbf{x}$ of some degree structure $\mathbf{D}$ over the reals can be thought of as defining a mapping from $\mathbf{D}$ to the standard model for second order arithmetic. Conversely (following Jockusch and Simpson [1976]), a *coding* of second order arithmetic involves *specifying* a collection of degrees, and relations on these degrees to represent addition and multiplication. If one can uniformly define the relationship between $\mathbf{x} \in \mathbf{D}$ and the code for the chosen representative $X \in \mathbf{x}$, then we say that $\mathbf{D}$ is *bi interpretable* with second order arithmetic. In local versions of the bi interpretability conjecture (arising from the work of Harrington and Slaman) one can replace the representative X with a number, namely the index for X in some canonical listing, with a consequent substitution of first order for second order arithmetic.

The *Bi interpretability Conjecture* is that $\mathcal{D}$ *is bi interpretable with second order arithmetic.* Bi interpretability and the rigidity of the standard model of second order arithmetic gives:

(1) *Rigidity of* $\mathcal{D}$.

And, since definitions in $\mathcal{D}$ can be read off from those in the standard model via the bi interpretation (if it were to exist), one immediately obtains a complete characterisation of the Turing definable relations:

(2) *The definable relations on* $\mathcal{D}$ *are exactly those given by the relations definable in second order arithmetic.*

Bi interpretability turns out to be technically more useful, but formally *equivalent* to Turing rigidity (Slaman and Woodin, private communication). The successes of the research programme based on bi interpretability are remarkable (see Nies, Shore and Slaman [1996] or Cooper [1997]). For instance, Slaman and Woodin have shown that recursion theoretic relativisation is sensitive to double jumps, in that only Turing degrees with identical double-jumps can have isomorphic cones above them. Although Martin [1968] showed that an assumption of projective determinacy was sufficient to guarantee an upper cone of bases of *elementarily equivalent* cones of $\mathcal{D}$.

Lerman's [1977] notion of *automorphism base* (a substructure which locally determines the global action of any automorphism), has successfully reduced rigidity of large structures to that of smaller, even local, ones. Examples of automorphism bases include (Jockusch and Posner [1981]) any comeager set $\boldsymbol{A} \subseteq \mathcal{D}$, the set of all noncomputable degrees *minimal* in $\mathcal{D}$, and (Slaman and Woodin) $\mathcal{E}$. When combined with results on definability, one gets particularly striking results.

Since, any automorphism of $\mathcal{D}$ must preserve every level of the arithmetical hierarchy of Turing degrees, and every level of the high/low hierarchies is invariant under all automorphisms of $\mathcal{D}$, there are some quite dramatic implications for Aut($\mathcal{D}$). For instance, if $\mathcal{E}$ or $\mathcal{D}(\leq \mathbf{0}')$ is rigid then so is $\mathcal{D}$. Which means that local structures contain a key to the Turing universe and its more material manifestations. Also, since there is in fact a *finite* automorphism base of c.e. degrees, Aut($\mathcal{D}$) is countable.

Further, known *local* automorphism bases provide automorphism bases for $\mathcal{D}$: such as (Lerman [1977]/Jockusch and Posner [1981]) every level of the high/low hierarchy (in particular, the low degrees), and (Ambos-Spies [ta]) every lower cone $\mathcal{E}(\leq \mathbf{a})$ $(\mathbf{a} \neq \mathbf{0})$. Finally, if we define a substructure $\mathcal{C}$ to be *rigid in* $\mathcal{D}$ if and only if $\psi \restriction \mathcal{C} =$ the identity for every $\psi \in \mathrm{Aut}(\mathcal{D})$, then we find that (Slaman and Woodin) $\mathcal{D}(\geq \mathbf{0}'')$ is rigid in $\mathcal{D}$.

The situation for the structures derived from the strong reducibilities is not very well known (see Odifreddi [ta]), although for none of them has there been shown to be a relatively rigid upper-cone counterpart to the classical part of our Universe. For those derived from the various truth-table reducibilities (*truth-table*, *bounded truth-table* and *weak truth-table*), there are as yet no fully worked out proofs of nonrigidity. Although the fact that there are known to be $2^{2^{\aleph_0}}$ automorphisms of the many-one degrees (ruling out any non-trivial invariant subsets) is an indicator for nonrigidity in relation to all the reducibilities intermediate between many-one and Turing reducibility.

5. Causal determinism restored

> *God does nothing by himself which he can do by another* – Isaac Newton
>
> (Jewish National and University Library, Jerusalem, Yahuda Manuscript Collection Var. I, Newton MS 15.5 (2, n.50), fol.67r.)

Given the present state of knowledge of the Turing universe, and the acceptance of it as a useful representation of the relationship between computable and incomputable aspects of the material universe, it is now possible to present a coherent and intuitively satisfying picture of the underlying causal structure of the Universe, together with a persuasive materialist basis for the associated epistemology, which immediately clarifies the numerous scientific and philosophical puzzles previously alluded to. This will eventually require us to be more specific about certain details of the modelling process.

But first, having discarded Laplace's clockwork Universe, it is necessary to re-examine the relationship between computability and determinism.

The nature of determinism

Allowing for the fact that the nature of Minkowski space-time dictates 'time slices' relative to 'initial' and 'subsequent' times which are appropriately chosen surfaces, one can start with a plausible formulation of determinism given by Bertrand Russell [1953], p. 398, which avoids any mention of mechanism or computability:

> A system is said to be 'deterministic' when, given certain data, $e_1, e_2, \ldots, e_n$, at times $t_1, t_2, \ldots, t_n$ respectively, concerning this system, if E_t is the state of the system at any time t, there is a functional relation of the form
>
> $$E_t = f(e_1, t_1, e_2, t_2, \ldots, e_n, t_n, t).$$
>
> The system will be 'deterministic throughout a given period' if t, in the above formula, may be any time within that period .... If the universe, as a whole, is such a system, determinism is true of the universe; if not, not.

However, one can then follow Russell (p. 401) in observing that:

> It follows that, theoretically, the whole state of the material universe at time t must be capable of being exhibited as a function of t. Hence our universe will be deterministic in the sense defined above. But if this be true, no information is conveyed about the universe in stating that it is deterministic. ... This, however, is plainly not what was intended.

So how does one escape the apparent falsity of Laplacian determinism, without ending up with a trivially true notion? The point is that having noted the existence of the functional relation, one does expect that E_t can be arrived at for given t other than by examining the state of the universe in question at time t. We do not conclude (Russell p. 401) that "the material universe *must* be subject to laws" — since we intuitively expect laws to involve some level of *coersion*, not merely recording. This involves some 'nice' description of E_t, which is logically simpler in some sense than the Universe itself. (There are other solutions to the problem, but these can usually be expressed in such terms.) And one would expect to arrive at such a description by identifying an appropriately close relationship between E_t and the known natural laws. The relationship between E_t, the boundary conditions, and the relevant laws, is assumed to be governed by the rule that all causal sequences permitted by the logical properties of the system are allowed to occur. The empirically familiar 'coarse grained' structure of reality, even at the quantum level, will follow from the appropriate choice of logical framework. Of course, if one has no theoretical explanation of the genesis of natural laws, then one is restricted to a sort of super-sophisticated taxonomy of laws, with all its Humean uncertainties. As Guth [1997] says in relation to the inductive extension of quantum-like behaviour to cosmogony:

> If the creation of the universe can be described as a quantum process, we would be left with one deep mystery of existence: What is it that determined the laws of physics?

— which leads us to an appropriate version of Penrose's [1987] 'strong determinism' (according to which, pp. 106-107, "all the complication, variety and apparent randomness that we see all about us, as well as the precise physical laws, are all exact and unambiguous consequences of one single coherent mathematical structure"), and the theoretical problem of how the relations of a physical system are subject to immanent definability. Such concerns are not new, elements of which can be traced back to to Leibniz's relational, as opposed to Newtonian, view of the Universe, and which find a particular expression (via a view[11] of the primacy of observation over fact) in C. S. Peirce (*The Architecture of Theories*, 1891) [1931-58]:

> To suppose universal laws of nature capable of being apprehended by the mind and yet having no reason for their special forms, but standing inexplicable and irrational, is hardly a justifiable position. Uniformities are precisely the sort of facts that need to be accounted for. Law is par excellence the thing that wants a reason. Now the only possible way of accounting for the laws of nature, and for uniformity in general, is to suppose them results of evolution.

We will argue below that the Turing model not only gives a logically concise description of the development of the Universe relative to its laws, but also of the form taken by those laws, resulting in a meaningful description of the level of determinism applying in the material universe and a bypassing of such epistemological difficulties. Whatever the relevance of the Turing model, it is surely an important but neglected task (of computability theory) to ascertain as closely as possible the computational complexity of the existing level of determinism of our Universe. Between the Laplacian 'clockwork universe' and the trivially deterministic one of Russell, there must lie a mathematically precise bound related to Russell's function.

The laws of nature

As we have seen, mathematical considerations suggest that mechanisms are as basic as material particles, and that at the subatomic level noninvariance may possibly be related to such ill-defined phenomena as arise from wave/particle confusion. But invariance appears to be an attribute of the (empirically detectable) fundamental forces of nature and the algorithmically describable laws constraining them. In any case, the abundance of Turing definable relations at the local level, but not of singletons, leads one to emphasise the link between natural laws and invariance, giving a more secure foundation for the notion of causal connection. Such connections of the

[11] cf. the Peircian maxim ([1931-58], vol. 5, paragraph 412): "if one can define accurately all the conceivable experimental phenomena which the affirmation or denial of a concept could imply, one will have a complete definition of the concept, and there is *absolutely nothing more* in it".

more fundamental kind have a clearly conceived, if mathematically sophisticated, genesis as materialisations of Turing defined mechanisms. While this link persists at the classical level even where no apparent algorithmic content to the causal connections exists. The difficulty in establishing empirically a relation so conceived puts the continuing debate about how to precisely describe what one means by a law (in particular, how to add a concise notion of *necessity* to Humean conjunctions of events) in a revealing context — and, as we shall see later, throws light on the relationship between *theory* and *observation*. However, the link between natural laws and invariance does not explain the *genesis* of *specific* laws. The existing mathematical theory of mechanism is as yet unadapted to deal with such basic uncertainties. Moreover (cf. the speculative cosmogony of, for example, Andrei Linde [1991], combining inflation and quantum theory), an element of arbitrariness in the definitions of the basic parameters of our Universe is consistent with a global determinism, in that one cannot exclude causally unconnected components of material existence, of which our Universe may be just one.

We notice that this model of causality is very much in the spirit of such Humean solutions as those of F. P. Ramsey [1978] (revived by David Lewis [1973]) in attempting to derive the precise counterpart to the intuitive content of the notion of a law via its context as a generalisation in some ideal systematisation of knowledge. Making the necessary translation between consequences of a deductive system and relations invariant in automorphic models, this is apparent from Ramsey (p. 138):

> ... [laws are] consequences of those propositions which we should take as axioms if we knew everything and organised it as simply as possible in a deductive system,

or Lewis (p. 73):

> ... a contingent generalization is a law of nature if and only if it appears as a theorem (or axiom) in each of the true deductive systems that achieves a best combination of simplicity and strength.

Moreover, the actual form of the Turing explanation provides explication of the notions of necessity of such rejecters of the Humean tradition as Armstrong [1983], Dretske [1977] and Tooley [1977] — and successfully addresses the criticisms (eg. Tooley [1977]) of the Ramsey–Lewis approach.

Of course, implicit in all this is a confirmation of the role of 'nomic necessity', particularly in the sense of Swoyer [1982], for whom laws express non-contingent relations. Earman [1986], p. 105, points out that:

> ... if there is nomic necessitation, its ultimate springs are most likely hidden from our view. The ultimate laws of nature, whatever they may be, will most likely involve universals whose instancings corre-

spond to states of affairs which are not directly observable and which are thus knowable only inferentially.

But this is wholly in line with the mathematical model.

See also Skyrms [1980] and Honderich [1988].

The picture which emerges from the theory of Turing automorphisms is one in which at neither the material nor the epistemological level can the basic causal relations guarantee a uniquely defined reality, but within which one can, from a realist perspective, explain the condensation of a so-called 'quasi-classical domain' in terms of Turing invariance. While the epistemological dimension is viewed as an integral part of the deterministic Universe, firmly based on the physical dimension, and algorithmically related to it via familiar proof-theoretic structure, but admitting qualitatively similar processes of systemic input as described for the material case. Intrinsic to this unity is the acceptance of human descriptions of the universe in a standard language as finitary approximations to limiting versions (existing in an appropriate infinitary language), no different in formal role to the rationals in relation to the descriptions of real numbers as infinitary decimals. Such a comprehensive relationship of model to that modelled is of course complex, but has the advantage of capturing intuitively based hierarchical structure independently of phenomenalist confusion. If such a scenario is accepted, there are a number of important consequences for science and for what are widely perceived to be its current crises.

Quantum versus classical reality

The first of these consequences is in relation to the search for a realist interpretation of quantum theory and a wider theoretical cohesion (encompassing, for example, what we know of gravity and the theory of general relativity). There is a conceptual difficulty in that the description of the constituent elements of the Universe in terms of reals involves an acceptance of *information content* (beyond those incidental details arising from the details of the specific presentation chosen) as the essential ingredient correlating with the underlying reality. And that in the quantum context there may be no counterpart to everyday experience leading one to expect that distinctions between entities can be achieved by reduction to a suitably basic level of information content. This brings brings Leibniz's formulation of the *identity of indiscernibles* (which roughly translates as the claim that objects must differ in some intrinsic, non-relational way to be distinct) into sharp focus — which in a letter to Samuel Clarke (see Alexander [1956]) is derived from another principle basic to realist interpretations of quantum theory, namely the *principle of sufficient reason*, which according to Leibniz himself (see Leibniz [1714], sections 31, 32) says that:

> ... there can be found no fact that is true or existent, or any true proposition, without there being a sufficient reason for its being so and not otherwise, although we cannot know these reasons in most cases.

The implication is that a well-developed cache of information is a prerequisite for any infrastructure which is of significance to a Universe whose laws are based on finitary transfer of data. In choosing the representation of a particular particle (assuming some form of atomism), one must require it to reflect its information content in relation to its global context. One may choose (consistently with the presentation) to represent information about the state of a larger body, in such a way that there will be a simple relationship between this and the conflated information content of its constituent parts. Ideally, one has a predilection for presentations of the Universe which reflect that information content in what we might describe as an *honest* way, relating to the boundary conditions in such a way that a particles actual evolved information content, relative to other constituent parts of the Universe, is essentially that of its representation. This is what Laplacian determinism would have provided, under fairly reasonable assumptions. Although the relativistic trend of physical theory first suggests one looks for presentations defined in terms of *relations* on the structure (cf. Smolin [1991]), this ignores the essentially hierarchical nature of information content. This is not to say that the information content does not exist independently of its history, but it does not seem possible to retrieve it, locally notated, without it.[12] So one looks to a Universe evolving according to processes based on definite mathematical principles. This means that the role of the causal structure in generating such information content requires that any presentation respects it, so that the range of allowable presentations is restricted by the available Turing automorphisms (as determined by Slaman and Woodin's finite automorphism base). The existence of non-trivial automorphisms indicate that there is no such thing as the hoped for honest presentation. This can be interpreted not just as indicating the existence of many indistinguishable but essentially different presentations, but as referring to possible material manifestations of the Turing model for basic causal structure in that matter can be regarded as a possible framework of labels for it. Since, in accordance with the principle of sufficient cause, nature appears to allow all histories not systemically excluded by the mathematical model, the observed effect is that of physically existing (although

[12]There is of course an extensive literature relating to historical determinism, reversibility of computation, and its consequences for the particulars of modelling scientific laws (to which we will not try to add). See, for example, Berlin [1969], Gandy [1980], Fredkin and Toffoli [1982], Margolus [1984] and Toffoli [1980].

the status of such existence varies according to different interpretations) parallel alternatives.

The particular form of the causal structure at the quantum level (ignoring for the moment the basal information content) will depend on the local structure of the Turing universe. This structure is not easily accessible (see for example Lerman [ta]). But as noted in section 4, the infinitary nature of Aut($\mathcal{D}$) translates, via known definabilities and automorphism bases, into that for Aut($\mathcal{E}$) and Aut($\mathcal{D}(\leq \mathbf{0}')$) (corresponding to the theoretically predicted, and empirically confirmed, multiplicity of available histories at the quantum level), while the definability of all levels of the Kleene-Post hierarchy corresponds to the structural regularities of theory. One does need to ask how much information content need be materially manifest for the theory to apply. Is it possible for nonrigidity to break down for particular local substructures of the Turing universe? If one restricts oneself to *invariant* substructures (and it is hard to see how any naturally occurring structure generated according to specific mathematical laws could be otherwise) one concludes that any nontrivial level of information content is likely to be accompanied by nonrigidity. Certainly any such structure containing $\mathcal{E}$ qualifies, as (by Ambos-Spies [ta]) do a number of basic definable substructures of $\mathcal{E}$. Although, at least in a formal sense, the existence or otherwise of *naturally realisable* substructures of $\mathcal{E}$ with pathological automorphism groups is still not known, there is no evidence that the material universe does not exhibit a full richness of noncomputable computably enumerable phenomena (relative to its initial conditions). In other words, it seems that according to all reasonable projections *one can view the material universe as forming an automorphism base for the Turing universe.* Present techniques seem sufficient to eliminate the possibility of definable computably enumerable singletons other than $\mathbf{0}$ or $\mathbf{0}'$, consistent with the association of the collapse of the wave function with entanglement with the classical level. Whether all such singletons are automorphism bases is not known.

The resulting explanation of Heisenberg's uncertainty principle in terms of systemic imperfections of invariance, extends to natural phenomena theoretically based on uncertainty, such as quantum fluctuations and the matter/anti-matter dichotomy. This limits the cosmogonical role of the former, in that invariance/noninvariance can only determine the form of an existing structure or phenomenon, suggesting that quantum fluctuations depend on a failure of definability of already present empty space as a basic prerequisite. So any idea (originating with Tryon [1973]) of creation of the Universe as *purely* such a quantum event must represent a victory of style over substance.

Returning to the question of how, specifically, the fundamental forces of nature emerged, speculation closely follows the current view provided by physics. What impresses itself upon us is that in the absence of *appropriate* (in a sense to be elaborated on) information content (in the very early universe, say) there would be insufficient basis for the differentiation of more than *one* such causal agent, manifesting itself according to the $\aleph_0$ available algorithmic materialisations and Turing automorphic images available. While a more developed information content would be expected to lead to a collapse in possibilities according to the emergent invariance, perceived as physically related to an underlying breakdown in symmetry. Further crystallisation of basic forces might emerge by a similar process, until, ultimately, all ambiguities are squeezed out of the system by the weight of growing organisation of information content, and the familiar structure of nature is in place.

One also obtains an extension of Penrose's 'cosmic censorship hypothesis' — "... namely the hypothesis that ... naked singularities do not occur", Penrose [1996], p. 27 — providing an explanation of the (non)-role of *any* mathematical singularities in nature. As Smolin [1997] reminds us: "Many people who work on quantum gravity have faith that the quantum theory will rescue us from the singularities" (proved by Penrose [1965] and Hawking and Penrose [1970] to be inevitably associated, under fairly simple and broad assumptions, with sufficiently large systems behaving according to the theory of general relativity). Echoing the Hartle-Hawking [1983] 'no boundary' cosmological model (best known for ingeniously avoiding the infinite regress of cause and effect familiar from most arguments for the existence of God), one is able, without discarding causality itself, to exclude *any* naturally occurring singularity by relating the level of definition of natural laws to the projected information content in closed systems. To do this, one needs to look more closely at the likely information content in the early universe, or more exactly, very close to the big bang. Leaving aside spontaneous creation (there appears to be little one can say about that, even when it comes disguised by a boundary-free context), the scenario is either a more structured creational one (which does not seem to be what one is dealing with), or extreme disorder (but with significant systemic regularities) inherited from a feature of a previous (or at least, logically connected) phase of a more extensively conceived universe. In view of what our own universe seems capable of passing on (ignoring more speculative projections), 'feature' will be interpreted as 'collapse'. Even within this refined context there is considerable scope for speculation — for example the cosmological evolutionary proposal of Smolin [1992], summarised in [1997], p. 88:

> What we are doing is applying [the] bounce hypothesis, not to the universe as a whole, but to every black hole in it. If this is true, then we live not in a single universe, which is eternally passing through the same recurring cycle of collapse and rebirth. We live instead

in a community of "universes", each one of which is born from an explosion following the collapse of a star to a black hole.

Potentially, an analysis of information content around a 'bounce' (with its corresponding quantum cosmology) can theoretically by-pass the extreme disorder implicit in the standard view of the 'big crunch', so putting such speculations on a firmer footing. But for the avoidance of space-time singularities, one need only extrapolate from the consequences of the inevitable homogenisation and local ramdomisation of information content near the big bang, with its consequent dissolution of the invariant relations on which the fundamental forces of nature depend. What is significant is not so much *loss* of information content in the vicinity of an incipient singularity (this is hard to quantify), but the decay of Turing invariant relations (or at the physical level, of the formations of nature) on which higher-order structure must be based. This can be formalised in terms of invariance within the class of random degrees (cf. Kučera [1990]), noting for instance that the 2-random degrees are known to avoid the cone above any nonzero degree below $\mathbf{0}'$. The eventual dismantling of the motive force for a singularity cannot avoid extreme compression and homogenisation of the antecedent information content, and this is necessary for the regularity of structure on which a new phase will depend. One seems to require at least the occupancy of a single atomic jump class. The picture that then emerges is of successive phases, involving structurally similar development, but potentially more *complex* in that it may be based on cumulatively more developed boundary conditions determined by previous phases. Theoretically, the underlying reason for this increase in complexity of the ensuing universe is to be found in disproofs of homogeneity in the Turing universe (see section 4). Further, the potential *relevance* of the increase is emphasised by Shore's [1981] demonstration that there are many Turing degrees $\mathbf{a}, \mathbf{b}$ for which $\mathcal{D}[\mathbf{a}, \mathbf{a}']$ and $\mathcal{D}[\mathbf{b}, \mathbf{b}']$ are not elementarily equivalent (for example $\mathcal{D}(\leq \mathbf{0}')$ is not elementarily equivalent to $\mathcal{D}[\mathbf{0}', \mathbf{0}'']$). While any analysis of the information content passed on by a collapsing universe depends on a much better understanding of the consequences of a failure of Church's thesis in a non-discrete physical system for the level its informational entropy.[13] (See David Layzer [1990] for a discussion of some of the issues involved.)

Another reassuring consequence of the Turing model is the theoretical counterpart to the classical appearance of reality above the quantum level provided by results on definability and rigidity in $\mathcal{D}$. The definability of the jump (and so in particular, of $\mathbf{a}'$ in $\mathcal{D}(\geq \mathbf{a})$, where $\mathbf{a}$ can be taken to be a member of the atomic jump class determined by the initial conditions of the

[13]The popular equation of 'more entropy' with 'less information' (see for example Ferris [1997], p. 92) must of course refer to *availability* of the information (according to the classic formulation of Shannon [1948]).

Universe), with its canonical relationship to the most basic naturally generated incomputability, provides a gateway to classical structure, entered via the background subatomic information content inherited from the early universe and intensified quantum contingency. Despite the apparent nonrigidity of $\mathcal{D}(\geq \mathbf{a}')$ in $\mathcal{D}(\geq \mathbf{a})$, the acquired information content implicit in the phase-transition from non-invariance to invariance can be relied on to lift the consequent phenomenological structure to the (rigid in $\mathcal{D}(\geq \mathbf{a})$) cone above $\mathbf{a}''$, thereby removing any possibility of a return to quantum ambiguity. (The phase-transition depends on the fact that any increment in local information content is *relative* to the context, and is expressed via a presentation which is *globally* modified by individual changes.) The apparent nonexistence of definable computably enumerable singletons other than $\mathbf{0}$ and $\mathbf{0}'$ is consistent with this picture.

How does the explanation of quantum nonlocality in terms of variable levels of underlying Turing invariance compare with existing, physically specific, 'realistic' interpretations? Of these the front-runner currently is the theory of 'decoherence' particularly associated with Gell-Mann and Hartle [1990], Griffiths [1984] and Omnès [1994], with its pleasing consequences for quantum cosmology. This seeks to augment the 'many-worlds' interpretation of Everett [1957] with an explanation, in terms of the logical structure of the Universe, of the origins of our 'quasi-classical domain' and of the 'branching' accompanying measurement which creates the inaccessibility to observation of the parallel worlds (re-termed *decoherent histories*) at the classical level.

According to the theory (for which we mainly refer to the version of Gell-Mann and Hartle), our observations can be made consistent with the 'many worlds' scenario by appropriately locating us within a particular structure of decohering alternatives, but are an inadequate guide to the specific *form* of the decoherence, except in so far as we can imperfectly project from the quantum level. This is described by Gell-Mann [1994] in terms of *coarse graining* (p. 144):

> For histories of the universe in quantum mechanics, coarse graining typically means following only certain things at certain times and only to a certain level of detail. A coarse-grained history may be regarded as a class of alternative fine-grained histories, all of which agree on a particular account of what is followed, but vary over all possible behaviors of what is not followed, what is summed over.

Then (Gell-Mann and Hartle [1990], p. 445):

> As observers of the universe, we deal with coarse grainings that are appropriate to our limited sensory perceptions, extended by instruments, communication, and records, but in the end characterized by a great deal of ignorance. Yet we have the impression that the

> universe exhibits a finer-grained set of decohering histories, independent of us, defining a sort of "classical domain", governed largely by classical laws, to which our senses are adapted to dealing with only a small part of it. No such coarse graining is determined by pure quantum theory alone. Rather, like decoherence, the existence of a quasiclassical domain in the universe must be a consequence of its initial condition and the Hamiltonian describing its evolution.

A quasiclassical domain "should be a set of alternative decohering histories, maximally refined consistent with decoherence, with its individual histories exhibiting as much as possible patterns of classical correlation in time". Disturbance by quantum events ensures that "[t]here are no classical domains, only quasiclassical ones". This leads on to a general consideration of maximal sets of alternative decohering histories (that is "those for which there are no finer-grained sets that are decoherent"), but then makes the special (or otherwise) status of the quasiclassical domain of which we seem to be a part problematic (see for example Dowker and Kent [1996]). Efforts to find a way out of the resulting confusion have led to an impression of desperation. The likelihood of decoherent coarse grained histories makes explanation of our quasiclassical domain centre (a variation on the anthropic principle) on the role of the observer (abstracted by Gell-Mann and Hartle as an IGUS, or 'information gathering and utilising system'), and leads to the 'lack of economy' complained of by Penrose [1989], p. 382, in the many worlds interpretation. Added to this, the theory does not appear to provide a full solution to the measurement problem.

However, if one looks more closely at accounts of the mechanism by which decoherence breaks down, one is struck by a remarkable convergence between these and local descriptions of the imposition of Turing invariance. Histories decohere in the absence of 'entanglement', expressed in terms of the interference term between histories. According to the Turing model of the causal structure of the universe, different automorphic images become possible in the absence of definability of substructure relative to what is rigid in $\mathcal{D}$. For the former, the emergence of a coarse grained alternative is related to cumulative entanglements affecting its corresponding fine grained components. There is no immediate reason for the cumulative breakdown of decoherence to define a unique convergent reality. This is easily recognised as an 'on the ground' description of the development, or otherwise, of invariance within a structure. Viewed at a phenomenological level, the full picture of decoherence between course grainings is complex. But empirically one is led to expect correlations — we are dealing with the *structure of nature* with all its observed regularities of behaviour. So one expects to abstract from the disparate phenomena a common mathematical framework, namely the causal structure captured by the Turing model. This holds out the possible renewal of the project based on known results of classical computability theory. From a recognition of the nature of the underlying mathematics of decoherence, one arrives at a precise explanation of why the coarse grained alternative we see has its special status, so answering what Omnès [1994], p. 504, calls "*the* problem, which is the existence of facts". What is missing from the decoherence scenario is now provided by the known features of the Turing universe, and the evidence for a quantum/classical dichotomy around the $\mathbf{0}'$, $\mathbf{0}''$ level, the rigidity in $\mathcal{D}$ of the causal structure above $\mathbf{0}''$ giving the required unique, familiar quasi-classical domain. One can even anticipate that the nature of the probabilities intrinsic to the measurement problem can be explained via a better knowledge of the automorphism group of the Turing universe.

There is also an apparent consistency with the other main contender, the reconstructed 'hidden-variables' theory of David Bohm. Developed in a sequence of papers and books, from Bohm [1952] to Bohm and Hiley [1993], the details of the theory — the wave function ψ taken to represent a real field (avoiding the apparent ambiguity of its reality status in the Gell-Mann and Hartle proposal) whose fluctuations originate at a classical sub-quantum-mechanical level involving hidden variables, with a nonlocal transfer of 'active information' (augmenting the 'pilot wave' idea of de Broglie [1927]) responsible for a sort of cosmological version of the 'global village' — are not transparently related to any notion of mathematical nonlocality. But within the philosophical constraints inherent in the approach, the picture developed (particularly of an *implicate* order, Bohm [1980]), becomes an informative metaphor for the more theoretically robust one. There are other proposals using hidden variables to restore exact determinism, for example that of Ghirardi, Rimini and Weber [1986], this time introducing at the quantum level a new and empirically undetected, randomly occurring, localising effect, which at macroscopic levels has the potential to fix the position of a solid object via state entanglement. So it would be too simplistic to dismiss all such theories as "a return to antiquated ideas", but the judgement (Omnés [1994], p. 401) that it

> ... seems difficult to accept that the deep and unexpected mathematical properties one found for the [operators] expressing classical properties [of] the decoherence effect do not contain a large part of truth. They rely upon so few assumptions and recover so many well-known features of reality that had remained unexplained before them that they have the kind of beauty Dirac coined as the mark of truth ... ,

is convincing. And the underpinning of the logical aspects within the Turing context of can only strengthen that impression.

Emergence and Entropy

Another area in which reductionist explanations have become augmented in recent years by those based on more global considerations is the biological sciences, with vigorous debate centering on such thorny problems as characterising the evolutionary process and the origins of life. A key concept has become that of 'emergence' in relation to the anti-entropic forces represented by life and its development, and more generally (see for example Holland [1998]). The problem is the recurring one of the breakdown of reductionism in the sciences, despite what appears to be a firm causal link between phenomena at different levels of knowledge. Recent analyses of the process whereby higher-order structure emerges via systemmic expressions of relatively well-understood local laws in biology are presaged by earlier ideas such as those of Lynn Margolis [1981] on 'mutualism', which as Steven Rose [1997] remarks, were once considered heretical, but (p. 229) have "now become the conventional wisdom of the textbooks". The key connection which has been made recently is between the emergence of order, in say the origins of life or in natural selection, and the mathematical phenomenon of attractors. See, for instance, Stuart Kauffman [1995], p. 26:

> The wonderful possibility, to be held as a working hypothesis, bold but fragile, is that on many fronts, life evolves toward a regime that is poised between order and chaos. The evocative phrase that points to this working hypothesis is this: life exists on the edge of chaos. Borrowing a metaphor from physics, life may exist near a kind of phase transition. ... Networks in the regime near the edge of chaos — this compromise between order and surprise — appear best able to coordinate complex activities and best able to evolve as well. It is a very attractive hypothesis that natural selection achieves genetic regulatory networks that lie near the edge of chaos.

Or again, Rose [1997], p. 166:

> The cellular web ... has a degree of flexibility which permits it to reorganize itself in response to injury or damage. Self-organization and self-repair are its essential autopoietic properties. These properties of stability and self-organization, which Stuart Kauffman has described a 'order for free', are the key to appreciating the fundamental irreducibility of living cells. Their metabolic organization is not merely the sum of their parts, and cannot be predicted simply by summing every enzyme reaction and substrate concentration that we can measure. For us to understand them, we have to consider the functioning of the entire ensemble.

But graphic and informative as these descriptions of the emergence of new relations are, it is the lack of a secure theoretical basis which makes such speculations "fragile". However, one is struck by the close parallels between the descriptions of self-organisation in nature, and the emergence of 'order for free', and ones conceptualisation of the process of emergence of Turing invariance. Once again, it is possible to make detailed correspondences, and there *is* an explanation, in the context of the Turing model.

Epistemological relativism

Corresponding to these consequences for the structure of matter, there are analogous ones for that of knowledge of the Universe. This is strikingly adumbrated (in ways not envisaged by the author of Turing's biography) in the following quote from Andrew Hodges [1983] (in relation to Hofstadter's [1979] views on "the significance of Gödel's incompleteness and Turing undecidability for the concept of Mind"), p. 540:

> Far more significant, in my view, is the limitation of human intelligence by virtue of its social embodiment – and this is a problem relegated to a marginal place in Hofstadter's work as in so many other accounts, though I have placed it at the centre of my own.[14] The study of Alan Turing's life does not show us whether human intelligence is limited, or not limited, by Gödelian paradoxes. It does show intelligence thwarted and destroyed by its environment.

As noted previously, incomputability in the objects of scientific investigation is relevant to the process of establishing truth in science, not just in that computability or otherwise determines the manner of discovery, but in that the process of discovery itself is an analysable physical phenomenon. To quote Bohm and Hiley [1993], p. 326:

> ... the view that our theories constitute appearances does not deny the independent reality of the universe as a whole. Rather it implies that even the appearances are part of this overall reality and make a contribution to it.

However Newton [1997] has a subtle variation on the theme that nature is deterministic and that it is just observation which is ill-defined, suggesting that it is the *language which is used to describe* observation which is inadequate. But this is analogous to saying invariance and Turing definability give rise to relevant distinctions — which they do not, by Slaman and Woodin (see the previous section). Anyway, instead of a classical reality sitting uncomfortably on top of a level of ambiguously defined micro-phenomena, one is now confronted with epistemological relativism, a failure of reductionism and a concomitant disunity of scientific theory. Similarly to before the emergent levels of definition enter again into the structural development via processes of systemic observation and awareness. What is familiar from

[14] Although oracle Turing machines get little more than a passing mention.

nature is the way one is able to establish a fairly comprehensive computable framework of relationships at a basic level, but whose explanatory power dissipates as the lines of causal connections extend — descriptions of the emergent chaos being effectively incomputable, even if the exact nature of the incomputability is not precisely characterised. Sitting on top of such turbulence, new regularities emerge involving new computable parameters and dynamic relations, giving rise to a more local theory, which may be less theoretically comprehensive, but maybe more important from an everyday perspective. The parallel with the structure of the Turing model is striking. Here again one sees (suggested by recent work) an onset of noninvariance, within which new structures become defined, but relative to which there is a recurrence of the lower-level formlessness.

The 'nonrigidity' of knowledge is already implicit, in a limiting form, in the so-called Duhem-Quine thesis. Also, Popper's [1959] qualification of the 'scientific method' is surely right in essentials, dealing as it does with the practical aspects of the consequently more subtle relationship between observation, hypothesis and theory, even if one persists in detecting scientific induction in the way confidence in a given theory is consolidated by a mixture confirmatory evidence (such as realised predictions) and *absence* of negatory evidence (and Popperian falsification — see Lakatos [1970] — being conditional on a competitive induction associated with *different* theories). Many relations on the Turing universe may be definable within the global context, but may not be computable, inductive extensions of observations. Others may be even less accessible to the scientific method. Related to this is the fact that a structure of knowledge involving unpredictability (already implicit in Gödel [1931], [1934], and made more relevant via the Turing model) means that one may have no way of judging whether a particular statement has empirical consequences or not, a fact commonly ignored — for instance, relevant to the above explanation for nonlocality in quantum mechanics, one can quote van Fraassen [1980], p. 95, on 'correlations in the behaviour of particles which have interacted in the past, but are now physically separated:

> These [hidden variable theories] do not predict exactly the same correlations — this is what makes these theories interesting to physics. So far, experiments appear to support quantum theories against those rivals. But the one response which is conspicuous by its absence is that an explanation of the correlations *must be found* which fits in exactly with quantum theory and does not affect its empirical content at all. Such metaphysical extensions of the theory (if indeed possible) would be philosophical playthings only. There are only two camps to the debate as far as physics is concerned: either this non-locality makes quantum theory pre-eminently suited to the

representation of the world (and we need to re-school our imaginations), or else quantum theory must be replaced by an empirically significant *rival*.

One also notes that observations may relate to definability in different ways, perhaps being repeatable, so inductively extendible, or having some individual significance in relation to such inductively (in the wider sense) derived frameworks. They may lead to 'theory-laden' facts, which again have an impact on the inductively based framework of theory. This means that although *reproducibility* of experiments depends on the underlying algorithmic content of the basic observations, it may be that the imaging of the phenomenon in question via the results of observation may exploit the full subtlety of the relationship between definability and observational data, with Husserl's [1913] notion of "eidetic intuition" being particularly relevant to epistemological manifestations of this relationship.

Particularly relevant to the extension of the model to the epistemological domain is Quine's [1953] cogently argued dismissal of any "fundamental cleavage between truths which are *analytic*, or grounded in meanings independently of matters of fact, and truths which are *synthetic*, or grounded in fact". For Quine (p. 44–45):

> Physical objects are conceptually imported ... as irreducible posits comparable, epistemologically, to he gods of Homer. ... Moreover, the abstract entities which are the substance of mathematics ... are another posit in the same spirit. Epistemologically these are myths on the same footing with physical objects and gods, neither better nor worse except for differences *in the degree to which they expedite our dealings with sense experiences*. (added italics)

While the exact mythological (or otherwise) status of the observations whereby objects are 'imported' is tangential to the discussion (although, for instance, one tends to be sceptical about mathematics as a cultural construct), it is clear that there is a conceptual convergence between this and the present picture, in that different categories of knowledge are associated with levels of irreducibility (independent of any particular atomist assumptions) based on qualitatively similar formation processes. As a result one can even speculate on the relevance of nonrigidity to the epistemological status of such well-known undecided statements of mathematics as the continuum hypothesis (CH). Even Gödel's [1964] approach to CH, ostensibly Platonist, seemed to admit a degree of relativism in that it involved contemplation of alternative models for ZFC, in as much as his description of the search for intuitively acceptable new axioms capable of deciding CH suggests an underlying consideration of new models in which such statements are true and which seem to realistically extend what is already believed to be true. It is a trivial observation that whatever the absolute status of mathematics, models are *theoretically* integral, so that our universe is symbiotically related to mathematical truth, in so far as it exists, via its modelling role. There arises the possibility that widely dispersed but apparently analogous phenomena may in fact be subject to a fundamental unifying principle.

So a first consequence of such nonrigidity is a partial confirmation of current relativistic views of human knowledge, but in a form which limits the extent of this relativism, and undermines its wilder inductive extensions. There is a difference between the conventionalism of Poincaré, for instance, and the more recent anti-realist views of scientific knowledge (see for example van Fraassen [1980]). There is a level of consistency of the former with Einstein's view of the theoretical results of science being in some sense *determined* (even if one does have underdetermination of theory by data to contend with), and anyway, the seeking out of conventions is not necessarily arbitrary and unrooted in an objective reality (cf. Newton [1997], p. 15). Again, one can differentiate between the pragmatism of say Peirce (the mathematics reflecting the subtlety of the argument), and the more radical, apparently anti-epistemological, versions of Schiller [1907] or Rorty [1979], [1982]. But while postmodernism can be termed (Gross and Levitt [1994], p. 87) an "instrument of revenge" against the tendency, derived from the earlier dominance of the logical positivism (the classical statement of which being A. J. Ayer [1936]), to constrain the humanities to the methodology of the empirical sciences, one cannot entirely dismiss the intuitions which underpin such analyses. Beyond the routine cataloguing of the dislocations of reality and discourse, and the tentative attempts (see for example Katherine Hayles [1990]) to enlist Gödel's theorem and chaos to explain the mismatch, there is potentially a solid theoretical explanation in terms of Turing nonrigidity and corresponding failures of definability in the real world. Implicit in the failure of bi interpretability is a fractured relationship between language and observation — there seems to be no alternative characterisation of the Turing definable relations available, while even at the local level (see Cooper [1997] and Nies, Shore and Slaman [1996]) definability in arithmetic and invariance are not comparable notions. On the other hand, it is difficult to envisage any consensus about what is certain, and what is culturally created (a contemporary battlefield as described by Gross and Levitt [1994]) without some sort of theoretical limitations on the inductive projections indulged in on both sides of the argument.

Secondly, one finds via the Turing model an echo of current doubts about the validity of reductionism in science, related to, for instance, the instrumentalism of Nancy Cartwright [1983] (there is no guarantee that received definitions of fundamental laws correspond exactly to their invariant status) or, more relevantly, the "epistemological pluralism" of Dupré [1993]. The Turing analysis does seem to give some credence to the disunity of science from a realist perspective. Reductionism is concisely described by Oppenheim and Putnam [1958], who

> ... propose the following levels: elementary particles, atoms, molecules, living cells, multicellular organisms, and social groups. ... Reduction consists in deriving the laws at each higher (reduced) level from the laws governing the objects at the next-lower (reducing) level. Such reduction, in addition to the knowledge of the laws at both the reducing and reduced levels, will also require so-called bridge principles (or bridge laws) identifying the kinds of objects at the reduced level with particular structures of the objects at the reducing level. Given the transitivity of such deductive derivation, the end point of this program will reveal the whole of science to have been derived from nothing but the laws of the lowest level and the bridge principles.

The problem is that a closer look at our *experience* of science does not entirely confirm this picture. Dupré (pp. 88–89), for instance, acknowledges reductionism's "great successes", in particular noting that: "Many of the greatest achievements of science depend essentially on insight into the structure of objects", but goes on to point out that "it does not follow that *all* scientific enquiry has to do with elucidation of structures". Consequently science, and human knowledge generally, often appears compartmentalised into diverse disciplines each with its own specific assumptions and technical basis. As David Bohm [1957], p. 133, observes:

> Any given set of quantities and properties of matter and categories of laws that are expressed in terms of these quantities and properties is in general applicable only within limited contexts, over limited ranges of conditions and to limited degrees of approximation, these limits being subject to better and better determination with the aid of further scientific research. Indeed, both the very character of the empirical data and the results of a more detailed logical analysis show that ... the possibility is always open that there may exist an unlimited variety of additional properties, qualities, entities, systems, levels, etc., to which apply correspondingly new kinds of laws of nature.

The resulting distinction between more basic and relatively local theories is very much in accord with the theoretically originating demarcation lines growing out of areas of chaos, with newly defined parameters emerging across boundaries, to be incorporated in new fields of study.

So it is possible to view the hierarchical nature of scientific theory as itself a response to the incomputabilities generated at successive levels, along with the surprising but theoretically explicable reattainment of algorithmic content (provided by a less fundamental, possibly, 'local' theory) at each level (comparable to the emergence of 'strange attractors' in chaos theory). The accumulation of *information* content at a particular level introduces new definabilities which become the raw material for the *algorithmic* content at the succeeding level. This description is closely echoed in Stephen Weinberg's [1992] quote (p. 61) from a talk by James Gleick:

> ... there are fundamental laws about complex systems, but they are new kinds of laws. They are laws of structure and organisation and scale, and they simply vanish when you focus on the individual constituents of a complex system — just as the psychology of a lynch mob vanishes when you interview individual participants.

So for example, the structure of material things and of our accompanying knowledge of them diminishes the *overall* importance of the search for a 'grand unified theory' (GUT) for the subatomic level, and exposes the inappropriateness of the term 'theory of everything' (TOE). There is also a partial confirmation of Dupré's argument for selective relativism (whereby one avoids the inductive extension of observed scientific inadequacy in certain contexts to an attack on the entire scientific project), in that there may well be areas in which the nature of the basic relations is associated with a very rapid accretion of complexity, so that the well-understood atomic dynamics tell one very little. Of course, one would argue that the information content (or chaos) of such contexts *does* define new parameters, which may become part of a higher level of analysis to which empiricism and logic *are* relevant, and that the definability of the jump does provide a link between the regularities at the different levels. But notice that the chaotic barrier to reduction (if the Turing model is accepted) is not just a practical one. The parameters at the succeeding level are basic to the resulting theory, but are computationally, in a fundamental sense, divorced from the lower level. (We are dealing with emergent Turing definabilities.)

Despite all the above, one notes that there is of course a reduction process at work — there is no breakdown in determinism — but it involves systemic, or globally originating, processes, so is imperfectly available to us epistemologically. It is also worth noting that natural laws may be related to definability in a less than precise way, in that the phenomenon is only approximated by the defined relationship (e.g. the laws of quantum mechanics enable bizarre exceptions, of vanishingly low probability, to higher level laws).

There is also (although this will get less attention than its real-world importance merits) an epistemological level corresponding to that of the material quasi-classical domain. Although contained within a relativistic context there is a potential convergence of agreed knowledge of the universe — a comprehensive world view is not just an aid to erroneous certainty.

However, on a technical note, it may be that one should distinguish mathematically between questions of epistemological *structure* and *practice*. Whereas parallel realities in nature, or differing but coexisting world views governing certain areas of human experience, may be acceptable, scientific activity depends on a more convergent framework. The computable basis for processes relative to historically emergent *partial* information can only be modelled successfully in the Turing context by allowing indeterministic oracle machines, and this is usually formulated (see Rogers [1967]) in terms of *enumeration reducibility*. A is said to be *enumeration reducible to* B ($A \leq_e B$) if and only if there is an algorithmic description of how to enumerate the information (i.e. the members) in A (in some order) given *any* enumeration of that in B. The resulting structure (captured in the *enumeration degrees* $\mathcal{D}_e$) forms a natural extension of the Turing model (under trivial notational adjustments). See Sorbi [1997] for a recent survey of results and techniques, or Cooper [1990] for the historical background. Recent work suggests that not only is the picture previously outlined, including the model of the familiar quasi-classical domain, to be found naturally embedded in the extended one, but there is a consistency of theory in that nonrigidity extends to $\mathcal{D}_e$ (although many other local and global characteristics may be very different). Moreover, there is a strong structural continuity in that (Sorbi [ta]) $\mathcal{D}$ forms an automorphism base for the enumeration degrees.

It would be hard to pass on without mentioning the relationship between free will and determinacy, and the traditional philosophical debates around the existence or otherwise of *origination* (the creation of new causal chains by free human choices), and the compatibility of determinism and moral freedom. Of course the renewed mathematical model of determinism rules out origination, including arguments[15] for it based on conjectural physical links between brain processes and quantum state vector reduction (cf. Penrose [1989], [1994], Stapp [1993]). Given determinism, other discussions are little changed in essentials, although determinism conceived less mechanistically has the effect of swinging the argument away from the incompatibilists (in regard to freedom and determinism) and the hard determinism of William James and his followers, according to which moral responsibility is illusory. Determinism can be viewed as assisting a complex and creative process of developing information content in which we and our judgements (perceived as free choices) are full participants. What is different is that, unlike in a clockwork universe, our participation does substantively change things. Moreover, the human mind involved in forming such judgements can be seen as involving a huge accretion of causal structure, with a subjective experience of free will based on the emergent form

[15]Anticipated in philosophical writings of Arthur Eddington [1928].

rather than on those of complexity and incipient disorder. The parallel between the way definability derives from complexity of information content, and our experience of mental creativity, suggests a process whereby great complexity and confusion, by a noncomputable causal connection, gives rise to unforseen structure. A concomitant of the functional complexity is the intimate relationship with a physical Universe of which our minds are microcosmic reflections (an aspect of mind central to the arguments of Dreyfus [1979] against the claims of classical AI). So the incidence of revolutionary new ideas depend not so much on more efficient (computable) mental processes but on their functional relationship to extreme confusion of mental content, and the level of systemic intolerance of formlessness. The lesson may be — avoiding the vagueness of, say, Bergson's [1907] *élan vital* underpinning his anti-scientism — that many evolutionary processes not only entail jumps in complexity of information content, but are essentially *creative*, in that the very same kind of collapse of alternative realities familiar from quantum theory are implicit in evolutionary selection. And it is this, even without direct physical links between quantum phenomena and brain processes, which may lead to recognition of the extraordinary aptness of the intuitions of Penrose and Stapp.

The existence of free will (as in the 'Free Will Defence') is intrinsic to particular responses to objections to theism (about which one might expect mathematics to have even less to say). But the basis of formal creation (viewed as a structure of invariance) in an informational contexture of hierarchical complexity can be interpreted theodistically. The aim of creation and development of a formal ideal provides a teleological context in which one can approach the problem of evil (prominent in both Western and Eastern philosophical discussions) in terms of necessity of complexity of structure (see, for example, Swinburne [1979], chs. 9–11, or John Hick [1978]). The teleological assumption is fairly minimal, given that, ultimately, the empty teleological context is not an option. In contrast, what makes the so-called 'anthropic principle' (the need for which the Turing model very successfully releases one from) so suspect is the strength and essential arbitrariness of the teleological premise.

6. Turing nonrigidity as paradigm shift

Not only does the Turing model, relating definability (in terms of inductively expanded observational data) and empirically based theories, provide a sympathetic context for Kuhn's description [1962] of how scientific knowledge (not a term Kuhn himself would have chosen) emerges, but the revision of 'the scientific paradigm' implicit in the model's acceptance can itself be seen as a Kuhnian paradigm shift.

Following on from the philosophical foundations laid down by Descartes, Galilei and others, it was the discoveries of Newton which established the scientific outlook which so dominated western society for the next two and a half centuries and still molds the thinking of countless numbers of working scientists. Having achieved the status of a paradigm (and so, according to Kuhn, 'declared invalid only if an alternate candidate is available to take its place'), it became more basic than any of the previously challenged examples described by Kuhn, its inadequacy reflected in many longstanding philosophical controversies, and eventually becoming a key ingredient in the current crisis in the way science is popularly regarded — much to the consternation and bemusement of the professional practitioners of science (see for instance Gross and Levitt [1994] or Newton [1997]). Constituent assumptions included the belief that observation (potentially) gives a *clear image* of a real world, and that a precisely defined reality is a normal state of affairs. In regard to the former, Kuhn (p. 121, third edition) talks of the need for, and lack of, "a viable alternative to the traditional epistemological paradigm ... initiated by Descartes and developed at the same time as Newtonian dynamics", observing how:

> Today research in parts of philosophy, psychology, linguistics, and even art history, all converge to suggest that the traditional paradigm is somehow askew.

As we have seen, the latter assumption presents us with an intractable anomaly involving our everyday experiences of the classical reality and the very successful descriptions of the quantum world. The Turing model both gives a framework within which to describe what nature achieves (via noninvariance) and one to structure (via definability) the process by which we try to gain knowledge of that achievement, and (Kuhn, p. 121) "... though the world does not change with a change of paradigm, the scientist afterward works in a different world."

Within computability theory there is already in progress what amounts to a minor paradigm shift. Nies, Shore and Slaman [1996] describe the development of coding techniques (for $\mathcal{E}$) as a vehicle for exploiting complexity of structure, and how "the ultimate expression of such coding procedures" became "embodied in the [bi interpretability] conjecture that crystallized the new paradigm of complexity as a route to characterization". In relation to this one should quote Kuhn (pp. 180–181, third edition):

> A revolution is for me a special sort of change involving a reconstruction of group commitments. But it need not be a large change, nor need it seem revolutionary to those outside a single community, consisting perhaps of fewer than twenty-five people.

This limited change associated with the discovery of Turing automorphisms is partly based on technical factors with their own associated Kuhnian incommensurabilities, but more significantly it provides further evidence of the extent to which the scientific paradigm has subtly infiltrated our culture. While the nature of the past twenty years' technical activity, the sweeping successes achieved within the bi interpretability programme of research, and the huge investment of time and energy, almost all on the side of definability, provide the infrastructure of the more restricted paradigm, the basic perceptions of the material world implicit in the larger one must inevitably colour one's expectations of its abstract counterpart, even if the link is not experienced in a particularly direct way. How typical (according to the Kuhnian scenario) of the establishment of a paradigm and its period of crisis that the research activity in the intervening period should have been so notable for its unconcern with foundational, philosophical and (even) applicational issues, and for its convergent technical direction.

Early local degree theory was motivated by the aim of characterising $\mathcal{D}$ and its more important substructures. The revelation of of what seemed increasingly pathological structure gave an air of hopelessness to this project, until renewed by Simpson [1976]. The resulting bi interpretability 'paradigm' supported an attempt, increasingly desperate, to relate this pathology to appropriate, more familiar, complexity of (mathematical) structure, and in the process to reduce the characterisation of the results of local degree theory to a small range of such results.[16] But the existence of nontrivial Turing automorphisms — not anticipated conceptually or technically — makes the nature of Turing definability *problematic*; one becomes even more aware of local degree theory as being *about* definability, rather than a reductionist tool in the struggle with pathology; while at the same time, the much wider ramifications connect the renewed project with very basic considerations of internal self-determination in the real world. The suggestion that "it is time for a new paradigm" (Nies, Shore and Slaman [1996]) becomes valid in a wider context.

Returning to the larger picture, one notes that radical as is the conceptual revision implicit in the logical explanation of the dichotomy between quantum and classical reality and of the associated appearance of breakdown in determinism, it is the parallel consequences for how we *know* the world (or otherwise) which entail a more basic renewal of thinking. Scientists (and mathematicians) have continued to be trapped in a clockwork universe, largely due to the dependence on the scientific method and the

[16]Of course, the commitment of researchers in the field to aspects of this programme varied. G. E. Sacks' well-known reaction to degree theoretic pathology was to focus on the importance of *techniques*, based on the intuition that there was something fundamental and relevant about the area in a wider sense. P. Odifreddi (private communication) pointed out (referring to the example of the many-one degrees $\mathcal{D}_m$, see Odifreddi [ta]) that even if the Turing structure *were* to be completely characterised — which seemed impossible — there would be no obvious consequences.

lack of a theoretical framework which is both consistent with the fruits of empiricism and able to extend to admit less well defined sensory evidence of limited reproducibility. Without such a framework, the sort of confusion surrounding the relationship of local to global, particularly in the arts and humanities, appears reminiscent of the alchemical confusion of the sciences before Newton. While the work of post-modernists, (post-)structuralists, and epistemological relativists of various shades, bears more than a passing resemblance to to the process of haphazard intuition and Aristotelian inference which then passed for 'natural philosophy'.

To quote Kuhn again, from the closing pages of his book, this time on the extent of epistemological relativism and the validity of a scientific enterprise based on the *nature of the world* (p. 173):

> What must nature, including man, be like in order that science be possible at all? Why should scientific communities be able to reach a firm consensus unattainable in other fields? Why should consensus endure across one paradigm change after another? And why should paradigm change invariably produce an instrument more perfect in any sense than those known before? ... It is not only the scientific community that must be special. The world of which that community is a part must also possess quite special characteristics, and we are no closer than we were at the start to knowing what these must be. That problem — What must the world be like in order that man may know it? — was not, however, created by this essay. On the contrary, it is as old as science itself, and it remains unanswered.

One answer, or at least the basis for an answer, has already been outlined above. Whether this can provide the basis for an update to the scientific paradigm, only time can tell. But there can be no doubt that the need for a reorganisation of the conceptual basis of our epistemological universe is no less pressing than that accompanying earlier crises in relation to the physical universe. Being realistic about such change, even when overdue, one can quote Darwin's [1872], p. 389 (in 1996 reprinted edn.), well-known comments from the concluding section of "The Origin of Species":

> Although I am fully convinced of the truth of the views given in this volume ..., I by no means expect to convince experienced naturalists whose minds are stocked with a multitude of facts all viewed, during a long course of years, from a point of view directly opposite to mine. ... [B]ut I look with confidence to the future, — to young and rising naturalists, who will be able to view both sides of the question with impartiality.

Or on a more pessimistic note, Max Planck [1949], pp. 33–34:

> ... a new scientific truth does not triumph by convincing its opponents and making them see the light, but rather because its oppo-

nents eventually die, and a new generation grows up that is familiar with it.

References

H. G. Alexander (ed.) [1956], *The Leibniz-Clarke Correspondence*, Manchester University Press, Manchester.

K. Ambos-Spies [ta], *Automorphism bases*, to appear.

D. Armstrong [1983], *What is a Law of Nature?*, Cambridge University Press, Cambridge.

A. Ashtekar, J. Lewandowski, D. Marolf, J. Mourão and T. Thiemann [1995], *Quantization of diffeomorphism invariant theories of connections with local degrees of freedom*, J. Math. Phys. **36**, 6456–6493.

A. Ashtekar and J. Stachel (eds.) [1991], *Conceptual Problems of Quantum Gravity*, Birkhäuser, Boston, Basel, Berlin.

A. Aspect, Dalibard and G. Roger [1982], *Experimental test of Bell's inequalities using time-varying analyzers*, Phys. Rev. Letters **49**, 1804–1807.

A. Aspect, P. Grangier and G. Roger [1982], *Experimental realization of Einstein-Podolsky-Rosen-Bohm gedanken experiment; a new violation of Bell's inequalities*, Phys. Rev. Letters **49**, 91.

A. J. Ayer [1936], *Language, Truth, and Logic*, Victor Gollancz, London; 2nd revised edn., 1946.

J. S. Bell [1964], *On the Einstein-Podolsky-Rosen paradox*, Physics **1**, 195–200; reprinted in J. S. Bell [1987], pp. 14–21.

J. S. Bell [1966], *On the problem of hidden variables in quantum mechanics*, Reviews of Modern Phys. **38**, 447–452; reprinted in J. S. Bell [1987], pp. 1–13.

J. S. Bell [1976], *Einstein-Podolsky-Rosen experiments*, in "Proceedings of the Symposium on Frontier Problems in High Energy Physics", Pisa, June 1976, pp. 33–45; reprinted in J. S. Bell [1987], pp. 81–92.

J. S. Bell [1987], *Speakable and Unspeakable in Quantum Mechanics: Collected papers on quantum philosophy*, Cambridge University Press, Cambridge, New York, Sydney.

P. A. Benioff [1982], *Quantum mechanical Hamiltonian models of Turing machines*, J. Stat. Phys. **29**, 515–546.

H.-L. Bergson [1907], *Creative Evolution*; translated by A. Mitchell, Macmillan, London, 1928.

I. Berlin [1969], in "Four Essays on Liberty", Oxford University Press, Oxford.

L. Blum, F. Cucker, M. Shub and S. Smale [1998], *Complexity and Real Computation*, Springer-Verlag, Berlin, Heidelberg, New York.

L. Blum and S. Smale [1993], *The Gödel incompleteness theorem and decidability over a ring*, in "From Topology to Computation: Proceedings of the Smalefest" (M. Hirsh, J. Marsden, M. Shub, eds.), Springer-Verlag, Berlin, Heidelberg, New York, pp. 321–339.

D. Bohm [1952], *A suggested interpretation of the quantum theory in terms of 'hidden' variables, I and II*, Phys. Rev. **85**, 166–193; reprinted in "Quantum Theory and Measurement" (J. A. Wheeler and W. H. Zurek, eds.), Princeton University Press, Princeton, NJ, 1983.

D. Bohm [1957], *Causality and Chance in Modern Physics*, Routledge and Kegan Paul, London.

D. Bohm [1980], *Wholeness and the Implicate Order*, Routledge, London, New York.

D. Bohm and B. J. Hiley [1993], *The Undivided Universe: An ontological interpretation of quantum theory*, Routledge, London, New York.

N. Cartwright [1983], *How the Laws of Physics Lie*, Oxford University Press, Oxford, New York.

J. L. Casti [1990], *Searching for Certainty: What Scientists can Know about the Future*, William Morrow, New York.

G. J. Chaitin [1987], *Algorithmic Information Theory*, Cambridge University Press, Cambridge, New York.

A. Church [1936], *A note on the Entscheidungsproblem*, J. Symbolic Logic **1**, 40–41 and 101–102.

A. Church [1957], *Application of recursive arithmetic to the problem of circuit synthesis*, Talks Cornell Summer Inst. in Symbolic Logic, Cornell, 3–50.

S. B. Cooper [1990], *Enumeration reducibility, nondeterministic computations and relative computability of partial functions*, in "Recursion Theory Week, Oberwolfach 1989" (K. Ambos-Spies, G. Müller, G. E. Sacks, eds.), Springer-Verlag, Berlin, Heidelberg, New York, pp. 57–110.

S. B. Cooper [1994], *Rigidity and definability in the non-computable universe*, in "Logic, Methodology and Philosophy of Science IX", Proceedings of the Ninth International Congress of Logic, Methodology and Philosophy of Science, Uppsala, Sweden, August 7–14,1991 (D. Prawitz, B. Skyrms and D. Westerstahl, eds.), North-Holland, Amsterdam, Lausanne, New York, Oxford, Shannon, Tokyo, pp. 209–236.

S. B. Cooper [1997], *Beyond Gödel's Theorem: The failure to capture information content*, in "Complexity, Logic and Recursion Theory" (A. Sorbi, ed.), Lecture Notes in Pure and Applied Mathematics, vol. 187, Marcel Dekker, New York, pp. 93–122.

S. B. Cooper [ta], *On a conjecture of Kleene and Post*, to appear.

C. Darwin [1872], *The Origin of Species by Means of Natural Selection or The Preservation of Favoured Races in the Struggle for Life*, 6th authorised edn., John Murray, London; reprinted (G. Beer, ed.) Oxford University Press, Oxford and New York, 1996.

M. Davis [1958], *Computability and Unsolvability*, McGraw-Hill, New York; reprinted by Dover Publications, New York, 1982.

L. de Broglie [1927], J. Physique, 6e série **82**, 225.

D. Deutsch [1985], *Quantum theory, the Church-Turing principle and the universal quantum computer*, Proc. Roy. Soc. (London) **A400**, 97–117.

B. J. Dobbs [1991], *The Janus Faces of Genius: The Role of Alchemy in Newton's Thought*, Cambridge University Press, Cambridge.

F. Dowker and A. Kent [1996], *On the consistent histories approach to quantum mechanics*, J. Stat. Phys. **82**, 1575–1646.

F. I. Dretske [1977], *Laws of Nature*, Phil. of Science **44**, 248–268.

H. L. Dreyfus [1979], *What Computers Can't Do*, Harper and Row, New York.

J. Dupré [1993], *The Disorder of Things: Metaphysical Foundations of the Disunity of Science*, Harvard University Press, Cambridge, Mass., and London.

J. Earman [1986], *A Primer On Determinism*, D. Reidel, Dordrecht, Boston, Lancaster, Tokyo.

A. S. Eddington [1928], *The Nature of the Physical World*, Cambridge University Press, Cambridge.

A. Einstein [1950], *Out of My Later Years*, Philosophical Library, New York.

A. Einstein, B. Podolsky and N. Rosen [1935], *Can quantum mechanical description of physical reality be considered complete?*, Phys. Rev. **47**, 777–780.

R. L. Epstein [1979], *Degrees of Unsolvability: Structure and Theory*, Lecture Notes in Mathematics No. 759, Springer-Verlag, Berlin, Heidelberg, New York.

Y. L. Ershov [1975], *The upper semilattice of numerations of a finite set*, Alg. Log. **14**, 258–284 (Russian); 14 (1975), 159–175 (English translation).

H. Everett, III [1957], *'Relative state' formulation of quantum mechanics*, Rev. Mod. Phys. **29**, 454–462; reprinted in "Quantum Theory and Measurement" (J. A. Wheeler and W. H. Zurek, eds.), Princeton University Press, Princeton, NJ.

S. Feferman [1957], *Degrees of unsolvability associated with classes of formalized theories*, J. Symbolic Logic **22**, 161–175.

S. Feferman *et al* (eds.) [1990], *Kurt Gödel, Collected Works, Vol. II: Publications 1938–1974*, Oxford University Press, New York, Oxford.

L. Feiner [1970], *The strong homogeneity conjecture*, J. Symbolic Logic **35**, 375–377.

T. Ferris [1997], *The Whole Shebang: A State-of-the-Universe(s) Report*, Weidenfeld & Nicolson, London.

R. P. Feynman [1982], *Simulating physics with computers*, Int. J. Theor. Phys. **21**, 467–488.

R. P. Feynman [1985], *QED: The Strange Theory of Light and Matter*, Princeton University Press, Princeton, NJ.

R. P. Feynman [1986], *Quantum mechanical computers*, Found. Phys. **16**, 507–531.

R. P. Feynman, R. B. Leighton and M. Sands [1965], *The Feynman Lectures on Physics, Vol. III*, Addison-Wesley, Reading, Mass..

J. D. Franson [1985], *Bell's Theorem and delayed determinism*, Phys. Rev. D **31**, 2529–2532.

E. Fredkin and T. Toffoli [1982], *Conservative logic*, Int. J. Theor. Phys. **21**, 219–253.

R. O. Gandy [1980], *Church's thesis and principles for mechanisms*, in "The Kleene Symposium", Proceedings of the Symposium held June 18–24, 1979 at Madison, Wisconsin, U.S.A. (K. J. Barwise, H. J. Keisler and K. Kunen, eds.), North-Holland, Amsterdam, New York, Oxford, pp. 123–148.

R. O. Gandy [1988], *The confluence of ideas in 1936*, in "The Universal Turing Machine: A Half-Century Survey" (R. Herken, ed.), Kammerer and Unverzagt, Hamburg.

M. Gell-Mann and J. B. Hartle [1990], *Quantum mechanics in the light of quantum cosmology*, in "Complexity, Entropy and the Physics of Information" (W. H. Zurek, ed.), Santa Fe Institute Studies in the Science of Complexity, vol. VIII, Addison-Wesley, Reading, Mass., pp. 425–458.

M. Gell-Mann [1994], *The Quark and the Jaguar: Adventures in the Simple and the Complex*, Freeman, New York.

R. Geroch and J. B. Hartle [1986], *Computability and physical theories*, Found. Phys. **16**, 533–550.

G. C. Ghirardi, A. Rimini and T. Weber [1986], *Unified dynamics for microscopic and macroscopic systems*, Phys. Rev. **D34**, 470–491.

K. Gödel [1931], *Über formal unentscheidbare Sätze der Principia Mathematica und verwandter Systeme I*, Monatsh. Math. Phys. **38**, 173–198.

K. Gödel [1934], *On undecidable propositions of formal mathematical systems*, mimeographed notes, in "The Undecidable. Basic Papers on Undecidable Propositions, Unsolvable Problems, and Computable Functions" (M. Davis, ed.), Raven Press, New York, 1965, pp. 39–71.

K. Gödel [1946], *Remarks before the Princeton bicentennial conference on problems in mathematics*, in S. Feferman *et al* [1990], pp. 150–153.

K. Gödel [1964], *What is Cantor's continuum problem?*, P. Benacerraf and H. Putnam (eds.), "Philosophy of Mathematics: selected readings", Prentice-Hall, Englewood Cliffs, NJ, pp. 258–273; reprinted in S. Feferman *et al* [1990], pp. 254–270.

K. Gödel [1972], *Some remarks on the undecidability results*, reprinted in S. Feferman *et al* [1990], pp. 305–306.

K. Gödel [1995], *The modern development of the foundations of mathematics in the light of philosophy*, in "Kurt Gödel, Collected Works, Vol. III: Unpublished Essays and

Lectures" (S. Feferman *et al*, eds.), Oxford University Press, New York, Oxford, 1995, pp. 374–387.

M. B. Green, J. H. Schwarz and E. Witten [1987], *Superstring Theory*, Cambridge University Press, Cambridge.

D. Greenspan [1973], *Discrete Models*, Addison-Wesley, Reading, Mass..

D. Greenspan [1980], *Arithmetic Applied Mathematics*, Pergamon Press, Oxford.

D. Greenspan [1982], *Deterministic computer physics*, Int. J. Theor. Phys. **21**, 505–523.

R. B. Griffiths [1984], *Consistent histories and the interpretation of quantum mechanics*, J. Statist. Phys. **36**, 219–272.

P. R. Gross and N. Levitt [1994], *Higher Superstition: The Academic Left and Its Quarrels with Science*, John Hopkins University Press, Baltimore.

A. Grzegorczyk [1955], *Computable functionals*, Fund. Math. **42**, 168–202.

A. Grzegorczyk [1957], *On the definitions of computable real continuous functions*, Fund. Math. **44**, 61–71.

A. H. Guth [1997], *The Inflationary Universe – The Quest for a New Theory of Cosmic Origins*, Addison-Wesley, New York, Harlow, England, Tokyo, Paris, Milan.

W. Haken [1973], *Connections between topological and group theoretical decision problems*, in "Word Problems: Decision problems and the Burnside problem in group theory", (W. W. Boone, F. B. Cannonito and R. C. Lyndon, eds.), Studies in Logic and the Foundations of Math., Vol. 71, North-Holland, Amsterdam, pp. 427–441.

L. A. Harrington and R. A. Shore [1981], *Definable degrees and automorphisms of* $\mathcal{D}$, Bull. Amer. Math. Soc. **4**, 97–100.

J. B. Hartle and S. W. Hawking [1983], *Wave function of the universe*, Phys. Rev. **D28**, 2960–2975.

S. W. Hawking [1977], *The Quantum Mechanics of Black Holes*, Scientific American; reprinted in S. W. Hawking, "Black Holes and Baby Universes and other essays", Bantam Books, Toronto, New York, London, Sydney, Aukland, 1993, pp. 91–103.

S. W. Hawking and R. Penrose [1970], *The singularities of gravitational collapse and cosmology*, Proc. Roy. Soc. (London) **A314**, 529–548.

N. K. Hayles [1990], *Chaos Bound: Orderly Disorder in Contemporary Literature and Science*, Cornell University Press, Ithaca, NY.

J. Hick [1978], *Evil and the God of Love*, Harper and Rowe, New York.

A. Hodges [1983], *Alan Turing: The Enigma*, Burnett Books and Hutchinson, London.

D. R. Hofstadter [1979], *Gödel, Escher, Bach: an Eternal Golden Braid*, Harvester Press, Hassocks, Sussex.

J. H. Holland [1998], *Emergence: Models, Metaphors, and Innovation*, Oxford University Press, New York, Oxford.

D. Hume [1739], *A Treatise of Human Nature*, ed. L. A. Selby-Bigge and P. H. Nidditch, Oxford University Press, Oxford, 1978.

D. Hume [1748], *An Enquiry Concerning Human Understanding*, ed. L. A. Selby-Bigge and P. H. Nidditch, sect. VII, Oxford University Press, Oxford, 1975.

E. Husserl [1913], *Ideas: General Introduction to Pure Phenomenology*, (tr. W. R. Boyce Gibson), George Allen and Unwin, London; and The Macmillan Company, New York, 1931.

W. James [1897], *The dilemma of determinism*, in "The Will to Believe and Other Essays in Popular Philosophy", Longmans, Green and co., New York, London; reprinted, Dover Publications, New York, 1956.

C. G. Jockusch, Jr. and D. Posner [1981], *Automorphism bases for degrees of unsolvability*, Israel J. Math. **40**, 150–164.

C. G. Jockusch, Jr. and R. A. Shore [1984], *Pseudo jump operators II: Transfinite iterations, hierarchies, and minimal covers*, J. Symbolic Logic **49**, 1205–1236.

C. G. Jockusch, Jr. and S. G. Simpson [1976], *A degree theoretic definition of the ramified analytical hierarchy*, Ann. Math. Logic **10**, 1–32.

S. Kauffman [1995], *At Home In The Universe: The Search for Laws of Self-Organisation and Complexity*, Viking/ Oxford University Press, London, New York, Toronto, Auckland.

S. C. Kleene and E. L. Post [1954], *The upper semi-lattice of degrees of recursive unsolvability*, Ann. of Math. (2) **59**, 379–407.

S. C. Kleene [1959], *Recursive functionals and quantifiers of finite types I*, Trans. Amer. Math. Soc. **91**, 1–52.

S. C. Kleene [1963], *Recursive functionals and quantifiers of finite types II*, Trans. Amer. Math. Soc. **108**, 106–142.

A. Kolmogorov and V. A. Uspenskii [1958], *On the definition of an algorithm*, A.M.S. transl. **29** (1963), 217–245, Usp. Mat. Nauk **13**, 3–28.

G. Kreisel [1965], *Mathematical logic*, in "Lectures on Modern Mathematics", Vol. III (T. L. Saaty, ed.), John Wiley & Sons, New York, pp. 95–195.

G. Kreisel [1967], *Mathematical logic: What has it done for the philosophy of mathematics?*, in "Bertrand Russell, Philosopher of the Century" (R. Schoenman, ed.), Allen and Unwin, London, pp. 201–272.

G. Kreisel [1970], *Church's Thesis: a kind of reducibility axiom for constructive mathematics*, in "Intuitionism and proof theory: Proceedings of the Summer Conference at Buffalo N.Y. 1968" (A. Kino, J. Myhill and R. E. Vesley, eds.), North-Holland, Amsterdam, London, pp. 121–150.

G. Kreisel [1971], *Some reasons for generalizing recursion theory*, in "Logic Colloquium '69: Proceedings of the Summer School and Colloquium in Mathematical Logic, Manchester, August 1969" (R. O. Gandy and C. E. M. Yates, eds.), North-Holand, Amsterdam, New York, pp. 139–198.

G. Kreisel [1974], *A notion of mechanistic theory*, Synthese **29**, 11–26.

A. Kučera [1990], *Randomness and generalizations of fixed point free functions*, in "Recursion Theory Week, Proceedings Oberwolfach 1989", (K. Ambos-Spies, G. Müller and G. E. Sacks, eds.), Springer, Berlin, pp. 245–254.

T. S. Kuhn [1962], *The Structure of Scientific Revolutions*, Third edition 1996, University of Chicago Press, Chicago, London.

R. A. La Budde [1980], *Discrete Hamiltonian mechanics*, Int. J. Gen. Syst. **6**, 3–12.

A. H. Lachlan [1972], *Recursively enumerable many-one degrees*, Alg. Log. **11**, 326–358 (Russian); 11 (1972), 186–202 (English translation).

D. Lacombe [1955a], *Extension de la notion de fonction récursive aux fonctions d'une ou plusieurs variables réelles, I*, C. R. Acad. Sc., Paris **240**, 2478–2480.

D. Lacombe [1955b], *Extension de la notion de fonction récursive aux fonctions d'une ou plusieurs variables réelles, II, III*, C. R. Acad. Sc., Paris **241**, 13–14, 151–153.

I. Lakatos [1970], *Falsification and the methodology of scientific research programmes*, in "Criticism and the Growth of Knowledge" (I. Lakatos and A. Musgrave, eds.), Cambridge University Press, Cambridge, pp. 91–195.

P. S. de Laplace [1819], *Essai philosophique sur les probabilités*, English trans. by F. W. Truscott and F. L. Emory, Dover, New York, 1951.

I. D. Lawrie [1990], *A Unified Grand Tour of Theoretical Physics*, Adam Hilger, Bristol, New York.

D. Layzer [1990], *Cosmogenesis: The Growth of Order in the Universe*, Oxford University Press, New York, Oxford.

G. W. Leibniz [1714], in L. E. Loemker (ed.), "Gottfried Wilhelm Leibniz: Philosophical Papers and Letters", Dordrecht, 1969.

M. Lerman [1977], *Automorphism bases for the semilattice of recursively enumerable degrees*, Notices Amer. Math. Soc. **24**, A-251, Abstract #77T-E10.

M. Lerman [1983], *Degrees of Unsolvability*, Perspectives in Mathematical Logic, Omega Series, Springer-Verlag, Berlin, Heidelberg, London, New York, Tokyo.

M. Lerman [ta], *Embedding partial lattices into the computably enumerable degrees*, to appear.

D. Lewis [1973], *Counterfactuals*, Harvard University Press, Cambridge, Massachusetts.

A. Linde [1991], *Inflation and quantum cosmology: The birth and early evolution of our Universe*, Phys. Scripts **T36**, 30–54.

B. Mandelbrot [1982], *The Fractal Geometry of Nature*, W. H. Freeman.

L. Margolis [1981], *Symbiosis in Cell Evolution*, W. H. Freeman, New York.

N. Margolus [1984], *Physics-like models of computation*, Physica **10D**, 81–95.

D. A. Martin [1968], *The axiom of determinateness and reduction principles in the analytical hierarchy*, Bull. Amer. Math. Soc. **74**, 687–689.

Ju. V. Matijasevič [1970], *Enumerable sets are Diophantine*, Dokl. Akad. Nauk. SSSR **191**, 279–282 (Russian); Sov. Math. Dokl. **11**, 354–357 (English translation).

W. McCulloch and W. Pitts [1943], *A logical calculus of the ideas immanent in nervous activity*, Bull. Math. Biophys. **5**, 115–133.

A. Nerode and R. A. Shore [1980], *Second order logic and first order theories of reducibility orderings*, in "The Kleene Symposium" (J. Barwise et al., eds.), North-Holland, Amsterdam, pp. 181–200.

A. Nerode and R. A. Shore [1980a], *Reducibility orderings: theories, definability and automorphisms*, Ann. Math. Logic **18**, 61–89.

R. G. Newton [1997], *The Truth of Science: Physical Theories and Reality*, Harvard University Press, Cambridge, Mass., and London.

A. Nies, R. A. Shore, T. A. Slaman [1996], *Definability in the recursively enumerable degrees*, Bull. Symbolic Logic **2**, 392–404.

A. Nies, R. A. Shore, T. A. Slaman [ta], *Interpretability and definability in the recursively enumerable degrees*, to appear.

P. Odifreddi [1989], *Classical Recursion Theory*, North-Holland, Amsterdam, New York, Oxford.

P. Odifreddi [1996], *Kreisel's Church*, in "Kreiseliana: About and Around Georg Kreisel" (P. Odifreddi, ed.), A. K. Peters, Wellesley, Mass.

P. Odifreddi [ta], *Reducibilities*, to appear in "The Handbook of Computability Theory" (E. Griffor, ed.), North-Holland, Amsterdam, New York, Oxford.

R. Omnès [1994], *The Interpretation of Quantum Mechanics*, Princeton University Press, Princeton, NJ.

E. Paliutin [1975], *Addendum to the paper of Ershov [1975]*, Alg. Log. **14**, 284–287 (Russian); 14 (1975) pp. 176–178 (English translation).

C. S. Peirce [1931-58], *The Collected Papers of C. S. Peirce (C. Hartshorne, P. Weiss and A. Burks, eds.)*, Harvard University Press, Cambridge, MA.

R. Penrose [1965], *Gravitational collapse and space-time singularities*, Phys. Rev. Lett. **14**, 57–59.

R. Penrose [1987], *Quantum physics and conscious thought*, in Quantum Implications: Essays in honour of David Bohm, (B. J. Hiley and F. D. Peat, eds.), Routledge & Kegan Paul, London, New York, pp. 105–120.

R. Penrose [1989], *The Emperor's New Mind: Concerning Computers, Minds, and the Laws of Physics*, Oxford University Press, Oxford, New York.

R. Penrose [1994], *Shadows of the Mind: A Search for the Missing Science of Consciousness*, Oxford University Press, Oxford, New York, Melbourne.

R. Penrose [1996], *Structure of spacetime singularities*, in "The Nature of Space and Time", by S. W. Hawking and R. Penrose, Princeton University Press, Princeton,

New Jersey, pp. 27–36.
L. S. Penrose and R. Penrose [1958], *Impossible objects: a special type of visual illusion*, British J. of Psychology **49**, 31–33.
M. K. E. L. Planck [1949], *Scientific Autobiography and Other Papers (tr. F. Gaynor)*, Williams and Norgate, New York, London.
K. R. Popper [1959], *The Logic of Scientific Discovery*, tr. of "Logik der Forschung", Vienna, 1934 (with the imprint '1935'), Hutchinson, London.
K. R. Popper [1983], *Realism and the Aim of Science*, Rowman and Littlefield, Totowa, N.J..
E. L. Post [1948], *Degrees of recursive unsolvability: preliminary report (abstract)*, Bull. Amer. Math. Soc. **54**, 641–642.
E. L. Post [1965], *Absolutely unsolvable problems and relatively undecidable propositions: Account of an anticipation*, (Submitted for publication 1941), in "The Undecidable. Basic Papers on Undecidable Propositions, Unsolvable Problems, and Computable Functions" (M. Davis, ed.), Raven Press, New York, 1965, pp. 340–433.
M. B. Pour-El and J. I. Richards [1983], *Noncomputability in analysis and physics*, Adv. Math. **48**, 44–74.
M. B. Pour-El and J. I. Richards [1989], *Computability in Analysis and Physics*, Springer-Verlag, Berlin, Heidelberg, New York, London, Paris, Tokyo.
W. van Orman Quine [1953], *Two dogmas of empiricism*, in "From a Logical Point of View", Harvard University Press, Cambridge, Massachusetts and London, pp. 20–46.
F. P. Ramsey [1978], *Foundations of Mathematics*, Humanities Press, Atlantic Highlands, New Jersey.
L. J. Richter [1979], *On automorphisms of the degrees that preserve jumps*, Israel J. Math. **32**, 27–31.
R. W. Robinson [1971], *Interpolation and embedding in the recursively enumerable degrees*, Ann. of Math. (2) **93**, 285–314.
H. Rogers, Jr. [1967a], *Some problems of definability in recursive function theory*, in "Sets, Models and Recursion Theory" (J. N. Crossley, ed.), Proceedings of the Summer School in Mathematical Logic and Tenth Logic Colloquium, Leicester, August–September, 1965, North-Holland, Amsterdam, pp. 183–201.
H. Rogers, Jr. [1967b], *Theory of Recursive Functions and Effective Computability*, McGraw-Hill, New York.
R. Rorty [1979], *Philosophy and the Mirror of Nature*, Princeton University Press, Princeton, NJ.
R. Rorty [1982], *Consequences of Pragmatism*, Harvester Press, Brighton.
S. Rose [1997], *Lifelines: Biology, Freedom, Determinism*, Allen Lane/ The Penguin Press, London, New York, Toronto, Auckland.
C. Rovelli and L. Smolin [1990], *Loop representations for quantum general relativity*, Nuclear Phys. **B331**, 80–152.
B. Russell [1953], *On the Notion of Cause, with Applications to the Free-Will Problem*, in "Readings in the Philosophy of Science" (H. Feigl and M. Brodbeck, eds.), Appleton-Century-Crofts, New York, pp. 387–407; reprinted from "Mysticism and Logic", George Allen & Unwin, pp. 180–205, and "Our Knowledge of the External World", W. W. Norton, London, 1929, pp. 247–256,.
G. E. Sacks [1963], *On the degrees less than* $\mathbf{0}'$, Ann. of Math. (2) **77**, 211–231.
G. E. Sacks [1964], *The recursively enumerable degrees are dense*, Ann. of Math. (2) **80**, 300–312.
G. E. Sacks [1966], *Degrees of Unsolvability*, (revised edition), Ann. of Math. Studies No. 55, Princeton University Press, Princeton, N.J.
G. E. Sacks [1985], *Some open questions in recursion theory*, in "Recursion Theory Week"

(H. D. Ebbinghaus, G. H. Müller and G. E. Sacks, eds.), Proceedings of a Conference held in Oberwolfach, West Germany, April 15–21, 1984, Lecture Notes in Mathematics No. 1141, Springer-Verlag, Berlin, Heidelberg, New York, Tokyo, pp. 333–342.

G. E. Sacks [1990], *Higher Recursion Theory*, Perspectives in Mathematical Logic, Springer-Verlag, Berlin, Heidelberg, New York, Tokyo.

F. C. S. Schiller [1907], *Studies in Humanism*, MacMillan, London, New York.

D. Scott [1975a], *λ-calculus and recursion theory*, Third Scandinavian Logic Symposium, (Kanger, ed.), North-Holland, Amsterdam, pp. 154–193.

D. Scott [1975b], *Data types as lattices*, Proc. Logic Conf., Kiel, Lecture Notes in Mathematics no. 499, Springer-Verlag, Heidelberg, Berlin, New York, pp. 579–651.

C. E. Shannon [1948], *A mathematical theory of communication*, Bell Syst. Tech. J. **27**, 379–423, 623–656.

R. Shaw [1981], *Strange attractors, chaotic behaviour, and information flow*, Z. Naturforsch. **36A**, 80–112.

R. Shaw [1984], *The dripping faucet as a model chaotic system*, The Science Frontier Express Series, Aerial Press, Santa Cruz, CA.

J. Shipman [1993], *Aspects of Computability in Physics,*, in the Proceedings of the 1992 Workshop on Physics and Computation, IEEE.

R. A. Shore [1981], *The theory of the degrees below* $\mathbf{0}'$, J. London Math. Soc. (2) **24**, 1–14.

R. A. Shore [1997], *Conjectures and Questions from Gerald Sacks' Degrees of Unsolvability*, Arch. Math. Logic **36**, 233–253.

S. G. Simpson [1977], *First-order theory of the degrees of recursive unsolvability*, Ann. of Math. (2) **105**, 121–139.

B. Skyrms [1980], *Causal Necessity*, Yale University Press, New Haven, CT.

T. A. Slaman [1991], *Degree structures*, in the Proceedings of the International Congress of Mathematicians, Kyoto, 1990, Springer-Verlag, Tokyo, pp. 303–316.

T. A. Slaman [1998], *Mathematical Definability*, in "Truth in Mathematics" (H. G. Dales and G. Oliveri, eds.), Oxford University Press, Oxford, New York, pp. 233–252.

T. A. Slaman and W. H. Woodin [1986], *Definability in the Turing degrees*, Illinois J. Math. **30**, 320–334.

P. Smolensky [1988], *On the proper treatment of connectionism*, Behavioral and Brain Sciences **11**, 1–74.

L. Smolin [1991], *Space and time in the quantum universe*, in A. Ashtekar and J. Stachel [1991], pp. 228–288.

L. Smolin [1992], *Did the Universe Evolve?*, Class. Quantum Grav. **9**, 173–191.

L. Smolin [1993], *What have we learned from non-perturbative quantum gravity?*, "General Relativity and Gravitation 1992: Proceedings of the Thirteenth International Conference on CRG, Cordoba, Argentina " (R. J. Gleiser, C. N. Kozameh and O. N. Moreschi, eds.), Institute of Physics Publications, Bristol.

L. Smolin [1997], *The Life of the Cosmos*, Weidenfeld and Nicolson, London.

R. I. Soare [1987], *Recursively Enumerable Sets and Degrees*, Springer-Verlag, Berlin, Heidelberg, London, New York.

R. I. Soare [1996], *Computability and recursion*, Bull. of Symbolic Logic **2**, 284–321.

A. Sorbi [1997], *The enumeration degrees of the Σ^0_2 sets*, in "Complexity, Logic and Recursion Theory" (A. Sorbi, ed.), Lecture Notes in Pure and Applied Mathematics, vol. 187, Marcel Dekker, New York, pp. 303–330.

A. Sorbi [ta], *Sets of generators and automorphism bases for the enumeration degrees*, to appear.

H. P. Stapp [1993], *Mind, Matter, and Quantum Mechanics*, Springer-Verlag, Berlin, Heidelberg, London, New York, Paris, Tokyo.

R. Swinburne [1979], *The Existence of God*, Clarendon Press, Oxford.

C. Swoyer [1982], *The Nature of Natural Laws*, Australasian J. of Phil. **60**, 203–223.

T. Toffoli [1980], *Reversible computing*, in "Automata, Languages and Programming" (De Bakker and Van Leeuwen, eds.), Springer-Verlag, Berlin, Heidelberg, London, New York, pp. 632–644.

T. Toffoli [1984], *Cellular automata as an alternative to (rather than an approximation of) differential equations in modelling physics*, in "Cellular Automata" (D. Farmer, T. Toffoli and S. Wolfram, eds.), North-Holland, Amsterdam, New York, Oxford, Tokyo, pp. 117–127.

M. Tooley [1977], *The Nature of Laws*, Canadian J. of Phil. **7**, 667–698.

E. P. Tryon [1973], *Is the Universe a Vacuum Fluctuation?*, Nature **246**, 396.

A. M. Turing [1936], *On computable numbers, with an application to the Entscheidungsproblem*, Proc. London Math. Soc. **42**, 230–265.

A. M. Turing [1939], *Systems of logic based on ordinals*, Proc. London Math. Soc. **45**, 161–228; reprinted in 'The Undecidable. Basic Papers on Undecidable Propositions, Unsolvable Problems, and Computable Functions' (M. Davis, ed.), Raven Press, New York, 1965, pp. 154–222.

A. M. Turing [1950], *Computing machinery and intelligence*, Mind **59**, 433–460; reprinted in 'Minds and Machines' (A. R. Anderson, ed.), Prentice-Hall, Englewood Cliffs, New Jersey, 1964, pp. 4–30.

B. C. van Fraassen [1980], *The Scientific Image*, Oxford University Press, Oxford, New York.

G. Y. Vichniac [1984], *Simulating physics with cellular automata*, in "Cellular Automata" (D. Farmer, T. Toffoli and S. Wolfram, eds.), North-Holland, Amsterdam, New York, Oxford, Tokyo, pp. 96–116.

J. von Neumann [1932], *Matematische Grundlagen der Quanten-mechanik*, Julius Springer-Verlag, Berlin; English tr.: Princeton University Press, Princeton, N.J., 1955.

Hao Wang [1993], *On physicalism and algorithmism: can machines think?*, Philosophia mathematica (Ser. III) **1**, 97–138.

S. Weinberg [1992], *Dreams of a Final Theory*, Pantheon, New York.

R. S. Westfall [1984], *Newton and Alchemy*, in "Occult and Scientific Mentalities in the Renaissance" (B. Vickers, ed.), Cambridge University Press, Cambridge, London, New York, Sydney, p. 315–335.

A Techniques-Oriented Survey of Bounded Queries

William Gasarch* Frank Stephan†

University of Maryland University of Heidelberg

Abstract The present work gives an overview on the field of bounded queries in recursion theory including the subfields of frequency computation and verboseness. The main topic is finding quantitative notions for the complexity of non-recursive sets in terms of the local complexity of computing the n-fold characteristic function. This work presents in particular the various proof methods popular in this field.

1 Introduction

Bounded queries is dedicated to studying the quantitative complexity of non-recursive sets. This complexity is measured by the amount of information that can be obtained about the n-fold characteristic function

$$x_1, x_2, \ldots, x_n \to C_n^A(x_1, x_2, \ldots, x_n) = (A(x_1), A(x_2), \ldots, A(x_n)).$$

While Turing degrees measure the overall complexity of a set, bounded queries measure the maximal complexity of this local n-fold characteristic function. If, for example, A is recursively enumerable, then the enumeration process for the whole set A allows also the enumeration locally of up to $n+1$ binary vectors such that one of them is the desired vector $C_n^A(x_1, x_2, \ldots, x_n)$. So recursively enumerable sets have a low local complexity since one can find the n-fold characteristic function by enumerating $n+1$ candidates instead of the 2^n ones necessary in the worst case.

*Dept. of C.S. and Inst. for Adv. Comp. Stud., University of MD., College Park, MD 20742, U.S.A., Email: gasarch@cs.umd.edu.

†Mathematisches Institut, Universität Heidelberg, Im Neuenheimer Feld 294, 69120 Heidelberg, Germany, EU, Email: fstephan@math.uni-heidelberg.de, supported by the Deutsche Forschungsgemeinschaft (DFG) grant Am 60/9-2.

Historically, the first approach to investigate C_n^A was to compute a vector which coincides with C_n^A on as many bits as possible. Rose [58] defined a set A to be (m,n)-*computable* iff one can everywhere predict at least m values of the n-fold characteristic function correctly.

> A is (m,n)-computable iff there is a total recursive function f which assigns to all distinct inputs $x_1, x_2, \ldots, x_n$ a binary vector $(y_1, y_2, \ldots, y_n)$ such that at least m of the equations $A(x_1) = y_1$, $A(x_2) = y_2, \ldots, A(x_n) = y_n$ hold.

Rose [58] asked for which frequencies m out of n does it hold that every (m,n)-computable set is already computable. Trahtenbrot [63, 64] answered the question by showing that every (m,n)-computable set is recursive iff $\frac{m}{n} > \frac{1}{2}$. Examples for non-recursive but still $(1,2)$-computable sets are (a) the semirecursive sets which are exactly those sets closed downward under some recursive linear ordering $\sqsubseteq$ and (b) the sets retraced by a total recursive function which can also be viewed as the infinite branches of some suitable recursive tree.

Independently, Beigel [3] and Gasarch [28] found another approach to measuring the complexity of C_n^A: they just counted the number of oracle queries necessary to compute any value of C_n^A either relative to A itself or relative to some optimally chosen oracle X. Furthermore, Hay [5, 31] came into the field by studying the complexity of the sets of Ershov's difference hierarchy [22, 23, 24]. The query complexity of A with respect to an optimal oracle X has also an alternative definition which does not mention oracles at all: A set A is (m,n)-verbose [4, 8] iff one can enumerate up to m possible values such that the n-fold characteristic function C_n^A is among them.

> A is (m,n)-verbose iff there is a partial recursive function ψ such that for every $x_1, x_2, \ldots, x_n$ there is a $k \in \{1, 2, \ldots, m\}$ with $\psi(x_1, x_2, \ldots, x_n, k)\downarrow\ = C_n^A(x_1, x_2, \ldots, x_n)$.

Beigel [3, 4] observed that C_n^A can be computed with h queries to a suitable oracle X iff A is $(2^h, n)$-verbose: Having a machine M computing C_n^A with h oracle queries, one just simulates all 2^h computation paths which are fully determined by the up to 2^h possible answers to the h queries. Then one enumerates the up to 2^h output vectors obtained by this process. For the converse direction, note that, for each $x_1, x_2, \ldots, x_n$, there is a $k \in \{1, 2, \ldots, 2^h\}$ such that $\psi(x_1, x_2, \ldots, x_n, k)$ computes $C_n^A(x_1, x_2, \ldots, x_n)$ and this k can be coded with h bits into a suitable oracle X at h places uniquely assigned to $(x_1, x_2, \ldots, x_n)$.

The advantage of the notion of (m,n)-verbose sets is that this notion is a bit more general than counting the queries to an optimal oracle: it permits also positive integers m which are not a power of 2. Only if the oracle queried is A itself or some other oracle linked to A does it still makes sense to count the queries.

Obviously every set is $(2^n, n)$-verbose. So the interesting question is which sets are (m, n)-verbose for some $m < 2^n$ and which values for m can be obtained. A connection to frequency computation is already given by the fact that every $(1, n)$-computable set is $(2^n - 1, n)$-verbose. The recursively enumerable sets are $(n + 1, n)$-verbose. Ershov's difference hierarchy [22, 23, 24] contains exactly those sets which are the set-theoretic difference of a constant number of recursively enumerable sets. One can show that each such set is $(kn+1, n)$-verbose for some k and all n. Also semirecursive sets are $(n + 1, n)$-verbose for all n. Beigel [3, 4] showed his famous Nonspeedup Theorem which states that every (n, n)-verbose set is already recursive, so the $(n + 1, n)$-verbose sets are the first nontrivial sets within the hierarchy of all these sets.

Related work is also dedicated to variants of the function C_n^A. Instead of looking at C_n^A directly, Owings [56] looked at the cardinality $\#_n^A$ which just maps $x_1, x_2, \ldots, x_n$ to $A(x_1) + A(x_2) + \ldots + A(x_n)$. Kummer [42] extended some initial results of Owings to the full Cardinality Theorem: If A is not recursive then there is no algorithm which enumerates for every input $x_1, x_2, \ldots, x_n$ up to n values such that the cardinality $\#_n^A(x_1, x_2, \ldots, x_n)$ is among these values. This no longer holds if only the last bit of the cardinality, Odd_n^A, is considered in place of $\#_n^A$, since this bit is from the range $\{0, 1\}$ and can be computed with one query from the truth-table cylinder of A. Nevertheless some non-trivial lower bounds where obtained for the complexity of computing Odd_n^A relative to A itself where A is semirecursive or recursively enumerable [10, 11].

The field of bounded queries is also studied within the framework of complexity theory. The recursion-theoretic notions can be transferred directly and the complexity-theoretic analogues obtained in this way have been investigated intensively [1, 2, 9, 12, 32, 33, 53, 55]. The complexity-theoretic results do not always parallel the recursion-theoretic ones. For example, there exist, for polynomial time computations, $(n, n+1)$-computable sets which are not computable. Neither Beigel's Nonspeedup Theorem nor Kummer's Cardinality Theorem can be transferred into the setting of polynomial time computation. This survey is dedicated to the recursion-theoretic part of bounded queries, the interested reader can look in the above cited references for information about bounded queries in complexity theory.

2 Trees of Consistent Strings

Trees are a suitable way to represent data consistent with auxiliary information about A. The main idea is to cut off all inconsistent branches as early as possible and then to show that the infinite branch corresponding to A satisfies some useful properties, for example, being the only infinite branch of T above some suitable node.

Trees are used to show Trahtenbrot's Theorem that every (m,n)-computable set is computable whenever $\frac{m}{n} > \frac{1}{2}$. Before going into the details of this theorem, the formal definition of (m,n)-computable sets is reincluded. The concept was invented and first investigated by Rose [58] and Trahtenbrot [63].

Definition 2.1 [58, 63] A set A is (m,n)-*computable* iff there is a total recursive function f which assigns to all distinct inputs $x_1, x_2, \ldots, x_n$ a binary vector $(y_1, y_2, \ldots, y_n)$ such that at least m of the equations $A(x_1) = y_1$, $A(x_2) = y_2, \ldots, A(x_n) = y_n$ hold. That is, the Hamming distance of the approximation $f(x_1, x_2, \ldots, x_n)$ and the original vector $C_n^A(x_1, x_2, \ldots, x_n)$ is at most $n - m$.

An example of $(1,2)$-computable sets is the semirecursive sets, introduced by Jockusch [35]: Each semirecursive set A possesses a recursive linear ordering $\sqsubseteq$ such that A is closed downward under this ordering $\sqsubseteq$. The set A is $(1,2)$-computed by

$$x_1, x_2 \to \begin{cases} (1,0) & \text{if } x_1 \sqsubseteq x_2; \\ (0,1) & \text{otherwise.} \end{cases}$$

One can easily show that semirecursive sets are also (m,n)-computable whenever $\frac{m}{n} \leq \frac{1}{2}$. This example shows that Trahtenbrot's result is optimal.

Theorem 2.2 [63, 64] *Let* $\frac{m}{n} > \frac{1}{2}$. *Then every* (m,n)-*computable set* A *is computable.*

Proof Let A be (m,n)-computable via the function f. Let the tree T contain all binary strings $\sigma \in \{0,1\}^*$ which satisfy, for all distinct $x_1, x_2, \ldots, x_n \in dom(\sigma)$, that at least m components of the n-bit vectors $f(x_1, x_2, \ldots, x_n)$ and $(\sigma(x_1), \sigma(x_2), \ldots, \sigma(x_n))$ coincide. Since f approximates A with frequency (m,n), the set A is an infinite branch of T in the sense that every string $A(0)A(1)\ldots A(x)$ is in T.

Now let B be a further infinite branch of T and assume that A and B differ at n places $x_1, x_2, \ldots, x_n$. Then both, the Hamming distance from $C_n^A(x_1, x_2, \ldots, x_n)$ to $f(x_1, x_2, \ldots, x_n)$ and that from $C_n^B(x_1, x_2, \ldots, x_n)$ to $f(x_1, x_2, \ldots, x_n)$, do not exceed $n - m$. So the Hamming distance of the vectors $C_n^A(x_1, x_2, \ldots, x_n)$ and $C_n^B(x_1, x_2, \ldots, x_n)$ is at most $2(n-m)$. But this value is smaller than n, which contradicts the fact that $C_n^A(x_1, x_2, \ldots, x_n)$ and $C_n^B(x_1, x_2, \ldots, x_n)$ have Hamming distance n. Therefore every two infinite branches of T differ on at most $n - 1$ places.

The Hamming distance of any further infinite branches B from A is defined and at most $n-1$, thus there is an infinite branch B whose Hamming distance from A is maximal. Let D be the finite set on which A and B disagree. The set B is also an infinite branch of the subtree

$$T' = \{\sigma \in T : (\forall x \in dom(\sigma) \cap D)\,[\sigma(x) = B(x)]\}$$

of T. Assume that C would be a further infinite branch of T' and $C(x) \neq B(x)$ for some x. By definition, $x \notin D$. So C and A differ on $D \cup \{x\}$ contrary to the choice of B as an infinite branch of T having maximal distance from A. Thus B is a unique infinite branch of T'. Now the following algorithm computes $B(x)$ for any x:

> Search for the first $t > x$ such that all $\sigma \in T' \cap \{0,1\}^t$ take only a unique value $y = \sigma(x)$ at x. Output this y as the value for $B(x)$.

This algorithm cannot output a value different from $B(x)$ since T' has at every length $t > x$ a string σ with $\sigma(x) = B(x)$. If the algorithm would not terminate for some given x, then there would for each length $t > x$ be a string $\sigma \in T'$ of length t with $\sigma(x) \neq B(x)$. In particular the subtree

$$T'' = \{\sigma \in T' : \sigma(x)\downarrow \Rightarrow \sigma(x) \neq B(x)\}$$

would be infinite. By König's Lemma [54, Theorem V.5.23], this subtree would also have an infinite branch C different from B contrary to the fact that B is the only infinite branch of T'. Thus B is computable and so is also the finite variant A of B. ∎

Jockusch and Soare [37, Theorem 2.5] showed that every infinite recursive tree has either an infinite computable branch or uncountably many infinite branches. Since a set has only countably many finite variants and all branches of the T above are finite variants of A, one can directly conclude that A is computable.

Call an infinite branch B of a tree T to be *isolated* iff there is an x such that every infinite branch C of T which coincides with B below x is identical to C. Isolated branches like B are recursive: The subtree T' of all nodes which coincide with B below x has only one infinite recursive branch and so by the result of Jockusch and Soare [37, Theorem 2.5] this infinite branch is recursive. The algorithm to compute B is the same as in Theorem 2.2.

3 Relativization to Minimal Pairs

It is often easier to deal with total than with partial recursive functions. Therefore it might be convenient to work in some relativized world where some given partial function can be replaced by a total extension. This idea becomes an essential part of a technique demonstrated in this section. The real power behind this idea is to work in two different relativized worlds such that every set A which is recursive in both relativized worlds is also recursive in the usual non-relativized one; such worlds are called minimal pairs. The relativized worlds are represented by two different oracles E_1 and E_2. Jockusch and Soare [37, Corollary 2.9] showed the existence of such oracles.

Theorem 3.1 [37] *There are oracles E_1 and E_2 such that E_1 and E_2 form a minimal pair with respect to Turing reducibility and every partial recursive function ψ with a finite range D has total D-valued extensions f_1 and f_2 such that f_1 is computable relative to E_1 and f_2 is computable relative to E_2.*

Now this technique of using relativizations to minimal pairs is demonstrated to prove Beigel's Nonspeedup Theorem and will then be extended to a proof for some results for Odd_n^A of a set A. Beigel's Nonspeedup Theorem can be formulated using the following notions of (m,n)-verbose and strongly (m,n)-verbose sets. The definition does not require the $x_1, x_2, \ldots, x_n$ to be distinct since both versions of the definition, the one which requires it and the one which does not, give the same concept.

Definition 3.2 [8] A set A is *(m,n)-verbose* iff there is a $\{0,1\}^n$-valued partial recursive function ψ such that for every $x_1, x_2, \ldots, x_n$ there is a $k \in \{1, 2, \ldots, m\}$ with $\psi(x_1, x_2, \ldots, x_n, k) \downarrow = C_n^A(x_1, x_2, \ldots, x_n)$. If ψ can be chosen to be total then A is called *strongly (m,n)-verbose.*

The technique of relativization can be used to give an alternative proof for Beigel's Nonspeedup Theorem [3, 4]. This alternative proof is done in two stages. First it is shown that every strongly (n,n)-verbose set is recursive and then this is extended to (n,n)-verbose sets using relativization.

Theorem 3.3 [3, 4] *If A is (n,n)-verbose then A is computable.*

Proof First take the case that A is strongly (n,n)-verbose. Let ψ be a total function which witnesses this fact according to Definition 3.2. Let T be the set of all strings which are consistent with ψ:

$$\begin{aligned}\sigma \in T \;\Leftrightarrow\; & (\forall x_1, x_2, \ldots, x_n \in dom(\sigma))\,(\exists k \in \{1, 2, \ldots, n\}) \\ & [\psi(x_1, x_2, \ldots, x_n, k) = (\sigma(x_1), \sigma(x_2), \ldots, \sigma(x_n))].\end{aligned}$$

Now A is an infinite branch of T and assume by way of contradiction that A is not isolated on T: then there are $x_1, x_2, \ldots, x_n$ such that $x_1 < x_2 < \ldots < x_n$ and infinite branches B_k which coincide with A below x_k and disagree with A at x_k. It follows that the first place where the vector $C_n^{B_k}(x_1, x_2, \ldots, x_n)$

disagrees with the vector $C_n^A(x_1, x_2, \ldots, x_n)$ is the k-th bit associated with x_k, thus all the $n+1$ different vectors $C_n^A(x_1, x_2, \ldots, x_n)$ and $C_n^{B_k}(x_1, x_2, \ldots, x_n)$ for $k = 1, 2, \ldots, n$ must be consistent with ψ at $x_1, x_2, \ldots, x_n$, a contradiction to the fact that ψ enumerates only n vectors. From this contradiction it follows that A is an isolated branch of T. Then A is computable as pointed out in and below the proof of Theorem 2.2.

Second, one has to extend the proof to the case of (n, n)-verboseness. The range of ψ is a finite set, namely $\{0,1\}^n$. Thus ψ has total extensions computable relative to E_1 and E_2. Arguing as in the first part, it follows that A is computable relative to E_1 and also relative to E_2. So the Turing degree of A must be below the degrees of E_1 and E_2. Since these form a minimal pair, the recursive degree is the only one satisfying this condition and A is computable. ∎

The next results deal with the set

$$\mathrm{Odd}_n^A = \{(x_1, x_2, \ldots, x_n) : A(x_1) + A(x_2) + \ldots + A(x_n) \text{ is odd}\}$$

and its query complexity measured in the number of queries needed to compute Odd_n^A relative to A. This complexity is just 1 if A is a truth-table cylinder. Therefore the more interesting results require restrictions on the choice of A. Here semirecursive and recursively enumerable sets A are considered.

There is a difference between parallel and serial queries. A computation uses parallel queries iff first all places for the queries are calculated and then the computation continues with the answers to all these queries. Serial queries do not have this constraint and so the place of the second query might depend on the answer to the first one and so on. The notion "B is computable relative to A with n parallel queries" is a generalization of Post's bounded truth-table reducibility [57] and is also called weak bounded truth-table reducibility. It was introduced by Friedberg and Rogers [26]. Kobzev [41] investigated the structure of weak bounded truth-table degrees inside Turing degrees and showed that every recursively enumerable Turing degree contains infinitely many different weak bounded truth-table degrees.

Definition 3.4 [3, 26, 57] A set B is *computable with n parallel queries relative to A* iff there is an algorithm which for given x first computes n places $f_1(x)$, $f_2(x)$, ..., $f_n(x)$ where B is queried and then computes $B(x)$ from the obtained data about B:

$$B(x) = \gamma(x, A(f_1(x)), A(f_2(x)), \ldots, A(f_n(x)))\downarrow .$$

This definition is equivalent to the one that B is *weakly bounded truth-table reducible to A with norm n*. If not only $f_1, f_2, \ldots, f_n$ but also γ is total, then one says that B is *bounded truth-table reducible to A with norm n* [54,

Definition III.8.1]. The definition can easily be extended by using a function in place of the set B and taking several variables instead of the single variable x.

Clearly Odd_n^A can be computed with n parallel queries to A for any set A. The next result shows that there are sets A for which this result is optimal, that is, for which Odd_{n+1}^A cannot be computed with n parallel queries to A.

Theorem 3.5 [10, 11] *Let A be semirecursive. If Odd_{n+1}^A is computable with n parallel queries to A then A is computable.*

Proof Assume that A is an initial segment of the recursive linear ordering $\sqsubseteq$ and Odd_{n+1}^A could be computed with n parallel queries to A. For given $x_1, x_2, \ldots, x_{n+1}$ let $y_1, y_2, \ldots, y_n$ be the places of these queries. Without loss of generality $x_h \sqsubseteq x_l$ and $y_h \sqsubseteq y_l$ if $h \leq l$. Let $\mathbf{v}_0, \mathbf{v}_1, \ldots, \mathbf{v}_{n+1}$ be the $n+2$ vectors consistent with $\sqsubseteq$, that is, $\mathbf{v}_i[h] = 1$ iff $h \leq i$ where $\mathbf{v}_i[h]$ denotes the h-th bit of $\mathbf{v}_i$. Note that $C_{n+1}^A(x_1, x_2, \ldots, x_{n+1})$ is among these vectors and that exactly i of the components $\mathbf{v}_i[h]$ are 1. Let $\mathbf{w}_0, \mathbf{w}_1, \ldots, \mathbf{w}_n$ be the corresponding vectors on $y_1, y_2, \ldots, y_n$. Now define

> $\mathbf{v}_i$ and $\mathbf{w}_j$ to be *consistent* iff there is a z such that $\mathbf{v}_i[k] = 1 \Leftrightarrow x_k \sqsubseteq z$ and $\mathbf{w}_j[h] = 1 \Leftrightarrow y_h \sqsubseteq z$ and the computation of $\mathrm{Odd}_{n+1}^A(x_1, x_2, \ldots, x_{n+1})$ terminates on the oracle answers $\mathbf{w}_j$ to the queries $y_1, y_2, \ldots, y_n$ with output c and $i \equiv c$ modulo 2.

It is easy to see the following four things. (a): It is impossible that on the one hand $\mathbf{v}_i$ and $\mathbf{w}_j$ are consistent and on the other hand $\mathbf{v}_{i+1}$ and $\mathbf{w}_j$ are consistent since the c depends only on $\mathbf{w}_j$ and thus either i or $i+1$ fails to satisfy that $i \equiv c$ modulo 2. (b): If on the one hand $\mathbf{v}_i$ and $\mathbf{w}_j$ are consistent and on the other hand $\mathbf{v}_k$ and $\mathbf{w}_j$ are consistent for some $k \geq i+2$ then $\mathbf{v}_{i+1}$ is not consistent with any vector $\mathbf{w}_h$ since $\mathbf{v}_{i+1}$ could only be consistent with $\mathbf{w}_j$ which fails because the c computed from $\mathbf{w}_j$ differs from $i+1$ modulo 2. (c): There is some vector $\mathbf{v}_i$ which is not consistent with any vector $\mathbf{w}_j$; this can be seen by combining (a) and (b) with the fact that there are $n+2$ vectors $\mathbf{v}_i$ and $n+1$ vectors $\mathbf{w}_j$. (d): The set of all $\mathbf{v}_i$ which are consistent with some $\mathbf{w}_j$ can be enumerated uniformly in $x_1, x_2, \ldots, x_{n+1}$.

So one has at most $n+1$ vectors which are consistent with some $\mathbf{w}_j$. On the other hand, $C_{n+1}^A(x_1, x_2, \ldots, x_{n+1})$ is among the vectors $\mathbf{v}_0$, $\mathbf{v}_1$, ..., $\mathbf{v}_{n+1}$ and is also consistent with one of the vectors $\mathbf{w}_j$ since Odd_{n+1}^A can be computed with n queries to A. It follows that one can enumerate all vectors $\mathbf{v}_i$ which are consistent with some $\mathbf{w}_j$ and obtains so up to $n+1$ vectors such that $C_{n+1}^A(x_1, x_2, \ldots, x_{n+1})$ is among them. So A is $(n+1, n+1)$-verbose and thus A is recursive. ∎

One can obtain a similar result for recursively enumerable sets using the technique of relativizing to a minimal pair.

Theorem 3.6 [10, 11] *Let A be recursively enumerable. If Odd^A_{n+1} is computable with n parallel queries to A then A is computable.*

Proof Let A_s be an enumeration of A. Define the partial function $\psi(x_1, x_2)$ as follows: If there is a stage s such that $x_1 \in A_s$ but $x_2 \notin A_s$ then let $\psi(x_1, x_2) = 1$. If there is a stage s such that $x_1 \notin A_s$ but $x_2 \in A_s$ then $\psi(x_1, x_2) = 2$. Let ψ be undefined otherwise.

Relative to E_i this function ψ has a total extension f. Now one has that whenever x_1 or x_2 is in A then also $x_{f(x_1,x_2)}$ is in A. Using Jockusch's original definition and the equivalence result [35, Definition 3.1 and Theorem 4.1], A is semirecursive relative to E_i.

By assumption Odd^A_{n+1} is computable with n parallel queries to A. This holds of course also relative to E_i. Since A is semirecursive relative to E_i, it follows by Theorem 3.5 that A is recursive relative to E_i. Since this holds for both sets E_i and these form a minimal pair, A is recursive. ∎

Similarly one can also show for semirecursive or recursively enumerable sets A that $\mathrm{Odd}^A_{2^n}$ cannot be computed with n serial queries relative to A unless A is recursive. The method is to show that relative to E_1 or E_2, one can turn the reduction with its n serial queries into a bounded truth-table reduction with $2^n - 1$ parallel queries. After that, one can proceed as before to show that A is recursive.

A further application of this technology is to extend Owings' first results [56] on the Cardinality Theorem to a full proof of this theorem: Owings showed that A is recursive if there is a total recursive function ψ such that $\#^A_n(x_1, x_2, \ldots, x_n)$ is equal to one of the values $\psi(x_1, x_2, \ldots, x_n, k)$ for $k \in \{1, 2, \ldots, n\}$. This gives an alternative to Kummer's proof [42]. Kummer and Stephan [45, Section 6] present a further variant very similar to the method used here to close the gap left by Owings.

4 Constructing Superterse Sets and Diagonalization

A set which is not (m, n)-verbose for any m, n with $m \leq 2^n - 1$ is called *superterse*. The notions verbose, terse and superterse relate to "modes of speech": A verbose speaker explains much, repeats often the same things and speaks redundantly. So the local information content is low and even small facts are told in a long story. A terse speaker tries to say as much as possible with as few words as possible. His speech cannot be compressed without losing facts. The difference between *terse* and *superterse* is that for a terse set A, the n-fold characteristic function cannot be computed relative to A itself with less than n queries, for a superterse set the n-fold characteristic function cannot be computed relative to any oracle with less than n queries. Originally, only those $(n+1,n)$-verbose sets were called *verbose* [4] for which in addition, for all n, $C^A_{2^n-1}$ can be computed with n queries to A itself. For example, semirecursive sets are verbose.

Superterse sets can also be characterized with respect to Kolmogorov complexity. The Kolmogorov complexity [51] of a binary vector $(y_1, y_2, \ldots, y_n)$ relative to some information η is the size of the shortest program σ (measured in bits for some underlying universal programming language) which computes the binary vector from the input η. There are two related notions of Kolmogorov complexity — one notion requires that only one of two different programs of the form σ and $\sigma\tau$ has an output, the other notion does not have any requirements with respect to program format — and there are also many acceptable numberings which could be used as the underlying universal programming languages. But the following characterization is robust in the sense that it holds for all of these variants of Kolmogorov complexity.

Proposition 4.1 *A is superterse iff for every n there are $x_1, x_2, \ldots, x_n$ such that the Kolmogorov complexity of $C^A_n(x_1, x_2, \ldots, x_n)$ relative to the input vector $(x_1, x_2, \ldots, x_n)$ is at least n.*

Proof Assume that A is not superterse. Then A is $(p(n), n)$-verbose for some polynomial p and all n [3, 8]. Beigel's proof can be reproduced by combining Theorem 7.6 (e) and Example 7.8 (a). This $(p(n), n)$-verboseness is even uniform in the sense that it can be achieved with the same function φ_e and input $n, x_1, x_2, \ldots, x_n$; the n itself is coded implicitly in the vector $(x_1, x_2, \ldots, x_n)$ so that one can compute n relative to this input vector. Given $(x_1, x_2, \ldots, x_n)$ one needs only to know for which $k \in \{1, 2, \ldots, p(n)\}$ the equation $C^A_n(x_1, x_2, \ldots, x_n) = \varphi_e(x_1, x_2, \ldots, x_n, k)\downarrow$ holds. The upper bound for k is the polynomial $p(n)$ and so one can code k with $O(\log(n))$ bits. Since e is a constant, there is some sufficiently large n such that the space to code e and the corresponding k is always below $n-1$ bits. This holds for all variants of Kolmogorov complexity and so the Kolmogorov complexity of $C^A_n(x_1, x_2, \ldots, x_n)$ relative to the input vector $(x_1, x_2, \ldots, x_n)$ is below $n-1$ bits for sufficiently large n.

For the converse direction let the Kolmogorov complexity of $C^A_n(x_1, x_2, \ldots, x_n)$ relative to the input $(x_1, x_2, \ldots, x_n)$ be always at most $n-1$ bits for some fixed n. That is, there is some universal partial recursive function ψ such that for all $x_1, x_2, \ldots, x_n$ there is some program σ of size at most $n-1$ bits with $\psi_\sigma(x_1, x_2, \ldots, x_n) = C^A_n(x_1, x_2, \ldots, x_n)$. Even in the worst case where every string represents a valid program, there are only $2^n - 1$ strings of length up to $n-1$ bits and one can translate the parameter σ of the Kolmogorov complexity into the parameter k of the definition of $(2^n-1, n)$-verboseness and translate the given function ψ to a partial recursive function

$\tilde{\psi}$ which assigns to input $x_1, x_2, \ldots, x_n, k$ just $\psi_\sigma(x_1, x_2, \ldots, x_n)$ for the k-th program σ. Then there is, for all $x_1, x_2, \ldots, x_n$, a k with $C_n^A(x_1, x_2, \ldots, x_n) = \tilde{\psi}(x_1, x_2, \ldots, x_n, k)$ and $\tilde{\psi}$ witnesses that A is not superterse. ∎

From this characterization it can be deduced that Kolmogorov random sets are a natural example for superterse sets. Beigel, Gasarch, Gill and Owings [4] give besides this example two more:

Every 1-generic set is superterse. Here a set A is 1-generic iff for every enumerable set W of strings there is an x such that either $A(0)A(1)\ldots A(x) \in W$ or no string $\sigma \in W$ extends $A(0)A(1)\ldots A(x)$, that is, all strings σ in W of length $x+1$ or more satisfy $A(y) \neq \sigma(y)$ for some $y \leq x$.

Every non-recursive truth-table cylinder is superterse. A truth-table cylinder is a set A such that every B which is truth-table reducible to A is also many-one reducible to A. Since every truth-table degree contains a truth-table cylinder and a semirecursive set which is not superterse and since the non-superterse sets are closed under bounded truth-table reduction, one has an alternative proof for Dëgtev's result that every non-recursive truth-table degree consists of at least 2 bounded truth-table degrees [4, 19, 62]: The truth-table cylinder is not btt-reducible to the semirecursive set.

Besides these natural examples for superterse sets, they can also easily be constructed explicitly. This construction is presented here since it is the basis for later, more complicated constructions.

Fact 4.2 [3, 4] *There is a superterse set.*

Proof One divides the natural numbers into disjoint intervals $I_{e,n}$ of length n. Now for each interval $I_{e,n} = \{x_1, x_2, \ldots, x_n\}$ one defines $A(x_1) = y_1$, $A(x_2) = y_2, \ldots, A(x_n) = y_n$ for the first $(y_1, y_2, \ldots, y_n) \in \{0,1\}^n$ which is different from those outputs $\varphi_e(x_1, x_2, \ldots, x_n, k)$ for $k = 1, 2, \ldots, 2^n - 1$ which are defined. It follows directly that A is not $(2^n - 1, n)$-verbose via φ_e for any n and e. So A is superterse. ∎

The set K is $(n+1, n)$-verbose since K is recursively enumerable. Nevertheless one can show that K fails to be strongly $(2^n - 1, n)$-verbose for every n and satisfies the notion corresponding to superterseness with respect to strong verboseness.

Theorem 4.3 *Some recursively enumerable sets, in particular K, are not strongly (m, n)-verbose for any n and $m < 2^n$.*

Proof Again one works with the intervals $I_{e,n}$ from above. Now one defines on a given interval $I_{e,n} = \{x_1, x_2, \ldots, x_n\}$ that

$$C_n^A(x_1, x_2, \ldots, x_n) = \begin{cases} (y_1, y_2, \ldots, y_n) & \text{if } \varphi_e(x_1, x_2, \ldots, x_n, k)\downarrow \text{ for all } k \in \{1, 2, \ldots, 2^n - 1\} \text{ and } (y_1, y_2, \ldots, y_n) \text{ is the first tuple in } \{0,1\}^n \text{ different from all these outputs;} \\ (0, 0, \ldots, 0) & \text{if } \varphi_e(x_1, x_2, \ldots, x_n, k)\uparrow \text{ for some } k \in \{1, 2, \ldots, 2^n - 1\}. \end{cases}$$

The set A is enumerable since either no element of $I_{e,n}$ is enumerated into A or all computations $\varphi_e(x_1, x_2, \ldots, x_n, k)$ terminate and then only those x_k are enumerated into A for which $y_k = 1$. It follows directly that A is not strongly $(2^n - 1, n)$-verbose via any function φ_e.

The set A is many-one reducible to K via some recursive function f. If now K would be strongly (m, n)-verbose for some m and n with $m < 2^n$ via some total function ψ, one would get that also A is strongly (m, n)-verbose via the function

$$x_1, x_2, \ldots, x_n, k \rightarrow \psi(f(x_1), f(x_2), \ldots, f(x_n), k)$$

contrary to the construction of A. So K is not strongly (m, n)-verbose for any n and $m < 2^n$. ∎

Friedberg [25] and Yates [67] introduced the permitting method which allows the combining of a diagonalizing construction with a framework to ensure that the constructed set A is computable relative to a given recursively enumerable set B.

Lachlan [48] constructed two recursively enumerable sets A and B such that A can be computed with one query relative to B but A is not truth-table reducible to B. This result can be obtained by combining results of Dëgtev [18] and Sacks [59]: Dëgtev [18] constructed a recursively enumerable and non-recursive set B such that all recursively enumerable sets which are truth-table reducible to B are either recursive or truth-table equivalent to B. Sacks [59] developed a method to split any given non-recursive but recursively enumerable set, in particular Dëgtev's set B, into a subset $A \subseteq B$ and the set $B - A$ such that A and $B - A$ are recursively enumerable and incomparable with respect to Turing reduction. The value $A(x)$ can be computed with 1 query to B: if $x \notin B$ then $x \notin A$ since $A \subseteq B$. If $x \in B$ then one enumerates A and $B - A$ until x shows up in one of these two sets. But A is not truth-table reducible to B since A is neither recursive nor truth-table equivalent to B.

The next theorem combines permitting, priority methods and the notion of strongly (m, n)-verbose sets to obtain that every non-recursive and recursively enumerable truth-table degree contains a set B — namely any semirecursive and recursively enumerable set B within the degree — such that there is a set A which is computable with one query to B but which is not bounded truth-table reducible to B. Since there are truth-table complete semirecursive sets, bounded truth-table reducibility cannot be replaced by truth-table for this result.

Theorem 4.4 *Let B be semirecursive and recursively enumerable but not computable. Then there is some A which is computable relative to B with only one query but which is not bounded truth-table reducible to B.*

Proof First the construction from Theorem 4.3 is adapted so that A is computable relative to B with one query. Then it is shown that every set which is bounded truth-table reducible to a semirecursive set is strongly $(2^n - 1, n)$-verbose for some n. Finally it is shown that A does not have this property and is hence not bounded truth-table reducible to B.

Now one works with intervals $I_{n,e,t}$ instead of $I_{n,e}$ from Theorem 4.3 where each interval $I_{n,e,t}$ has the length n. The parameter t is used to permit a diagonalization on this interval against the e-th function as a witness for A being strongly $(2^n - 1, n)$-verbose. If φ_e is total, then one of the permitted diagonalizations will succeed and so rule out the witness φ_e. On the interval $I_{n,e,t} = \{x_1, x_2, \ldots, x_n\}$ the procedure does the following:

$$C_n^A(x_1, x_2, \ldots, x_n) = \begin{cases} (y_1, y_2, \ldots, y_n) & \text{if } t \in B \text{ and the computation} \\ & \varphi_e(x_1, x_2, \ldots, x_n, k) \text{ converges} \\ & \text{for all } k \in \{1, 2, \ldots, 2^n - 1\} \\ & \text{before } t \text{ is enumerated into } B \\ & \text{and } (y_1, y_2, \ldots, y_n) \text{ is the first} \\ & \text{tuple in } \{0,1\}^n \text{ different from} \\ & \text{all these outputs;} \\ (0, 0, \ldots, 0) & \text{otherwise.} \end{cases}$$

The construction gives directly that the resulting set A is recursively enumerable. Furthermore, A can be computed with one query relative to B: If $t \notin B$ then A is 0 on the whole interval $I_{n,e,t}$. If $t \in B$ then one counts the time to enumerate t into B and checks whether all the computations $\varphi_e(x_1, x_2, \ldots, x_n, k)$ terminate within this time. If so, one diagonalizes them; if not, one again takes A to be 0 on the whole interval.

Assume now that A is btt-reducible to B with norm m and let $n = m + 5$. Then C_n^A can be computed with nm parallel queries to B. Since B is semirecursive, one can compute for B up to $nm+1$ vectors such that one of them contains C_{nm}^B on the queried places. Now one can evaluate the btt-reduction with each of these vectors as supposed answers and one receives up to $nm+1$ possibilities for C_n^A. Since $n \geq m$ and $n \geq 5$, one has that $mn + 1 \leq n^2 + 1 < 2^n$.

So the $nm + 1$ possibilities do not cover all possible characteristic vectors at these n places and A is strongly $(2^n - 1, n)$-verbose.

Assume that φ_e is a total function which might be a witness for the fact that A is strongly $(2^n - 1, n)$-verbose. Let $f(t)$ be the time to compute all vectors $\varphi_e(x_1, x_2, \ldots, x_n, k)$ for $k = 1, 2, \ldots, 2^n - 1$ on the interval $I_{n,e,t} = \{x_1, x_2, \ldots, x_n\}$. If every $t \in B$ would be enumerated into B within $f(t)$ computation-steps then B would be recursive since one could check the membership to B by simulating the enumeration process during these $f(t)$ steps. Thus there is a t which is enumerated into B after more than $f(t)$ steps. Now for this t, the diagonalization on the interval $I_{e,n,t}$ is permitted and thus φ_e does not witness that A is strongly $(2^n - 1, n)$-verbose. So A is not strongly $(2^n - 1, n)$-verbose for any n and therefore A is also not btt-reducible to B. ∎

Though not intended, this construction is still similar to the Sacks splitting method. The set $\tilde{B}$ given as the union of all $I_{e,n,t}$ with $t \in B$ is many-one equivalent to B. Now $A \subseteq \tilde{B}$ and an analysis of the enumeration of A shows that also $\tilde{B} - A$ is recursively enumerable. So A is to a certain extent obtained by splitting a set "very similar to B".

5 Counting Mind Changes

In the previous section it was shown that for certain recursively enumerable sets B there are recursively enumerable sets A which are computable relative to B with only one query but not bounded truth-table reducible to B. The next result shows that on the other hand there are sets B for which the notions of computing with n parallel queries and bounded truth-table reduction with norm n coincide. This coincidence is true for sets of hyperimmune free Turing degree [54, Section V.5] — one can equip the reduction with a clock which majorizes the computation time relative to the correct oracle — and it is also true for 2-generic sets, that is, for sets which are 1-generic in the world relativized to K. Within this section, it is shown that also some recursively enumerable sets have this property.

Beigel, Gasarch and Hay [5, Theorem 6.9 and Corollary 6.10] did the original construction for K, but the underlying method is more general. In fact it works for every recursively enumerable set B which satisfies the relation that every further recursively enumerable set which is Turing reducible to B is also many-one reducible to B. The Turing degrees of such sets are called *m-topped*. Downey and Jockusch [21] showed that there are m-topped degrees besides those of $\emptyset$ and K; these degrees are low_2 but not low.

Theorem 5.1 [5, 21] *Let B be a recursively enumerable set having greatest many-one degree among the recursively enumerable sets of its Turing degree.*

Then the statements (a) *and* (b) *are equivalent for any m and any recursively enumerable set A. If in addition $m = 2^n - 1$ then all four statements are equivalent.*
(a) *A is computable relative to B with m parallel queries.*
(b) *A is btt-reducible to B with norm m.*
(c) *A is computable relative to B with n serial queries.*
(d) *A is computable relative to B with n serial queries via a reduction which converges also if some of the oracle answers are false.*

Proof The main tool for this proof is the set of all mind changes a Turing machine makes relative to B. So for given machine M, set B and enumeration B_s, the set B_M of mind changes is defined as

$$(x,k) \in B_M \Leftrightarrow (\exists s_0, s_1, \ldots, s_k \text{ with } s_0 < s_1 < \ldots < s_k)$$
$$(\forall l < k)\,[M_{s_l}^{B_{s_l}}(x)\downarrow \neq M_{s_{l+1}}^{B_{s_{l+1}}}(x)\downarrow\,].$$

Since one wants to obtain some information about a reduction from A to B, one can without loss of generality assume that M^B is total and $M_s^{B_s}(x)$ is either undefined or takes the value 0 or 1 at every stage s. Therefore for every x one can compute the value $b(x)$ taken at the first stage s_0 where $M_{s_0}^{B_{s_0}}(x)$ is defined. Then one knows that for even indices l the value $M_{s_l}^{B_{s_l}}(x)$ is $b(x)$ and for odd indices l the value is $1 - b(x)$.

It is easy to see that B_M is Turing reducible to B. It follows that B_M is many-one reducible to B by the choice of B. So queries to B_M can be transformed effectively into queries to B. Now one uses these observations to prove the Theorem. The directions (b $\Rightarrow$ a) and (d $\Rightarrow$ c) follow directly from the definition.

(a $\Rightarrow$ b): The places of the m queries are fixed and independent of B. Thus any mind change can only be due to enumerating one queried element into B. Since there are m places, there are at most m mind changes. The number of mind changes can be deduced by the m parallel queries about $(x,1), (x,2), \ldots, (x,m)$ to B_M which can be realized by queries to the corresponding values of B. If the number of mind changes is even, then $A(x) = b(x)$, if it is odd, then $A(x) = 1 - b(x)$. This evaluation terminates for all oracle answers, thus A is bounded truth-table reducible to B with norm m.

(c $\Rightarrow$ b): Without loss of generality, M makes on every converging path to every oracle exactly n queries: if M makes less than n queries then one can add some dummy queries, if M wants to make somewhere the $n+1$-st query then one aborts the computation path and outputs some default value since this path is incorrect. One uses now at every stage the oracle B_s; if the answer to the n serial oracle queries are $(y_1, y_2, \ldots, y_n)$ at some stage and $(z_1, z_2, \ldots, z_n)$ at some later stage and in both cases the computations converge but have a different result then these two vectors of answers must differ and there is a first l with $y_l \neq z_l$. Now this difference can only be due to the fact that the x' queried at this place is enumerated into B. It follows that $y_l = 0$ and $z_l = 1$, in particular $(y_1, y_2, \ldots, y_n)$ is lexicographically before $(z_1, z_2, \ldots, z_n)$. Therefore the vector of the answers to the queries increases in lexicographic order and so there are at most $m = 2^n - 1$ mind changes. Thus one knows the number of mind changes by asking queries for $(x, 1), (x, 2), \ldots, (x, m)$ to B_M which can be realized by queries to the corresponding values of B. So again A is bounded truth-table reducible to B with norm m.

(b $\Rightarrow$ d): Using implicitly part (a $\Rightarrow$ b) one can assume that $A(x)$ is calculated using the queries to $(x, 1), (x, 2), \ldots, (x, m)$ to B_M. One knows that $B_M(x, l) \geq B_M(x, k)$ whenever $l \leq k$ unless the oracle answers are incorrect. This allows the use of binary search to find a maximal l such that $B_M(x_l) = 1$; this l takes one of the values $0, 1, 2, \ldots, m$ since $(x, m+1), (x, m+2), \ldots \notin B_M$ and $(x, 0) \in B_M$. The binary search needs n queries. In the case of incorrect oracle answers, one ends up after n queries also with some l, but the l might be incorrect. In the case of correct oracle answers, l is just the number of mind changes. If l is even then $A(x) = b(x)$, if l is odd then $A(x) = 1 - b(x)$. Since this last step can formally also be done for any incorrect l obtained from incorrect oracle answers, the reduction terminates also for incorrect oracle answers. ∎

6 On Autoreducible Sets, Non-Uniform Case-Distinction and Priority Methods

Trahtenbrot introduced the notion of autoreducible set [65]. A set is autoreducible iff there is an algorithm which computes $A(x)$ without querying A at x. Formally, one can also use the oracle $A \cup \{x\}$ instead of A so that a query to x is useless since it receives the default answer 1. So one obtains the following definition:

Definition 6.1 [65] A is autoreducible iff there is a partial recursive function ψ using an oracle with $A(x) = \psi^{A \cup \{x\}}(x)\downarrow$ for all x.

On the one hand, many types of sets occurring in the world of bounded queries can easily be shown to be autoreducible. On the other hand, it is often easier to construct a non-autoreducible set with some additional property than to diagonalize the given property directly. So one can for example use the construction of non-autoreducible sets to obtain sets which are also not frequency-computable. In this section, this proof-method is used to construct a recursively enumerable set A for which Odd_m^A cannot be computed with less than m serial queries to A [4, Theorem 11].

This is not true for semirecursive sets where, if $m = 2^n - 1$, only n serial queries are necessary, that is, the number of serial queries needed is logarithmic in m. So only some sets might be so difficult that Odd_m^A requires always m serial queries to A. Since Odd_m^A can be computed directly from C_m^A, one has that also C_m^A needs m serial queries to A. Sets with this property are called *terse* [3, 4].

As already said, the existence of a terse recursively enumerable set is mirrored back to Trahtenbrot's result [65] on the existence of a recursively enumerable set which is not autoreducible. Kummer and Stephan [46] obtained the result that every non-terse set is autoreducible. The proof follows the technique of non-uniform case-distinction. Other applications of these proof-technique are that an $(n, 2n)$-computable set cannot be the join of two non-recursive sets of different Turing degree [46, Theorem 2.9], that $(n, 2n)$-computable sets are not strongly hyperimmune [46, Theorem 4.6] and that frequency computable sets are not part of an inseparable pair of recursively enumerable sets [46, Theorem 3.2].

Kaufmann [38] did further research on the question to which degree such case-distinctions are still effective. For example, she showed that given a program e which (m, n)-computes some set where $\frac{m}{n} > \frac{1}{2}$ one can compute a list of $n + 1 - m$ programs such that one of them computes up to finitely many errors the characteristic function of a set A which is (m, n)-computable via this program e. But there is no effective procedure which does the same outputting only $n - m$ programs.

Theorem 6.2 [46] *Let A be not autoreducible. Then, for all n, Odd_n^A cannot be computed with less than n serial queries to A. In particular, A is terse.*

Proof Assume that for some given set A there is an n such that Odd_{n+1}^A is computable with n serial queries relative to A. Let this n be minimal; n cannot be 0 since otherwise A would be computable.

Assume that there is an x_1 such that for all $x_2, \ldots, x_{n+1}$ the computation of $\mathrm{Odd}_{n+1}^A(x_1, x_2, \ldots, x_{n+1})$ makes a query to A at x_1. Then one can compute $\mathrm{Odd}_n^A(x_2, x_3, \ldots, x_{n+1})$ with $n - 1$ queries to A by computing $\mathrm{Odd}_{n+1}^A(x_1, x_2, \ldots, x_{n+1})$ and replacing that of the n queries which goes to x_1 by the fixed constant $A(x_1)$. This contradicts to the assumption that Odd_n^A cannot be computed with $n - 1$ serial queries to A and so this case does not occur.

So for every x_1 there are some $x_2, \ldots, x_{n+1}$ such that $\mathrm{Odd}_{n+1}^A(x_1, x_2, \ldots, x_{n+1})$ can be computed with some queries to A where the computation does not query x_1 itself. Such a computation and $x_2, \ldots, x_{n+1}$ can be found using queries to A at any places except at x_1 by parallel search. After computing $\mathrm{Odd}^A_{n+1}(x_1, x_2, \ldots, x_{n+1})$ one queries A at $x_2, \ldots, x_{n+1}$ and obtains so the value for $A(x_1)$: x_1 is in A iff $\mathrm{Odd}^A_{n+1}(x_1, x_2, \ldots, x_{n+1}) + A(x_2) + \ldots + A(x_{n+1})$ is odd. So one obtains an autoreduction for A. The theorem holds since its contrapositive is proven. ∎

The proof can easily adapted to obtain similar results. Kummer and Stephan [46, Theorem 2.10] also showed that every frequency computable set is autoreducible. The same holds also for every strongly $(2^n - 1, n)$-verbose set since these sets are all frequency computable. Thus, the next result of Trahtenbrot [65] combined with the previous theorem gives not only a proof for the existence of recursively enumerable terse sets but also an alternative, though more complicated, proof for the existence of recursively enumerable sets which are not frequency computable. The proof uses priority methods combined with finite injury as the construction of an incomplete recursively enumerable degree by Soare [61, Chapter VII].

Theorem 6.3 [65] *There is a recursively enumerable set which is not autoreducible.*

Proof The construction of A meets the following requirements:

$$(e) \quad (\exists x)\,[A(x) \neq \varphi_e^{A\cup\{x\}}(x)].$$

The term "Priority Argument" means that the construction always deals with the condition (e) of highest priority which needs attention at some stage and which satisfies (e) in a way such that no already satisfied condition of higher priority becomes again unsatisfied. Here (e') has higher priority than (e) iff $e' < e$. The term "Finite Injury" means that there are only finitely many stages where a requirement (e) may be unsatisfied or even be destroyed in order to satisfy arguments of higher priority.

Let $\varphi_{e,s}^{A_s\cup\{x\}}(x)$ denote that the first s computation steps of the e-th machine with oracle $A_s \cup \{x\}$ are simulated where every query to some y needs y computation steps and therefore the computation queries only places $y \leq s$. Furthermore, every requirement is linked to a marker $x_{e,s}$ which denotes the place where the requirement (e) is intended to be satisfied. The $x_{e,s}$ move only finitely often to a greater place and converge to places $x_{e,\infty}$. For them it holds that

$$(e) \quad A(x_\infty) \neq \varphi_e^{A\cup\{x_\infty\}}(x_\infty).$$

The construction is done in stages. At stage 0 one initializes $x_{e,0} = e$ and $A_0 = \emptyset$. To compute A_{s+1} from A_s one does at stage $s+1$ the following:

Find the smallest e such that

- $e \leq s$ and $x_{e,s} \leq s$;
- $A_s(x_{e,s}) = 0$;
- $\varphi_{e,s}^{A_s \cup \{x_{e,s}\}}(x_{e,s}) \downarrow = 0$.

If such an e does not exist then do nothing ($x_{e',s+1} = x_{e',s}$ for all e' and $A_{s+1} = A_s$) else enumerate $x_{e,s}$ into A ($A_{s+1} = A_s \cup \{x_{e,s}\}$) and move all markers $x_{e',s}$ for $e' > e$ to places where they cannot destroy the computation ($x_{e',s+1} = x_{e',s}$ for $e' \leq e$ and $x_{e',s+1} = e' + s$ for $e' > e$).

For the verification, one first shows inductively that all x_e move only finitely often. Assume that this is true for all $e' < e$; since this condition is void for $e = 0$ it is not necessary to look at the base case. Now there is a stage s such that no $x_{e'}$ with $e' < e$ moves after stage s. Then, there is also a stage $s' \geq s$ such that $A(x_{e',s'}) = A_{s'}(x_{e',s'})$ for all $e' < e$. Since x_e only moves when some $x_{e',s'}$ with $e' < e$ is enumerated into A, it follows that $x_{e,s'}$ does not move any more and $x_{e,\infty} = x_{e,s'}$. So every x_e moves only finitely often.

Now assume by way of contradiction that some requirement (e) is never satisfied. Then in particular

$$A(x_{e,\infty}) = \varphi_e^{A \cup \{x_{e,\infty}\}}(x_{e,\infty}) \downarrow .$$

Now the contradiction is shown for both possible cases $A(x_{e,\infty}) = 1$ and $A(x_{e,\infty}) = 0$.

(a): $A(x_{e,\infty}) = 1$. Then there is a stage $s+1$ where $x_{e,\infty}$ is enumerated into A. So $x_{e,\infty} = x_{e',s}$ at that stage for some e'. It holds that $e' = e$: On the one hand when x_e moved at some stage $s'+1$ onto this position $x_{e,\infty}$ then $x_{e,s'+1} \geq s'$. On the other hand $s \geq x_{e,\infty}$ since only values below s are enumerated into A at stage $s+1$. So $s' \leq s$ and $e' = e$; formally this argumentation does not cover the case when x_e never moves and $x_{e,\infty} = x_{e,0}$ but then it is directly clear that $e' = e$. The computation $\varphi_{e,s}^{A_s \cup \{x_{e,s}\}}(x_{e,s})$ has converged to 0 at that stage s and all markers $x_{e'}$ with $e' > e$ are from that stage on outside the places where this computation has queried. Furthermore, also no element below $x_{e,s}$ was enumerated into A after stage s since otherwise x_e would have moved again and $x_{e,\infty} \neq x_{e,s}$. So case (a) does not occur.

(b): $A(x_{e,\infty}) = 0$. Then for almost all stages $s > x_{e,\infty}$ it holds that $x_{e,s} = x_{e,\infty}$ and $\varphi_{e,s}^{A \cup \{x_{e,s}\}}(x_{e,s}) \downarrow = 0$. For all these s, the algorithm to compute A_{s+1} would either enumerate $x_{e,s}$ or some $x_{e',s}$ with $e' < e$ into A. The first does not happen since otherwise $A(x_{e,\infty}) = 1$, the second does not happen since otherwise $x_{e,s}$ would move to some place larger than $x_{e,\infty}$. So case (b) also does not occur.

It follows that all requirements (e) are satisfied in the limit and that A is not autoreducible by any φ_e.

Since every computable set is autoreducible, A is not recursive. But A is recursively enumerable since the algorithm only enumerates elements into A but never takes any elements from A out again. ∎

So one obtains the result that there is a recursively enumerable terse set. Beigel, Gasarch, Gill and Owings [4] showed with a construction a bit more complicated than the one for the non-autoreducible set that every non-recursive recursively enumerable Turing degree has a recursively enumerable terse set. This result cannot be obtained by the method used here since there is a non-recursive recursively enumerable Turing degree whose recursively enumerable sets are all autoreducible [49, 50].

Theorem 6.4 [4, Theorem 11] *Every non-recursive recursively enumerable Turing degree contains a recursively enumerable terse set.*

7 Local Methods and Coding Theory

Dëgtev [20] studied the inclusion problem of frequency computation with a special emphasis on recursively enumerable sets. The inclusion problem is

$$\{(m, n, h, k) : \text{ every } (m, n)\text{-computable set is } (h, k)\text{-computable}\}.$$

Dëgtev introduced the following local combinatorial notion which is related to the inclusion problem.

Definition 7.1 [20] A vector set $V \subseteq \{0,1\}^k$ with $k \geq n$ is (m, n)-admissible iff for every projection W of V onto n coordinates there is a vector $\mathbf{v}$ which coincides with every vector in $\mathbf{w} \in W$ in at least m components. This definition can be extended to $k < n$: A vector set $V \subseteq \{0,1\}^k$ with $k < n$ is (m, n)-admissible iff $m - (n - k) \leq 0$ or V is $(m - (n - k), k)$-admissible.

Dëgtev [20] showed that if every (m, n)-admissible vector set is (h, k)-admissible then every (m, n)-computable set is (h, k)-computable and obtained the converse for $m = 1$ and $n \geq 2$: If there is some $(1, n)$-admissible vector set V which is not (h, k)-admissible then some $(1, n)$-computable set A is not (h, k)-computable. If furthermore V contains the vectors $(0, 0, \ldots, 0)$ and $(1, 1, \ldots, 1)$ then A can be chosen recursively enumerable. But the converse does not hold in general, at least not if $\frac{m}{n} > \frac{1}{2}$. The vector set $\{(0,0,0), (0,0,1), (0,1,0), (1,0,0)\}$ is $(2, 3)$-admissible but not $(1, 1)$-admissible while every $(2, 3)$-computable set is also recursive, that is, $(1, 1)$-computable, by Trahtenbrot's Theorem [63].

The next result gives an example of how Dëgtev's criterion can be used to show that some $(1, 3)$-computable set A fails to be $(2, 5)$-computable. The construction uses priority arguments with finite injury in order to diagonalize on the one hand any function which might (2, 5)-compute A and preserve on the other hand that A is (1, 3)-computable.

Theorem 7.2 [20] *There is a recursively enumerable set which is* (1, 3)*-computable but not* (2, 5)*-computable.*

Proof Let the vector set V contain all vectors $(y_1, y_2, y_3, y_4, y_5)$ such that either $y_1 + y_2 + y_3 + y_4 + y_5 \in \{0, 1, 5\}$ or exactly two neighbouring y_i and y_j are 1 where y_5 and y_1 are also viewed to be neighbours, that is, $(1, 0, 0, 0, 1) \in V$, too. This vector set V is (1, 3)-admissible but not (2, 5)-admissible. It contains the vectors $(0, 0, 0, 0, 0)$ and $(1, 1, 1, 1, 1)$ which is essential for getting a recursively enumerable set. Besides (1, 3)-computability the construction meets the following requirements:

(e) If φ_e is total then there is an interval I_x such that $\varphi_e(I_x)$ and $C_5^A(I_x)$ have Hamming distance 4 or 5.

The interval I_x can just be fixed to $\{5x+1, 5x+2, 5x+3, 5x+4, 5x+5\}$. As in Theorem 6.3 the construction uses movable markers. But now, the marker $x_{e,s}$ does not stand for a single point but for an interval $I_{x_{e,s}}$. Again the construction is done in stages. At stage 0 one initializes $x_{e,s} = e$ and $A_0 = \emptyset$. One now explicitly needs a flag which states which requirements are satisfied; all requirements are set to be unsatisfied at stage 0. To compute A_{s+1} from A_s one does at stage $s + 1$ the following:

Find the smallest e such that

- $e \leq s$ and $x_{e,s} \leq s$;
- According to its flag, (e) is not satisfied at stage s;
- $\varphi_{e,s}(I_x)\downarrow = (y_1, y_2, y_3, y_4, y_5)$ for some y_1, y_2, y_3, y_4, y_5.

If such an e does not exist then do nothing ($x_{e',s+1} = x_{e',s}$ for all e', $A_{s+1} = A_s$ and do not change the flags) else update as follows:

- Set the flag for (e) to be satisfied. The flags for requirements (e′) remain untouched if $e' < e$ and are reset to unsatisfied if $e' > e$.
- Find a vector $(z_1, z_2, z_3, z_4, z_5) \in V$ which has Hamming distance 4 or 5 from $(y_1, y_2, y_3, y_4, y_5)$ and enumerate exactly the $5x_{e,s} + k$ with $z_k = 1$ into A_{s+1}.
- Enumerate all not yet enumerated elements from the intervals I_x into A_{s+1} which satisfy $x_{e,s} < x \leq s$, that is, let A_{s+1} contain all elements $5x_{e,s} + 6$, $5x_{e,s} + 7$, ..., $5s + 4$, $5s + 5$ plus those from A_s and those from the previous item.

- Move for $e' > e$ all markers to new places: $x_{e',s+1} = s + e' - e$. The markers for $e' \leq e$ do not change their place: $x_{e',s+1} = x_{e',s}$.

As in Theorem 6.3 it is verified that every marker moves only finitely often. Also one can see that if φ_e is defined and outputs an element from $\{0,1\}^5$ on the five inputs from the interval given by $x_{e,\infty}$ then (e) is eventually satisfied and so A is not $(2,5)$-computable via φ_e. If φ_e is partial or outputs somewhere some illegal output, this is also true anyway. So A is not $(2,5)$-recursive. Furthermore, it follows from the construction that A is recursively enumerable. Note that on every interval I_x the characteristic function of A is in V: either no element is enumerated and it equals $(0,0,0,0,0)$ or by the second case in the construction some vector from V is chosen explicitly or by the third case in the construction it equals $(1,1,1,1,1)$ which is also in V. Now it remains to verify that A is $(1,3)$-computable. The function to witness this is constructed by taking the first possible case in the following case-distinction where without loss of generality $x_1 < x_2 < x_3$.

$$f(x_1,x_2,x_3) = \begin{cases} (1,1,1) & \text{if some } x_k \in A_{x_3+1}; \\ (y_1,y_2,y_3) & \text{if } x_1,x_2,x_3 \in I_x \text{ and } V \text{ projected} \\ & \text{onto the corresponding three} \\ & \text{coordinates is } (1,3)\text{-admissible} \\ & \text{via } (y_1,y_2,y_3); \\ (0,0,1) & \text{otherwise.} \end{cases}$$

In the first two cases it is clear that one of the components is correct. In the third case, one knows that x_1 and x_3 belong to different intervals I_x and $I_{x'}$; note that $x < x'$ by $x_1 < x_3$. Furthermore, both are not enumerated into A_s for some $s \leq x'$. So it follows whenever x_1 becomes enumerated into $A_{s'}$ for $s' > x'$ then also every element in $I_{x'}$ and in particular x_3. Thus whenever $A(x_1) \neq 0$ then $A(x_3) = 1$ and so one component is correct also in the third case. ∎

One can generalize the proof from above to show that, in the case $\frac{m}{n} \leq \frac{1}{2}$, there is an (m,n)-computable set which is not (h,k)-computable if there is a vector set $V \subseteq \{0,1\}^k$ which is $(m-l, n-2l)$-admissible for $l = 1,2,\ldots,m$ but not (h,k)-admissible [44]. As the example of the $(2,3)$-computable sets already showed, the criterion (m,n)-admissible alone does not characterize the inclusion problem. But it characterizes the uniform inclusion problem where *the (m,n)-computable sets are uniformly (h,k)-computable* iff there is an algorithm which translates every program e into a program e' such that whenever A is (m,n)-computable via φ_e then A is also (h,k)-computable via $\varphi_{e'}$.

Theorem 7.3 *The (m,n)-computable sets are uniformly (h,k)-computable iff every (m,n)-admissible vector set $V \subseteq \{0,1\}^k$ is (h,k)-admissible.*

Proof Assume that every (m,n)-admissible vector set $V \subseteq \{0,1\}^k$ is (h,k)-admissible. Given the function φ_e which (m,n)-computes A it is now shown how $\varphi_{e'}$ is constructed.

There is a tree T such that $\sigma \in T$ iff σ is consistent with φ_e, that is, iff for all distinct $x_1, x_2, \ldots, x_n \in dom(\sigma)$ and for $(y_1, y_2, \ldots, y_n) = \varphi_e(x_1, x_2, \ldots, x_n)$ at least m of the equations $\sigma(x_1) = y_1$, $\sigma(x_2) = y_2$, ..., $\sigma(x_n) = y_n$ hold. Now for each $t \geq n$ the vector set $\{0,1\}^t \cap T$ is (m,n)-admissible. For distinct $x_1, x_2, \ldots, x_k$ and $t = x_1 + x_2 + \ldots + x_k + n$ one can compute $V = \{0,1\}^t \cap T$ and the projection W of V onto $x_1, x_2, \ldots, x_k$. Both sets, V and W, are (m,n)-admissible and so W is (h,k)-admissible. There is a vector $(y_1, y_2, \ldots, y_k)$ which coincides with every vector in W in at least h components. So $\varphi_{e'}$ outputs for input $x_1, x_2, \ldots, x_k$ the first such vector $(y_1, y_2, \ldots, y_k) \in \{0,1\}^k$ found. Since every string $A(0)A(1)\ldots A(t)$ is in T it follows that $C_k^A(x_1, x_2, \ldots, x_k)$ is in W and thus agrees with $\varphi_{e'}$ in at least h components. So A is (h,k)-computable via $\varphi_{e'}$.

For the converse direction, assume that V is some (m,n)-admissible but not (h,k)-admissible subset of $\{0,1\}^t$ — it is sufficient to consider $t = k$ but this direction works for all t. Then also the vector set $W = V \times \{0\}^\infty$ is (m,n)-admissible in the sense that every projection onto n coordinates is (m,n)-admissible and there is a single computable function φ_e such that every $A \in W$ — where the set A is identified with the infinite vector given by its characteristic function — is (m,n)-computable by φ_e. On the other hand there cannot be any single function $\varphi_{e'}$ which (h,k)-computes every $A \in W$ since the restriction of $\varphi_{e'}$ to the first t components would witness that V is (h,k)-admissible contrary to its choice. So since there is no solution e', it is also not possible to find it uniformly in e. ∎

The negative result that whenever some (m,n)-admissible vector set V is not (h,k)-admissible then one can, given some φ_e, not always find some $\varphi_{e'}$ looks a bit artificial since it is principally unsolvable and seems to be repairable with some finite case-distinction and knowledge about A. But Case, Kaufmann, Kinber and Kummer [13] showed that it is impossible to close the gap. They used for this negative result some notion from learning theory which might also be formalized as follows: there is no limiting recursive process which has access to A as an oracle and translates a program which (m,n)-computes A into one which (h,k)-computes A. Putting all these things together one obtains the following theorem.

Theorem 7.4 [13] *The following statements are equivalent:*

(a) *Every (m,n)-admissible vector set is (h,k)-admissible.*

(b) *Every (m,n)-admissible vector set $V \subseteq \{0,1\}^k$ is (h,k)-admissible.*
(c) *For each e there is an e' such that every set which is (m,n)-computable via φ_e is also (h,k)-computable via $\varphi_{e'}$.*
(d) *Every (m,n)-computable set is uniformly (h,k)-computable, that is, the e' in* (c) *can be computed from e.*
(e) *There is machine M such that every $e_n = M(e, A(0)A(1)\ldots A(n))$ is defined and whenever φ_e (m,n)-computes A then almost all e_n are equal to an e' where $\varphi_{e'}$ (h,k)-computes A.*

So the uniform inclusion and uniform equality problem can be studied by analyzing admissible vector sets. A further motivation for studying these vector sets is that the criterion "every (m,n)-admissible vector set $V \subseteq \{0,1\}^k$ is (h,k)-admissible" coincides with the statement "every polynomial time (m,n)-computable set is polynomial time (h,k)-computable". That is, admissible vector sets characterize also the polynomial time inclusion problem for frequency computation.

For every d,k,m there are $(m,m+d+1)$-admissible vector sets which are not $(k,k+d)$-admissible so that uniform equality can only hold between parameters having the same difference $d=n-m$. Kinber [40] showed that a vector set is $(2,3)$-admissible iff it is $(m,m+1)$-admissible for every $m \geq 2$ and so gave a nontrivial example of such an equality. Kinber gave an example of a $(3,5)$-admissible vector set which is not $(4,6)$-admissible. Hinrichs and Wechsung [32] extended the study of (m,n)-admissible sets and showed that, for all d, on the one hand there is, for all $m < 2^d$, some $(m,m+d)$-admissible vector set which is not $(m+1,m+d+1)$-admissible and on the other hand there is an m' such that, for all $m \geq m'$, every $(m,m+d)$-admissible vector set is $(m+1,m+d+1)$-admissible. The smallest possible value of this m' is still unknown for all $d \geq 3$.

Definition 7.5 [8, 20, 47] Let $V \subseteq \{0,1\}^k$ be a vector set and $\frac{m}{n} \leq \frac{1}{2}$ for the first two definitions.

The vector set V is *strongly (m,n)-admissible* iff for every t with $t < m \wedge n-t \leq k$ and for every projection W onto $n-t$ coordinates there is a vector $\mathbf{v}$ such that at least m of the components of $\mathbf{v}$ are 0 and for every $\mathbf{w} \in W$ at least $m-t$ of the components of $\mathbf{v}$ and $\mathbf{w}$ coincide [47].

The vector set V is *special (m,n)-admissible* iff for every s,t with $s+t<m \wedge n-s-t \leq k$ and for every projection W onto $n-s-t$ coordinates there is a vector $\mathbf{v}$ such that at least $m-s$ of the components of $\mathbf{v}$ are 0, at least $m-t$ of the components of $\mathbf{v}$ are 1 and for every $\mathbf{w} \in W$ at least $m-s-t$ of the components of $\mathbf{v}$ and $\mathbf{w}$ coincide [47].

The vector set V is *(m,n)-good* iff there is a sequence $m_1, m_2, \ldots, m_n$ such that every projection of V onto h coordinates with $1 \leq h \leq \min\{k,n\}$

contains at most m_h elements and $m_s + m_t \leq m_{s+t} + 1$ for all s, t with $s + t \leq n$ [8].

The next theorem characterizes several inclusion problems combinatorically.

Theorem 7.6 [8, 47, 52] (a) *Every (m,n)-computable set is (h,k)-computable iff every strongly (m,n)-admissible vector set $V \subseteq \{0,1\}^k$ is (h,k)-admissible.*
(b) *Every recursively enumerable (m,n)-computable set is (h,k)-computable iff every special (m,n)-admissible vector set $V \subseteq \{0,1\}^k$ is (h,k)-admissible.*
(c) *Every (m,n)-computable set is strongly (h,k)-verbose iff every (m,n)-computable set is (h,k)-verbose iff every strongly (m,n)-admissible vector set $V \subseteq \{0,1\}^k$ is (h,k)-good.*
(d) *Every strongly (m,n)-verbose set is (h,k)-computable iff every (m,n)-good vector set $V \subseteq \{0,1\}^k$ is (h,k)-admissible.*
(e) *Every (m,n)-verbose set is (h,k)-verbose iff every strongly (m,n)-verbose set is strongly (h,k)-verbose iff every (m,n)-good vector set $V \subseteq \{0,1\}^k$ is (h,k)-good. If there is some (m,n)-good vector set $V \subseteq \{0,1\}^k$ which is not (h,k)-good then there is a strongly (m,n)-verbose set which is not (h,k)-verbose.*

This theorem gives implicitly a decision procedure for the inclusion problem since there are only finitely many vector sets $V \subseteq \{0,1\}^k$ and it is sufficient to test the conditions for them all. Therefore part (a) of the theorem implies McNicholl's result of the recursiveness of the inclusion problem.

Theorem 7.7 [52] *The inclusion problem of frequency computation is computable.*

Until now it was not possible to find examples where the criteria in Theorem 7.6 (a) and (b) differ from the straightforward generalization of Dëgtev's result which would give for m, n with $\frac{m}{n} \leq \frac{1}{2}$: The class of the (recursively enumerable) (m,n)-computable sets is contained in the class of the (h,k)-computable sets iff every (m,n)-admissible vector set $V \subseteq \{0,1\}^k$ is (h,k)-admissible.

The characterizations in terms of conditions on finite sets allow the use of finite combinatorics and coding theory for obtaining some information on the inclusion problems. The following two examples illustrate this method.

Example 7.8 (a) *Every $(2^k - 1, k)$-verbose set is $(p(n), n)$-verbose for all n where $p(n) = 1 + n + \frac{n(n-1)}{2} + \frac{n(n-1)(n-2)}{6} + \ldots + \frac{n!}{(n-k-1)! \cdot (k-1)!}$ is the sum of the first k binomial coefficients for n.*
(b) *For every $m \geq 2$ there is a $(1, 2^m - m)$-computable set which is not $(2, 2^m - 1)$-computable.*

Proof The first result (a) is based on a result of Sauer which implies that every $(2^k - 1, k)$-good vector set is also $(p(n), n)$-good for all n [60].

The second result (b) is obtained using Hamming codes. These codes code n bits with m parity check-bits such that 1 error can be detected and corrected if $m + n + 1 \leq 2^m$. For arbitrary m, let $n = 2^m - m - 1$. Let $V \subseteq \{0,1\}^{n+m}$ be the set of all code words; V has 2^n elements and is thus $(1, n+1)$-admissible: Every projection W of V has at most 2^n elements and thus some vector $\mathbf{u} \in \{0,1\}^{n+1}$ is not in W. Let $\mathbf{v}$ be the complement of $\mathbf{u}$, that is, $\mathbf{v}[h] = 1 - \mathbf{u}[h]$ for all h, then $\mathbf{u}$ is the only vector which does not agree with $\mathbf{v}$ in one component and every vector $\mathbf{w} \in W$ agrees with $\mathbf{v}$ in at least 1 component. Since the Hamming code corrects one error, the Hamming spheres of radius 1 generated by the code elements are disjoint. Each such sphere contains the center and $n+m$ elements around, that is $n+m+1 = 2^m$ elements in total. So the 2^n Hamming spheres cover together 2^{n+m} vectors, that is the whole vector set $\{0,1\}^{n+m}$. Given any vector $\mathbf{u} \in \{0,1\}^{n+m}$ there is a vector $\mathbf{v} \in V$ such that the complement of $\mathbf{u}$ belongs to the Hamming sphere of radius 1 generated by $\mathbf{v}$. Thus this complement coincides with $\mathbf{v}$ in at least $n + m - 1$ components and thus $\mathbf{v}$ agrees with $\mathbf{u}$ in at most 1 component. It follows that V is not $(2, n+m)$-admissible. Using $n + m + 1 = 2^m$, one can deduce that there is a $(1, 2^m - m)$-computable set which is not $(2, 2^m - 1)$-computable. ∎

The second result was obtained while attacking the still unsolved problem whether for each n there is a $(1, n)$-computable set which is not $(2, 2n - 1)$-computable [44]. Other explicit non-inclusions are: Some recursively enumerable $(n, 2n)$-computable sets are not $(1, n)$-computable [20], some recursively enumerable $(n, 2n + m + 1)$-computable sets are not $(k, 2k + m)$-computable for any k [47]. Dëgtev [20] found already all straightforward inclusions like that every (m, n)-computable set is also $(m - k, n - k)$-computable for $k < m$ and $(hm, hn + l)$-computable for all h, l. Also some nontrivial inclusions hold, for example every $(2, 4)$-computable set is also $(3, 6)$-computable [47].

Related to the inclusion problem is the equality problem, that is the problem to determine all (m, n, h, k) such that a set is (m, n)-computable iff it is (h, k)-computable. The equality problem for frequency computation and also for verboseness have explicit solutions not depending on difficult combinatorial properties.

Theorem 7.9 [6, 47] (a) *The classes of (m, n)-computable and (h, k)-computable sets coincide iff $(m, n) = (h, k)$ or $\frac{m}{n} > \frac{1}{2} \wedge \frac{h}{k} > \frac{1}{2}$.*
(b) *The classes of (m, n)-verbose and (h, k)-verbose sets coincide iff $(m, n) = (h, k)$ or $m \geq 2^n \wedge h \geq 2^k$ or $m \leq n \wedge h \leq k$ or the fractions $q = \frac{m}{n}, \frac{h}{k}$ are both between 1 and 2 and have the same $l \in \{0, 1, 2, \ldots\}$ with $\frac{2l+1}{l+1} < q \leq \frac{2l+3}{l+2}$.*

Beigel, Gasarch and Kinber [6] considered also the following generalization of frequency computation: A set A has k-enumerable frequency (m, n) iff there is a partial recursive function ψ which assigns to every $x_1, x_2, \ldots, x_n$ and $h \in \{1, 2, \ldots, k\}$ a tuple $(y_1, y_2, \ldots, y_n)$ such that one of these up to k tuples agrees with $C_n^A(x_1, x_2, \ldots, x_n)$ on at least m components. For example every non-recursive set A has 2-enumerable but not 1-enumerable frequency $(3, 5)$: The characteristic function on any five inputs contains either 3 times a 0 or 3 times a 1, thus one enumerates for any input the two vectors $(0, 0, 0, 0, 0)$ and $(1, 1, 1, 1, 1)$ and knows that one of them coincides with the 5-fold characteristic function in three components. On the other hand, A does not have 1-enumerable frequency $(3, 5)$ since A is not recursive and therefore also not $(3, 5)$-computable. The next theorem summarizes the results.

Theorem 7.10 [5] *Below are listed four classes of sets such that for every set in the class the set has k-enumerable frequency (m, n) iff the corresponding condition on k is satisfied.*
(a) *A (m, n)-recursive: $k \geq 1$;*
(b) *A semirecursive but not computable: $k \geq \frac{n+1}{2(n-m)+1}$;*
(c) *$A = K$: $k \geq \frac{n+1}{n-m+1}$;*
(d) *A superterse: k Hamming spheres of radius $n - m$ cover $\{0, 1\}^n$.*

Proof (a): If A is (m, n)-computable via f this frequency is clearly also 1-enumerable by enumerating just the output of f. The converse direction is void since having k-enumerable frequency (m, n) requires enumerating at least 1 vector.

(b): Let A be semirecursive. Then one can compute a sequence $\mathbf{v}_0, \mathbf{v}_1, \ldots,$ $\mathbf{v}_n$ such that each vector $\mathbf{v}_{i+1}$ differs from $\mathbf{v}_i$ by turning one 0 into 1 and $C_n^A(x_1, x_2, \ldots, x_n)$ is among them. The Hamming distance between vectors $\mathbf{v}_i$ and $\mathbf{v}_j$ is just $|i - j|$.

If $k \geq \frac{n+1}{2(n-m)+1}$ then one enumerates, for $h = 1, 2, \ldots, k$, the vector $\mathbf{v}_i$ with $i = \min\{n, (2(n - m) + 1)h - (n - m) - 1\}$. One can verify that each vector $\mathbf{v}_j$ has Hamming distance at most $n - m$ from some vector enumerated. So does $C_n^A(x_1, x_2, \ldots, x_n)$ and A has k-enumerable frequency (m, n).

For the converse direction, assume that A has k-enumerable frequency (m, n). If the Hamming distance of two vectors $\mathbf{v}_i$ and $\mathbf{v}_j$ to some vector $\mathbf{u}$ is not above $n - m$ then the Hamming distance between $\mathbf{v}_i$ and $\mathbf{v}_j$ is at most $2(n - m)$. Thus for any given vector $\mathbf{u}$ there are at most $2(n - m) + 1$ vectors $\mathbf{v}_i$ which have Hamming distance up to $n - m$ from $\mathbf{u}$. It follows that A is $(k(2(n - m) + 1), n)$-verbose: Since A has k-enumerable frequency (m, n), some enumeration process enumerates up to k vectors $\mathbf{u}$ such that for each such $\mathbf{u}$ one can enumerate up to $2(n - m) + 1$ of the vectors $\mathbf{v}_i$ that coincide with $\mathbf{u}$ in at least m components. It follows that the n-fold characteristic function of A is among these $k(2(n-m)+1)$ vectors. Since A is not recursive, $k(2(n-m)+1) \geq n+1$ and $k \geq \frac{n+1}{2(n-m)+1}$ by Beigel's Nonspeedup Theorem.

(c): If A_s is a recursive enumeration of the set A then one can k-enumerate the frequency (m,n) by outputting all vectors $C_n^{A_s}(x_1, x_2, \ldots, x_n)$ whose number of 1s is divisible by $n-m+1$. If $C_n^A(x_1, x_2, \ldots, x_n)$ has $h(n-m+1)+l$ many 1s then one knows that the Hamming distance to the last output vector is just l and so at most $n-m$. Thus every recursively enumerable set has k-enumerable frequency (m,n) if $k \geq \frac{n+1}{n-m+1}$.

Here is a proof-sketch for the converse direction: By direct diagonalization as in Theorem 4.3 one can show that there is a recursively enumerable set A such that A does not have k-enumerable frequency (m,n) whenever $k < \frac{n+1}{n-m+1}$. One can also show that whenever A is one-one reducible to a set B and B has k-enumerable frequency (m,n) so does A. Since every recursively enumerable set is one-one reducible to K [54, Section III.7] it follows that K also does not have k-enumerable frequency (m,n).

(d): If there are k spheres of radius $n-m$ which cover the vector set $\{0,1\}^n$ then one can for any input enumerate just the centers of these spheres and obtains a witness that any given set has k-enumerable frequency (m,n) via this process.

For the converse direction assume that A would have k-enumerable frequency (m,n) via ψ. For given $x_1, x_2, \ldots, x_n$ one can now enumerate for each $h \in \{1,2,\ldots,k\}$ all tuples with Hamming distance at most $n-m$ from some tuple $\psi(x_1, x_2, \ldots, x_n, h)$ provided that this is defined. By assumption, the n-fold characteristic function $C_n^A(x_1, x_2, \ldots, x_n)$ is always among these tuples. Since A is superterse there must be $x_1, x_2, \ldots, x_n$ such that all tuples from $\{0,1\}^n$ are enumerated. So each tuple in $\{0,1\}^n$ has at most Hamming distance $n-m$ from some tuple $\psi(x_1, x_2, \ldots, x_n, h)$ and so these up to k tuples define up to k Hamming spheres of radius $n-m$ which cover $\{0,1\}^n$. ∎

Part (d) of the theorem gives only a combinatorial criterion. This criterion, the number of Hamming spheres of radius r which are necessary to cover $\{0,1\}^n$ has been studied intensively — see [14, 15, 16, 34, 66] for an overview of these results — but no easy formula has been found to express it.

8 The Cardinality Theorem

Owings [56] was motivated from Beigel's Nonspeedup Theorem [3, 4] to search for a parallel result with respect to Cardinality: Whenever one can enumerate for $x_1, x_2, \ldots, x_n$ up to n values such that one of them is the cardinality of

A on this input then A is recursive. Formally, one defines for any partial recursive function φ_e the class $\mathcal{A}_{e,n}$ which contains all sets A such that for each $x_1, x_2, \ldots, x_n$ there is $k \in \{1, 2, \ldots, n\}$ such that $\varphi_e(x_1, x_2, \ldots, x_n, k)\downarrow = \#_n^A(x_1, x_2, \ldots, x_n)$ — the class $\mathcal{A}_{e,n}$ is empty if φ_e is not a suitable operator. Owings [56] showed the intended result only for total functions: Whenever φ_e is total then $\mathcal{A}_{e,n}$ contains only recursive sets. Kummer [30, 42] improved then this result to the Cardinality Theorem.

Theorem 8.1 [42] *If one can for each distinct $x_1, x_2, \ldots, x_n$ enumerate up to n values such that the local cardinality $\#_n^A(x_1, x_2, \ldots, x_n)$ is among these n values, that is, if $A \in \mathcal{A}_{e,n}$ for some e and n, then A is computable.*

Kummer's proof for this is quite interesting since it brought two new ideas into the field of bounded queries: the idea of recursively enumerable trees and the application of Ramsey Theory to these trees [17]. The main three steps for his proof are:

- One constructs from φ_e the recursively enumerable tree T of all σ which are consistent with φ_e.
- One shows that T has the combinatorial property that for some m no full binary tree of degree m is embeddable into T.
- One shows that from this non-embeddability result it follows that every infinite branch of T and therefore also every set in $\mathcal{A}_{e,n}$ is recursive.

The first step is quite straightforward and uses the following recursively enumerable tree T derived from φ_e:

$$\begin{aligned} \sigma \in T \;\Leftrightarrow\; & (\forall \text{ distinct } x_1, x_2, \ldots, x_n \in dom(\sigma))\,(\exists k \in \{1, 2, \ldots, n\}) \\ & [\sigma(x_1) + \sigma(x_2) + \ldots + \sigma(x_k) = \varphi_e(x_1, x_2, \ldots, x_n, k)\downarrow]. \end{aligned}$$

Now it remains to formalize and explain the second and third step. For the second step, only a proof-sketch is given, Kummer [42] gives the full proof within Lemmas 2 and 3.

The second and third step need two definitions. First, β_k is the full binary tree of having $2^k - 1$ nodes including the leaves. Second, the tree β_k *is embeddable into T above σ* if there is an order preserving mapping g from all binary strings of length up to k to nodes in T such that $g(\lambda) = \sigma\tau$ for some τ where λ stands for the empty string. Order preserving means that $g(\eta a)$ is of the form $g(\eta)a\tau$ for all strings $\eta \in \{0,1\}^*$ shorter than k and all $a \in \{0,1\}$, in particular, η is a prefix of τ iff $g(\eta)$ is a prefix of $g(\tau)$ for all nodes $\eta, \tau \in \beta_k$. The tree $\beta_0 = \{\lambda\}$ is embeddable above every $\sigma \in T$.

Proposition 8.2 *If T is derived from φ_e then there is a k such that the full binary tree β_k is not embeddable into T.*

Proof sketch Using some Ramsey style theorems for finite trees, Kummer's main idea is to show [42, Lemma 3] that for each n there is a some k such that whenever β_k is embeddable into T then one can find places $x_1, x_2, \ldots, x_n$ (not necessarily in ascending order) and finite branches $\sigma_0, \sigma_1, \ldots, \sigma_n$ in T such that $\sigma_m(x_h) \downarrow = 0$ if $m < h$ and $\sigma_m(x_h) \downarrow = 1$ if $m \geq h$. Then one obtains that $\sigma_m(x_0) + \sigma_m(x_1) + \ldots + \sigma_m(x_n) = m$. So one has that for $x_1, x_2, \ldots, x_n$ every cardinality m is realized by some σ_m which contradicts that by construction of T only n out of $n+1$ cardinalities can be realized by some $\sigma \in T$. ∎

Proposition 8.3 *Assume that some full binary tree β_k is not embeddable into T. Then every infinite branch of T is recursive.*

Proof Some full binary tree β_k is not embeddable into T by assumption, but β_0 is embeddable into T above every $\sigma \in T$.

So if the set A represents an infinite path through T then there is a maximal h such that β_h is embeddable above every node $A(0)A(1)\ldots A(x)$. Furthermore, there is some y such that the tree β_{h+1} is not embeddable above $A(0)A(1)\ldots A(y)$. These two facts allow the computation of $A(x)$ as follows:

> On input x search for the first σ extending $A(0)A(1)\ldots A(y)$ such that $x \in dom(\sigma)$ and β_h is embeddable into T above σ. Having this σ, output $\sigma(x)$.

Since A is an infinite path, the computation always terminates. If now the computation gives for some x some wrong value then there must be a σ' extending $A(0)A(1)\ldots A(y)$ which itself is not a prefix of A such that β_h is embeddable above σ'. By choice of h, β_h is also embeddable above $\sigma = A(0)A(1)\ldots A(|\sigma'|)$. The strings σ and σ' have some longest common prefix σ'' which extends $A(0)A(1)\ldots A(y)$. Now one can embed the tree β_{h+1} above σ'' in contradiction to the choice of y by assigning $g(\lambda) = \sigma''$ and embed the two subtrees β_h with roots 0 and 1 above σ and σ'. This cannot occur and therefore the above algorithm is correct. ∎

The gap between the results of Owings and Kummer might also be overcome by applying the technique of relativization presented in Section 3. But as shown in [45], Kummer's technique can also be used to prove several generalizations of the Cardinality Theorem for which Owings' technique [56] apparently does not suffice (even if restricted to total functions). The transition from $\mathcal{A}_{e,n}$ to a characteristic function for a set $A \in \mathcal{A}_{e,n}$ is not uniform.

Kaufmann [38] had the idea to measure the degree of uniformity of such a transition by the quantity of programs which must be computed from e

such that at least one of these programs agrees with some set in $\mathcal{A}_{e,n}$ modulo finitely many errors.

Theorem 8.4 [38, Theorems 5.1.1 and 5.1.2]
(a) *There is an algorithm that computes m programs from e such that whenever $\mathcal{A}_{e,n}$ is not empty and whenever φ_e is total then one of these m programs computes — modulo finitely many errors — some set in $\mathcal{A}_{e,n}$ where m is the parameter of the first full binary tree β_m not embeddable into any recursive tree obtained from any total function $\varphi_{e',n}$ such that $\mathcal{A}_{e',n}$ is not empty.*
(b) *In general, given the class $\mathcal{E}$ of all recursive trees T such that β_m cannot be embedded into T, then one can compute from a characteristic index e of T indices $a_1, a_2, \ldots, a_m$ such that one of them computes an infinite branch of T modulo finitely many errors. The number m of the indices $a_1, a_2, \ldots, a_m$ is optimal and cannot be replaced by $m-1$.*
(c) *Let m be an arbitrarily large constant. Then there is no algorithm which computes m programs from e whenever φ_e enumerates a tree T where T has an infinite branch but β_2 is not embeddable into T such that one of these m programs computes — modulo finitely many errors — some infinite branch of T.*

Motivated by this result, Kaufmann and Kummer [38, 39] studied the general question: given a program (or enumeration) for a tree T, how difficult is it to find a program for the characteristic function of some infinite branch of T. Since this does not work for arbitrary trees, one looks at classes of trees T which satisfy some additional requirements as for example that T has at each level at most k different nodes ($|\{0,1\}^t \cap T| \leq k$ for all t). For these special cases there are algorithms which output for recursive indices of trees k programs and for recursively enumerable indices $2k-1$ programs such that one of them computes — modulo finitely many errors — some infinite branch of T provided that T is not finite. Case, Kaufmann, Kinber and Kummer [13] applied these ideas also to the field of learning theory. Kummer and Ott [43] considered then the general model of learning infinite branches of a tree while knowing a program for the characteristic function of the tree.

An application of Kummer's Cardinality Theorem is to detect a minimum number k such that there is a non-recursive set A with k-enumerable frequency (m,n). For semirecursive sets A the condition $k \geq \frac{n+1}{2(n-m)+1}$ is necessary and sufficient. The proof for this lower bound in Theorem 7.10 (b) uses that every Hamming sphere of radius $n-m$ can cover only up to $2(n-m)+1$ tuples which are consistent with the linear ordering connected to the semirecursive set. In the general case, such an ordering does not exist. But one can still see that every Hamming sphere of radius $n-m$ covers up to $2(n-m)+1$ cardinalities: that one induced by its center and the $n-m$ next ones below and above. So the calculations of the lower bound can be preserved and Beigel, Gasarch and Kinber [6, Theorem 3.1] obtained the following result.

Theorem 8.5 [6] *If a non-recursive set has k-enumerable frequency (m, n) then $k \geq \frac{n+1}{2(n-m)+1}$.*

9 Nondeterministic Queries

The non-deterministic variant of Turing reductions does not give any extra power in recursion theory. But this is no longer true if one restricts the size and number of queries polynomially as done in complexity theory — there are A and B such that A can be computed relative to B only with non-deterministic queries. Beigel, Gasarch and Owings [7] investigated to which extent this is also true for a constant bound on the number of queries and no bound on their size.

Definition 9.1 [7] A function f is *computable with n non-deterministic queries to A* iff there is a partial recursive function ψ such that for every x there are $x_1, x_2, \ldots, x_n$ such that $f(x) = \psi(x, x_1, x_2, \ldots, x_n, C_n^A(x_1, x_2, \ldots, x_n)) \downarrow$ and whenever $\psi(x, z_1, z_2, \ldots, z_n, C_n^A(z_1, z_2, \ldots, z_n))$ converges to some value y for some $z_1, z_2, \ldots, z_n$ then $f(x) = y$.

A set A is *n-subjective* iff every function f which is computable relative to A is also computable with n non-deterministic queries to A. A set A is *objective* iff it is not n-subjective for any n.

Note that the program of ψ can be made deterministic, even if it is specified non-deterministic in the sense that it converges on some path for the correct $x_1, x_2, \ldots, x_n$ and $C_n^A(x_1, x_2, \ldots, x_n)$ but may diverge on other paths. But it is still required that whenever ψ has auxiliary input of the form $z_1, z_2, \ldots, z_n, C_n^A(z_1, z_2, \ldots, z_n)$ then ψ either diverges or converges to the correct value on every computation path. The next result allows to characterize the n-subjective sets using the fixed function C_{n+1}^A.

Theorem 9.2 [7] *A set A is n-subjective iff C_{n+1}^A can be computed with n non-deterministic queries to A.*

Proof If every f computable relative to A can be computed with n non-deterministic queries, then this also holds for the fixed function C_{n+1}^A. For the converse direction let f be any function computed by a machine M with arbitrarily many queries to A and let x be any input.

Now one guesses non-deterministically a sequence $D_0, D_1, \ldots, D_m$ of finite sets such that $f(x)$ can be computed relative to A with queries from the set D_0, such that $|D_m| = n$ and such that, for every set D_k with $k < m$, there are $x_1, x_2, \ldots, x_{n+1} \in D_k$ for which $C_{n+1}^A(x_1, x_2, \ldots, x_{n+1})$ can be computed with n non-deterministic queries to A at some arguments $y_1, y_2, \ldots, y_n$ which satisfy in addition $D_{k+1} = (D_k - \{x_1, x_2, \ldots, x_{n+1}\}) \cup \{y_1, y_2, \ldots, y_n\}$.

From this construction it follows that the characteristic function of A on each set D_k can be computed from knowing A on D_{k+1}. Furthermore, each set D_{k+1} has less elements than D_k and thus one obtains after sufficiently many steps a set D_m having n elements. So one can evaluate $f(x)$ by knowing C_n^A on this set D_m and so f is computable with n non-deterministic queries. ∎

So one has a real dichotomy: either every function recursive in A is already computable with n non-deterministic queries relative to A for some n or every function C_n^A needs already n non-deterministic queries for all n.

There are quite natural examples for n-subjective and objective sets. Obviously 0-subjective sets and recursive sets coincide. The following sets are 1-subjective: truth-table cylinders, retraceable sets, recursively enumerable semirecursive sets and their complements. In general, semirecursive sets are only 2-subjective. Random sets and 1-generic sets are objective. One can also improve the result that whenever C_{n+1}^A is computable with n queries to A then A is autoreducible from deterministic to non-deterministic queries. The proof of Theorem 6.2 also holds for non-deterministic queries instead of deterministic ones and so whenever Odd_{n+1}^A can be computed with n non-deterministic queries to A then A is autoreducible. Since Odd_{n+1}^A is easier to compute than C_{n+1}^A one obtains the following theorem.

Theorem 9.3 *Every n-subjective set is autoreducible.*

Jockusch and Paterson [36] showed that recursively enumerable degrees and degrees above the degree of K do contain sets which are not autoreducible. So these degrees contain also objective sets [7, Theorem 25]. In addition, Beigel, Gasarch and Owings [7, Theorem 24] showed that the objective set in a recursively enumerable Turing degree can also be chosen to be recursively enumerable. But they left open whether every non-recursive Turing degree contains an objective set [7, Section 6]. Stephan [62] gave a negative answer to this question. He constructed a retraceable set A such that A is bounded truth-table reducible to every B in the same Turing degree with norm 2. This result can be used to show that every B in the Turing degree of A is 2-subjective.

Theorem 9.4 [62] *There is a non-recursive Turing degree containing only 2-subjective sets.*

Proof Stephan [62] showed that there is a set A retraced by a total recursive function which is btt-reducible with norm 2 to every set B within the same Turing degree. The fact that A is retraced by a total recursive function is equivalent to the existence of a uniformly recursive procedure $u \to A_u$ with $A_u = A \cap \{0, 1, \ldots, u\}$ for all $u \in A$.

Now let B be any set Turing equivalent to A. It is shown that B is 2-subjective: Let x_1, x_2, x_3 be arbitrary inputs and let M be a deterministic Turing machine which computes B relative to A. Now a non-deterministic machine N with one query to u works as follows.

$$N(x_1, x_2, x_3, u, A(u)) = \begin{cases} (y_1, y_2, y_3) & \text{if } A(u) = 1 \text{ and, for } i = 1, 2, 3, \\ & M^{A_u}(x_i)\downarrow = y_i \text{ without oracle} \\ & \text{queries to } A_u(x) \text{ for } x > u; \\ \uparrow & \text{otherwise.} \end{cases}$$

The oracle answers are correct if the computation of M asks the oracle A_u only below u. Since A is infinite, one can take u so large that $M^A(x_i)$ does not make any query above u and obtains then the correct answer. So one only needs to show that the queries to A can be replaced by queries to B using a btt-reduction from A to B. Let f_1, f_2, g describe the btt-reduction in the sense that $A(u) = g(u, B(f_1(u)), B(f_2(u)))$.

$$N'(x_1, x_2, x_3, v, w, C_2^B(v, w)) = \begin{cases} (y_1, y_2, y_3) & \text{if there is } u \text{ such that} \\ & f_1(u) = v,\ f_2(u) = w, \\ & g(u, B(v), B(w)) = 1 \\ & \text{and, for } i = 1, 2, 3, \\ & M^{A_u}(x_i)\downarrow = y_i \text{ with-} \\ & \text{out oracle queries to} \\ & A_u(x) \text{ for } x > u; \\ \uparrow & \text{otherwise.} \end{cases}$$

So the non-deterministic computation guesses the u which is used by N, verifies with two queries to B that $u \in A$ and proceeds then as the original N. It follows that B is 2-subjective since N' computes C_3^B with two non-deterministic queries to B. ∎

10 Open Problems

The equality problem for (m, n)-computable and (m, n)-verbose sets has been solved explicitly. But the corresponding inclusion problems have only combinatorial characterizations which are difficult to evaluate. Several explicit inclusions or non-inclusions are unknown, for example, whether every $(1, n)$-computable set is $(2, m)$-computable for some $n > 1$ and $m < 2n$. Furthermore, it is unknown whether the the inclusion problems for recursively enumerable sets and for all sets are really different. Although the combinatorial criteria are different for specific V they still might induce the same inclusion structure. Until now, no concrete example of (m, n) and (h, k) is known such that the inclusion holds for recursively enumerable but not for all sets.

For the case of the uniform equalities, that is the set of all (m, n, h, k) such that functions (m, n)-computing and (h, k)-computing a set A can be translated effectively into each other, no explicit characterization is known. One knows that the difference $d = n - m$ and $d' = k - h$ must be equal if uniform equality holds. For every d there are infinitely many m such that $(m, m + d)$-computable sets are uniformly $(m + 1, m + d + 1)$-computable and vice versa. The condition $m \geq 2^d$ is necessary for these m, but for $d \geq 3$, it is unknown which is the first m such that this uniform equality holds.

There are a few unsolved problems which do not relate to finite combinatorics. Given $n \geq 3$ it is unknown which is the minimum m such that there is a strongly (m, n)-verbose p-complete set [46]. The same question is open for strongly hyperimmune sets. Gasarch [27] asked whether every non-recursive Turing degree contains a cardinality terse set, that is, a set where $\#_n^A$ cannot be computed with less than n queries to A for every n.

Acknowledgment We would like to thank Peter Fejer, Andrew Lee and Maciek Smuga-Otto for proofreading and helpful discussion. We are also grateful to the anonymous referee for several comments and suggestions.

References

[1] A. Amir, R. Beigel and W. Gasarch. Some connections between bounded query classes and non-uniform complexity. *Proceedings of the Fifth Annual Structure in Complexity Theory Conference*, IEEE Computer Society Press, 232–243, 1990

[2] A. Amir and W. Gasarch. Polynomial terse sets. *Information and Computation*, 77:37–56, 1988.

[3] R. Beigel. *Query-Limited Reducibilities.* PhD thesis, Stanford University, 1987. Also available as Report No. STAN-CS-88-1221.

[4] R. Beigel, W. Gasarch, J. T. Gill and J. C. Owings. Terse, superterse and verbose sets. *Information and Computation*, 103:68–85, 1993.

[5] R. Beigel, W. Gasarch and L. Hay. Bounded query classes and the difference hierarchy. *Archive for Mathematical Logic*, 29:69–84, 1989.

[6] R. Beigel, W. Gasarch and E. Kinber. Frequency computation and bounded queries. *Theoretical Computer Science*, 163:177–192, 1996.

[7] R. Beigel, W. Gasarch and J. C. Owings. Nondeterministic bounded query reducibilities. *Annals of Pure and Applied Logic*, 41:107–118, 1989.

[8] R. Beigel, M. Kummer and F. Stephan. Quantifying the amount of verboseness. *Information and Computation*, 118:73–90, 1995.

[9] R. Beigel, M. Kummer and F. Stephan. Approximable sets. *Information and Computation*, 120:304–314, 1995.

[10] R. Beigel, W. Gasarch, M. Kummer, G. Martin, T. McNicholl and F. Stephan. On the query complexity of sets. *Proceedings of the Twentyfirst International Symposium on Mathematical Foundations of Computer Science* – MFCS '96, Springer LNCS 1113, 206–217, 1996.

[11] R. Beigel, W. Gasarch, M. Kummer, G. Martin, T. McNicholl and F. Stephan. On the complexity of Odd_n^A. Manuscript, 1997.

[12] J. Cai and L. A. Hemachandra. Enumerative counting is hard. *Information and Computation*, 82(1):34–44, July 1989.

[13] J. Case, S. Kaufmann, E. Kinber and M. Kummer. Learning recursive functions from approximations. *Journal of Computer and System Sciences*, 55:183–196, 1997.

[14] G. Cohen and P. Frankl. Good coverings of Hamming spaces with spheres. *Discrete Mathematics*, 56:125–131, 1989.

[15] G. Cohen, M. Karpovsky and H. Mattson. Covering radius — survey and recent results. *IEEE Transactions on Information Theory*, 31:338–343, 1985.

[16] G. Cohen, A. Lobstein and N. Sloane. Further results on the covering radius of codes. *IEEE Transactions on Information Theory*, 32:680–694, 1986.

[17] W. Deuber. A generalization of Ramsey's theorem for regular trees. *Journal of Combinatorial Theory* (*Series B*), 18:18–23, 1975.

[18] A. N. Dëgtev. tt- and m-degrees. *Algebra and Logic* 12:143–161, 1973, in Russian, 12:78–89, 1973, translation.

[19] A. N. Dëgtev. Three theorems on tt-degrees. *Algebra and Logic* 17:270–281, 1978, in Russian, 17:187–194, 1978, translation.

[20] A. N. Dëgtev. On (m, n)-computable sets. In *Algebraic Systems*, edited by D. I. Moldavanskij, Ivanova State University, 1981, in Russian.

[21] R. Downey and C. Jockusch. T-degrees, jump classes and strong reducibilities. *Transactions of the AMS*, 301:103–120, 1987.

[22] Yu. Ershov. A hierarchy of sets I. *Algebra and Logic*, 7(1):47–74, 1967, in Russian, 7:25–43, 1968, translation.

[23] Yu. Ershov. A hierarchy of sets II. *Algebra and Logic*, 7(4):38–47, 1967, in Russian, 7:212–232, 1968, translation.

[24] Yu. Ershov. A hierarchy of sets III. *Algebra and Logic*, 9:34–51, 1970, in Russian, 7:20–31, 1970, translation.

[25] R. Friedberg. The fine structure of degrees of unsolvability of recursively enumerable sets. In *Summaries of Cornell University Summer Institute for Symbolic Logic, Communications Research Division, Inst. for Def. Anal.*, Princeton, 404–406, 1957.

[26] R. Friedberg and H. Rogers. Reducibilities and completeness for sets of integers. *Zeitschrift für mathematische Logik und Grundlagen der Mathematik*, 5:117–125, 1959.

[27] W. Gasarch. Bounded queries in recursion theory: A survey. *Proceedings of the sixth Annual Conference on Structure in Complexity Theory*, IEEE Computer Society Press, 62–78, 1991.

[28] W. Gasarch. *A hierarchy of functions with applications to recursive graph theory.* Technical Report 1651, University of Maryland, Department of Computer Science, 1985.

[29] W. Gasarch and G. Martin. *Bounded queries in recursion theory.* Manuscript.

[30] V. Harizanov, M. Kummer and J. C. Owings, Jr. Frequency computation and the cardinality theorem. *Journal of Symbolic Logic*, 57:682–687, 1992.

[31] L. Hay. Letters to W. Gasarch, 1985.

[32] M. Hinrichs and G. Wechsung. Time bounded frequency computations. *Information and Computation*, 139:234–257, 1997.

[33] A. Hoene and A. Nickelsen. Counting, selecting and sorting by query-bounded machines. *Proceedings of the Tenth Annual Symposium on Theoretical Aspects in Computer Science* — STACS '93, Springer LNCS, 665:196–205, 1993.

[34] I. Honkala. Modified bounds for covering codes. *IEEE Transactions on Information Theory*, 37:351–365, 1991.

[35] C. G. Jockusch. Semirecursive sets and positive reducibility. *Transactions of the AMS*, 131:420–436, May 1968.

[36] C. G. Jockusch and M. Paterson. Completely autoreducible degrees. *Zeitschrift für Mathematische Logik und Grundlagen der Mathematik*, 22:571–575, 1976.

[37] C. G. Jockusch and R. I. Soare. Π_1^0 classes and degrees of theories. *Transactions of the AMS*, 173:33–56, 1972.

[38] S. Kaufmann. *Quantitative Aspects in Computability Theory.* Shaker Verlag, Aachen, 1998, in German.

[39] S. Kaufmann and M. Kummer. On a quantitative notion of uniformity. *Fundamentae Informatica*, 25:59–78, 1996.

[40] E. B. Kinber. *Frequency computable functions and frequency enumerable sets.* Candidate dissertation, Riga, 1975, in Russian.

[41] G. Kobzev. Recursive enumerable bw-degrees. *Matematicheskie Zametki*, 21(6):839–846, in Russian, 21:473–477, 1977, translation.

[42] M. Kummer. A proof of Beigel's cardinality conjecture. *Journal of Symbolic Logic*, 57:677–681, 1992.

[43] M. Kummer and M. Ott. Learning branches and learning to win closed games. *Proceedings of the Ninth Annual ACM Conference on Computational Learning Theory* — COLT'96, 280-291, 1996.

[44] M. Kummer and F. Stephan. *Some Aspects of Frequency Computation.* Technical Report 21/91, Institute for Logic, Complexity and Deductive Systems, University of Karlsruhe, 76128 Karlsruhe, Germany.

[45] M. Kummer and F. Stephan. Effective search problems. *Mathematical Logic Quarterly*, 40:224–236, 1994.

[46] M. Kummer and F. Stephan. Recursion theoretic properties of frequency computation and bounded queries. *Information and Computation*, 120:59–77, 1995.

[47] M. Kummer and F. Stephan. The power of frequency computation. *Proceedings of the Tenth International Conference on Fundamentals of Computation Theory* — FCT '95, Springer LNCS, 969:323-332, 1995.

[48] A. H. Lachlan. Some notions of reducibility and productiveness. *Zeitschrift für mathematische Logik und Grundlagen der Mathematik*, 11:17–44, 1965.

[49] R. E. Ladner. Mitotic recursively enumerable sets. *Journal of Symbolic Logic*, 38:199–211, 1973.

[50] R. E. Ladner. A completely mitotic nonrecursive recursively enumerable degree. *Transactions of the American Mathematical Society*, 184:479–507, 1973.

[51] M. Li and P. Vitányi. *An Introduction to Kolmogorov Complexity and Its Applications*. Second Edition, Springer, Heidelberg, 1997.

[52] T. McNicholl. *The Inclusion Problem for Generalized Frequency Classes*, Ph.D. Dissertation, The George Washington University, Washington D.C., 1995.

[53] A. Nickelsen. On polynomially D-verbose sets. *Proceedings of the Fourteenth Annual Symposium on Theoretical Aspects of Computer Science — STACS '97*, Springer LNCS, 1200:307–318, 1997.

[54] P. Odifreddi. *Classical Recursion Theory.* North-Holland, Amsterdam, 1989.

[55] M. Ogihara. Polynomial-time membership comparable sets. *Proceedings of the Ninth Annual Structure in Complexity Theory Conference*, 2–11, 1994.

[56] J. C. Owings, Jr. A cardinality version of Beigel's Nonspeedup Theorem. Journal of Symbolic Logic 54:761–767, 1989.

[57] E. Post. Recursively enumerable sets of positive integers and their decision problems, *Bulletin of the American Mathematical Society*, 50:284–316, 1944.

[58] G. Rose. An extended notion of computability. In *Abstracts of the International Congress for Logic, Methodology and Philosophy of Science*, page 14, 1960.

[59] G. E. Sacks. On the degrees less than $\mathbf{0}'$. *Annals of Mathematics*, 77:211–231, 1963.

[60] N. Sauer. On the densities of families of sets. *Journal of Combinatorial Theory Series A*, 13:145–147, 1972.

[61] R. I. Soare. *Recursively Enumerable Sets and Degrees.* Perspectives in Mathematical Logic. Springer, Heidelberg, 1987.

[62] F. Stephan. On the structures inside truth-table degrees. *Forschungsberichte Mathematische Logik 29 / 1997, Mathematisches Institut, Universität Heidelberg*, Heidelberg, 1997.

[63] B. A. Trahtenbrot. On the frequency computability of functions. Algebra and Logic 2:25–32, 1963, in Russian.

[64] B. A. Trahtenbrot. Frequency computation. *Proceedings Steklov Inst. Math.*, 133:223–234, 1973.

[65] B. A. Trahtenbrot. On autoreducibility. *Doklady Acad. Nauk* 192:1224-1227, 1970, in Russian, 11:814–817, 1972, translation.

[66] G. V. Wee. Improved sphere bounds on the covering radius of codes. *IEEE Transactions on Information Theory*, 34:237–245, 1988.

[67] C. E. M. Yates. Three theorems on the degree of recursively enumerable sets. Duke Mathematical Journal 32:461–468, 1965.

Relative Categoricity in Abelian Groups

Wilfrid Hodges

Queen Mary & Westfield College
London E1 4NS

Abstract

We consider structures A consisting of an abelian group with a subgroup A^P distinguished by a 1-ary relation symbol P, and complete theories T of such structures. Such a theory T is (κ, λ)-categorical if whenever A, B are models of T of cardinality λ with $A^P = B^P$ and $|A^P| = \kappa$, there is an isomorphism from A to B which is the identity on A^P. We state all true theorems of the form: If T is (κ, λ)-categorical then T is (κ', λ')-categorical.

1 Introduction

Relative categoricity has been for a long time one of the least popular branches of stability theory. In 1986 Saharon Shelah published a formidable paper [11] giving a deep classification under a weak form of the generalised continuum hypothesis, and concluding 'There are no particular problems (especially if you have read §4)'. I am not sure whether his paper contains any results that one could explain to a non-specialist. Since Shelah's paper there has been nothing like the Paris breakthrough that suddenly made classical stability theory all the rage in 1979. To the best of my knowledge nobody has made a serious attempt at a geometric theory outside the context of covers ([1], [2], [6]). Let me add a morsel of temptation: relative categoricity was probably the first area where first-order stability methods were applied to unstable theories.

In the early 1980s I hoped to find a way into relative categoricity by looking at a family of theories that we think we understand reasonably well, namely abelian groups with distinguished subgroups. There was never much chance that the results would be typical, since we have too many special benefits in this case; the underlying theory is Horn and all its completions are stable. In the event I stubbed my toe on one particular detail and turned aside to other work. Two years ago in St Petersburg Anatolii Yakovlev helped me over that detail, and now I think the results are in reasonably good shape. The model-theoretic methods are antediluvian, but as I said, we are still waiting for people to apply new tools to these questions. The proofs of the two-cardinal case (the one case that goes in principle beyond Shelah's paper [11]) are too long to give here, and for this case I do little more than state the results. For those who want it, a full account is online (at least until further publication) at `ftp.qmw.ac.uk/pub/preprints/hodges/relcatab.tex`.

My thanks to Ian Hodkinson, with whom I discussed this work as it developed in the 1980s.

2 Relative categoricity

Throughout this paper, T is a complete theory in a countable first-order language $L(P)$, one of whose symbols is a 1-ary relation symbol P; L is the language got by dropping P from $L(P)$; and for every model A of T, the set P_A of elements of A which satisfy the formula $P(x)$ is the domain of a substructure A^P of the reduct $A|L$. We call A^P the *P-part* of A. By a (κ, λ)-*structure* (or a (κ, λ)-*model* if we are talking about models of T) we mean an $L(P)$-structure A with $|A^P| = \kappa$ and $|A| = \lambda$.

We say that T is (κ, λ)-*categorical* if T has (κ, λ)-models, and whenever A, B are any two such models with $A^P = B^P$, there is an isomorphism from A to B over A^P (i.e. which is the identity on A^P).

We say that T is *relatively categorical* if whenever A, B are any two models of T with $A^P = B^P$, there is an isomorphism from A to B over A^P.

Lemma 2.1 *If T is (κ, λ)-categorical and A is a (κ, λ)-model of T, then every automorphism of A^P extends to an automorphism of A.*

Proof. Let α be an automorphism of A^P. Construct a structure B and an isomorphism $\gamma : A \to B$ so that $A^P = B^P$ and γ extends α^{-1}. By assumption there is an isomorphism $\beta : A \to B$ over A^P. Then $\gamma^{-1}\beta$ is an automorphism of A extending α. □

We say that T has the *reduction property* if for every formula $\phi(\bar{x})$ of $L(P)$ there is a formula $\phi^*(\bar{x})$ of L such that if A is any model of T and $\bar{a}$ a tuple of elements of A^P, then

$$A \models \phi(\bar{a}) \quad \Leftrightarrow \quad A^P \models \phi^*(\bar{a}).$$

The next result is in some sense a model-theoretic version of Lemma 2.1.

Lemma 2.2 *Suppose $\omega \leq \kappa \leq \lambda$ and T is (κ, λ)-categorical.*

(a) *(Pillay and Shelah [9]) If $\kappa = \lambda$ then T has the reduction property.*

(b) *If every model A of T has a direct sum decomposition $A_1 \oplus A_2$ where $A^P \subseteq A_1$ and A_2 is infinite, then T has the reduction property. (The notion of 'direct sum' should be one which allows a Feferman-Vaught theorem; see below.)* □

3 Group pairs

For the rest of this paper we specialise to the case where structures A, B, C, D etc. are abelian groups whose P-part is a subgroup. For brevity we call such structures *group pairs*. Thus L is the first-order language of abelian groups and T is a complete theory of group pairs.

A formula is said to be *positive primitive*, or more briefly *p.p.*, if it has the form $\exists \bar{x} \bigwedge_{i \in I} \phi_i$ where each ϕ_i is atomic. The same proof as for modules (e.g. [5] section A1) gives:

Theorem 3.1 *For every formula $\phi(\bar{x})$ of $L(P)$ there is a boolean combination $\phi'(\bar{x})$ of sentences and p.p. formulas of $L(P)$, which is equivalent to $\phi(\bar{x})$ in all group pairs. Every complete theory of group pairs is stable.* □

Corollary 3.2 *If T has a (κ, λ)-model with $\omega \leq \kappa < \lambda$, then T has a (κ', λ')-model whenever $\omega \leq \kappa' \leq \lambda'$.*

Proof. This follows from Shelah [10] Conclusion V.6.14(2) (noting the assumption on his p. 223 that T is stable). □

We use standard abelian group notation such as $A[p]$, following Fuchs [3]. An element a of A is *p^k-divisible* if $a = p^k b$ for some element b of A. We shall abbreviate 'of finite exponent' to 'bounded'. We say that a group pair A is *bounded over P* if the group A/A^P is bounded. Noting that if $A \equiv B$ and A is bounded over P then so is B, we can say also that the complete theory $\mathrm{Th}(A)$ is *bounded over P*; likewise with other notions that depend only on the elementary equivalence class of a structure. We say that two subgroups of a group are *disjoint* if they meet only in 0, and that T is *disjoint from P* if $A^P = \{0\}$ in every model A of T.

4 Feferman-Vaught arguments

In the class of group pairs, we can form direct sums $A = \bigoplus^P_{i \in I} A_i$. The definition is the same as for abelian groups, except that we also require that for any element $a = \sum_{i \in I} a_i$ with a_i in A_i,

$$a \in A^P \Leftrightarrow \text{ for all } i \in I, a_i \in A_i^P.$$

Right to left follows from the fact that P picks out a subgroup. Left to right doesn't; we look for a criterion which guarantees that it does hold.

Lemma 4.1 *Let A be a group pair, and let A_i $(i \in I)$ be subgroups of A such that $A = \bigoplus_{i \in I} A_i$ as abelian groups. Suppose also that there are a subset J of I and an element j_0 of J such that $A^P \subseteq \bigoplus_{i \in J} A_i$, and $A_j \subseteq A^P$ for all $j \in J \setminus \{j_0\}$. Then $A = \bigoplus^P_{i \in I} A_i$ as group pairs.*

Proof. Suppose $a = \sum_{i \in I} a_i$, $a \in A^P$. Then $a = (\sum_{j_0 \neq i \in J} a_i) + a_{j_0}$. By assumption each a_i $(j_0 \neq i \in J)$ is in A^P. So a_{j_0} is in A^P too. Also when $i \notin J$, $a_i = 0 \in A^P$. □

Henceforth all the direct sums that we consider will be abelian group direct sums which satisfy the criterion of this lemma, so that they are also group pair direct sums.

Note that a direct sum $A \oplus^P B$ is in fact a direct product $A \times B$, so that the Feferman-Vaught theorem applies (e.g. [5] section 9.6). The Feferman-Vaught theorem tells us among other things that if $B_1 \equiv C_1$ and $B_2 \equiv C_2$ then $B_1 \oplus B_2 \equiv C_1 \oplus C_2$; and similarly with $\preccurlyeq$ for $\equiv$.

By the Feferman-Vaught theorem, if A_1 and A_2 have complete theories T_1 and T_2 respectively, then the theory $T = \mathrm{Th}(A_1 \oplus A_2)$ depends only on T_1 and T_2; so we can write T as $T_1 \oplus T_2$. This notation will always imply that T_1 and T_2 are complete theories. Note that in general not every model of the theory $T = T_1 \oplus T_2$ will have the form $B_1 \oplus B_2$ with B_i a model of T_i $(i = 1, 2)$; when every model of T does have such a decomposition, we say that the decomposition $T_1 \oplus T_2$ is *precise*. Most of the conclusions of this paper depend on showing that a (κ, λ)-categorical theory of group pairs has a precise decomposition of a certain form.

Lemma 4.2 *Suppose T has a model A such that A/A^P is not bounded. Then for any model B of T and any cardinal κ, $B \preccurlyeq B \oplus \mathbb{Q}^{(\kappa)}$.*

Proof. If A/A^P is unbounded then the same holds for every model B of T; so we can assume B is A. By a compactness argument using the fact that A/A^P is unbounded, there is $A' \succcurlyeq A$ where A' has a direct summand $\mathbb{Q}$ which is disjoint from A'^P. Hence we can write $A' = A'_1 \oplus \mathbb{Q}$ as an abelian group direct sum with $A'^P \subseteq A'_1$, so that by Lemma 4.1 it is also a group pair direct sum. By the Szmielew invariants ([5] section A2) or more simply the upward Löwenheim-Skolem theorem, $\mathbb{Q} \preccurlyeq \mathbb{Q} \oplus \mathbb{Q}^{(\kappa)}$ and so by Feferman-Vaught,

$$A \preccurlyeq A' = A'_1 \oplus \mathbb{Q} \preccurlyeq A'_1 \oplus \mathbb{Q} \oplus \mathbb{Q}^{(\kappa)} = A' \oplus \mathbb{Q}^{(\kappa)}.$$

But also, by Feferman-Vaught again,

$$A \subseteq A \oplus \mathbb{Q}^{(\kappa)} \preccurlyeq A' \oplus \mathbb{Q}^{(\kappa)}.$$

So $A \preccurlyeq A \oplus \mathbb{Q}^{(\kappa)}$. □

We write $\mathbb{Q}_P$ for $\mathbb{Q}$ with every element satisfying P.

Lemma 4.3 *Suppose T has a model A such that A^P is not bounded. Then for every model B of T and every cardinal κ, $B \preccurlyeq B \oplus \mathbb{Q}_P^{(\kappa)}$.*

Proof. As in the previous lemma we can take B to be A. By assumption there is A' such that $A \preccurlyeq A'$ and $\mathbb{Q}_P$ is a subgroup of A'^P and hence of A'. So as an abelian group direct sum, $A' = A'_1 \oplus \mathbb{Q}_P$; but again the decomposition meets the criterion of Lemma 4.1, so that this is a group pair direct sum. The previous argument completes the proof. □

Theorem 4.4 *Assume $\omega \leq \kappa < \lambda$.*

(a) If T is (κ, κ)-categorical then T is bounded over P.

(b) If T is (κ, λ)-categorical and has unbounded P-part, then T is (κ', λ)-categorical for every κ' with $\omega \leq \kappa' \leq \kappa$.

Proof. (a) Take a (κ, κ)-model A of T, and suppose that A/A^P is unbounded. Let B be $A \oplus \mathbb{Q}^{(\omega)}$, which is also a (κ, κ)-model of T. Let D be a subgroup of B which contains $\mathbb{Q}^{(\omega)}$, is divisible torsion-free and disjoint from B^P, and is maximal with these properties. Then we can write B as $B_1 \oplus D$ with $B^P \subseteq B_1$. Now split into two cases according as B_1/B_1^P is bounded or not. If it is unbounded, then by Lemma 4.2, $B_1 \preccurlyeq B \preccurlyeq B_1 \oplus \mathbb{Q}^{(\kappa)}$, and the first and third of these group pairs are non-isomorphic (κ, κ)-models with the same P-part, contradicting (κ, κ)-categoricity. On the other hand if B_1/B_1^P is bounded, compare $B_2 = B_1 \oplus \mathbb{Q}$ with $B_3 = B_1 \oplus \mathbb{Q}^{(\kappa)}$; they are both (κ, κ)-models with the same P-part, but they are not isomorphic since the dimensions of the $\mathbb{Q}$-vector spaces $\mathbb{Q} \otimes (B_2/B_2^P)$ and $\mathbb{Q} \otimes (B_3/B_3^P)$ are different.

(b) (Yakovlev) By Corollary 3.2, T has (κ', λ)-models. Let A, B be (κ', λ)-models with $A^P = B^P$. By assumption A^P is unbounded, so $A \preccurlyeq A \oplus \mathbb{Q}_P^{(\kappa)}$, and this latter is a (κ, λ)-model A'. Likewise $B \preccurlyeq B \oplus \mathbb{Q}_P^{(\kappa)} = B'$. Also $A'^P = A^P \oplus \mathbb{Q}_P^{(\kappa)} = B'^P$. So by assumption there is an isomorphism $\alpha' : A' \to B'$ which is the identity on A'^P. Writing π_1 for projection onto the first factor, define $\alpha : A \to B$ by

$$\alpha(a) = \pi_1(\alpha'(a)).$$

Since α' takes the second factor of A'^P to the second factor of B'^P, α is an isomorphism from A to B. □

When $\kappa < \lambda$, we also need analogues of Lemma 4.2 which tell us when direct factors of the forms $\mathbb{Z}(p^\infty)$ or $\mathbb{J}_p$ can appear. The outcome is that (κ, λ)-categoricity always implies that A/A^P is the direct sum of a divisible group and a bounded group; and moreover if T is not bounded over P then A^P is also the direct sum of a divisible group and a bounded group. I omit details.

5 Algebraic facts

We shall need a method for extending an isomorphism $\gamma : A^P \to A'^P$ to an isomorphism $\beta : A \to A'$. The method that I used in the 1980s was model-theoretic. Then Roger Villemaire [13] helpfully pointed out that one of my assumptions said simply that certain Ulm-Kaplansky invariants are zero, so that one could use the back-and-forth method of Kaplansky and Mackey [7] (an improvement of Ulm's method), at least for countable groups. For our case we need two generalisations. The first is to groups of any cardinality; happily we have this through the work of Paul Hill reported in Fuchs [4] section 79ff. The second generalisation is that our base group is arbitrary; it need not even be torsion. So we need a version of Hill's argument which is 'fibred' over an arbitrary base group. (But in other respects our situation is much more restricted than Hill's.)

Let A be an abelian group and p a prime. For each ordinal k and ∞ we define $p^k A$ by induction on k:

$$p^0 A = A.$$

$$p^{k+1} A = p(p^k A).$$

$$p^\delta A = \bigcap_{k<\delta} p^k A \quad (\delta \text{ limit or } \infty).$$

The *p-height* of an element a of A, $\mathrm{ht}^p_A(a)$, is the least ordinal k such that $a \in p^k A \setminus p^{k+1} A$, or ∞ if there is no such ordinal k.

If B is a subgroup of A and p is a prime, we say that an element a of A is *p-proper over* B if the maximum value of $\mathrm{ht}^p_A(c)$ as c ranges over the coset $a + B$ is $\mathrm{ht}^p_A(a)$. Adapting a term of Hill, we say that B is *p-nice* in A if the heights $\mathrm{ht}^p_A(a)$ as a ranges over $A \setminus B$ form a finite set (which is necessarily a finite initial segment of the natural numbers, possibly together with ∞). We say that B is a *very nice* subgroup of A if B is a p-nice subgroup of A for every prime p.

Note that if B is a very nice subgroup of A then for every prime p and every element a of A, the coset $a + B$ contains an element which is p-proper over B.

Lemma 5.1 *If A/B is a bounded group, then B is a very nice subgroup of A.*

Proof. For each prime p the p-th component $(A/B)_p$ of A/B is bounded and hence a direct sum of cyclic p-groups of order $\leq p^m$ for some finite m. Now if $p^k c = a + b$ with $a \notin B$ and $b \in B$, then $p^k(c + B) = (a + B)$ is a non-zero element of $(A/B)_p$, and hence $k < m$. □

Let A be a group and B a subgroup of A. We say that A is *tight* over B if A/B is torsion and for each prime p the following hold: for each finite k,

$$(p^k A)[p] \subseteq p^{k+1}A + B,$$

and

$$(p^\infty A)[p] \subseteq B.$$

Note that if B is very nice in A, then tightness implies that no element of $A[p]$ outside B is p-proper over B. (Under the same assumption, it also implies that the Ulm-Kaplansky invariants of A over B are all zero.) Note also that if T is bounded over P, then either all or no models of T are tight over P.

Lemma 5.2 *Let A be a group with a subgroup B, and suppose that A/B is bounded. Then A is tight over B if and only if*

> *There is no non-trivial finite subgroup D of A such that $B \cap D = 0$ and $(D + B)/B$ is pure in A/B.*

Proof. Let p be a prime and k a natural number.

First suppose that A is tight over B. Let D be a non-trivial finite subgroup of A such that $B \cap D = \{0\}$. Decomposing D as a direct sum of cyclic groups of prime power order, let one of the non-trivial summands be $\langle a \rangle$, say of order p^{k+1}. Put $b = p^k a$. Then $p^k b$ lies in $(p^k A)[p]$. So by tightness there is $c \in A$ such that $p^{k+1}c - b \in B$. Hence $b + B$ is p^{k+1}-divisible in A/B, implying that $(D + B)/B$ is not pure in A/B.

Second, assume the condition and let a be an element of A such that $p^{k+1}a = 0$; so p^k is a typical element of $(p^k A)[p]$. If $p^k a \in B$ then $p^k a \in p^{k+1} + B$ trivially. Suppose on the other hand that $p^k a \notin B$. By the condition, $\langle a + B \rangle / B$ is not pure in A/B. Since $\langle a + B \rangle / B$ is a p-group, it is q-divisible for every prime $q \neq p$, and hence $p^k a + B$ must be p^{k+1}-divisible in A/B, so that some element of $p^k a + B$ is p^{k+1}-divisible in A. Thus A is tight over B, as required. □

Theorem 5.3 (Extension Theorem) *Let A, A' be abelian groups with very nice subgroups B, B' respectively. Suppose that both A/B and A'/B' are torsion groups, and that A (resp. A') is tight over B (resp. B). Let $\gamma : B \to B'$ be* *an isomorphism which preserves p-heights in the sense of A, A' for all primes p. Then γ extends to an isomorphism from A to A'.*

Proof sketch. The proof starts by decomposing A as a pushout of extensions A_p of B, where for each prime p, A_p/B is p-torsion; and likewise with A'. It suffices to lift γ to each A_p separately. From this point on, follow the Kaplansky-Mackey argument closely (Fuchs [4] p. 62f). Choose an element a of $A_p[p]$ which is p-proper over B, and use γ to match up a with an element a' of A' such that $\gamma(pa) = pa'$. The fact that both A and A' are tight over B, B' allows one to match up a with an element a' in A', extending γ to an isomorphism from $B_1 = B + \langle a \rangle$ to $B_1' = B' + \langle a' \rangle$ which preserves heights in A, A'. Iterate this procedure to form increasing chains of subgroups B_i and B_i', taking unions at limits. □

The Extension Theorem takes care of the part of a model A which is tight over A^P. For the rest we need the following lemma.

Lemma 5.4 *Let A be a group pair which is a model of T and which has the decomposition $A = C \oplus D$ where $A^P \subseteq C$. Suppose that A/A^P is bounded and C is tight over A^P. Then the Szmielew invariants of D, and hence the complete first-order theory of D, can be read off from T.*

Proof. Consider for example the Ulm invariants:

$$\frac{(p^k A)[p]}{(p^{k+1}A + p^k A^P)[p]} = \frac{(p^k C)[p]}{(p^{k+1}C + p^k A^P)[p]} \oplus \frac{(p^k D)[p]}{(p^{k+1}D)[p]}$$

which by the tightness of C over A^P

$$= \frac{(p^k D)[p]}{(p^{k+1}D)[p]}.$$

□

The results of this section deliver the following structure theorem. It serves us well for the case of (κ, κ)-categoricity; when $\kappa < \lambda$, other methods are needed as well.

Theorem 5.5 (Structure Theorem) *Suppose A is bounded over P. Then $A = A_1 \oplus A_2$, where A_1 is a tight extension of A^P. This decomposition is unique in the sense that if $A = A_1' \oplus A_2'$ where A_1' is a tight extension of A^P, then $A_2 \cong A_2'$ and the identity on A^P extends to an isomorphism from A_1 to A_1'.* □

6 (κ, κ)-categoricity

Lemma 6.1 *Let T be a complete theory of group pairs, and let (a)–(c) be the statements:*

(a) T is (κ, κ)-categorical for some infinite cardinal κ.

(b) T has the decomposition $T_1 \oplus T_2$ where T_1 is bounded over P and tight over P, T_1^P has infinite models and T_2 is disjoint from P and bounded.

(c) As (b), and the given decomposition is precise and unique with the stated properties.

Then (a) implies (b), and (b) is equivalent to (c).

Proof. (a) $\Rightarrow$ (b): Assuming (a), Theorem 4.4(a) tells us that T is bounded over P. So by the Structure Theorem (Theorem 5.5), $T = T_1 \oplus T_2$ where T_1 is bounded over P and tight over P, and T_2 is disjoint from P. If $A_1 \models T_1$, $A_2 \models T_2$ and $A = A_1 \oplus A_2$, then $A_2 \subseteq A/A^P$, so that A_2 is bounded and hence T_2 is bounded. Since T has (κ, κ)-models, T_1^P has infinite models.

(b) $\Leftrightarrow$ (c): Clearly (c) implies (b). Assuming (b), suppose $T = T_1 \oplus T_2 = U_1 \oplus U_2$ are two decompositions as in (b), and let $A = A_1 \oplus A_2$ and $B = B_1 \oplus B_2$ be models of T witnessing these decompositions. Adding new relation symbols to name the factors, we apply the Robinson joint consistency lemma (e.g. [5] Theorem 6.6.1) to find a model C of T with decompositions $C = C_1 \oplus C_2 = C_1' \oplus C_2'$ witnessing the two decompositions of T. By Theorem 5.5, C_1 and C_1' are isomorphic over C^P so that $T_1 = \mathrm{Th}(C_1) = \mathrm{Th}(C_1') = U_1$. Also $T_2 = U_2$ by Lemma 5.4. This proves the uniqueness of the decomposition. But every model of T is bounded over P, so by Theorem 5.5 again it has a decomposition as in (b); this proves preciseness. □

Our main results on (κ, κ)-categoricity are the following two theorems. The first is a kind of Morley theorem. (It is not true for relatively categorical theories in general; see Shelah and Hart [12].)

Theorem 6.2 *Let T be a complete theory of group pairs which has models A with A^P infinite. Then the following are equivalent.*

(a) T is relatively categorical.

(b) T is (κ, κ)-categorical for some uncountable κ.

(c) T has the reduction property, and $T = T_1 \oplus T_2$ where T_1 is bounded over P and tight over P, T_1^P has infinite models and T_2 has finite models.

Proof. (a) $\Rightarrow$ (b): Trivial.

(b) $\Rightarrow$ (c): Assume κ is uncountable and T is (κ, κ)-categorical. By Lemma 2.2(a), T has the reduction property. By Lemma 6.1, T has a precise decomposition $T_1 \oplus T_2$ where T_1 is bounded over P and tight over P, T_1^P has infinite models and T_2 is bounded. Let $A = A_1 \oplus A_2$ be a corresponding decomposition of a (κ, κ)-model of T. If A_2 is infinite then A_2 is elementarily equivalent to a structure B of cardinality ω and a structure C of cardinality κ. By Feferman-Vaught, $A_1 \oplus B$ and $A_1 \oplus C$ are also (κ, κ)-models of T. But these two structures are not isomorphic, since otherwise Theorem 5.5 would make B isomorphic to C. So T_2 has finite models.

(c) $\Rightarrow$ (a): Assume (c), and let A and B be models of T with $A^P = B^P$. By (b) $\Rightarrow$ (c) of Lemma 6.1, both A and B have decompositions $A = A_1 \oplus A_2$, $B = B_1 \oplus B_2$ witnessing the decomposition of T in (c). By the reduction property, the identity map α on A^P preserves finite p-heights from A to B for each prime p, and hence also from A_1 to B_1. So by Theorem 5.5, α extends to an isomorphism β from A_1 to B_1. Since A_2 and B_2 are elementarily equivalent and finite, they are also isomorphic; whence β extends to an isomorphism from A to B over A^P. □

Theorem 6.3 *Let T be a complete theory of group pairs which has models A with A^P infinite. Then the following are equivalent:*

(a) T is (ω, ω)-categorical.

(b) T has the reduction property and T/T^P is bounded.

(c) T has the reduction property, and $T = T_1 \oplus T_2$ where T_1 is bounded over P and tight over P, T_1^P has infinite models and T_2 is bounded.

Proof. (a) $\Rightarrow$ (b) $\Rightarrow$ (c) are already proved in Lemma 2.2(a) and the proof of Lemma 6.1. (c) $\Rightarrow$ (a): The argument of the previous theorem applies, noting that infinite bounded abelian groups are ω-categorical. □

7 (κ, λ)-categoricity with $\kappa < \lambda$

If $\kappa < \lambda$ then Shelah [11] is no longer available as a guide. But for abelian groups the facts are straightforward when κ is finite, and one can refer to Macintyre [8] for an algebraic description.

Theorem 7.1 *Let n be finite.*

(a) T is (n, ω)-categorical if and only if T has the reduction property and every model A of T is a bounded group in which the subgroup A^P has order n.

(b) *If λ is uncountable, then T is (n, λ)-categorical if and only if T has the reduction property and every model A of T is an uncountably categorical abelian group in which the subgroup A^P has order n.* $\square$

When the cardinals are infinite, the groups in question are harder to describe but the results seem as neat as we could hope for. The best heuristic is to look for the same kind of precise decomposition that applies in the case of (κ, κ)-categoricity, extending Lemma 5.4 and Theorem 5.5 as necessary. The fact that in general T is not bounded over P means that properties of cotorsion groups play a large role.

Theorem 7.2 *Suppose $\omega \leq \kappa < \lambda$, and T is (κ, λ)-categorical. Then T is (κ', λ')-categorical whenever $\omega \leq \kappa' < \lambda'$.*

Proof sketch. By the methods of section 4 we show that every model A of T has the form $A = C \oplus D$ where $A^P \subseteq C$, C is torsion and tight over A^P, A^P is very nice in C, $|C| = |A^P|$ and D is the direct sum of a divisible group and a bounded group; exactly one homogeneous direct summand of D has infinite rank. The tightness of C over A^P allows us to read off the Szmielew invariants of the torsion part of D from the first-order theory of A, just as in Lemma 5.4.

Now suppose κ', λ' are as in the theorem, and A, A' are (κ', λ')-models of T with $A^P \cong A'^P$. We have decompositions $A = C \oplus D$, $A' = C' \oplus D'$ as above. There are two cases. The first is that T is unbounded over P. In this case both D and D' must be direct sums of $\mathbb{Q}^{(\lambda)}$ and torsion groups in which each homogeneous direct summand has finite rank. The second case is that T is bounded over P. Then D is a torsion group, and exactly one of its homogeneous direct summands has rank λ while the rest have finite rank. In either case D and D' have isomorphic torsion parts and isomorphic torsion-free parts, and hence $D \cong D'$.

It remains to show that C is isomorphic to C' over their P-parts. By Lemma 2.2(b), T has the reduction property, and it follows that the isomorphism from A^P to A'^P is p-height-preserving in A, A' (and hence also in C, C') for all primes p. The Extension Theorem, Theorem 5.3, does the rest. $\square$

References

[1] G. Ahlbrandt and M. Ziegler, What's so special about $(\mathbb{Z}/4\mathbb{Z})^{(\omega)}$?, *Archive for Math. Logic* 31 (1991) 115-132.

[2] D. M. Evans, A. Pillay and B. Poizat, Le groupe dans le groupe, *Algebra i Logika* 29 (1990) 368–378.

[3] L. Fuchs, *Infinite Abelian Groups I*, Academic Press, New York 1970.

[4] L. Fuchs, *Infinite Abelian Groups II*, Academic Press, New York 1973.

[5] W. Hodges, *Model Theory*, Cambridge Univ. Press, Cambridge 1993.

[6] W. Hodges and A. Pillay, Cohomology of structures and some problems of Ahlbrandt and Ziegler, *J. London Math. Soc.* 50 (1994) 1–16.

[7] I. Kaplansky and G. W. Mackey, A generalization of Ulm's theorem, *Summa Brasil. Math.* 2 (1951) 195–202.

[8] A. Macintyre, On ω_1-categorical theories of abelian groups, *Fundamenta Math.* 70 (1971) 253–270.

[9] A. Pillay and S. Shelah, Classification over a predicate I, *Notre Dame J. Formal Logic* 26 (1985) 361–376.

[10] S. Shelah, *Classification Theory and the Number of Non-isomorphic Models*, North-Holland, Amsterdam 1978.

[11] S. Shelah, Classification over a predicate II, in *Around Classification Theory of Models*, Lecture Notes in Mathematics 1182, Springer, Berlin 1986, pp. 47–90.

[12] S. Shelah and B. Hart, Categoricity over P for first order T or categoricity for $\phi \in L_{\omega_1\omega}$ can stop at $\aleph_k$ while holding for $\aleph_0, \ldots, \aleph_{k-1}$. *Israel J. Math.* 70 (1990) 219–235.

[13] R. Villemaire, Abelian groups $\aleph_0$-categorical over a subgroup, *J. Pure Appl. Algebra* 69 (1990) 193–204.

Computability and Complexity Revisited

Neil D. Jones*
DIKU, University of Copenhagen, Denmark
E-mail, web: neil@diku.dk, http://www.diku.dk/people/NDJ.html

Abstract

A programming approach to computability and complexity theory yields proofs of central results that are sometimes more natural than the classical ones; and some new results as well. This paper contains some high points from the recent book [9], emphasising what is different or novel with respect to more traditional treatments. Topics include:

- Kleene's *s-m-n* theorem applied to compiling and compiler generation.
- Proof that *constant time factors do matter*: for on a natural computation model, problems solvable in linear time have a proper hierarchy, ordered by coefficient values.
- Characterisations in programming terms of the classes LOGSPACE and PTIME. These are intrinsic: without externally imposed space or time computation bounds.
- Kleene's Second Recursion theorem: an efficient implementation.
- Results on which problems possess optimal algorithms, including: Levin's Search theorem (first time in book form) and Blum's Speedup.
- Boolean program problems complete for PTIME, NPTIME, PSPACE.

1 Introduction

The recent book [9] differs didactically and foundationally from traditional treatments of computability and complexity theory. Its didactic aim is to teach the main theorems of these fields to computer scientists. Its approach is to exploit as much as possible the readers' programming intuitions, and to motivate subjects by their relevance to programming concepts. Foundationally it differs by using, instead of the natural numbers $\mathbb{N}$, binary trees as data (as in the programming language Lisp). Consequently programs *are*

*This research was partially supported by the Danish Natural Science Research Council (*DART* project), and the Esprit *Atlantique* project.

data, avoiding the usual complexities and inefficiencies of encoding program syntax as Gödel numbers.

2 The WHILE and I languages

We use a simple programming language WHILE, which is in essence a small subset of LISP or Pascal. Why just this language? Because WHILE seems to have just the right mix of expressive power and simplicity for our purposes.

Expressive power is important when writing programs that deal with programs as data. The WHILE language's data structures (binary trees of atomic symbols) are particularly well suited to this, giving simple and clear constructions. On the other hand, classical approaches based on the natural numbers necessitate schemes for assigning *Gödel numbers* to encode program texts and fragments, and numerical functions to build and decompose Gödel numbers.[1]

Simplicity of the computational formalism is also essential to prove theorems about programs and their behaviour. This rules out the use of larger, more powerful languages, since proofs about programs would simply be too complex to be easily understood. We go in the opposite direction: when proving theorems about programs, we use an equivalent but still simpler language I, that is identical to WHILE but limited to programs with one variable and one atom.

2.1 Syntax of WHILE data and programs

Values in the set $\mathbb{D}_A$ of data values are built up from a fixed finite set A of so-called *atoms* by finitely many applications of the pairing operation. It will not matter too much exactly what A contains, except that we will identify one of its elements, `nil` for several purposes.

A value $\mathtt{d} \in \mathbb{D}_A$ is a binary tree with atoms as leaf labels. A tree can either be a single atom, e.g., `marilyn`, or a pair $(\mathtt{d}_1 . \mathtt{d}_2)$ which has $\mathtt{d}_1$ as left subtree (also called the "head"), and $\mathtt{d}_2$ as right subtree (also called the "tail"). An example, written in "fully parenthesised form": `((a.((b.nil).c)).nil)`.

Definition 2.1 Let $A = \{\mathtt{a}_1, \ldots, \mathtt{a}_n\}$ be some finite set. Then

1. $\mathbb{D}_A$ is the smallest set satisfying $\mathbb{D}_A = (\mathbb{D}_A \times \mathbb{D}_A) \cup A$. The *pairing* operation "." yields value $(\mathtt{d}_1 . \mathtt{d}_2)$ when applied to values $\mathtt{d}_1$, $\mathtt{d}_2$.

2. The *size* $|d|$ of a value $\mathtt{d} \in \mathbb{D}_A$ is defined as follows: $|\mathtt{d}| = 1$ if $\mathtt{a} \in A$, and $1 + |\mathtt{d}_1| + |\mathtt{d}_2|$ if $\mathtt{d} = (\mathtt{d}_1 . \mathtt{d}_2)$.

[1] While this can be done (as Gödel himself showed using the Chinese Remainder Theorem), the technical constructions involved are messy, and the program simulations achieved are not tight enough to establish time hierarchy results.

A compact linear notation for values: Unfortunately it is hard to read deeply parenthesised structures (one has to resort to counting), so we will use a more compact "list notation" taken from the Lisp and Scheme languages, in which

()	stands for	nil
(d_1 d_2 $\cdots$ d_n)	stands for	(d_1.(d_2.$\cdots$(d_n.nil)$\cdots$))

The syntax of WHILE programs is given by the "informal syntax" part of figure 2.1. (Ignore the "concrete syntax" for now.) Programs have only one input/output variable X1, and to manipulate tree structures built by cons from atoms. Operations hd and tl (head, tail) decompose such structures, and atom=? tests atoms for equality. In tests, nil serves as "false," and anything else serves as "true." We often write false for nil, and true for (nil.nil).

Syntactic category:		Informal syntax:	Concrete syntax:
P : Program	::=	read X1; C; write X1	C
C : Command	::=	Xi := E	(:= nil^i E)
	|	C1; C2	(; C1 C2)
	|	if E then C1 else C2	(if E C1 C2)
	|	while E do C	(while E C)
E : Expression	::=	Xi	(var nil^i)
	|	D	(quote D)
	|	cons E1 E2	(cons E1 E2)
	|	hd E	(hd E)
	|	tl E	(tl E)
	|	atom=? E1 E2	(atom=? E1 E2)
D : Data-value	::=	A | (D.D)	A | (D.D)
A : Atom	::=	nil | ...	nil | ...

Figure 2.1: Program syntax: informal and concrete

In general A will be fixed, so to reduce notation we write $\mathbb{D}$ instead of $\mathbb{D}_A$. An example, the following program, **reverse**:

```
read X1;
  X2 := nil;
  while X1 do { X2 := cons (hd X1) X2;  X1 := tl X1 };
  X1 := X2;
write X1
```

satisfies [[reverse]](a.(b.(c.nil))) = (c.(b.(a.nil))) or, in the compact notation, [[reverse]](a b c) = (c b a). In examples we will use a more relaxed syntax, e.g., the input and output variables need not both be X1.

Definition 2.2 The language I is identical to WHILE, with two exceptions: $A = \{\texttt{nil}\}$, so data values have only one atom; and programs have only one variable X.

2.2 Semantics of WHILE and I programs

Informally, the net effect of *running a program* p is to compute a partial function $[\![\texttt{p}]\!]^{\texttt{WHILE}}$ from $\mathbb{D}$ to $\mathbb{D}$, so the result (if defined) of running program p on input d is expressed $[\![\texttt{p}]\!]^{\texttt{WHILE}}(\texttt{d})$. If the language is understood from context, the superscript may be dropped.

A slight abuse of notation to have a way to express nontermination: we write $[\![\texttt{p}]\!](\texttt{d}) = \bot$ to mean that $[\![\texttt{p}]\!](\texttt{d})$ is undefined, and $[\![\texttt{p}]\!]^{\texttt{WHILE}} : \mathbb{D} \to \mathbb{D}_\bot$ in case $[\![\texttt{p}]\!]^{\texttt{WHILE}}$ is (as usual) a partial function.

Control structures are *sequential composition* C1;C2, the *conditional* if E then C1 else C2, and the *while loop* while E do C. In tests, any non-nil value serves as "true." A formal definition of the semantics may be found in [9]. For brevity, running time is only informally defined. This definition is natural, provided the data-sharing implementation techniques used in Lisp and other functional languages is used.

Definition 2.3 The running time $time_{\texttt{p}}(\texttt{d}) \in \{0, 1, 2, \ldots\} \cup \{\bot\}$ is obtained by counting 1 every time any of the following is performed while computing $[\![\texttt{p}]\!](\texttt{d})$ as defined in the semantics: a variable or constant reference; an operation hd, tl, cons, or := is applied; or a test in an if or while command. Its value is $\bot$ if the computation does not terminate.

2.3 Two simple constructions

Two examples illustrate the simplicity of program manipulation. Both use the concrete syntax of Figure 2.1, which we now explain.

When a program is used as data in $\mathbb{D}$, it is represented as shown in the "concrete syntax" column of Figure 2.1. Notation: $\texttt{nil}^i$ stands for (nil ... nil), the list of i nil's. As an example, the reverse program in concrete syntax is the following element of $\mathbb{D}$:

```
(; (:= nil¹ (quote nil))
   (while (var nil¹)
          (; (:= (var nil²) (cons (hd (var nil¹))
                                  (var nil²)     ) )
             (:= nil¹ (tl (var nil¹)))
 ) )      )
```

2.3.1 The halting problem

Theorem 2.4 There is no WHILE-program q deciding the halting problem, meaning that for any WHILE-program p and input d

$$[\![\mathtt{q}]\!](\mathtt{p.d}) = \begin{cases} \mathtt{true} & \text{if } [\![\mathtt{p}]\!](\mathtt{d}) \neq \bot \\ \mathtt{false} & \text{if } [\![\mathtt{p}]\!](\mathtt{d}) = \bot \end{cases}$$

Proof Suppose such a q exists. It must have form: `read X; C; write X`. Construct the following program r from q:

```
read X;
X := cons X X;          (* Does program X stop on input X? *)
C;                      (* Apply program q to answer this *)
if X then  while X do X := X (* Loop if X stops on input X *)
     else  X := nil;         (* Terminate if it does not *)
write X
```

Consider the input X = r. First, suppose that $[\![\mathtt{r}]\!](\mathtt{r}) \neq \bot$. Then control in r's computation on input r must reach the **else** branch above (else r would loop on r). But then X = **false** holds after command C, so $[\![\mathtt{r}]\!](\mathtt{r}) = \bot$ by the assumption that q decides the halting problem. This is contradictory. Conclusion: $[\![\mathtt{r}]\!](\mathtt{r}) = \bot$. By similar reasoning, $[\![\mathtt{r}]\!](\mathtt{r}) = \bot$ is also impossible.

The only unjustified assumption was existence of q, so this must be false.

2.3.2 The *s-m-n* theorem

A *program specialiser* spec is given a subject program p together with part of its input data, s. Its effect is to construct a new program $\mathtt{p_s} = [\![\mathtt{spec}]\!](\mathtt{p.s})$ which, when given p's remaining input d, will yield the same result that p would have produced given both inputs.

Definition 2.5 Program spec is a *specialiser* for WHILE-programs if $[\![\mathtt{spec}]\!](\_)$ is total, and if for any p ∈ WHILE-*programs* and s, d ∈ WHILE-*data*

$$[\![\mathtt{p}]\!](\mathtt{s.d}) = [\![[\![\mathtt{spec}]\!](\mathtt{p.s})]\!](\mathtt{d})$$

Theorem 2.6 (Kleene's *s-m-n* theorem)[2]. There is a program specialiser for WHILE-programs.

[2]Traditional version: $\phi_p^{m+n}(x_1,\ldots,x_m,y_1,\ldots,y_n) = \phi^n_{s^m_n(p,x_1,\ldots,x_m)}(y_1,\ldots,y_n)$. This version is much simpler, partly because of the linear notation $[\![\_]\!]$ instead of $\phi_i^n(\_)$, and partly because of the fact that $I\!D$ has a built-in pairing operation. Thus $m = n = 1$ is fully general, so s_n^m for all m, n are not necessary, being replaced by the single one program spec.

Proof Program p has form: `read X; Body; write X`. Given known input s, consider program p_s = `read X; X := cons s X; Body; write X`. This is obviously (albeit somewhat trivially) correct since p_s, when given input d, will first assign the value (s.d) to X, and then apply p to the result. It suffices to see how to construct $p_s = [\![\texttt{spec}]\!](\texttt{p.s})$. A program to transform input s and the concrete syntax of program p into p_s is easily constructed.

3 Compilation and Interpretation

These concepts provide a natural bridge between the worlds of recursion theory and computer science. Briefly: most of the traditional computability theory constructions are compilations; and the well-known *universal function* is just the function computed by a "self-interpreter." The concepts take on more life for computer scientists if expressed in a context of several programming languages, so we first generalise a bit:

3.1 Programming languages more generally

Definition 3.1 A *programming language* L consists of

1. Two sets, L-*programs* and L-*data*, and two distinct elements **true**, **false** ∈ L-*data*.
2. L's *semantic function*: for every p ∈L-*programs*, a corresponding (partial) input-output function $[\![\texttt{p}]\!]^{\texttt{L}}(\_) : \texttt{L-}\mathit{data} \to \texttt{L-}\mathit{data}_\bot$.
3. L's *running time function*: for every p ∈L-*programs*, a corresponding (partial) time-usage function $\mathit{time}^{\texttt{L}}_{\texttt{p}}(\_) : \texttt{L-}\mathit{data} \to I\!N \cup \{\bot\}$ such that for any p ∈ L-*programs* and d ∈ L-*data*, $[\![\texttt{p}]\!]^{\texttt{L}}(\texttt{d}) = \bot$ if and only if $\mathit{time}^{\texttt{L}}_{\texttt{p}}(\texttt{d}) = \bot$

Definition 3.2 L-program p *decides* a subset $A \subseteq$ L-*data* if for any d ∈ L-*data*

$$[\![\texttt{p}]\!](\texttt{d}) = \begin{cases} \texttt{true} & \text{if } \texttt{d} \in A \\ \texttt{false} & \text{if } \texttt{d} \in \texttt{L-}\mathit{data} \setminus A \end{cases}$$

3.2 Compilation

Suppose we are given three programming languages: a *source language* S, a *target language* T, and an *implementation language* L, and that S-*data* = T-*data*, S-*programs* ⊆ L-*data*, T-*programs* ⊆ L-*data*.

Definition 3.3 An L-program comp is a *compiler* from S to T if $[\![\texttt{comp}]\!](\_)$ is total, and for every p ∈ S-*programs* and d ∈ $I\!D$,

$$[\![p]\!]^S(d) = [\![[\![comp]\!]^L(p)]\!]^T(d)$$

3.3 Interpretation

Self-interpreters, under the name *universal programs*, play (and have played since the 1930's) a central role in theorems of both complexity and computability theory. Generalising to several languages, an interpreter `int` ∈ L-programs for a language S has a pair of inputs: a *source program* p ∈ S-programs, and the source program's input data d ∈ $I\!D$.

Definition 3.4 Program `int` is an *interpreter* for S *written in* L if for every p ∈ S-programs and every d ∈ $I\!D$,

$$[\![p]\!]^S(d) = [\![int]\!]^L(p.d)$$

3.3.1 An interpreter `i` for 1-variable WHILE programs

We now give an example interpreter written in language WHILE. This is nearly a self-interpreter, except that for the sake of simplicity we restrict it to programs containing only one variable X. The interpreter, called `i`, is in Figure 3.1, where the STEP macro is in Figure 3.2. The interpreter will use atoms `while`, `:=`, etc. as in the concrete syntax of Figure 2.1, together with some additional ones for programming convenience, e.g. "`do_hd`" and "`do_while`". Every value manipulated by `i` will thus lie in $I\!D_{A_1}$, where

$$\begin{aligned} A_1 \;=\; & \{\texttt{nil}, \texttt{var}, \texttt{quote}, \texttt{cons}, \texttt{hd}, \texttt{tl}, \texttt{atom=?}, :=, ;, \texttt{while}, \texttt{var}, \\ & \texttt{do\_cons}, \texttt{do\_hd}, \texttt{do\_tl}, \texttt{do\_asgn}, \texttt{do\_while}\} \end{aligned}$$

Theorem 3.5 `i` as defined in Figure 3.1 is an interpreter written in WHILE for the language $\texttt{WHILE}^{1var}$, which is identical to WHILE except that programs are restricted to one variable X.

To aid compactness and readability, the STEP part of `i` is given by a set of transitions in Figure 3.2, i.e., rewrite rules (Cd, s, v) ⇒ (Cd′, s′, v′) describing transitions from one state of form

(Cd, s, v) = (*Code-stack, Computation-stack, Value*)

to the next state. Expression evaluation is based on the following *net effect property*. Suppose the value of expression E is d, provided that the current value of variable X is v. Then (E.Cd, s, v) $\Rightarrow^+$ (Cd, d.s, v) where $\Rightarrow^+$ means "can be rewritten in one or more steps to". A similar net effect property characterises execution of commands.

```
read PD;                        (* Input is (program.data) *)
  Cd := cons (hd PD) nil;       (* Control stack = (program.nil) *)
  Val := tl PD;                 (* The value of X = data *)
  Stk := nil;                   (* Computation stack is initially empty *)
  while Cd do STEP;             (* Repeat while control stack nonempty *)
write Val
```

Figure 3.1: Interpreter i *for* $\texttt{WHILE}^{1var}$

Blank entries in Figure 3.2 correspond to values that are neither referenced nor changed in a rule. The rules can easily be programmed in the WHILE language. For example, the three while transitions could be programmed as in Figure 3.3.

3.3.2 Self-interpretation of WHILE and I

We will call an interpreter for a language L which is written in the same language L a *self-interpreter* or *universal program*, and we will often use name u for a universal program. By Definition 3.4, u must satisfy $[\![\mathtt{p}]\!]^{\mathtt{L}}(\mathtt{d}) = [\![\mathtt{u}]\!]^{\mathtt{L}}(\mathtt{p.d})$ for every L-program p and data value d.

The interpreter i just given is *not* a self-interpreter due to the restriction to just one variable in the interpreted programs. Extension to a full self-interpreter is a straightforward programming exercise.

Theorem 3.6 There exists a self-interpreter for the WHILE language.

There also exists a self-interpreter u for the minimal language I. This fact will have interesting complexity consequences.

Theorem 3.7 There exists a self-interpreter u for the I language.

To prove this we first need a concrete syntax for I-programs; that for WHILE will not do, since it uses more than the one atom nil. Specifically, programs must be represented as elements of $\mathbb{D}_{\{\mathtt{nil}\}} = \{\mathtt{nil}\} \cup \mathbb{D}_{\{\mathtt{nil}\}} \times \mathbb{D}_{\{\mathtt{nil}\}}$.

Proof of Theorem 3.7 (sketch). Choose a one-to-one encoding $c : A_1 \to \mathbb{D}_{\mathtt{nil}}$, and extend it to $c : \mathbb{D}_{A_1} \to \mathbb{D}_{\mathtt{nil}}$ in the natural way. Construct self-interpreter u0 (a WHILE-program) from the interpreter i of Theorem 3.5 by modifying its code so all references to data from input I-program $\mathtt{p} \in \mathbb{D}_{A_1}$ instead deal with the encoded program data from $c(\mathtt{p}) \in \mathbb{D}_{\{\mathtt{nil}\}}$. The effect is that for any I-program p and input $\mathtt{d} \in \mathbb{D}_{\{\mathtt{nil}\}}$

$$[\![\mathtt{u0}]\!]^{\mathtt{WHILE}}(c(\mathtt{p}).\mathtt{d}) = [\![\mathtt{i}]\!]^{\mathtt{WHILE}}(\mathtt{p.d}) = [\![\mathtt{p}]\!]^{\mathtt{I}}(\mathtt{d})$$

Program u0 uses only one atom, but has several variables. Finally, obtain I-program u from u0 by packing its several variables into one using "cons."

Cd	Stk	Val	⇒	Cd	Stk	Val
nil			⇒	nil		
(quote D).Cd	S		⇒	Cd	D.S	
(var nil).Cd	S	v	⇒	Cd	v.S	v
(hd E).Cd			⇒	E.do_hd.Cd		
do_hd.Cd	(T.U).S		⇒	Cd	T.S	
do_hd.Cd	nil.S		⇒	Cd	nil.S	
(tl E).Cd			⇒	E.do_tl.Cd		
do_tl.Cd	(T.U).S		⇒	Cd	U.S	
do_tl.Cd	nil.S		⇒	Cd	nil.S	
(cons E1 E2).Cd			⇒	E1.E2.do_cons.Cd		
do_cons.Cd	U.T.S		⇒	Cd	(T.U).S	
(; C1 C2).Cd			⇒	C1.C2.Cd		
(:= X E).Cd			⇒	E.do_asgn.Cd		
do_asgn.Cd	w.S	v	⇒	Cd	S	w
(if E C1 C2).Cd			⇒	E.do_if.C1.C2.Cd		
do_if.C1.C2.Cd	(T.U).S		⇒	C1.Cd	S	
do_if.C1.C2.Cd	nil.S		⇒	C2.Cd	S	
(while E C).Cd			⇒	E.do_while. (while E C).Cd		
do_while. (while E C).Cd	(T.U).S		⇒	C.(while E C).Cd	S	
do_while.C1.Cd	nil.S		⇒	Cd	S	

Figure 3.2: The STEP *macro, expressed by rewriting rules*

4 More than you ever expected from the *s-m-n* theorem

4.1 Partial evaluation: program specialisation in practice

The program specialiser of Theorem 2.6 was very simple, and the programs it outputs are slightly slower than the ones from which they were derived. (This was also true of Kleene's original construction.) On the other hand, program specialisation can be done so as to yield *efficient* specialised programs. This is known in the programming language community as *partial evaluation*; see [10] for a thorough treatment and a large bibliography.

Applications of program specialisation include *compiling* (done by specialising an interpreter to its source program), and *generating compilers from*

```
if hd hd Cd = while then (* Set up iteration *)
   Cd := (cons (hd tl hd Cd) (cons do_while Cd)) else
if hd Cd = do_while then
   if hd Stk then          (* Do body if test is true *)
   { Cd := (hd tl tl tl Cd) . (tl Cd);
     Stk := tl Stk } else (* Else exit while *)
   { Stk := tl Stk; Cd := tl tl Cd } else ...
```

Figure 3.3: WHILE *code for rewrite rules to implement "*while*"*

interpreters, by using the specialiser to specialise itself to a given interpreter. Surprisingly, this can give quite efficient programs as output.

A simple but nontrivial example of partial evaluation Consider Ackermann's function, with program:

```
a(m,n) = if m =? 0 then n+1 else
         if n =? 0 then a(m-1,1)
         else a(m-1,a(m,n-1))
```

Computing a(2,n) involves recursive evaluations of a(m,n) for m = 0, 1 and 2, and various values of n. A partial evaluator can evaluate expressions m=?0 and m-1, and function calls of form a(m-1,...) can be unfolded. We can now specialise function a to the values of m, yielding a less general program that is about twice as fast:

```
a2(n) = if n =? 0 then 3 else a1(a2(n-1))
a1(n) = if n =? 0 then 2 else a1(n-1)+1
```

4.2 Compiling and compiler generation by the Futamura projections

This section shows an application of the *s-m-n* theorem in computer science to compiling and, more generally, to generating program generators. For simplicity we elide the name of the language L that is the implementation, input, and output language of the specialiser.

The starting point is an interpreter program int for some programming language S. By definition this satisfies $[\![\texttt{source}]\!]^{S}(\texttt{input}) = [\![\texttt{int}]\!](\texttt{source.input})$ for any S-program source and data input $\in \mathbb{D}$.

First Futamura projection: $\texttt{target} = [\![\texttt{spec}]\!](\texttt{int.source})$. This shows that *a specialiser can compile.* Correctness is to show that target is a program in the specialiser's output language which is equivalent to S-program source, i.e., $[\![\texttt{source}]\!]^{\texttt{S}} = [\![\texttt{target}]\!]$:

$$
\begin{array}{rcll}
[\![\texttt{source}]\!]^{\texttt{S}}(\texttt{input}) & = & [\![\texttt{int}]\!](\texttt{source.input}) & \text{Def'n of interpreter} \\
& = & [\![[\![\texttt{spec}]\!](\texttt{int.source})]\!](\texttt{input}) & \text{Def'n of specialiser} \\
& = & [\![\texttt{target}]\!](\texttt{input}) & \text{Def'n of } \texttt{target}
\end{array}
$$

Second Futamura projection: $\texttt{comp} = [\![\texttt{spec}]\!](\texttt{spec.int})$. This shows that *a specialiser can generate a compiler.* Correctness is to show that comp is a compiler from S to the specialiser's output language. It constructs the just-mentioned target program from the source program:

$$
\begin{array}{rcll}
\texttt{target} & = & [\![\texttt{spec}]\!](\texttt{int.source}) & \text{Definition of target} \\
& = & [\![[\![\texttt{spec}]\!](\texttt{spec.int})]\!](\texttt{source}) & \text{Definition of specialiser} \\
& = & [\![\texttt{comp}]\!](\texttt{source}) & \text{Definition of } \texttt{comp}
\end{array}
$$

Third Futamura projection: $\texttt{cogen} = [\![\texttt{spec}]\!](\texttt{spec.spec})$. This shows that *a specialiser can generate a compiler generator*[3]. Correctness is to show that cogen constructs the just-mentioned compiler from the interpreter int:

$$
\begin{array}{rcll}
\texttt{comp} & = & [\![\texttt{spec}]\!](\texttt{spec.int}) & \text{Definition of comp} \\
& = & [\![[\![\texttt{spec}]\!](\texttt{spec.spec})]\!](\texttt{int}) & \text{Definition of specialiser} \\
& = & [\![\texttt{cogen}]\!](\texttt{int}) & \text{Definition of } \texttt{cogen}
\end{array}
$$

Perhaps the most surprising thing about this equational reasoning is that it also works well in practice: A variety of partial evaluators (= program specialisers = *s-m-n* functions) have been constructed, and they give good results in practical applications [10].

Speedups from self-application.

Each of program execution, compilation, compiler generation, and compiler generator generation can be done in two different ways:

$$
\begin{array}{lclcl}
\texttt{output} & = & [\![\texttt{int}]\!](\texttt{source.input}) & = & [\![\texttt{target}]\!]^{\texttt{S}}(\texttt{input}) \\
\texttt{target} & = & [\![\texttt{spec}]\!](\texttt{int.source}) & = & [\![\texttt{comp}]\!](\texttt{source}) \\
\texttt{comp} & = & [\![\texttt{spec}]\!](\texttt{spec.int}) & = & [\![\texttt{cogen}]\!](\texttt{int}) \\
\texttt{cogen} & = & [\![\texttt{spec}]\!](\texttt{spec.spec}) & = & [\![\texttt{cogen}]\!](\texttt{spec})
\end{array}
$$

The exact timings vary according to the design of spec and int, and with the implementation language L. Nonetheless, we have observed in practical computer experiments that *in each case the rightmost run is often about 10*

[3] It transforms interpreters into compilers.

times faster than the leftmost. Moral: self-application can generate programs that run faster!

5 Constant time factors *do* matter

The Turing machine constant speedup theorem is traditionally one of the first learned in complexity courses – and one that gives practically oriented students a bad impression of theory: While the proof is not too complex, what it says is extremely counterintuitive, going against daily programming experience.

In effect it says that *from any Turing machine program, one may construct an equivalent one that runs twice as fast* (asymptotically, and if its running time is superlinear). Unfortunately, the construction is useless for speeding up real programs, as it in essence involves changing to a double-density tape.

The theorem's truth is an immediate consequence of the use of an "unfair time measure": one Turing machine state transition is counted as taking one time unit, *regardless of the size of the Turing machine's tape alphabet.*

We show a more satisfying result (at least from a programmer's perspective): a proper hierarchy exists among problems that can be solved by I programs in linear time. In particular there exist constants $0 < a < b$ and set A such that the question $\mathtt{d} \in A$? can be decided in time $a \cdot |\mathtt{d}|$ but not in time $b \cdot |\mathtt{d}|$, *regardless of how clever one is* at programming and/or algorithm design. Such an absolute result is rare in computer science, and attracts the students' attention.

This result is false for Turing machines; its status for the full `WHILE` language is an open question.

5.1 Time-bounded complexity classes

Definition 5.1 Given programming language L and $f : I\!N \to I\!N$, we define

$$
\begin{array}{lcl}
\text{TIME}^{\mathtt{L}}(f) & = & \{A \subseteq I\!D \mid \ A \text{ is decided by some L-program p, such that} \\
 & & \qquad\qquad time^{\mathtt{L}}_{\mathtt{p}}(\mathtt{d}) \leq f(|\mathtt{d}|) \ \text{ for all } \mathtt{d} \in \mathtt{L}\text{-}data\} \\
\text{LINTIME}^{\mathtt{L}} & = & \bigcup_k \text{TIME}^{\mathtt{L}}(\lambda n \,.\, kn) \\
\text{PTIME}^{\mathtt{L}} & = & \bigcup\{\text{TIME}^{\mathtt{L}}(f) \mid f : I\!N \to I\!N \text{ is a polynomial}\}
\end{array}
$$

Theorem 5.2 $\text{PTIME}^{\mathtt{WHILE}} = \text{PTIME}^{\mathtt{I}}$, and $\text{LINTIME}^{\mathtt{WHILE}} = \text{LINTIME}^{\mathtt{I}}$ (when restricted to input values containing only the atom `nil`).

This equivalence is easy to show by data encodings and packing several variables into one using `cons`.

Theorem 5.3 Constant factor hierarchy: There is a b such that for all $a \geq 1$ there is a set $A \subseteq \mathbb{D}$ which is in $\text{TIME}^{\text{L}}(a \cdot b \cdot n)$, but not in $\text{TIME}^{\text{L}}(a \cdot n)$.

The key to this result is the existence of an "efficient" interpreter for I, as seen in Definition 5.4. The proof diagonalises (as for undecidability of the halting problem), using a time-bounded extension of the self-interpreter of Theorem 3.7 to stay within the required time bounds. The current framework makes this substantially simpler than traditional time-hierarchy proofs.

5.2 Interpretation overhead and "efficiency"

Let `int` be an interpreter for S written in L. Assuming both an L-machine and an S-machine are at one's disposal, interpretation is usually rather slower than direct execution of S-programs. In practice, an interpreter's running time on inputs `p` and `d` typically satisfies

$$time^{\text{L}}_{\texttt{int}}(\texttt{p.d}) \leq a_{\texttt{p}} \cdot time^{\text{S}}_{\texttt{p}}(\texttt{d})$$

for all `d`. Here $a_{\texttt{p}}$ is a "constant" independent of `d`, but it may depend on the source program `p`. Often $a_{\texttt{p}} \doteq c + f(\texttt{p})$, where constant c represents the time taken for "dispatch on syntax" and recursive calls of the evaluation or command execution functions; and $f(\texttt{p})$ represents the time for variable access.

Definition 5.4 An interpreter `int` (for S written in L) is *efficient* if there is a constant a such that for all $\texttt{d} \in \mathbb{D}_A$ and every S-program `p`

$$time^{\text{L}}_{\texttt{int}}(\texttt{p.d}) \leq a \cdot time^{\text{S}}_{\texttt{p}}(\texttt{d})$$

Constant a is here quantified *before* `p`, so the slowdown caused by an efficient interpreter is independent of `p`.

Theorem 5.5 The interpreter `u` for I written in I from Theorem 3.7 is efficient according to Definition 5.4

5.3 An efficient timed universal program

Definition 5.6 An I-program `tu` is an *efficient timed universal program* if there is a constant k such that for all $\texttt{p} \in$ I-*programs*, $\texttt{d} \in \mathbb{D}$ and $n \geq 1$:

1. If $time_{\texttt{p}}(\texttt{d}) \leq n$ then $[\![\texttt{tu}]\!](\texttt{p . d . nil}^n) = ([\![\texttt{p}]\!](\texttt{d})\texttt{.nil})$
2. If $time_{\texttt{p}}(\texttt{d}) > n$ then $[\![\texttt{tu}]\!](\texttt{p . d . nil}^n) = \texttt{nil}$
3. $time_{\texttt{tu}}(\texttt{p.d.nil}^n) \leq k \cdot \min(n, time_{\texttt{p}}(\texttt{d}))$.

The effect of $[\![\texttt{tu}]\!]$(`p.d.nil`^{n}) is to simulate `p` for $\min(n, time_\texttt{p}(\texttt{d}))$ steps. If $time_\texttt{p}(\texttt{d}) \le n$, i.e., `p` terminates within n steps, then `tu` produces a non-`nil` value containing `p`'s result. If not, the value `nil` is yielded, indicating "time limit exceeded."

We first construct an efficient timed interpreter `tt` for `I`, but in the form of a `WHILE`-program. The idea is to take the interpreter seen before for one-variable `WHILE` programs, and add some extra code and an extra input, a *time bound* of the form `nil`^{n} stored in a variable `Cntr`, so obtaining a program `tt`. Every time the simulation of one operation of program input `p` on data input `d` is completed, the time bound is decreased by 1. Here is `tt`:

```
read X;                      (* X = (p . d . nilⁿ) *)
Cd := cons (hd X) nil;       (* Code to be executed *)
Val := hd (tl X);            (* Initial value of simulated X *)
Cntr := tl (tl X);           (* Time bound *)
Stk := nil;                  (* Computation stack *)
while Cd do
 if Cntr
 then { if hd (hd Cd) ∈ {quote, var, do_hd, do_tl,
                             do_cons, do_asgn, do_while}
        then Cntr := tl Cntr;
        STEP; X := cons Val nil;}
 else {Cd := nil; X := nil};
write X
```

Theorem 5.7 There exists an efficient timed universal program `tu`.

Proof Build `tu` by modifying `tt` as done in the proof of Theorem 3.7.

5.4 The linear-time hierarchy

This result shows there is a constant b such that for any $a \ge 1$ there is a decision problem which cannot be solved by any I-program that runs in time bounded by $a \cdot n$, *regardless of how clever* one is at programming, or at problem analysis, or both. On the other hand, the problem *can* be solved by an I-program in time $a \cdot b \cdot n$ on inputs of size n.

Theorem 5.8 There is a constant b such that for all $a \ge 1$, there is a set A in $\text{TIME}^\texttt{I}(a \cdot b \cdot n)$ that is not in $\text{TIME}^\texttt{I}(a \cdot n)$.

Proof First define program `diag` informally:

```
read X;
Timebound := nil^{a·|X|};
Arg := cons X (cons X Timebound);
X := tu Arg; (* Use tu to run X on X for up to a·|X| steps *)
if hd X then X := false else X := true;
write X
```

Claim: the set $A = \{\texttt{d} \mid [\![\texttt{diag}]\!]^{\texttt{L}}(\texttt{d}) = \texttt{true}\}$ is in $\text{TIME}^{\text{I}}(a \cdot b \cdot n)$ for an appropriate b, but is not in $\text{TIME}^{\text{I}}(a \cdot n)$. Further, b is independent of a.

We now analyse the running time of program diag on input p. Since a is fixed, $\texttt{nil}^{a\cdot|\texttt{d}|}$ can be computed in time $c \cdot a \cdot |\texttt{d}|$ for some c and any d. From Definition 5.6, there exists k such that the timed universal program tu of Theorem 5.7 runs in time $time_{\texttt{tu}}((\texttt{p.d.nil}^n)) \leq k \cdot min(n, time_{\texttt{p}}(\texttt{d}))$. Thus on input p, the command "X := tu Arg" takes time at most

$$k \cdot min(a \cdot |\texttt{p}|, time_{\texttt{p}}(\texttt{p})) \leq k \cdot a \cdot |\texttt{p}|$$

so program diag runs in time at most

$$c \cdot a \cdot |\texttt{p}| + k \cdot a \cdot |\texttt{p}| + e$$

where c is the constant factor used to compute $a \cdot |\texttt{X}|$, k is from the timed universal program, and e accounts for the time beyond computing Timebound and running tu. Now $|\texttt{p}| \geq 1$ so

$$c \cdot a \cdot |\texttt{p}| + k \cdot a \cdot |\texttt{p}| + e \leq a \cdot (c + k + e) \cdot |\texttt{p}|$$

which implies that $A \in \text{TIME}^{\text{I}}(a \cdot b \cdot n)$ with $b = c + k + e$.

Now suppose for the sake of contradiction that $A \in \text{TIME}^{\text{I}}(a \cdot n)$. Then there exists a program p which also decides membership in A, and does it quickly, satisfying $time_{\texttt{p}}(\texttt{d}) \leq a \cdot |\texttt{d}|$ for all $\texttt{d} \in I\!\!D$. Consider cases of $[\![\texttt{p}]\!](\texttt{p})$ (yet another diagonal argument). Then $time_{\texttt{p}}(\texttt{p}) \leq a \cdot |\texttt{p}|$ implies that tu has sufficient time to simulate p to completion on input p. By Definition 5.6, this implies

$$[\![\texttt{tu}]\!](\texttt{p.p.nil}^{a\cdot|\texttt{p}|}) = ([\![\texttt{p}]\!](\texttt{p})\texttt{.nil})$$

If $[\![\texttt{p}]\!](\texttt{p})$ is **false**, then $[\![\texttt{diag}]\!](\texttt{p}) = \texttt{true}$ by construction of diag. If $[\![\texttt{p}]\!](\texttt{p})$ is true, then $[\![\texttt{diag}]\!](\texttt{p}) = \texttt{false}$. Both cases contradict the assumption that p and diag both decide membership in A. The only unjustified assumption was that $A \in \text{TIME}^{\text{I}}(a \cdot n)$, so this must be false, completing the proof.

This construction has been carried out in detail on the computer by Hesselund and Dahl, who establish a stronger result in [4]: that $\text{TIME}^{\text{I}}(201 \cdot a \cdot n + 48)$ properly includes $\text{TIME}^{\text{I}}(a \cdot n)$.

Further, for any non-zero "time constructible" $T(n)$ (using the natural definition), there is a b such that $\text{TIME}^{\text{I}}(b \cdot T(n)) \setminus \text{TIME}^{\text{I}}(T(n))$ is non-empty.

6 Gödel's incompleteness theorem

Gödel's theorem is often presented as "any proof system of a certain minimal complexity is either incomplete or inconsistent." We sidestep the problem of dealing with the multiplicity of proof systems familiar in mathematical logic by generalising it to a "inference system."

We introduce a tiny logical language DL, in which one can make statements about values in $\mathbb{D}$, and then prove that no inference system, by our definition, can generate exactly the set of true statements in DL.

This implies that any inference system which only allows true statements of DL to be proven (i.e., is consistent) cannot generate all true statements of DL (i.e., incomplete). Conclusion: "the full truth" of DL statements cannot be ascertained by means of axioms and rules of logical deduction.

Definition 6.1 An *inference system* $\mathcal{I}$ consists of

1. One finite set of *predicate names* $P, Q, \ldots, Z$ and one of *inference rules* $R_1, R_2, \ldots, R_m$.

2. For each inference rule R_r, a *type*: $R_r : P_1 \times \ldots \times P_k \rightarrow P$ where $P, P_1, \ldots, P_k$ are predicate names.

3. Each inference rule R_r of type $P_1 \times \ldots \times P_k \rightarrow P$ is a decidable *inference relation*: $R_r \subseteq \mathbb{D}^k \times \mathbb{D}$.

Definition 6.2 A *proof tree* of an inference system $\mathcal{I}$ is defined as usual. The set $Thms_P^{\mathcal{I}}$ of all theorems of form $P(\mathtt{d})$, where P is a predicate name, is defined by

$$Thms_P^{\mathcal{I}} = \{\mathtt{d} \mid \mathcal{I} \text{ has a proof tree with root } P(\mathtt{d})\}$$

Theorem 6.3 $Thms_P^{\mathcal{I}}$ is a recursively enumerable set for any inference system $\mathcal{I}$ and predicate P.

The logical language DL for $\mathbb{D}$. An abstract syntax of DL is given by a grammar defining *terms*, which stand for values in $\mathbb{D}$, and *statements*, which are assertions about relationships among terms.

Terms: `T ::= nil | (T.T) |` x_0 `|` x_1 `| ...`
Statements: `S ::= T=T++T |` $\neg$ `S | S` $\wedge$ `S |` $\exists x_i$ `S`

The symbol ++ stands for the "append" operation on list values. Logical operators $\vee, \Rightarrow, \forall$, etc. can be defined as usual, and equality $\mathtt{T} = \mathtt{T}'$ is syntactic sugar for $\mathtt{T} = \mathtt{T}'$`++nil`. Statements are interpreted in the natural way, for example the relation "x is a sublist of y" could be expressed by:

$$\exists u \exists v \exists w (y = w{+}{+}v \wedge w = u{+}{+}x)$$

Definition 6.4 $\mathcal{T}$ is the set of *true closed statements* of DL

Definition 6.5 A predicate $P \subseteq \mathbb{D}^n$ is *representable in* DL if there is a statement $\mathtt{S}(x_1, \ldots, x_n)$ such that

$$P = \{(\mathtt{d}_1, \ldots, \mathtt{d}_n) \in \mathbb{D}^n \mid \mathit{Substitute}(\mathtt{S}, (x_1, \ldots, x_n), (\mathtt{d}_1, \ldots, \mathtt{d}_n)) \in \mathcal{T}\}$$

Lemma 6.6 If set $A \subseteq \mathbb{D}^n$ is representable in DL, then so is $\overline{A} = \mathbb{D}^n \setminus A$.

Theorem 6.7 For any I-program p, the set dom($[\![\mathtt{p}]\!]$) is representable in DL

Proof of this is by a short induction on program syntax.

Theorem 6.8 (Gödel's incompleteness theorem.) $\mathcal{T}$ is not recursively enumerable.

Proof Consider the set $\mathit{HALT} = \mathrm{dom}([\![\mathtt{u}]\!])$ where u is the universal program (self-interpreter) for I programs. By Theorem 6.7 it is representable in DL. By Lemma 6.6, $\overline{\mathit{HALT}}$ is representable by some DL-statement $\mathtt{F}(x)$, so

$$\overline{\mathit{HALT}} = \{(\mathtt{p.d}) \mid \mathit{Substitute}(\mathtt{F}, x, (\mathtt{p.d})) \in \mathcal{T}\}$$

Suppose $\mathcal{T}$ were recursively enumerable. Then there must exist a program q such that $\mathcal{T} = \mathrm{dom}([\![\mathtt{q}]\!])$. But then for any I-program p and input d, we have

$$(\mathtt{p.d}) \in \overline{\mathit{HALT}} \text{ iff } [\![\mathtt{q}]\!](\mathit{Substitute}(\mathtt{F}, x, (\mathtt{p.d}))) \neq \bot$$

This would imply that $\overline{\mathit{HALT}}$ is recursively enumerable, which is false.

Corollary 6.9 For any inference system $\mathcal{I}$ and predicate name P:

$$\text{If } \mathit{Thms}_P^{\mathcal{I}} \subseteq \mathcal{T} \text{ then } \mathit{Thms}_P^{\mathcal{I}} \neq \mathcal{T}$$

In effect this says that if any inference system proves only true DL statements, then it cannot prove all of them. In other words there is and always will be a difference between *truth* (at least for DL) and *provability* by inference systems. This captures one essential aspect of Gödel's incompleteness theorem. In comparison with the original proof, and others seen in the literature, this one uses surprisingly little technical machinery (though it admittedly builds on the nontrivial difference between recursive and recursively enumerable sets).

Gödel's original work began with a logical system containing Peano arithmetic, and applied diagonalisation to construct a witness: an example of a statement S which is true, but which cannot be provable. Gödel's original witness is (intuitively) true since it in effect asserts "there is no proof in this system of S." Our version indeed uses diagonalisation, but on I programs instead, and to prove that the problem $\overline{HALT}$ is not recursively enumerable.

7 Levin's optimal search theorem

A great many familiar problems are searches. Consider the predicate

$$R(\mathcal{F}, \theta) \equiv \text{truth assignment } \theta \text{ makes formula } \mathcal{F} \text{ true}$$

The problem to find θ (if it exists) when given only $\mathcal{F}$ is the familiar and apparently intractable *satisfiability problem.* As is well known, it is much easier to check truth of $R(\mathcal{F}, \theta)$.

Definition 7.1 A *witness function* for a binary predicate $R \subseteq \mathbb{D} \times \mathbb{D}$ is a function $f : \mathbb{D} \to \mathbb{D}$ such that:

$$\forall \mathtt{x} (\exists \mathtt{y} \,.\, R(\mathtt{x}, \mathtt{y})) \Rightarrow R(\mathtt{x}, f(\mathtt{x}))$$

A *brute-force search* program for finding a witness immediately comes to mind. Given $x \in \mathbb{D}$ we just enumerate elements $\mathtt{y} \in \mathbb{D}$, checking one after the other until a witness pair $(\mathtt{x}, \mathtt{y}) \in R$ has been found[4]. It is quite obvious that this strategy can yield an extremely inefficient program, since it may waste a lot of time on wrong candidates until it finds a witness. Levin's theorem states a surprising fact: for many interesting problems there is another brute-force search strategy that not only is efficient, but *optimal* up to constant factors. The difference is that Levin's strategy generates and tests not *solutions*, but *programs.*

[4] If R is decidable, this is straightforward. If semi-decidable but not decidable, a "dovetailing" of computations can be used to test $(\mathtt{x}, \mathtt{d}_0) \in R?$, $(\mathtt{x}, \mathtt{d}_1) \in R?$, ... in parallel.

Theorem 7.2 *Levin's Search Theorem.* Let $R \subseteq \mathbb{D} \times \mathbb{D}$ be a recursively enumerable binary predicate, so $R = \text{dom}([\![\texttt{r}]\!])$ for some program r. Then there is a WHILE program opt such that $f = [\![\texttt{opt}]\!]$ is a witness function for R. Further, for every program q that computes a witness function f for R:

$$time_{\texttt{opt}}(\texttt{x}) \leq a_{\texttt{q}}(time_{\texttt{q}}(\texttt{x}) + time_{\texttt{r}}(\texttt{x}.f(\texttt{x})))$$

for all x, where $a_{\texttt{q}}$ is a constant that depends on q but not on x. Further, the program opt can be effectively obtained from r.

Sketch of proof of Levin's theorem. Without loss of generality we assume that when program r is run with input (x.y), if $(\texttt{x}, \texttt{y}) \in R$ it gives y as output. Otherwise, it loops forever. Enumerate $\mathbb{D} = \{\texttt{d}_0, \texttt{d}_1, \texttt{d}_2, \ldots\}$ effectively (it can be done in constant time per new element). Build program opt to compute as follows:

1. A "main loop" to generate all finite trees. At each iteration one new tree is added to list $\texttt{L} = (\texttt{d}_n \ldots \texttt{d}_1\texttt{d}_0)$. Tree $\texttt{d}_n$ for $n = 0, 1, 2, \ldots$ will be treated as the command part of the n-th I program $\texttt{p}_n$.

2. Iteration n will process programs $\texttt{p}_k$ for $k = n, n-1, \ldots, 1, 0$ as follows:

   (a) Run $\texttt{p}_k$ on input x for a "time budget" of 2^{n-k} steps.

   (b) If $\texttt{p}_k$ stops on x with output y, then run r on input (x.y), so $\texttt{p}_k$ and r together have been executed for at most 2^{n-k} steps.

   (c) If $\texttt{p}_k$ or r failed to stop, then replace k by $k-1$, double the time budget to 2^{n-k+1} steps, and reiterate.

3. If running $\texttt{p}_k$ followed by r terminates within time budget 2^{n-k}, then output $[\![\texttt{opt}]\!](\texttt{x}) = \texttt{y}$ and stop; else continue with iteration $n+1$.

Thus the programs are being interpreted concurrently, every one receiving some "interpretation effort." We stop once any one of these programs has both *solved our problem and been checked*, within its given time bounds. Note that opt will loop in case no witness is found. The following table showing the time budgets of the various runs may aid the reader in following the flow of the construction and correctness argument.

The keys to "optimality" of opt are the efficiency of the self-interpreter STEP operation, plus a policy of allocating time to the concurrent simulations so that the total time will not exceed, by more than a constant factor, the time of the program that finishes first.

Time budget	p_0	p_1	p_2	p_3	p_4	p_5	ldots
$n = 0$	1	-	-	-	-	-	...
$n = 1$	2	1	-	-	-	-	...
$n = 2$	4	2	1	-	-	-	...
$n = 3$	8	4	2	1	-	-	...
$n = 4$	16	8	4	2	1	-	...
$n = 5$	32	16	8	4	2	1	...
$n = 6$	64	32	16	8	4	2	...
...	...						

Suppose $\mathtt{q} = \mathtt{p}_k$ computes a witness function f. At iteration n, program $\mathtt{p}_k$ and the checker $\mathtt{r}$ are run for 2^{n-k} steps. Therefore (assuming $R(\mathtt{x}.f(\mathtt{x}))$ is true) the process above will not continue beyond iteration n, where

$$2^{n-k} \geq time_{\mathtt{p}_k}(\mathtt{x}) + time_{\mathtt{r}}(\mathtt{x}.f(\mathtt{x}))$$

A straightforward time analysis yields

$$time_{\mathtt{opt}}(\mathtt{x}) \leq c2^k(time_{\mathtt{q}}(\mathtt{x}) + time_{\mathtt{r}}(\mathtt{x}.f(\mathtt{x})))$$

as required, where c is not excessively large. It must be admitted, however, that the constant factor $c2^k$ is enormous. The interesting thing is that it exists at all, i.e., that the construction gives, from an asymptotic viewpoint, the best possible result.

8 Programming characterisations of LOGSPACE and PTIME

The data construction operation of a `WHILE` is "`cons`", which combines two binary trees into a single, larger one. If this operation is unavailable, however, the data decomposition operations `hd`, `tl` still provide a "read-only" access to the `WHILE`-program's input. Although limited, such "cons-free" programs are not trivial. The following theorem, proven in [9], characterises their computational power exactly.

Theorem 8.1 A set $A \subseteq \{0,1\}^*$ is decidable by a LOGSPACE-bounded Turing machine program if and only if it is decidable by a `WHILE` program without `cons` operations.

An interesting aspect: this is an *intrinsic* characterisation of LOGSPACE, with no time or storage bounds given a priori. A similarly intrinsic characterisation also exists for PTIME:

Theorem 8.2 A set $A \subseteq \{0,1\}^*$ is decidable by a polynomial time-bounded WHILE or Turing machine program if and only if it is decidable by a *recursive* WHILE program without cons operations.

A recursive WHILE program, as defined in [9], is a collection of mutually recursive procedures (paramerless, and communicating by means of global variable declarations). A construction to show essentially the same result was done by Cook in [2], expressed in terms of "auxiliary push-down automata." Our program-oriented version seems a more natural formulation.

A puzzling observation. A recursive WHILE program can run in exponential time – even though such programs can decide all and only sets in PTIME. Thus even though a polynomial-time problem is being solved, the solver can run in superpolynomial time.

The results above can be interpreted as saying that recursive programs without "cons" are capable of simulating imperative ones with "cons;" but at a formidable cost in computing time (exponentially larger). In essence, we have shown that the heap can be replaced by the stack, but at a very high time cost. It is not known, however, *whether this cost is necessary.*

9 Complete problems for PTIME, NPTIME, and PSPACE via Boolean programs

Program analysis is in general undecidable due the halting problem's recursive unsolvability, and Rice's general result that all nontrivial extensional program properties are undecidable. On the other hand, *finite-memory* programs [12, 13] do have decidable properties since their entire state spaces can be computed. A series of theorems relating well-known complexity classes to finite program analysis appears in [9].

Definition 9.1 (The language BOOLE) A *boolean program* is an input-free program $\texttt{p} = \texttt{I}_1 \ldots \texttt{I}_m$ where each instruction I and expression E is of form given by:

$$
\begin{array}{lcl}
\texttt{I} & ::= & \texttt{X := E} \mid \texttt{I}_1\texttt{; I}_2 \mid \texttt{goto}\ \ell \mid \texttt{if E then I}_1\ \texttt{else I}_2 \\
\texttt{E} & ::= & \texttt{X} \mid \texttt{true} \mid \texttt{false} \mid \texttt{E}_1 \vee \texttt{E}_2 \mid \texttt{E}_1 \wedge \texttt{E}_2 \mid \neg\ \texttt{E} \mid \texttt{E}_1 \Rightarrow \texttt{E}_2 \mid \texttt{E}_1 \Leftrightarrow \texttt{E}_2 \\
\texttt{X} & ::= & \texttt{X0} \mid \texttt{X1} \mid \ldots
\end{array}
$$

Theorem 9.2 The problem of deciding membership in the following subsets of BOOLE-*programs* is characterised as follows:

1. $\{$ p $\in$ BOOLE-*programs* $\mid$ $[\![$p$]\!]$ = **true** $\}$: complete for PSPACE (polynomial space)

2. $\{$goto-free p $\in$ BOOLE-*programs* $\mid$ $[\![$p$]\!]$ = **true** $\}$: complete for PTIME (polynomial time)

3. $\{$goto-free p $\in$ BOOLE-*programs* $\mid$ $[\![$q;p$]\!]$ = **true** for some q $\}$: complete for NPTIME (nondeterministic polynomial time)

10 Related Work

The book [15] by Kfoury, Moll and Arbib has similar aims, but [9] goes further in two respects: it covers complexity theory as well as computability; and it demonstrates the advantages that come from structuring *both programs and data.* [15] deals with structured programs, but uses the natural numbers as data.

Paul Voda is redeveloping recursion theory on the basis of Lisp-like data structures. A book is forthcoming; and [18] is a recent article.

Several other works have given exact characterizations of complexity classes in terms of programming languages or other formal languages. Most are, however, rather more technical than the LOGSPACE and PTIME characterizations above.

Early results: Meyer and Ritchie's "loop language" characterization of the elementary functions [17] has a similar approach to that of this paper, but at a (much) higher complexity level. Another early result was the discovery that spectra of first-order logic are identical with NEXPTIME [14], further developed actively as finite model theory, see Immerman [7] and many others.

More recently, Bellantoni and Cook's characterisation [1] of PTIME using "tiered recursion" has attracted much interest. Girard, Scedrov and Scott gave a linear logic characterization of PTIME in [5]. Leivant and Marion [16], and Hillebrand and Kanellakis [6] have characterized several complexity classes by variants of the lambda-calculus.

References

[1] S. Bellantoni and S. Cook, A new recursion-theoretic characterization of the polytime functions. Computational Complexity **2** (1992), 97–110.

[2] S. A. Cook, Characterizations of pushdown machines in terms of time-bounded computers. Journal of the ACM **18** (1971), 4–18.

[3] Torben Amtoft Hansen, Thomas Nikolajsen, Jesper Larsson Träff, and N. D. Jones. Experiments with implementations of two theoretical constructions. In *Lecture Notes in Computer Science 363*, pages 119–133. Springer Verlag, 1989.

[4] C. Dahl and M. Hessellund. Determining the constant coefficients in a time hierarchy. Technical report, Department of Computer Science, University of Copenhagen, feb 1994.

[5] J.-Y. Girard, A. Scedrov, P. Scott, Bounded linear logic: a modular approach to polynomial time computability, *Theoretical Computer Science*, vol. 97, pp. 1–66, 1994.

[6] G. Hillebrand, P. Kanellakis, On the expressive power of simply typed and let-polymorphic lambda calculi, *Logic in Computer Science*, IEEE Computer Society Press, pp. 253–263, 1996.

[7] N. Immerman, Relational queries computable in polynomial time, *Information and Computation*, vol. 68, pp. 86-104, 1986.

[8] N. D. Jones. Constant time factors *do* matter. In Steven Homer, editor, *STOC '93. Symposium on Theory of Computing*, pages 602–611. ACM Press, 1993.

[9] N. D. Jones, *Computability and Complexity from a Programming Perspective.* The MIT Press, 1997.

[10] N.D. Jones, C. Gomard, P. Sestoft. *Partial Evaluation and Automatic Program Generation.* Prentice-Hall International, 1993.

[11] N. D. Jones. LOGSPACE and PTIME characterized by programming languages. *Theoretical Computer Science*, 1998.

[12] N. D. Jones and S. Muchnick. Even simple programs are hard to analyze. *Journal of the Association for Computing Machinery*, 24(2):338–350, 1977.

[13] N. D. Jones and S. Muchnick. Complexity of finite memory programs with recursion. *Journal of the Association for Computing Machinery*, 25(2):312–321, 1978.

[14] N. D. Jones, A. Selman, Turing machines and the spectra of first-order formulae with equality, *Journal of Symbolic Logic*, vol. 39, no. 1, pp. 139–150, 1974.

[15] A.J Kfoury, R.N. Moll, and M.A. Arbib. *A Programming Approach to Computability.* Texts and monographs in Computer Science. Springer-Verlag, 1982.

[16] D. Leivant, J-Y. Marion, Lambda calculus characterizations of ptime, *Fundamenta Informaticae*, vol. 19, pp. 167–184, 1993.

[17] A. Meyer, D. Ritchie, A classification of the recursive functions, *Zeitschrift MLG*, vol. 18, pp. 71–82, 1972.

[18] Paul Voda. Subrecursion as Basis for a Feasible Programming language. In Steven Homer, editor, *Logic in Comp. Science September 94*. Lecture Notes in Computer Science 933, Springer Verlag 1995.

Effective Model Theory: The Number of Models and Their Complexity[1]

Bakhadyr Khoussainov[2]
Cornell University, Ithaca NY 14853
University of Aukland, Aukland New Zealand

Richard A. Shore[3]
Cornell University, Ithaca NY 14853

Abstract

Effective model theory studies model theoretic notions with an eye towards issues of computability and effectiveness. We consider two possible starting points. If the basic objects are taken to be theories, then the appropriate effective version investigates decidable theories (the set of theorems is computable) and decidable structures (ones with decidable theories). If the objects of initial interest are typical mathematical structures, then the starting point is computable structures. We present an introduction to both of these aspects of effective model theory organized roughly around the themes of the number and types of models of theories with particular attention to categoricity (as either a hypothesis or a conclusion) and the analysis of various computability issues in families of models.

1. Basic Notions

The lectures on which this paper is based were intended to be a brief introduction to effective model theory centered around one set of issues: the number of models of specified type and, in particular, the notion of categoricity. For more general

[1] This paper is primarily based on the short course on effective model theory given by the second author at the ASL summer meeting, Logic Colloquium '97, at the University of Leeds. Much of the material was also presented by the authors in invited talks at a special session of the ASL annual meeting, University of California, Irvine, USA, 1995; the Discrete Mathematics, Theoretical Computer Science, and Logic Conference, Victoria University, Wellington, New Zealand, 1996; the Special Session on Feasible Mathematics of the AMS annual meeting, Orlando, USA, 1996; the annual meeting of the ASL, University of Wisconsin, Madison, 1996, USA; and the Special Session on Computable Mathematics and its Applications of the AMS annual meeting, Baltimore, January 1998.

[2] Partially supported by ARO through MSI, Cornell University, DAAL03-91-C0027.

[3] Partially supported by NSF Grants DMS-9204308, DMS-9503503, INT-9602579 and ARO through MSI, Cornell University, DAAL03-91-C0027.

introductions we refer the reader to *The Handbook of Recursive Algebra* (Ershov et al. [1998]), especially the articles by Harizanov [1998] and Ershov and Goncharov [1998]. This *Handbook* also contains other useful survey papers on aspects of effective model theory and algebra and an extensive bibliography. The one most closely related to the theme of this paper is Goncharov [1998]. Another interesting survey is Millar [1999] in *The Handbook of Computability Theory* (Griffor [1999]). Two books in progress on the subject are Ash and Knight [1999] and Harizanov [2000]. These are all good sources for material and references. An extensive and very useful bibliography prepared by I. Kalantari [1998] can also be found in Ershov et al. [1998].

One might well begin with the question of what effective model theory is about. Of course, it is about investigating the subjects of model theory with an eye to questions of effectiveness. What then is model theory about and what does one mean by effectiveness? As for model theory we simply quote from two standard texts (to which we also refer the reader for the terminology, notation and results of classical model theory). Chang and Keisler [1990] say "Model theory is the branch of mathematical logic which deals with the connection between a formal language and its interpretations, or models." Hodges [1993] says "Model theory is the study of the construction and classification of structures within specified classes of structures." We can take these two definitions as expressing two views of the proper subject of model theory. The first starts with formal languages and so we may say with theories. (We take a *theory* T to be simply a set of sentences in some (first-order) language L, called the *language of* T. We say that a theory T is *complete* if $T \vdash \sigma$ or $T \vdash \neg\sigma$ for every sentence σ of L.) The second starts with mathematical structures. One might think of these views as, respectively, logical and algebraic. They lead to a basic dichotomy in the approach to effective model theory. Should we "effectivize" theories or structures. Of course, the answer is that we should investigate both approaches and their interconnections. As for what one means by "effectiveness", there are many notions ranging from ones in computer science to ones of descriptive set theory that have some claim to being versions of effectiveness. Most, if not all, of them can be reasonably called in to analyze different model theoretic questions. In this paper, we limit ourselves to what we view as the primary notion of effectiveness: Turing computability (or, equivalently, recursiveness). Thus we are lead to formal definitions of the two basic notions of our subject, effective theories and structures.

Definition 1.1 A theory T is *decidable* if the theorems of T form a computable set. A structure $\mathcal{A}$ (for a language L) with *underlying set* (or *domain*) A is *decidable* if $Th(\mathcal{A}, a)_{a \in A}$, the *complete* (or *elementary*) *diagram* of $\mathcal{A}$, i.e. the set of all sentences (with constant symbols for each element of A) true in $\mathcal{A}$, is computable. $\mathcal{A}$ is *computable* if $D(\mathcal{A}, a)_{a \in A}$, the (*atomic*) *diagram* of $\mathcal{A}$, i.e. the set of all atomic sentences or their negations (again with constant symbols for each element of A) true in $\mathcal{A}$, is computable.

For those whose basic object of interest, or at least starting point, consists of theories, the decidable theories are the natural effective objects of study. In line with standard model theoretic usage a structure whose complete theory has some property P is often said to also have property P and so we have decidable structures. This is the "logical" point of view. On the other hand, the algebraist or general mathematician usually starts with structures. From this point of view, the effective objects are the computable structures. After all, when one thinks of what a computable group should be one thinks that it should be a group structure for which the group operation is computable and similarly for all other typical algebraic structures. One does certainly not assume that even the word problem, let alone the complete diagram, is computable.

Note that we are deliberately avoiding all issues of coding or Gödel numbering. There are two common approaches to this issue. The Eastern, and especially the Russian, school favors numerations. One starts with a classical structure $\mathcal{A}$ and provides a *numeration* (or *enumeration*), that is a map ν from the natural numbers $\mathbb{N}$ onto the underlying set A of the structure $\mathcal{A}$. The *numerated* (or *enumerated*) structure $\langle \mathcal{A}, \nu \rangle$ is called *constructive* if the (appropriately coded) atomic diagram of $\mathcal{A}$, with constant symbols i for $i \in \mathbb{N}$ interpreted as $\nu(i)$, is computable (recursive). $\langle \mathcal{A}, \nu \rangle$ is *strongly constructive* if the complete diagram of $\mathcal{A}$ with constant symbols i for $i \in \mathbb{N}$ interpreted as $\nu(i)$ is computable. These notions essentially correspond to what we call computable and decidable structures, respectively.

An established Western approach is to say that all elements are natural numbers, all sets are subsets of $\mathbb{N}$ and all functions are functions from $\mathbb{N}$ to $\mathbb{N}$. In this view, languages are Gödel numbered, structures consist of a set of numbers and relations and functions on that set. The formal definitions of computable or recursive for subsets of, and functions on, $\mathbb{N}$ are then simply applied directly to theories and structures. We adopt what might be viewed as a less formal version of the second approach along the lines followed in Shoenfield [1971] and now, we think, prevalent in thinking (if not always in writing) about computability. Given that we are not considering issues raised by the theory of enumerations, we see no reason to explicitly code objects as numbers. After all, we now "know" what effective and computable mean not only for numbers but for all kinds of data structures from strings to arrays on arbitrary finite alphabets. Thus we talk about a computable language without the formalities of Gödel numbering and so about computable theories, types, etc. Similarly, we have computable structures, lists of names for their elements, diagrams and theories. These may or may not "be" sets of, or functions on, $\mathbb{N}$. Any reader who prefers explicit Gödel numbering is certainly able to make the appropriate translations. (We may at times, however, resort to indices to clarify certain uniformity issues.) For those interested in the issues related specifically to numerations we refer the reader to Ershov [1977].

Of course, the notions of effectiveness associated with Turing computability only make sense in the countable setting.

- All languages, sets, structures and the like are assumed to be countable unless explicitly stated otherwise.

Even so, not all sets or structures are computable. Classically, one typically identifies isomorphic structures. Of course, this eliminates all issues of effectiveness and so is often not appropriate here. We will have to distinguish between classically isomorphic models. The following definitions of presentations and presentability help us make these distinctions.

Definition 1.2 A structure $\mathcal{A}$ is *computably (decidably) presentable* if $\mathcal{A}$ is isomorphic to a computable (decidable) structure $\mathcal{B}$ which we call a *computable (decidable) presentation* of $\mathcal{A}$.

Before launching into theorems and analyses, we present a few examples of decidable or computable theories and structures. These theories and structures will serve as examples for many of the notions and results we consider below. Proofs for many of the facts we cite about these structures can be found in Chang and Keisler [1990, 3.4].

Example 1.3 Our language here is that of (linear) orders with one binary predicate $\leq$. We consider two theories *DeLO*, dense linear orderings with no first or last element and *DiLO*, discrete linear orderings with first but no last element. *DeLO* is axiomatisable, $\aleph_0$ *-categorical* (i.e. all countable models are isomorphic) and so complete and decidable. *DiLo* is axiomatisable and complete and so decidable but not $\aleph_0$-categorical. The standard structures associated with these theories are $\mathbb{Q}$ and $\mathbb{N}$, respectively, with their natural orderings. Both are decidable. As *DeLO* is $\aleph_0$-categorical every model (remember we are considering only countable structures) is isomorphic to $\mathbb{Q}$ and so decidably presentable. *DeLO* has effective quantifier elimination and so every computable model is actually decidable. On the other hand, not every model of *DiLO* is even computably presentable nor is every computable model decidable as we shall see below (for example, in Proposition 6.1). (To see that not every model of *DiLO* is computably presentable, note that at the cost of a couple of jumps we can form the quotient of a given *DiLO* by the equivalence relation of being finitely far apart. This procedure can produce an arbitrary ordering with first element. If the quotient ordering is not arithmetic, the original model can't be computably presented.)

Example 1.4 The next theory we mention is ACF_0, algebraically closed fields of characteristic 0. The language is that of field theory with $0, 1, +$ and $\times$. ACF_0 is axiomatisable, $\aleph_1$*-categorical*, i.e. all models of cardinality $\aleph_1$ are isomorphic, and so complete and decidable. ACF_0 also has effective quantifier elimination and so here too every computable model is actually decidable. Even though ACF_0 is not $\aleph_0$-categorical, every model is decidably presentable and below we prove a general theorem establishing this fact (Theorem 5.2).

Example 1.5 Finally, we briefly discuss *PA*, Peano Arithmetic or if one prefers any suitable finitely axiomatized subtheory such as Robinson's Q [1950]. The language has $0, 1, +$ and $\times$ with the usual axioms. Of course PA is axiomatisable but, by Gödel's incompleteness theorem, it is neither complete nor decidable. It is not $\aleph_0$-categorical (the compactness theorem provides nonstandard models not isomorphic to $\mathbb{N}$). No model is decidable (again by the incompleteness theorem) and only the standard model $\mathbb{N}$ is computably presentable.

Proposition 1.6 (Tennenbaum [1959]) *No nonstandard model of PA or even of Robinson's Q is computably presentable.*

Proof sketch. We assume that one has developed the theory T in question enough to, say prove unique factorization into primes and that the standard universal partial computable function is representable in that there is a formula $F(e, x, s, i)$ such that, for each e, x, s, i in $\mathbb{N}$, $\phi_{e,s}(x) = i$ if and only if $T \vdash F(e, x, s, i)$. (We do not bother to differentiate between a number and the numeral representing it.) One now shows that T proves the simple fact that
$(*)\forall s \exists y \forall e([F(e, e, s, 0) \rightarrow p_e | y \wedge p_{2e+1} \nmid y] \wedge [\neg F(e, e, s, 0) \rightarrow p_e \nmid y \wedge p_{2e+1} | y])$
where $p_e | y$ is a formula saying that the eth prime divides y.

Now let $\mathcal{A}$ be any nonstandard model of T, s any nonstandard element of $\mathcal{A}$ and y the element of $\mathcal{A}$ guaranteed by $(*)$. We define the function f on $\mathbb{N}$ by $f(e) = 1$ if $\mathcal{A} \models p_e | y$ and $f(e) = 0$ if $\mathcal{A} \models p_{2e+1} | y$. Clearly $f(e)$ is computable from the atomic diagram of $\mathcal{A}$ by searching for an element z such that $\mathcal{A} \models p_e \times z = y$ or $\mathcal{A} \models p_{2e+1} \times z = y$. (One must exist by $(*)$.) However, f is clearly not computable. Indeed, f is *diagonally noncomputable* : $\forall e(f(e) \neq \phi_e(e))$. Thus $\mathcal{A}$ is not computably presentable. □

2. The Effective Completeness Theorem

A common theme in model theory is the investigation of questions about when given theories have models with specified properties. Typical examples include characterizing when theories have atomic, prime, universal, homogeneous or saturated models. Other questions involve models of various ranks or dimension, with or without indiscernibles or even more ambitiously attempts to characterize all the models of a given theory. In effective model theory one naturally wants to know when theories have decidable or computable models of each type or even to attempt to characterize the decidable or computable models of a given theory. We will investigate a few examples of such questions. We begin with the issue of when a theory has a model at all – Gödel's completeness theorem.

Theorem 2.1 (Completeness Theorem) *If a theory T is consistent it has a model.*

We present one effective analog of the completeness theorem for decidable theories with a proof modeled on Henkin's proof of the classical completeness theorem. This method of construction is simple but basic for many results in both classical and effective model theory and we will see several variants latter on.

Theorem 2.2 (Effective Completeness Theorem) *If a theory T is consistent and decidable then it has a decidable model.*

Proof. We assume that the classical Henkin construction is known and so provide only a sketch so that we can check its effective content. Let L_c be the language L of T extended by infinitely many new constants c_i and let σ_e be a (computable) list of the sentences of L_c. We construct an increasing sequence of finite sets Ψ_s of sentences of L_c (with $\wedge\Psi_s = \psi_s$) consistent with T with union Ψ as in the Henkin proof of the completeness theorem. We need to satisfy the requirements P_e for each $e \in \mathbb{N}$:

- $P_e : \sigma_e \in \Psi$ or $\neg\sigma_e \in \Psi$ and if σ_e is of the form $\exists x\theta(x)$ and in Ψ then $\theta(c_i) \in \Psi$ for some i.

Construction: At stage s ask if σ_s is consistent with $T \cup \Psi_s$. If so put σ_s into Ψ_{s+1} and, if σ_s is $\exists x\theta(x)$, also put $\theta(c_i)$ into Ψ_{s+1} for some as yet unmentioned c_i. If σ_s is not consistent with $T \cup \Psi_s$ put $\neg\sigma_s$ into Ψ_{s+1}.

Verifications: Obviously, Ψ is complete and the standard argument shows that it is consistent. As usual the elements of the desired model $\mathcal{M}$ are the equivalence classes of the c_i under the equivalence relation $\equiv$ given by $c_i \equiv c_j$ iff $(c_i = c_j) \in \Psi$ and the relations and functions on $\mathcal{M}$ are determined in the natural way by the formulas in Ψ.

The only issue for us now is the effectiveness of the construction. First we note that one can verify that if T is decidable then Ψ is computable. The only question we must answer at stage s is if σ_s is inconsistent with $T \cup \Psi_s$. This is equivalent to whether or not $\psi_s \rightarrow \neg\sigma_s$ (with new free variables z_i substituted in for the constants c_i appearing in Ψ_s or σ_s) is a theorem of T. As T is decidable the answer to these questions is a computable function of s. Thus the equivalence relation $c_i \equiv c_j$ is computable. (Just look at Ψ_{s+1} where $c_i = c_j$ is σ_s.) So the equivalence classes form a computable set (the domain of $\mathcal{M}$) and the relations and functions on $\mathcal{M}$ are determined by Ψ. Indeed, as usual, a sentence σ is true in $\mathcal{M}$ if and only if $\sigma \in \Psi$ and so $\mathcal{M}$ is decidable as required. □

One theme in effective model theory that we will not pursue investigates the question of how hard it is (say in terms of Turing degree or levels of the (hyper)arithmetic hierarchy) to construct models of a given type when it is not possible to produce decidable or even computable ones. We consider the completeness theorem as our only example. In the construction above the only noneffective step was deciding if σ_s is consistent with $T \cup \Psi_s$. As one can always answer this question computably in T' (the Turing jump of T), every consistent theory T has a model computable, indeed decidable, in T'.

Corollary 2.3 *If T is consistent then there is a model $\mathcal{M}$ of T such that the elementary theory of $\mathcal{M}$, $Th(\mathcal{M}, m)_{m \in M}$, is computable in T' (and so Δ^0_2 in T). Indeed by the low basis theorem, there is always one with $T' \leq_T (Th(\mathcal{M}, m)_{m \in M})'$.*

Proof. The first assertion follows immediately from the construction and discussion above. For the second, instead of a single Ψ we build a binary tree (of choices of σ_s (and Henkin axioms as appropriate) or $\neg\sigma_s$). We terminate any path that becomes inconsistent when we find a proof of inconsistency from T. This produces an infinite binary tree computable in T (the particular Ψ constructed above is an infinite path through this tree). The low basis theorem (Jockusch and Soare [1972]) says that there is an infinite path P through the tree with $P' \leq_T T'$. As above we can construct the desired model (and its complete diagram) computably in P as required. □

- For the sake of convenience we assume from now on that all theories are consistent.

We can now say (in some sense) when a theory T has a decidable model.

Corollary 2.4 *A complete theory T has a decidable model if and only if it is decidable. An arbitrary theory T has a decidable model if and only if it has a decidable complete extension.*

Proof. If $\mathcal{M}$ is a model of T and T is complete then the set of theorems of T is simply the intersection of $Th(\mathcal{M}, m)_{m \in M}$ with the sentences of the language L of T and so T is decidable if $\mathcal{M}$ is decidable. Even if T is not complete, if $\mathcal{M}$ is a decidable model of T then this set is a decidable complete extension of T. The other (if) direction of both assertions in the Corollary follow from Theorem 2.2. □

We will not in general assume that theories are complete. However, finite models have little interest from the viewpoint of Turing computability.

- We assume from now on that all theories have only infinite models.

Now that we "know" when a theory T has a decidable model, we might well ask how many decidable models a theory can have. For now we identify models up to classical isomorphisms and so we might better ask how many decidably presentable models can a theory have. The issues of identifying computable models only when there is a computable isomorphism between them will be taken up in § 6-7.

If T is incomplete then every decidable complete extension has a decidable model by Theorem 2.2 and, of course, models of distinct extensions are not isomorphic. Moreover, every decidable model of T is a model of some complete decidable extension of T. Thus if one is interested in the number of decidably presentable models of a theory, it suffices to consider only complete decidable theories. We begin with the possibility that there is only one as in our example $DeLO$ of a decidable $\aleph_0$ -categorical theory.

Proposition 2.5 *If a theory T is $\aleph_0$-categorical then the following conditions are equivalent:*

1. T is decidable.
2. T has a decidable model.
3. All models of T are decidably presentable.

Proof. As $\aleph_0$-categoricity implies completeness, the equivalences all follow directly from the hypothesis, definitions and Theorem 2.2. □

Now, it is a remarkable classical theorem due to Vaught [1961] that no complete theory has exactly two (isomorphism types of) models. The effective analog for decidable models is, however, false.

Theorem 2.6 (Millar [1979], Kudaibergenov [1979]) *There is a decidable theory T with exactly two (isomorphism types of) decidably presentable models.*

Proof sketch. Let f be a partial computable function whose range is $\{0, 1\}$ and which does not have a total computable extension. Consider the (computably enumerable but computably inseparable) sets $M_0 = \{x|f(x) = 0\}$ and $M_1 = \{x|f(x) = 1\}$. Let $f_0 \subset f_1 \subset \ldots$ be an effective approximation to f such that $k \notin dom(f_s)$ for all $k > s$.

The language of T contains infinitely many unary and binary predicates P_i and R_i, respectively, where $i \in \omega$. Consider first the theory T_0 whose axioms are the following set of statements:

1. $\forall x P_0(x) \& \forall y (P_{i+1}(y) \rightarrow P_i(y))$, where $i \in \omega$.
2. If $R_k(x, y)$, then $x \neq y$ and $P_k(x) \& P_k(y)$.
3. If $x \neq y$, $P_s(x) \& P_s(y)$ and $f_s(k) = 0$, then $R_k(x, y)$.
4. If $x \neq y$, $P_s(x) \& P_s(y)$ and $f_s(k) = 1$, then $\neg R_k(x, y)$.

One can check that the following four properties hold of T_0:

1. T_0 has a decidable model completion T. Moreover T has a unique 1–type (Definition 3.1) p such that $P_k(z) \in p$ for all $k \in \omega$.
2. If a model $\mathcal{A}$ of T has at least two elements realizing p, then $\mathcal{A}$ is not decidably presentable.
3. If a model $\mathcal{A}$ of T has fewer than two elements realizing p, then $\mathcal{A}$ is decidably presentable.
4. If $\mathcal{A}_1$ and $\mathcal{A}_2$ are models of T with the same finite number of elements realizing p, then $\mathcal{A}_1$ and $\mathcal{A}_2$ are isomorphic.

These properties show that T has exactly two decidably presentable models. □

The above proof can easily be generalized:

Corollary 2.7 *For each $n \leq \omega$, there exists a theory with exactly n nonisomorphic decidable models.* □

As for our examples above, an analysis of the structure of models of $DiLO$ as in Chang and Keisler [1990, 3.4] easily implies that there are countably many distinct decidable models. The same is true for ACF_0 as we shall see in Theorem 5.2.

Although the natural effective version of Vaught's theorem fails, the proof (properly effectivized) can be used to give a similar result for decidable models (Theorem 4.4 below). We first need to study another aspect of the question of how many decidable models a theory T can have: When are each of the classically studied types of models such as prime, atomic or saturated models of a decidable theory decidably presentable?

3. Decidable Prime Models

We begin our study of specific types of models with prime and atomic models. They will play a crucial role in the next two sections.

Definition 3.1 An *n-type* Γ or $\Gamma(x_1, \ldots, x_n)$ of a theory T is a set of formulas with n free variables in the language of T which is consistent with T such that $\sigma(x_1, \ldots, x_n)$ or $\neg\sigma(x_1, \ldots, x_n)$ belongs to Γ for each such formula. An n-type $\Gamma(x_1, \ldots, x_n)$ of a theory T is *principal* if there is a formula $\theta(x_1, \ldots, x_n)$ such that $T \vdash \theta(x_1, \ldots, x_n) \to \sigma(x_1, \ldots, x_n)$ for every $\sigma \in \Gamma$. In this case we say that $\theta(x_1, \ldots, x_n)$ is a *complete formula* that *generates* Γ.

Definition 3.2 A model $\mathcal{A}$ of a theory T in the language L is a *prime* model of T if it can be elementarily embedded into every model of T. $\mathcal{A}$ is *atomic* if every n-tuple of elements from A satisfies a complete formula $\theta(x_1, \ldots, x_n)$ of L. (Each of these models is unique (up to isomorphism) if it exists.)

The notions of prime and atomic coincide for countable models and so we motivate our characterization of decidable prime models by two classical characterizations.

Theorem 3.3 *A complete theory T in a language L has a prime model if and only if every formula of L consistent with T is a member of a principal type over T.*

Theorem 3.4 *A complete theory T in a language L has an atomic model if and only if every formula of L consistent with T can be extended to a complete formula.*

As the notions of atomic and prime coincide (for countable models), each of these theorems provides a characterization of the theories with prime models. We now consider what might be the appropriate effective versions of these theorems. In one direction, note that every type realized in a prime model of T is principal and all principal types are realized in every model of T. Thus, if T has a decidable prime model, not only is every formula consistent with T a member of a principal type (and so completable) but there is a uniformly computable list of these principal types given by the ones realized in the decidable prime model.

The classical theorems at first glance suggest that this condition might be sufficient. We should use this list of computable types to construct the model. However, an additional possible uniformity is suggested by each classical characterization. The characterization of prime models suggests that we might need to be able to go uniformly effectively from formulas to (indices for) principal types containing them. The characterization of atomic models suggests that one might need to be able to go uniformly effectively from formulas to generating formulas for the principal types containing them. Although the two classical versions are equivalent these two effective versions are not. The first is clearly necessary as given a formula ψ consistent with T and a decidable prime model $\mathcal{A}$ we can computably find an n-tuple of elements of $\mathcal{A}$ satisfying ψ. The set of formulas satisfied by this n-tuple $\mathcal{A}$ is then a computable principal type containing ψ. It turns out that this condition is also sufficient. The second condition clearly implies the first and so is sufficient but not, as it turns out, necessary.

Theorem 3.5 (Harrington [1974]; Goncharov and Nurtazin [1973]) *A complete decidable theory T has a decidable prime model if and only if there is a computable function taking each formula to (an index for) a computable principal type containing it.*

Proof. We construct the desired model by a priority argument reminiscent of that for the Sacks splitting theorem for computably enumerable sets [1963] but instead producing a Henkin construction that restricts the types realized to the principal ones.

Let σ_e list the formulas of L_c the language of T extended by new constants c_i. We construct in stages a sequence of finite sets $\Psi_s(c_1, \ldots, c_{n_s})$ of sentences consistent with T with union Ψ as in the proof of Theorem 2.2. Again we let $\psi_s = \wedge\Psi_s$. At each stage s of the construction $\Gamma_{e,s}$ will be a principal e-type containing the formula $\exists y_{e+1}, \ldots, \exists y_{n_s} \psi_s(x_1, \ldots, x_e, y_{e+1}, \ldots, y_{n_s})$. Our goal is to satisfy the requirements P_e of Theorem 2.2 as well as new ones Q_e that guarantee that the model constructed is prime by making sure that only principal types are realized. We satisfy Q_e by making sure that $\Gamma_{e,s}$ is eventually constant and so that $[c_1], \ldots [c_n]$ satisfies the principal type $\Gamma_e (= \lim_s \Gamma_{e,s})$. (We denote the equivalence class of c_i in the model built from the constants as in Theorem 2.2 by $[c_i]$.)

- $P_e : \sigma_e \in \Psi$ or $\neg\sigma_e \in \Psi$ and if σ_e is of the form $\exists x\theta(x)$ and in Ψ then $\theta(c_i) \in \Psi$ for some i.
- $Q_e : \langle [c_1], \ldots [c_e] \rangle$ realizes a principal type $\Gamma_e = \lim_s \Gamma_{e,s}$.

Construction: At stage s, if only one of σ_s and $\neg\sigma_s$ is consistent with $T \cup \Psi_s$ put it into Ψ_{s+1}. Suppose it is ρ that is put into Ψ_{s+1} and so $T \vdash \psi_s \rightarrow \rho$. As $\exists y_{e+1}, \ldots, \exists y_{n_s} \psi_s(x_1, \ldots, x_e, y_{e+1}, \ldots, y_{n_s})$ is in $\Gamma_{e,s}$ which is a complete type over T, and $T \vdash \psi_s \rightarrow \rho$, $\exists y_{e+1}, \ldots, \exists y_{n_{s+1}} \psi_{s+1}(x_1, \ldots, x_e, y_{e+1}, \ldots, y_{n_{s+1}})$ is also in $\Gamma_{e,s}$. So we can let $\Gamma_{e,s+1}$ be $\Gamma_{e,s}$ for all e. If both σ_s and $\neg\sigma_s$ are consistent with $T \cup \Psi_s$, the problem is that adding σ_s (or $\neg\sigma_s$) to Ψ_s to form Ψ_{s+1} may make $\exists y_{e+1}, \ldots, \exists y_{n_{s+1}} \psi_s(x_1, \ldots, x_e, y_{e+1}, \ldots, y_{n_{s+1}})$ not be a member of $\Gamma_{e,s}$ for various numbers e. This would force us to change our choice of the type realized by $\langle [c_1], \ldots, [c_e] \rangle$ and so make $\Gamma_{e,s+1} \neq \Gamma_{e,s}$. We view this as an injury to requirement Q_e (which requires that $\Gamma_{e,s}$ eventually stabilize). As in the Sacks splitting theorem we act so as to minimize the priority of the first requirement injured.

More precisely, we let ψ^0_{s+1} be $\psi_s \wedge \sigma_s$ and ψ^1_{s+1} be $\psi_s \wedge \neg\sigma_s$. We let $e_{i,s}$ (for $i = 0, 1$) be the least $e \leq s$ such that $\exists y_{e+1}, \ldots, \exists y_{n_{s+1}} \psi^i_{s+1}(x_1, \ldots, x_e, y_{e+1}, \ldots, y_{n_{s+1}})$ is not in $\Gamma_{e,s}$. (If none exists, $e_{i,s} = s$.) If $e_{0,s} \leq e_{1,s}$ let $\psi_{s+1} = \psi^1_{s+1}$ and otherwise let $\psi_{s+1} = \psi^0_{s+1}$. Let $e_s = \min\{e_{0,s}, e_{1,s}\}$. For $e \leq e_s$ we can let $\Gamma_{e,s+1} = \Gamma_{e,s}$ as for such e, $\exists y_{e+1}, \ldots, \exists y_{n_{s+1}} \psi_{s+1}(x_1, \ldots, x_e, y_{e+1}, \ldots, y_{n_{s+1}}) \in \Gamma_{e,s}$. For $e > e_s$ we redefine $\Gamma_{e,s+1}$ as the first in our uniformly computable list of principal types which contains $\exists y_{e+1}, \ldots, \exists y_{n_s+1} \psi_{s+1}(x_1, \ldots, x_e, y_{e+1}, \ldots, y_{n_s+1})$.

If we have put $\exists x \theta(x)$ into Ψ, we put $\theta(c_i)$ in as well for some unused c_i. This clearly does not require any change in the $\Gamma_{e,s+1}$ already defined.

Verifications: As T is decidable and the types on our list are uniformly computable, the construction is clearly computable. We clearly satisfy the P_e requirements and so construct a decidable model $\mathcal{M}$ as in Theorem 2.2. As all sentences σ_i involving only $c_1, \ldots, c_e$ that are put into Ψ_s at stage s belong to the principal type $\Gamma_{e,s}$, if we can show that $\lim_s \Gamma_{e,s}$ exists for each e (and is say Γ_e) then we will have shown that, in $\mathcal{M}$, $\langle [c_1], \ldots [c_e] \rangle$ realizes the principal type Γ_e as required to guarantee that $\mathcal{M}$ is a prime model of T.

We prove by induction on e that there is a stage t_e such that $e_s > e$ for all $s \geq t_e$ and so $\Gamma_{e,s} = \Gamma_{e,t_e}$ for all $s > t_e$. Suppose that t_{e-1} exists. We need to show that e_s is greater than e for all sufficiently large s. Now, by the definition of t_{e-1}, $e \leq e_s$ for every $s > t_{e-1}$ and so by the choice of Ψ_{s+1} in the construction, $\Gamma_{e,s} = \Gamma_{e,t_e} = \Gamma_e$ for all $s > t_e$. As Γ_e is principal, some $\sigma(x_1, \ldots, x_e)$ is a generator and so by some stage $t \geq t_{e-1}$ we have added σ to Ψ_t. We claim that $e_s > e$ for every $s > t$. Consider σ_s for any $s > t$. The only way e_s could be e is if both σ_s and $\neg\sigma_s$ are consistent with $T \cup \Psi_s$ but $\exists y_{e+1}, \ldots, \exists y_{n_{s+1}} \psi^i_{s+1}(x_1, \ldots, x_e, y_{e+1}, \ldots, y_{n_{s+1}})$ is not in $\Gamma_{e,s}$ for $i = 0$ or 1. As $\exists y_{e+1}, \ldots, \exists y_{n_{s+1}} \psi_s(x_1, \ldots, x_e, y_{e+1}, \ldots, y_{n_{s+1}}) \rightarrow \sigma$ and σ is complete this would mean that

$$\exists y_{e+1}, \ldots, \exists y_{n_{s+1}} \psi_s(x_1, \ldots, x_e, y_{e+1}, \ldots, y_{n_{s+1}}) \rightarrow \neg\exists y_{e+1}, \ldots, \exists y_{n_{s+1}} \sigma_s(x_1, \ldots, x_e, y_{e+1}, \ldots, y_{n_{s+1}})$$

or that

$$\exists y_{e+1}, \ldots, \exists y_{n_{s+1}} \psi_s(x_1, \ldots, x_e, y_{e+1}, \ldots, y_{n_{s+1}}) \rightarrow \neg\exists y_{e+1}, \ldots, \exists y_{n_{s+1}} \neg\sigma_s(x_1, \ldots, x_e, y_{e+1}, \ldots, y_{n_{s+1}})$$

so that σ_s or $\neg\sigma_s$, respectively, would be inconsistent with Ψ_s contrary to our assumption. Thus t is the required stage t_e. □

We finish this section with an alternative version of Theorem 3.5 and some remarks about various uniformity conditions.

Corollary 3.6 *A complete decidable theory T has a decidable prime model if and only if T has a prime model and the set of all principal types of T is uniformly* *computable.*

Proof. The only if direction of this Corollary is clearly implied by the Theorem. Suppose then that T has a prime model and the set of principal types of T is uniformly computable. As T has a prime model, every formula ψ is a member of a principal type and so the search among those in the given set for one containing ψ terminates and provides the computable function required in the theorem. □

The effective uniformity in the listing of the computable principal types is necessary as an explicit hypothesis:

Theorem 3.7 (Millar[1978]) *There is a complete decidable theory T all of whose types are computable with a prime model but no decidable (or even computable) one.*

Finally, we show that the possible alternate version of Theorem 3.5 that asks for a computable way to go from a formula to a completion is false and so "uniformly atomic" is stronger than "uniformly prime" even for decidable $\aleph_1$-categorical theories.

Proposition 3.8 *There is a (complete) decidable $\aleph_1$-categorical theory T with a decidable prime model but with no computable function taking formulas to complete extensions.*

Proof. The language of T has infinitely many unary predicates R_i. The axioms of T say that the cardinality of each R_i is exactly 2 and that R_i and R_j are disjoint for distinct i and j except for some *designated* triples $\langle i, j, k\rangle$ such that R_k consists of one element from each of R_i and R_j. Moreover, no two distinct designated triples have any entry in common. The actual list of axioms for T is thus determined by the list of designated triples. This list will be defined recursively to diagonalize against each possible computable partial function θ_e which might be a candidate for a function taking formulas to complete extensions. Thus T will be axiomatisable. It is also $\aleph_1$ -categorical. (The part of the model consisting of elements in any R_i is uniquely determined by the axioms. The rest just consists of $\aleph_1$ many elements not in any R_i.) Thus T is complete and decidable.

The list of designated triples is effectively enumerated in increasing order (and so is computable) by waiting to diagonalize each θ_i at the formula $R_{2i}(x)$. If $\theta_i(R_{2i}(x))$ converges at stage s, we choose j, k larger than any number mentioned already and designate the triple $\langle 2i, 2j+1, 2k+1\rangle$. In particular, if $\theta_i(R_{2i}(x))$ is the generating formula $\theta(x)$ (which implies $R_{2i}(x)$) then θ cannot mention R_{2k+1}. We claim that T can prove neither that $\theta(x)$ implies $R_{2k+1}(x)$ nor that it implies $\neg R_{2k+1}(x)$ and so θ_i is not a function taking formulas to complete extensions. To see that no information about $R_{2k+1}(x)$ can be implied by $\theta(x)$ consider the theory T' gotten by restricting T to the language L' which is L without the predicate R_{2k+1}. T' is clearly also $\aleph_1$ -categorical and consistent with $\theta(x)$. Let $\mathcal{A}$ be a model of T' and a an element realizing $\theta(x)$. Let b be the other element of R_{2i} in $\mathcal{A}$ and c and

d the elements of R_{2j+1}. (R_{2i} and R_{2j+1} are disjoint by construction.) We can easily expand $\mathcal{A}$ to a model of T by interpreting R_{2k+1} as either $\{a, c\}$ or $\{b, d\}$. Thus $\theta(x)$ cannot imply either $R_{2k+1}(x)$ or $\neg R_{2k+1}(x)$. □

4. Saturated Models and the Number of Decidable Models

Definition 4.1 A model $\mathcal{A}$ of a theory T in the language L is a *saturated* model of T if it realizes every type of T with finitely many parameters from A. (If it exists, the saturated model of T is unique.)

The characterization of decidable theories with decidable saturated models is somewhat easier than for prime ones.

Theorem 4.2 (Morley [1976], Millar [1978], Goncharov [1978a]) *A decidable theory T has a decidable saturated model if and only if the types of T are uniformly computable.*

Proof sketch. If T has a decidable saturated model $\mathcal{A}$ then the types of T are uniformly computable as we can simply list the n-tuples from A and, for each of them the set of formulas it satisfies. For the other direction, we can use the uniformly computable list of types to do an effective Henkin construction. As the construction proceeds, we designate new constants to realize each potential type over previously introduced constants. As all the potential types over new constants are given uniformly computably as restrictions to a subset of their free variable of ones on our given list this procedure can be effectively organized. Roughly speaking, the plan is to continue to make the designated constants realize the appropriate type until an inconsistency is reached. We can check for inconsistencies with previously assigned types since they are all uniformly computable. We use a priority ordering to guarantee that, despite the need to cancel attempts at realizing certain potential types, each actual type over the constants introduced is in fact realized. Thus the model constructed is saturated as required. □

By Millar [1978], the explicit assumption of uniformity is necessary even if one assumes that the decidable theory T has a saturated model and all its types are computable. Millar [1978, p. 63] suggests that the proof of this results can be modified to show that there is no connection between the decidability of the saturated and prime models (when both exist). We now show that, in fact, if there is a decidable saturated model then there is a decidable prime model.

Proposition 4.3 (Ershov [1980, 381-382], see also Goncharov [1997, Theorem 3.4.4]) *If a complete theory T has a decidable saturated model then it has a decidable prime model.*

Proof. As T has a decidable model it is itself decidable by Corollary 2.4. As it has a decidable saturated model, Theorem 4.2 gives us a uniformly computable list Γ_e of all the types of T. By Theorem 3.5, it suffices to prove that, given any formula ϕ consistent with T, we can go effectively to a principal type Γ containing ϕ. We begin with the first type Γ_{n_0} on our list containing $\phi = \phi_0$. We proceed recursively to extend ϕ to ϕ_i and define a type Γ_{n_i} containing ϕ_i. Given ϕ_i, Γ_{n_i} and σ_i (from the list of all formulas with the same number of free variables as ϕ), we ask if both σ_i and $\neg\sigma_i$ are consistent with $T \cup \{\phi_i\}$. If not, $\phi_{i+1} = \phi_i$ and $n_{i+1} = n_i$. If so, we find the first e_0 and e_1 such that $\phi_i \wedge \sigma_i \in \Gamma_{e_0}$ and $\phi_i \wedge \neg\sigma_i \in \Gamma_{e_1}$, respectively. We let n_{i+1} be the larger of e_0 and e_1 and let ϕ_{i+1} be $\phi_i \wedge \sigma_i$ or $\phi_i \wedge \neg\sigma_i$ accordingly. It is clear that the sequence n_i is nondecreasing as at step i of the construction if e_0 and e_1 are defined then one of them is n_i and we always take the larger. As this procedure is effective, $\{\phi_i | i \in \omega\}$ generates a computable type Γ containing ϕ. If n_i is not eventually constant, Γ would be a type of T not equal to any Γ_e for a contradiction. Once n_i has stabilized say at n we can define e_0 and e_1 at only finitely many stages s as each time we do so we extend ϕ_s and eliminate one possible Γ_j for $j < n$ from future consideration. Thus ϕ_i also eventually stabilizes say at ϕ_e. It is now clear that ϕ_e generates the type Γ_n which is therefore the required principal type containing ϕ. □

We now see what the proof of Vaught's theorem that a complete theory cannot have exactly two models gives us.

Corollary 4.4 *If a complete but not $\aleph_0$-categorical theory T has a decidable saturated model then it has at least three decidable models.*

Proof. Let $\mathcal{A}$ be a decidable saturated model of T. By Proposition 4.3, T has a decidable prime model $\mathcal{B}$. As T is not $\aleph_0$-categorical, the decidable saturated model $\mathcal{A}$ of T is not a prime model and so $\mathcal{A}$ and $\mathcal{B}$ are not isomorphic. Thus $\mathcal{A}$ realizes a nonprincipal (but computable) type $\Gamma(\overline{x})$. $\mathcal{A}$ can clearly be expanded to a saturated model of $T \cup \Gamma(\overline{c})$ by properly interpreting the constants $\overline{c}$ and so $T \cup \Gamma(\overline{c})$ has a decidable saturated model and hence a decidable prime model $\mathcal{C}$ by Proposition 4.3. Of course, the restriction of $\mathcal{C}$ is a decidable model of T. As in the proof of Vaught's theorem (as in Chang and Keisler [1990, Theorem 2.3.15]), this model cannot be isomorphic to either $\mathcal{A}$ or $\mathcal{B}$. □

On the other hand, if a decidable theory T has no decidable prime model (and so no decidable saturated model) then it has infinitely many decidable prime models. To see this, we quote a simple case of Millar's effective omitting types theorem.

Theorem 4.5 (Millar [1983]) *If T is a decidable theory and $\{\Gamma_i | i < n\}$ a finite set of computable nonprincipal types of T then there is a decidable model of T omitting every (i.e. not realizing any) Γ_i.*

Corollary 4.6 *If a decidable theory T does not have a decidable prime model then T has infinitely many decidable models.*

Proof. By Theorem 2.2, T has a decidable model $\mathcal{A}$. As $\mathcal{A}$ is not a prime model it realizes some nonprincipal type Γ_1. By Theorem 4.5, there is a decidable model $\mathcal{A}_1$ of T omitting Γ_1. As $\mathcal{A}_1$ is not prime, it realizes a nonprincipal type Γ_2 distinct from Γ_1 by construction. We now get a decidable $\mathcal{A}_2$ omitting both Γ_1 and Γ_2. Continuing in this way we get an infinite sequence Γ_i of computable nonprincipal types of T and decidable nonisomorphic models $\mathcal{A}_i$ of T as required. (Each $\mathcal{A}_i$ realizes Γ_{i+1} but not Γ_j for any $j \leq i$.) □

Another variation on the question of how many decidable models a decidable theory can have asks when is every model of T decidably presentable. One obvious necessary condition is that all types in T are computable. (Every type is realized in some model and only computable types can be realized in a decidable model.) Thus, in particular, T can have only countably many types. This condition is not sufficient and the problem remains open in general. There are a couple of partial answers. The answer is simple for $\aleph_0$-categorical theories and is supplied by Proposition 2.5. The nicest result is for $\aleph_1$-categorical theories to which we now turn.

5. $\aleph_1$-Categorical Theories

If a theory T is $\aleph_1$-categorical (and so complete) but not $\aleph_0$-categorical then the Baldwin-Lachlan theorem [1971] supplies us with a full classification of the models of T in terms of a well defined notion of dimension. There are countably many models $\mathcal{A}_i$ of T and they are arranged in a liner order of type $\omega + 1$ with respect to elementary embedding ascending with increasing dimension:

$$\mathcal{A}_0 \preceq \mathcal{A}_1 \preceq \mathcal{A}_2 \preceq \ldots \preceq \mathcal{A}_n \preceq \ldots \preceq \mathcal{A}_\infty.$$

$\mathcal{A}_0$, the model of dimension zero is the prime model of T and $\mathcal{A}_\infty$, the unique model of infinite dimension, is the saturated model of T. The model $\mathcal{A}_i$ for $i > 0$ is the model of dimension i.

The classic example of an $\aleph_1$ but not $\aleph_0$- categorical theory is ACF_0. Here the dimension of a model is its transcendence degree over the prime field $\mathbb{Q}$. $\mathcal{A}_0$, the prime model, is the algebraic closure of $\mathbb{Q}$. $\mathcal{A}_\infty$, the saturated model, is the algebraic closure of the rationals extended by infinitely many transcendental elements. Each $\mathcal{A}_i$ for $i > 0$ is the algebraic closure of $\mathbb{Q}$ extended by i many transcendentals.

The general problem we wish to address is the following:

Question 5.1 *If T is $\aleph_1$ but not $\aleph_0$-categorical theory when (and which of) its models are decidably or computably presentable?*

5.1 Decidable Models of $\aleph_1$-Categorical Theories

Of course, if T is $\aleph_1$-categorical and so complete, it has a decidable model if and only if it is itself decidable (Theorem 2.4). Actually, the decidability of T is enough to guarantee that every model is decidably presentable:

Theorem 5.2 (Harrington[1974], Khisamiev [1974]) *If T is $\aleph_1$-categorical and decidable then every model of T is decidably presentable.*

Proof. We first use the results of Baldwin and Lachlan [1971] to show that we can reduce the problem to that of the existence of decidable prime models for a decidable theory T. (All the model theoretic facts we cite in this proof can be found in Baldwin and Lachlan [1971].)

As T is $\aleph_1$-categorical, there is a principal n-type $\Gamma(x_1, \ldots, x_n)$ such that $T' = T \cup \Gamma(c_1, \ldots, c_n)$ (with c_i new constants) has a *strongly minimal formula*, i.e. a formula $\phi(x)$ of L' (the language L of T expanded by new constants c_i) such that for every model $\mathcal{A}$ of T' and every formula $\psi(x)$ of L', exactly one of $\{a \in A | \mathcal{A} \models \phi(a) \wedge \psi(a)\}$ and $\{a \in A | \mathcal{A} \models \phi(a) \wedge \neg\psi(a)\}$ is finite. Of course, T' is $\aleph_1$-categorical. Note that as T is decidable and Γ is principal, T' is also decidable ($T' \vdash \phi \Leftrightarrow \phi \in \Gamma \Leftrightarrow T \vdash \theta \rightarrow \phi$ where θ is a generator of Γ). As all models of T can be extended to ones of T', we can assume for the proof of our theorem that T has a strongly minimal formula ϕ.

Now each model of an $\aleph_1$-categorical theory T with a strongly minimal formula ϕ is the prime model of an extension T' of T by constants d_i satisfying a type Δ which says that $\phi(d_i)$ holds for each i and that the d_i are algebraically independent, i.e. there is no formula $\psi(x, \overline{y}) \in \Delta$ such that for some n, $\exists^{\leq n} x(\phi(x) \wedge \psi(x, \overline{y})) \in \Delta$. (In fact, the cardinality of the set of d_i is the dimension of the model and uniquely determines it.) Again T' is clearly $\aleph_1$-categorical. We must verify that it is also decidable, i.e. Δ is computable. We prove by induction on the number n of d_i that the corresponding types Δ_n and theories $T_n = T \cup \Delta_n(d_1, \ldots, d_n)$ are uniformly decidable. (They are complete by definition.) For $n+1$, consider any formula $\psi(x, d_1, \ldots, d_n)$. In each model $\mathcal{A}$ of T_n exactly one of $\{a \in A | \mathcal{A} \models \phi(a) \wedge \psi(a, d_1, \ldots, d_n)\}$ and $\{a \in A | \mathcal{A} \models \phi(a) \wedge \neg\psi(a, d_1, \ldots, d_n)\}$ is finite by the strong minimality of ϕ. By compactness, there is then an $m \in \mathbb{N}$ such that $T_n \vdash \exists^{\leq m} x(\phi(x) \wedge \psi(x, \overline{d}))$ or $T_n \vdash \exists^{\leq m} x(\phi(x) \wedge \neg\psi(x, \overline{d}))$. As T_n is decidable, we can search for and find such an m for ψ or $\neg\psi$. The other is in Δ, i.e. if $T_n \vdash \exists^{\leq m} x(\phi(x) \wedge \psi(x, \overline{d}))$ then $\neg\psi(x, \overline{d}) \in \Delta$ and if $T_n \vdash \exists^{\leq m} x(\phi(x) \wedge \neg\psi(x, \overline{d}))$ then $\psi(x, \overline{d}) \in \Delta$. Thus each T_n and $T_\infty = \cup T_n$ is decidable and the models of T are precisely the prime models of these theories. To prove our theorem it therefore suffices to show that each of these theories has a decidable prime model.

By Theorem 3.5, it suffices to show that if T is a decidable $\aleph_1$-categorical theory with a strongly minimal formula ψ then there is a computable function taking any formula $\sigma(\overline{x})$ to a computable principal type Γ_σ containing σ.

Given σ, we construct a computable type Γ in stages e by starting with σ and adding on each σ_e in turn if it is consistent with what we have put in Γ so far and, if σ_e is $\exists y(\psi(y) \wedge \theta(y, \overline{x}))$, we also add in $\exists y(\psi(y) \wedge \theta(y, \overline{x}) \wedge \phi(y))$ for some algebraic ϕ, i.e. one such that $T \vdash \exists^{\leq n} y(\psi(y) \wedge \phi(y))$ for some $n \in \omega$. Of course, if σ_e is not consistent with what we have so far we add on $\neg\sigma_e$. The point here is that if $\exists y(\psi(y) \wedge \theta(y, \overline{x}))$ is consistent with what we have so far then the formula gotten by adding it on is realized in the prime model of T say by $\overline{c}$. Now that model has only algebraic realizations of ψ and so whatever element witnessed $\exists y(\psi(y) \wedge \theta(y, \bar{c}))$ is algebraic and so also satisfies some algebraic formula ϕ. Thus $\exists y(\psi(y) \wedge \theta(y, \bar{x}) \wedge \phi(y))$ can be consistently added on as desired.

We claim that Γ is principal and so the required Γ_σ. Consider the prime model $\mathcal{A}$ of $T \cup \Gamma(\bar{c})$ and any $a \in A$ such that $\mathcal{A} \models \psi(a)$. As $\mathcal{A}$ is a prime model of $T \cup \Gamma(\bar{c})$, a realizes a principal type over $T \cup \Gamma(\bar{c})$ generated say by $\theta(y, \bar{c})$. If a is not algebraic then for every formula ϕ and every $n \in \omega$, $T \cup \Gamma(\bar{c}) \vdash \theta(y, \bar{c}) \to [\phi(y) \to \neg\exists^{\leq n} y(\psi(y) \wedge \phi(y))]$. On the other hand, as $\mathcal{A} \models \psi(a) \wedge \theta(a, \bar{c})$, $\exists y(\psi(y) \wedge \theta(y, \bar{x})) \in \Gamma$ and so by construction $\exists y(\psi(y) \wedge \theta(y, \bar{x}) \wedge \phi(y)) \in \Gamma$ for some ϕ such that $T \vdash \exists^{\leq n} y(\psi(y) \wedge \phi(y))$ for some n for a contradiction. Thus $\mathcal{A}$ has only algebraic solutions of ψ, i.e. it is the model of dimension 0, and so $\mathcal{A}$ is actually the prime model of T. As Γ is realized in $\mathcal{A}$, it must be principal over T as required.

(This last argument is attributed to Lachlan in Harrington [1974]. Harrington's own proof is also instructive. It begins with the observation that the function taking a formula σ to its rank as defined in Baldwin [1973] can be seen to be a computable map from formulas into $\mathbb{N}$ by the arguments presented in that paper. Thus, given a formula σ consistent with T, we may computably define a type $\Gamma = \cup\Gamma_e$ containing σ by putting in, for each e in turn, either σ_e or $\neg\sigma_e$ so as to always preserve consistency and to reduce the rank of $\bigwedge \Gamma_e$ if possible. Eventually, the rank must stabilize and so we produce a principal type Γ containing σ.) □

5.2 Computable Models

We now turn to the question of which models of an $\aleph_1$-categorical but not $\aleph_0$-categorical theory T are computably presentable if T is not decidable. It is easy to find such a theory with no computable models by coding a noncomputable set S into every model. (For example, extend ACF_0 by adding on new unary predicates P_i and, for each $i \in \omega$, axioms $\forall x(P_i(x) \to x = 0)$ and $P_i(0)$ if $i \in S$ but $\neg P_i(0)$ if $i \notin S$.) Thus the question is, if T has a computable but no decidable model, which of the models $\mathcal{A}_i$ of T can or cannot be computable. Only a few facts are known.

Theorem 5.3 (Goncharov [1978], Kudaibergenov [1980]) *For every $n \in \mathbb{N}$ there is an $\aleph_1$-categorical but not $\aleph_0$ -categorical theory T such that $\mathcal{A}_0, \ldots, \mathcal{A}_n$ are all computably presentable but not $\mathcal{A}_i$ for $i > n$.*

Proof. Fix $n \in \mathbb{N}$. The language for the required theory T will consist of a unary predicate P_k and an n-ary predicate R_k for each $k \in \mathbb{N}$. The axioms for T will code a computably enumerable but not computable set $B = \cup B_s$ into each model of dimension greater than n while maintaining the possibility that the models of dimension less than or equal to n are computably presentable.

Axioms:

- The P_k are nested downward with respect to k and exactly one element drops out at each k, i.e. for each $k \in \mathbb{N}$ we have the following axioms:

  * $\forall x(P_{k+1}(x) \to P_k(x))$
  * $\exists! x(P_k(x) \wedge \neg P_{k+1}(x))$
- For each $k \in \mathbb{N}$ we wish to require that

$$R_k(x_1, \ldots, x_n) \Leftrightarrow \bigwedge\{x_i \neq x_j | i \neq j\} \wedge \exists s(k \in B_s \wedge x_1, \ldots, x_n \in P_s).$$

  We enforce this requirement by the following axioms:
  * $R_k(x_1, \ldots, x_n) \to x_i \neq x_j$ for $i \neq j$.
  * For each $s \in \mathbb{N}$ and $k \in B_s$: $\bigwedge\{x_i \neq x_j | i \neq j\} \wedge x_1, \ldots, x_n \in P_s \to R_k(x_1, \ldots, x_n)$.
  * For each $s \in \mathbb{N}$ and $k \notin B_s$: $\bigvee\{x_i \notin P_s | i \leq n\} \to \neg R_k(x_1, \ldots, x_n)$.

Verifications: It is easy to see that the cardinality of $\cap P_s^{\mathcal{A}}$ uniquely determines the isomorphism type of any model $\mathcal{A}$ of T and that all models $\mathcal{A}$ of size $\aleph_1$ have $\aleph_1$ many elements in $\cap P_s^{\mathcal{A}}$. Thus T is $\aleph_1$-categorical. Indeed, the cardinality of $\cap P_s^{\mathcal{A}}$ is the dimension of $\mathcal{A}$.

We claim that a model $\mathcal{A}$ of T is computably presentable if and only if there are fewer than n distinct elements in $\cap P_s^{\mathcal{A}}$. For one direction, suppose that there are distinct $c_1, \ldots, c_n$ in $\cap P_s^{\mathcal{A}}$. In that case, $k \in B \Leftrightarrow \mathcal{A} \models R_k(c_1, \ldots, c_n)$ and so $\mathcal{A}$ cannot be computably presentable as B is not computable.

For the other direction, we wish to construct a computable model $\mathcal{A}$ of T with $m < n$ many elements $c_1, \ldots, c_m$, in $\cap P_s^{\mathcal{A}}$. We let the other elements of the desired model be the natural numbers and we put i in P_k if and only if $i \geq k$. We now only have to computably define the predicates R_k. Given distinct elements $a_1, \ldots, a_n$ from A, not all of them are from among the c_i and so we can effectively find an s and indeed the smallest s such that one of them is not in P_s. We then let $R_k(a_1, \ldots, a_n)$ hold if and only if $k \in B_{s-1}$. This clearly defines a computable model $\mathcal{A}$ of T with $|\cap P_s^{\mathcal{A}}| = m$ as required. □

Thus any initial segment of the models of T can be the computably presentable ones. The obvious questions arise as to what else is possible.

Question 5.4 *Which subsets of $\omega + 1$ can be the set of computably presentable models of an $\aleph_1$-categorical but not $\aleph_0$-categorical theory T with a computable model? In particular, must the prime model always be computably presentable? Must the saturated model be computably presentable if all the others are?*

The following theorem answers the two specific questions asked. All other instances of the general question are open.

Theorem 5.5 (Khoussainov, Nies and Shore [1997]) *There are $\aleph_1$ -categorical but not $\aleph_0$-categorical theories T_1 and T_2 such that*

i) All models of T_1 except the prime one are computably presentable.
ii) All models of T_2 except the saturated one are computably presentable.

Proof (For T_1). Given $S \subset \omega$ we construct a structure $\mathcal{A}_S$ of signature $L =$

$(P_0, P_1, P_2, \ldots)$, where each P_i is a binary predicate symbol having the following properties:

- The theory T_S of the structure $\mathcal{A}_S$ is $\aleph_1$ - but not $\aleph_0$–categorical and A_S is the prime model of T_S.
- Each nonprime model A of T_S has a computable presentation if and only if S is Σ_2^0.
- A computable prime model provides S with a certain recursion-theoretic property but there exists a Σ_2^0–set which does not have this property.

The building blocks of our structures $\mathcal{A}_S$ will be finite structures that we call n-cubes and now define by induction on n.

Definition 5.6 A 1–*cube* C_1 is a structure $(\{a, b\}, P_0)$ such that $P_0(x, y)$ holds in C_1 if and only if $(x = a$ and $y = b)$ or $(y = a$ and $x = b)$. Given two disjoint n-cubes we get an $n + 1$-*cube* as an expansion of their union by letting P_n be an isomorphism between the n-cubes. An ω–*cube* is an increasing union of n–cubes, $n \in \omega$ with signature $(P_0, P_1, P_2, \ldots)$

Definition 5.7 If $S \subseteq \omega$, $\mathcal{A}_S$ is the disjoint union of n -cubes for $n \in S$ and $T_S = Th(\mathcal{A}_S)$.

Lemma 5.8 *If S is infinite, then T_S is $\aleph_1$- but not $\aleph_0$ –categorical and the model with no ω-cubes is its prime model.*

Proof. It is easy to see that the model $\mathcal{A}_S$ satisfies the following conditions which are all expressible by a set of axioms in the language L:

1. $\forall x \exists y P_0(x, y)$ and for each n, P_n defines a partial one-to-one function. (We abuse notation by also denoting this partial function by P_n.)
2. For all $n \neq m$ and for all x, $P_n(x) \neq P_m(x)$.
3. For each n and for all x if $P_n(x)$ is defined, then $P_0(x)$, $P_1(x)$, ..., $P_{n-1}(x)$ are also defined.
4. For all n, m and for all x if $P_n(x)$ and $P_m(P_n(x))$ are defined, then $P_m(P_n(x)) = P_n(P_m(x))$.
5. For all $k, n > n_1 \geq n_2 \geq \ldots \geq n_{k-1} \geq n_k$, $\forall x(P_{n_1}(\ldots(P_{n_k}(x)\ldots) \neq P_n(x))$.
6. For each $n \in \omega$, $n \in S$ if and only if there exists exactly one n–cube which is not contained in an $n+1$–cube.

Let $\mathcal{M}$ be a model which satisfies all the above statements. For each $n \in S$, $\mathcal{M}$ must have an n–cube which is not contained in an $n + 1$–cube. If an $x \in M$ does not belong to any n–cube for $n \in S$, then x is in an ω–cube. Thus any two models which satisfy this list of axioms are isomorphic if and only if they have the same number of ω–cubes. In particular, if $\mathcal{M}_1$ and $\mathcal{M}_2$ are models of T_S of cardinality $\aleph_1$, each has $\aleph_1$ many ω–cubes (as each cube is countable). Thus $\mathcal{M}_1$ and $\mathcal{M}_2$ are isomorphic and T_S is an $\aleph_1$- but not $\aleph_0$–categorical theory. It is clear that the prime model is the one with no ω-cubes. □

Lemma 5.9 *Each nonprime model of T_S is computably presentable if and only if S is Σ_2^0.*

Proof. If $\mathcal{M}$ is a model of T_S, $s \in S$ if and only if $\mathcal{M} \models \exists x \exists y \forall z (P_s(x,y) \& \neg P_{s+1}(x,z))$. Thus if $\mathcal{M}$ is computably presentable S is Σ_2^0. For the other direction, note that it suffices to construct a computable model $\mathcal{M}_1$ with one ω-cube when $S \in \Sigma_2^0$. (We can computably add on more ω-cubes as desired.) We build $\mathcal{M}$ by putting in an n-cube when, according to the Σ_2^0 representation of S as $\{n | \exists x \forall y H(x,y,n)\}$, we seem to have a witness x that $n \in S$. When the witness fails, we merge this n-cube into the ω-cube that we are building. More formally, at stage 0 we start to build a substructure $\mathcal{B}$ that will be an m-cube for some m at every stage s and will at the end of the construction be an ω-cube. At stage s, we first put into $\mathcal{M}$ an n-cube for each $n < s$ for which we do not have one and associate the cube with the first number x that has not yet been associated with n. Then, we merge $\mathcal{B}$ and the existing n-cubes for those $n < s$ for which there is a $y < s$ such that $H(x,y,n)$ fails for the x currently associated with n into an m-cube for some m larger than any number yet used in the construction. Clearly the substructure $\mathcal{B}$ becomes the only ω-cube of $\mathcal{M}$. Moreover, for $n \in \omega$, there is an n-cube in the final structure $\mathcal{M}$ if and only if $\exists x \forall y H(x,y,n)$, i.e. if and only if $n \in S$ as required. □

We now provide the recursion theoretic property of S that is guaranteed by the existence of a computable prime model of T_S (but not by any of the other models being computably presentable).

Definition 5.10 A function f is *limitwise monotonic* if there exists a computable function $\phi(x,t)$ such that $\phi(x,t) \leq \phi(x,t+1)$ for all $x,t \in \omega$, $\lim_t \phi(x,t)$ exists for every $x \in \omega$ and $f(x) = \lim_t \phi(x,t)$.

Lemma 5.11 *If the prime model of T_S is computably presentable then S is the range of a limitwise monotonic function.*

Proof. Suppose $\mathcal{M}$ is a computable prime model of T_S. Define $\phi(x,s)$ for each $x \in M$ and $s \in \mathbb{N}$ as the largest $n < s$ such that $P_n(x,y)$ holds for some $y < s$. It is clear that $\phi(x,s)$ is monotonic in s. As every $x \in M$ is in an n-cube for some n, $\phi(x,s)$ is equal to this n for all sufficiently large s. □

Lemma 5.12 *There exists a Δ_2^0 set A which is not the range of any limitwise monotonic function.*

Proof. Let $\phi_e(x,t)$ be a list of all candidates for representations of limitwise monotonic functions f_e. At stage s we define a finite set A_s so that $A(y) = lim_s A_s(y)$ exists for all y (and hence A is Δ_2^0). We also satisfy the following requirements to guarantee that A is not the range of a limitwise monotonic function.

R_e : If $f_e(x) = lim_t \phi_e(x,t) < \omega$ for all x, then $range(f_e) \neq A$.

The strategy to satisfy a single R_e works as follows: At stage s, pick a witness m_e, enumerate m_e into A (i.e. set $A_s(m_e) = 1$). Now R_e is satisfied (since m_e

remains in A) unless at some later stage t_0 we find an x such that $\phi_e(x, t_0) = m_e$. If so, R_e ensures that $A(\phi_e(x,t)) = 0$ for all $t \geq t_0$. Thus, either $f_e(x) \uparrow$ or $f_e(x) \downarrow$ and $f_e(x) \notin A$.

Keeping $\phi_e(x,t)$ out of A for all $t \geq t_0$ can conflict with a lower priority ($i > e$) requirement R_i since it maybe the case that $m_i = \phi_e(x, t')$ for some $t' > t_0$. However, if $f_e(x) \downarrow$, then from some point on there is only one number that R_e prevents from being a candidate for m_i. If $f_e(x) \uparrow$, then the restriction is transitory, i.e. as $\phi_e(x,t)$ is monotonic in t each candidate for m_i is eventually released and never prevented from being chosen as the final value of m_i.Thus each lower priority R_i will eventually be able to choose a witness m_i that it will never have to change because of the actions of R_e. In this way, every requirement can be satisfied by a typical finite injury priority argument. □

Proof sketch (For T_2). We take a Π^0_2 set S defined by $k \in S \Leftrightarrow \forall n \exists m H(n,m,k)$ which is not Σ^0_2. (H is some computable predicate on $\mathbb{N}^3$.) We now code S into a computable structure $\mathcal{A}$ with unary predicates P_i and predicates $R_{k,s}$ of arity k for $i, k, s \in \mathbb{N}$. The relevant properties of $\mathcal{A}$ that can be guaranteed by axioms in this language are as follows:

- The $P_i^{\mathcal{A}}$ form a descending chain of sets with one element dropping out at each i.
- The $R^{\mathcal{A}}_{k,s}$ code the approximation $H(n,m,k)$ to $k \in S$ by requiring that if j is least such that $\forall n \leq s \exists m \leq j(H(n,m,k))$ and $x_1, \ldots, x_k \in P_j$ are distinct for $i \leq k$ then $R_{k,s}(x_1, \ldots, x_k)$ holds and not otherwise. (In particular, if $k \notin S$ then for some s_0 we have axioms saying that $R_{k,s}(x_1, \ldots, x_k)$ does not hold for any $s \geq s_0$ and any $x_1, \ldots, x_k$.)

The theory T_S of $\mathcal{A}_S$ is $\aleph_1$- but not $\aleph_0$-categorical with the dimension of a model $\mathcal{A}$ being once again determined by the cardinality of $\cap P_i^{\mathcal{A}}$. The intuition is that the more elements there are in $\cap P_i^{\mathcal{A}}$ for a model $\mathcal{A}$ of T_S, the more of the Π^0_2 -approximation to S that we can "recover" from the diagram of $\mathcal{A}$. In particular, if $\mathcal{A}$ is the saturated model of T_S, $\cap P_i^{\mathcal{A}}$ is infinite and S is Σ_2 in $\mathcal{A}$: $k \in S \Leftrightarrow \exists x_1, \ldots x_k \in A[(\forall i)(\mathcal{A} \models P_i(x_1) \wedge \ldots P_i(x_k)) \wedge (\forall s)(\mathcal{A} \models R_{k,s}(x_1, \ldots x_k))]$. As S is not Σ^0_2, the saturated model of T_S is not computably presentable. For each $t < \omega$, however, we can (nonuniformly) build a computable model $\mathcal{A}_t$ of T_S with t many elements in $\cap P_i^{\mathcal{A}_t}$. The information needed is $S \cap (t+1)$ and, for each $k \leq t$ which is not in S the least n for which there is no m such that $H(n,m,k)$ holds. □

All the theorems in this subsection about computable models of $\aleph_1$ -categorical theories use infinite signatures. Not too much is known about the existence of such structures and theories in finite signatures or for ones that are extensions of standard algebraic theories. One interesting example is Herwig, Lempp and Ziegler [1999] who have established Theorem 5.3 for $n = 0$ with T an extension of the theory of groups in the standard signature.

6. Computable Dimension and Categoricity

Until now we have taken the classical approach of identifying models up to classical isomorphism. However, it is not obvious that even two computable (or decidable) models that happen to be isomorphic should be identified when one is interested in effective procedures. There could well be (and indeed, as we shall see, there are) structures with presentations $\mathcal{A}$ and $\mathcal{B}$ such that the two presentations have different effective properties. For example, there are computable presentations of $\langle \mathbb{N}, \leq \rangle$ on which the successor function is not computable.

Proposition 6.1 *There is a computable presentation $\mathcal{A} = \langle A, \leq_A \rangle$ of $\langle \mathbb{N}, \leq \rangle$ such that the successor function on $\mathcal{A}$ is not computable.*

Proof. $\mathcal{A}$ will consist of the even numbers in their usual order plus an infinite set of odd numbers determined and placed in the ordering by a procedure designed to guarantee that no computable function ϕ_e is the successor function on $\mathcal{A}$. At stage s we check, for each $e < s$, if $\phi_e(2e)$ has converged at stage s and is equal to $2e+2$. If so we put $2s+1$ into A and place it between $2e$ and $2e+2$. It is obvious that A is computable and that $\phi_e(2e)$ is not the successor of $2e$ in $\mathcal{A}$ for any e. □

The natural approach to the issue raised by such examples is to identify structures or presentations only when there is a computable isomorphism between them. Of course, this only makes sense when the structures themselves are computable.

- Henceforth all structures will be computable.

Definition 6.2 $\mathcal{A}$ is *computably isomorphic* to $\mathcal{B}$, $\mathcal{A} \cong_c \mathcal{B}$, if there is a computable $f : A \to B$ which is an isomorphism. We also say then that $\mathcal{A}$ and $\mathcal{B}$ are *of the same computable isomorphism type.*

Definition 6.3 The *(computable) dimension* of a structure $\mathcal{A}$ is number of its computable isomorphism types. $\mathcal{A}$ is *computably categorical* if its computable dimension is 1, i.e. every $\mathcal{B}$ isomorphic to $\mathcal{A}$ is computably isomorphic to $\mathcal{A}$.

Note that in a computably categorical structure $\mathcal{A}$ every definable relation that is computable in any presentation of $\mathcal{A}$ is computable in every presentation of $\mathcal{A}$ and so for such structures the effectiveness of definable properties is independent of the presentation.

Example 6.4 $\mathbb{Q}$ (the rationals) with its usual linear order is computably categorical: The standard back and forth argument showing that the theory of dense linear orderings without endpoints is countably categorical is effective and so produces computable isomorphisms between any two such orderings.

Example 6.5 $\mathbb{N}$ as a model or PA or indeed as a structure with only the successor function $s(x)$ (given as $x+1$ in the language of arithmetic) is computably categorical: Given any $\mathcal{B}$ isomorphic to $\mathbb{N}$, one defines the required computable $f : \mathbb{N} \to \mathcal{B}$

by recursion. $f(0)$ is the first element of $\mathcal{B}$ and if $f(n)$ is defined as $b \in B$ then $f(n+1) = s^{\mathcal{B}}(b)$. However, it is easy to see from Proposition 6.1 that $\langle \mathbb{N}, \leq \rangle$ is not computably categorical. (If s is the successor function on $\mathbb{N}$ and $f : \mathbb{N} \rightarrow A$ were a computable isomorphism into the $\mathcal{A}$ of Proposition 6.1, fsf^{-1} would be a computable successor function on $\mathcal{A}$.)

Example 6.6 Every finitely generated structure is computably categorical by the natural generalization of the preceding argument for $\langle \mathbb{N}, s \rangle$.

Example 6.7 $\overline{\mathbb{Q}}$, the algebraic closure of the rationals and so the prime model of ACF_0, is computably categorical but $\widetilde{\mathbb{Q}}$, the countable saturated model of ACF_0 (i.e. the algebraic closure of the rationals extended by infinitely many transcendentals) has computable dimension ω (Corollary 6.12).

All of these examples have dimension 1 or ω but, actually, every $n \leq \omega$ is possible.

Theorem 6.8 (Goncharov [1980a]) *For each n, $1 \leq n \leq \omega$ there is a structure of dimension n.*

Goncharov uses a priority argument to construct families of uniformly computably enumerable sets with (in a precise sense) exactly n many distinct enumerations and then codes them into structures so as preserve the dimension. We will see other approaches to these results in Theorem 6.22 and Corollary 7.16. Although there are interesting codings of these families into familiar types of mathematics structures such as groups and rings (see § 9), we do not know of any "natural" structures with dimension n for $1 < n < \omega$. Indeed, for many classes of structures it is possible to prove that they are computably categorical or have dimension ω. In most of these cases it is actually possible to characterize the structures that are computably categorical.

Theorem 6.9 (Goncharov [1973], LaRoche [1977], Remmel [1981], Goncharov and Dzgoev [1980]) *A Boolean algebra is computably categorical if it has finitely many atoms. If not, it has dimension ω.*

Theorem 6.10 (Remmel [1981a], Goncharov and Dzgoev [1980]) *A linear order is computably categorical if it has only finitely many pairs of adjacent elements. If not, it has dimension ω.*

We can deduce a similar result on algebraically closed fields from a general theorem about computable categoricity among decidable presentations of a structure.

Theorem 6.11 (Nurtazin [1974]) *Suppose $\mathcal{A}$ is a decidable structure. If there are finitely many elements $\bar{c} \in A$ such that $(\mathcal{A}, \bar{c})$ is the prime model of the theory $Th(\mathcal{A}, \bar{c})$ and the set of complete formulas of this theory is computable, then any* *two decidable presentations of $\mathcal{A}$ are computably isomorphic. On the other hand, if there are no such $\bar{c}$, then there are infinitely many decidable presentations of $\mathcal{A}$ no two of which are computably isomorphic.*

Corollary 6.12 (Nurtazin [1974]; Metakides and Nerode [1979]) *An algebraically closed field of finite transcendence degree over its prime field is computably categorical. One of infinite transcendence degree has dimension ω.*

Proof. Let T be the theory of algebraically closed fields of characteristic 0. As T has quantifier elimination every computable model $\mathcal{A}$ of T is decidable. (Given a sentence with quantifiers (in the expanded language with constants for elements of A) find the quantifier free equivalent. Its truth can be decided by the computability of $\mathcal{A}$. As T is $\aleph_1$-categorical every model $\mathcal{A}$ is the prime model of $T' = T \cup \Gamma(\bar{c}) \cup \Delta(\bar{d})$ for a computable principal type Γ providing the theory with a strongly minimal formula and the type Δ of a sequence of transcendentals (independent elements) as described in the proof of Theorem 5.2. (Actually, for this particular T, Γ is not needed as it is already strongly minimal.) The sequence $\bar{d}$ is finite if and only if the transcendence degree of $\mathcal{A}$ over its prime field is finite. In particular if the transcendence degree is infinite, there is no finite sequence as required and so $\mathcal{A}$ would have infinite computable dimension. On the other hand, if the sequence is actually finite, we can effectively decide if a given formula $\phi(\bar{d}, \bar{x})$ is an atom. As in the proof of Theorem 5.2, we can go effectively to a computable principal type Γ of T' containing $\phi(\bar{d}, \bar{x})$. For this particular theory, however, we can enumerate the complete formulas. (In characteristic 0, they just say that (for some ordering of the x's), each x in turn satisfies some irreducible polynomial over the previous ones.) We can thus find such a generating formula γ in Γ and then ask if $\phi \longrightarrow \gamma$. If so ϕ is complete and not otherwise. (Metakides and Nerode [1979] give a direct proof of this Corollary.) □

An important program is thus to characterize or at least classify computably categorical structures and theories whose models are computably categorical. One major success along these lines is the characterization by Goncharov [1975] of computably categorical structures whose two quantifier theory is decidable in terms of Scott families.

Definition 6.13 A **Scott family** for a structure $\mathcal{A}$ is a computable sequence

$$\phi_0(\bar{a}, x_1, \ldots, x_{n_0}), \phi_1(\bar{a}, x_1, \ldots, x_{n_1}), \ldots,$$

of $\exists$-formulas, i.e. prenex ones with only existential quantifiers, satisfiable in $\mathcal{A}$, where $\bar{a}$ is a finite tuple of elements from $\mathcal{A}$, such that every n-tuple of elements from $\mathcal{A}$ satisfies one these formulas and any two tuples satisfying the same formula from the above sequence can be interchanged by an automorphism of $\mathcal{A}$.

Definition 6.14 A structure $\mathcal{A}$ is *n-decidable* (for $n \in \mathbb{N}$) if the set of prenex sentences of $Th(\mathcal{A}, a)_{a \in A}$ with $n - 1$ alternations of quantifiers is computable. So, for example, $\mathcal{A}$ is 1-*decidable* if the set of prenex sentences of $Th(\mathcal{A}, a)_{a \in A}$ with either only existential or only universal quantifiers is decidable.

Proposition 6.15 *If a structure $\mathcal{A}$ has a Scott family, then $\mathcal{A}$ is computably categorical.*

Proof. Let $\phi_0(\bar{a}, x_1, \ldots, x_{n_0}), \phi_1(\bar{a}, x_1, \ldots, x_{n_1}), \ldots$ be a Scott family for $\mathcal{A}$, where $\bar{a} = (a_0, \ldots, a_{m-1})$. Let $\mathcal{A}_1$ and $\mathcal{A}_2$ be computable presentations of $\mathcal{A}$. We define a mapping $f : \mathcal{A}_1 \rightarrow \mathcal{A}_2$ by stages. We can assume that for each $j \in \{0, \ldots, m-1\}$, a_j^i is the element in A_i corresponding to the constant a_j. At even stages we define images of elements from $\mathcal{A}_1$, at odd stages we define preimages of elements from $\mathcal{A}_2$.

Stage 0. Set $f_0 = \{(a_0^1, a_0^2), \ldots, (a_{m-1}^1, a_{m-1}^2)\}$.

Stage 2k>0. We can suppose that the function f_{2k-1} has been defined. Assume that $f_{2k-1} = \{(a_0^1, a_0^2), \ldots, (a_{m-1}^1, a_{m-1}^2), (b_1, d_1), \ldots, (b_s, d_s)\}$ and that f_{2k-1} can be extended to an isomorphism from $\mathcal{A}_1$ to $\mathcal{A}_2$. Let b be the first number in A_1 not in the domain of f_{2k-1}. Consider the tuple $(b_1, \ldots, b_s, b)$. Find an i such that $\phi_i(\bar{a}, b_1, \ldots, b_s, b)$ holds in $\mathcal{A}_1$. Hence $\exists x \phi_i(\bar{a}, d_1, \ldots, d_s, x)$ holds in $\mathcal{A}_2$. Find the first $d \in A_2$ for which $\phi_i(\bar{a}, d_1, \ldots, d_s, d)$ holds. Extend f_{2k-1} by letting $f_{2k} = f_{2k-1} \bigcup \{(b, d)\}$.

Stage 2k+1. We define f_{2k+1} similarly so as to put the least element of A_2 not yet in the range of f_{2k} into that of f_{2k+1}.

Finally, let $f = \bigcup_{i \in \omega} f_i$. Clearly, f is a computable isomorphism. □

Theorem 6.16 (Goncharov [1975]) *If $\mathcal{A}$ is 2-decidable then it is computably categorical if and only if it has a Scott family.*

Of course, the if direction of this Theorem follows from the preceding Proposition. For the other direction, one uses a priority argument to build a $\mathcal{B}$ and a Δ_2^0 isomorphism between $\mathcal{A}$ and $\mathcal{B}$. Attempts are made to make sure that no ϕ_e is an isomorphism between $\mathcal{A}$ and $\mathcal{B}$. If one of the attempts fails, the construction builds a Scott family for $\mathcal{A}$. (See Ash and Knight [1999] for the details of an ingenious but relatively simple proof.)

Note that the definition of computable categoricity is on its face a Π_1^1 property. This theorem gives a Σ_1^1 equivalent (having a Scott family). Actually, the property of having a Scott family can easily be seen to be arithmetic as the requirement for an isomorphism can be replaced by the existence of a set of finite partial isomorphisms with the back and forth property. Thus, for 2-decidable structures, Theorem 6.16 gives a characterization that is significantly simpler than the underlying definition of computable categoricity.

We now turn to the specific issue of persistence of computable categoricity under expansions by constants that will turn out to be a route into various results and examples of the sorts listed above. In particular, it will lead us to a proof that the existence of a Scott family is not necessary for computable categoricity.

6.1 Persistence of Computable Categoricity

Classically, it is an easy consequence of the Ryll-Nardzewski Theorem that having a countably categorical theory is *persistent*, i.e. preserved under expansions by finitely many constants.

Theorem 6.17 *If $Th(\mathcal{A})$, the theory of a structure $\mathcal{A}$, is countably categorical then so is the theory of any expansion of $\mathcal{A}$ by finitely many constants.*

The natural question for computable categoricity has been considered by Millar, Goncharov and others. It is posed as the Millar-Goncharov problem in Ershov and Goncharov [1986]:

Question 6.18 (Millar,Goncharov) *Is computable categoricity persistent, i.e. if $\mathcal{A}$ is computably categorical is also every expansion of $\mathcal{A}$ by finitely many constants?*

It is not hard to see that if a structure $\mathcal{A}$ has a Scott family $\phi_i(\bar{a}, x_1, \dots, x_{n_i})$ then every expansion by finitely many constants $c_1, \dots c_m$ also has one. We simply slightly modify the original Scott family. (Essentially, one replaces each formula $\phi_i(\bar{a}, x_1, \dots, x_{n_i})$ by $\phi_i(\bar{a}, c_1, \dots, c_m, x_1, \dots x_{n_i - m})$ and then lists only the satisfied formulas. Then, one can easily check that the sequence $\psi_0, \psi_1, \dots$ is a Scott family for the expanded structure $(\mathcal{A}, c_1, \dots, c_m)$.) Thus Theorem 6.16 gives us an answer when $\mathcal{A}$ is 2-decidable.

Corollary 6.19 (Goncharov [1975]) *If $\mathcal{A}$ is 2-decidable then the expansion of $\mathcal{A}$ by finitely many constants is also computably categorical.*

Millar has improved this result by one quantifier by a quite different proof. So, roughly speaking, it suffices to be able to solve systems of equalities and inequalities.

Theorem 6.20 (Millar [1986]) *If $\mathcal{A}$ is 1-decidable then the expansion of $\mathcal{A}$ by finitely many constants is also computably categorical.*

Proof (*Hirschfeldt*). Suppose we are given $\mathcal{A}$ and $\mathcal{B}$ isomorphic, computably categorical and 1-decidable with $\langle \mathcal{A}, a \rangle \cong \langle \mathcal{B}, b \rangle$. We will build $\mathcal{C}$ via a Henkin construction, a sequence g_s of partial isomorphisms from $\mathcal{C}$ to $\mathcal{B}$ and, for each potential isomorphism $\Phi_e : \mathcal{C} \to \mathcal{A}$, a partial map $h_e : \mathcal{C} \to \mathcal{B}$ such that

- either there is an e such that h_e is total and $h_e \Phi_e^{-1}$ is an isomorphism from $\langle \mathcal{A}, a \rangle$ to $\langle \mathcal{B}, b \rangle$,
- or $g = \lim_s g_s$ exists and is an isomorphism from $\mathcal{C}$ to $\mathcal{B}$ but no Φ_e is an isomorphism from $\mathcal{C}$ to $\mathcal{A}$.

As the second alternative contradicts the hypothesis that $\mathcal{A}$ is computably categorical, we will have the desired computable isomorphism between $\langle \mathcal{A}, a \rangle$ and $\langle \mathcal{B}, b \rangle$. In the construction we actually act, when we can, to guarantee that Φ_e is not an isomorphism from $\mathcal{C}$ to $\mathcal{A}$ (and so we do not have to worry about it). Thus we let R_e be the requirement that Φ_e is not an isomorphism from $\mathcal{C}$ to $\mathcal{A}$. As the construction proceeds, we say that R_e is satisfied (or not) depending on whether we have a certain type of witness to Φ_e's not being an isomorphism from $\mathcal{C}$ to $\mathcal{A}$.

For convenience, we assume that the domain of each model considered here is $\mathbb{N}$. Let $\{\theta_n\}_{n\in\omega}$ be an effective list of all atomic sentences in the language of $\mathcal{A}$ expanded by adding a constant $\mathbf{i}$ for each $i \in \omega$. By θ_n^0 and θ_n^1 we mean $\neg\theta_n$ and θ_n, respectively.

For any conjunction Γ of literals containing no constant $\mathbf{i}$ for $i > m$ and partial computable function Φ with computable domain, we let $f(k) = \mathbf{n}$ if $\Phi(k) \downarrow= n$, $f(k) = x_k$ if $\Phi(k) \uparrow$, and denote by $\Gamma[\Phi]$ the formula $\exists x_0 \cdots \exists x_m \Gamma(\mathbf{0}/f(0), \ldots \ldots, \mathbf{m}/f(m))$. So, for example, if θ_n is the sentence $P(\mathbf{0}, \mathbf{1}, \mathbf{2}, \mathbf{3})$ and $\Phi = \{\langle 1, 7\rangle, \langle 3, 5\rangle\}$, then $\theta_n^1[\Phi] = \exists x_0 \exists x_1 \exists x_2 \exists x_3 P(x_0, \mathbf{7}, x_2, \mathbf{5})$, while, on the other hand, $\theta_n^0[\Phi] = \exists x_0 \exists x_1 \exists x_2 \exists x_3 \neg P(x_0, \mathbf{7}, x_2, \mathbf{5})$.

We note a few immediate consequences of this definition. In what follows, ε will always be either 0 or 1.

Proposition 6.21

1. If $\mathcal{M} \models \Gamma[\Phi]$ and Φ is an extension of Ψ then $\mathcal{M} \models \Gamma[\Psi]$.
2. If $\mathcal{M} \models \Gamma[\Phi]$ and $S \supset \operatorname{dom} \Phi$ then there is an extension Ψ of Φ with domain S such that $\mathcal{M} \models \Gamma[\Psi]$.
3. If $\mathcal{M} \models \Gamma[\Phi]$ and $\mathcal{M} \models \neg((\Gamma \wedge \theta_n^\varepsilon)[\Phi])$ then $\mathcal{M} \models (\Gamma \wedge \theta_n^{1-\varepsilon})[\Phi]$.
4. Let $a_0, \ldots, a_m$ and $b_0, \ldots, b_m$ be two sequences of natural numbers. If $\mathcal{M} \models \neg(\Gamma[\{\langle n, a_n\rangle \mid n \leq m\}])$ and $\mathcal{N} \models \Gamma[\{\langle n, b_n\rangle \mid n \leq m\}])$ then $\langle \mathcal{M}, a_0, \ldots, a_n\rangle \not\cong \langle \mathcal{N}, b_0, \ldots, b_n\rangle$.
5. Suppose that Φ is total, $\operatorname{dom} \Psi = \{0, \ldots, r-1\}$, $\mathcal{M} \models \theta[\Phi \restriction r]$ and $\mathcal{N} \models \neg(\theta[\Psi])$ for some literal θ. Then there is a total computable f which is the identity on $\{0, \ldots, r-1\}$ such that $\mathcal{M} \models \theta[\Phi \circ f]$. Let $\theta' = \neg\theta[f]$. Then $\mathcal{M} \models \neg(\theta'[\Phi])$ and $\mathcal{N} \models \neg((\neg\theta')[\Psi])$. □

We now describe our construction.

Construction. At each stage s, we define partial computable functions g_s and $h_{i,s}$, $i \in \omega$. We also construct the atomic diagram $\Delta_\mathcal{C}$ of $\mathcal{C}$ by adding on one of the literals θ_s^0 or θ_s^1 at each stage s. We use the following notations: Γ_s is the conjunction of all the literals in $\Delta_\mathcal{C}$ at the end of stage s; $z_{e,s}$ is the least number such that $\Phi_{e,s}(z_{e,s}) = a$, if one exists, $z_{e,s} = 0$ otherwise; $r_{e,s} = \sup((\bigcup_{i<e} \operatorname{dom} h_{i,s}) \cup \{z_{i,s} \mid i \leq e\} \cup \{e\})$.

We say that a stage s is e-expansionary if $\Phi_{e,s}$ is injective, $\Phi_{e,s}(z_{e,s}) \downarrow= a$, $\{0, \ldots, r_{e,s}\} \subseteq \operatorname{dom} \Phi_{e,s}$, $\operatorname{dom} \Phi_{e,s} \supsetneq \{0, \ldots, \sup(\operatorname{dom} \Phi_{e,s-1})\}$, and $\operatorname{rng} \Phi_{e,s} \supsetneq \{0, \ldots, \sup(\operatorname{rng} \Phi_{e,s-1})\}$. (Thus, if there are infinitely many e -expansionary stages, Φ_e is total, injective, and surjective.)

We begin at $s = 0$ with $\Gamma_0 = \emptyset$, $g_0 = \emptyset$ and $h_{e,0} = \emptyset$ for each $e \in \omega$. We assume by induction that $\mathcal{B} \models \Gamma_s[g_s]$ and for each $e \in \omega$, $\mathcal{B} \models \Gamma_s[h_{e,s}]$. At stage $s+1$ we find the least $e \leq s$, if any, such that R_e is not satisfied and one of the following conditions holds.

1. For some ε, $\mathcal{B} \models (\Gamma_s \wedge \theta_s^\varepsilon)[g_s \restriction r_{e,s}+1]$ and $\mathcal{A} \models \neg((\Gamma_s \wedge \theta_s^\varepsilon)[\Phi_{e,s}])$.
2. Not 1 and for some ε,
   (a) $\mathcal{B} \models (\Gamma_s \wedge \theta_s^\varepsilon)[g_s \restriction r_{e,s}+1]$,
   (b) $\mathcal{B} \models (\Gamma_s \wedge \theta_s^\varepsilon)[h_{e,s}]$, and
   (c) $s+1$ is an e-expansionary stage.
3. Not (1 or 2 a and b), and for some ε,
   (a) $\mathcal{B} \models (\Gamma_s \wedge \theta_s^\varepsilon)[g_s \restriction r_{e,s}+1]$,
   (b) $\mathcal{B} \models \neg((\Gamma_s \wedge \theta_s^{1-\varepsilon})[g_s \restriction r_{e,s}+1])$, and
   (c) $\mathcal{B} \models \neg((\Gamma_s \wedge \theta_s^\varepsilon)[h_{e,s}])$.

If such an e exists, we say that e is active at stage $s+1$. Let $r = r_{e,s}+1$. For each $i > e$, let $h_{i,s+1} = \varnothing$. For each $i < e$, let $h_{i,s+1} = h_{i,s}$. Declare all R_i, $i > e$, to be unsatisfied.

If 1 or 3 holds we must abandon the current attempt at the isomorphism h and so let $h_{e,s+1} = \varnothing$. If 1 holds, we have a witness to fact that Φ_e is not an isomorphism from $\mathcal{C}$ to $\mathcal{A}$ and we declare R_e to be satisfied.

If 2 holds, there are two cases. If $h_{e,s} = \varnothing$, we restart our definition of h_e using the assumed isomorphism between $\langle \mathcal{A}, a\rangle$ and $\langle \mathcal{B}, b\rangle$: Find the least tuple $\langle a_0, \dots, a_{r-1}\rangle$ of distinct numbers such that $a_{z_{e,s}} = b$ and if we define $h_{e,s+1}$ to be the partial function mapping each $n < r$ to a_n, then

1. $\mathcal{B} \models \Gamma_{s+1}[h_{e,s+1}]$ and
2. for all $t \leq s$ and $\delta \in \{0,1\}$, $\mathcal{B} \models (\Gamma_t \wedge \theta_t^\delta)[h_{e,s+1}] \Rightarrow \mathcal{A} \models (\Gamma_t \wedge \theta_t^\delta)[\Phi_{e,s} \restriction r_{e,s}+1]$,

and define $h_{e,s+1}$ in this manner. (Such a tuple exists because, since R_e is not satisfied, $\mathcal{A} \models \Gamma_{s+1}[\Phi_{e,s}]$, so that $\mathcal{A} \models \Gamma_{s+1}[\{\langle z_{e,s}, a\rangle\}]$, and $\langle \mathcal{A}, a\rangle \cong \langle \mathcal{B}, b\rangle$.)

If $h_{e,s} \neq \varnothing$, we extend h_e so as to keep h_e and $h_e\Phi_e^{-1}$ looking like isomorphisms. If $|\mathrm{dom}\, h_{e,s}|$ is even, let k be the least number not in $\mathrm{rng}\, h_{e,s}$, let n be a number larger than any previously appearing in the construction, and define $h_{e,s+1} = h_{e,s} \cup \{\langle n, k\rangle\}$. If $|\mathrm{dom}\, h_{e,s}|$ is odd, let p be the least number not in $\mathrm{dom}\, h_{e,s}$, let m be such that $\mathcal{B} \models \Gamma_{s+1}[h_{e,s} \cup \{\langle p, m\rangle\}]$, and let $h_{e,s+1} = h_{e,s} \cup \{\langle p, m\rangle\}$.

If no such e exists, let ε be such that $\mathcal{B} \models (\Gamma_s \wedge \theta_s^\varepsilon)[g_s]$ and let $r = \max(\mathrm{dom}\, g_s)+1$. For each $i \in \omega$, let $h_{i,s+1} = h_{i,s}$.

In any case, we continue to extend the diagram $\Delta_\mathcal{C}$ and the isomorphism g. We add θ_s^ε to $\Delta_\mathcal{C}$ and let $\Gamma_{s+1} = \Gamma_s \wedge \theta_s^\varepsilon$. If $|\mathrm{dom}(g_s \restriction r)|$ is even, let k be the least number not in $\mathrm{rng}(g_s \restriction r)$, let n be a number larger than any previously appearing in the construction, and let $g_{s+1} = g_s \restriction r \cup \{\langle n, k\rangle\}$. If $|\mathrm{dom}(g_s \restriction r)|$ is odd, let p be the least number not in $\mathrm{dom}(g_s \restriction r)$, let m be such that $\mathcal{B} \models \Gamma_{s+1}[g_s \restriction r \cup \{\langle p, m\rangle\}]$, and set $g_{s+1} = g_s \restriction r \cup \{\langle p, m\rangle\}$.

Notice that, whichever case holds, $\mathcal{B} \models \Gamma_{s+1}[g_{s+1}]$ and for each $e \in \omega$, $\mathcal{B} \models \Gamma_{s+1}[h_{e,s+1}]$, which are the induction hypotheses needed for the next stage of the construction.

Verifications. Since at each stage $s+1$ we added either θ_s or its negation to $\Delta_\mathcal{C}$, $\Delta_\mathcal{C}$ is the atomic diagram of a structure $\mathcal{C}$. Because $\mathcal{A}$ and $\mathcal{B}$ are 1-decidable, the construction is effective and so C is computable.

Suppose first that there is an e such that R_e is active infinitely often and let e be the least such number. We wish to show that $h_e\Phi_e^{-1}$ is the desired computable isomorphism from $\langle\mathcal{A}, a\rangle$ to $\langle\mathcal{B}, b\rangle$. Let s_0 be a stage such that no R_i is active for $i < e$ at any stage $t \geq s_0$. It follows from the definition of $r_{e,s}$ that there exists an $s_1 \geq s_0$ such that $r_{e,t} = r_{e,s_1}$ for all $t \geq s_1$. Let $r_e = r_{e,s_1}$. It follows from the definition of g_s that there exists $s_2 \geq s_1$ such that $g_t \upharpoonright r_e + 1 = g_{s_2} \upharpoonright r_e + 1$ for all $t \geq s_2$. As R_e is active infinitely often it is never satisfied after stage s_2. So condition 1 never holds after this stage. Thus $\mathcal{A} \vDash \Gamma_s[\Phi_e]$ for every $s \geq s_2$, and hence Φ_e is an isomorphism from C to $\mathcal{A}$.

We claim that it is not possible for condition 3 to hold infinitely often. Suppose otherwise. Let $s_3 \geq s_2$ be such that $\operatorname{dom}\Phi_{e,s_3} \supseteq \{0, \ldots, r_e\}$. Inspecting the way $h_{e,s+1}$ is defined when case 2 holds and $h_{e,s} = \varnothing$, we see that there is an $s \geq s_3$ such that $h_{e,s+1} = \{\langle n, a_n\rangle \mid n \leq r_e\}$ for a tuple $\langle a_0, \ldots, a_{r_e}\rangle$, $a_{z_{e,s}} = b$, such that for all $t > s$,

1. $\mathcal{B} \vDash \Gamma_{t+1}[\{\langle n, a_n\rangle \mid n \leq r_e\}]$ and
2. $\mathcal{B} \vDash (\Gamma_t \wedge \theta_t^{1-\varepsilon})[\{\langle n, a_n\rangle \mid n \leq r_e\}] \Rightarrow \mathcal{A} \vDash (\Gamma_t \wedge \theta_t^{1-\varepsilon})[\Phi_e \upharpoonright r_e + 1]$.

Such a tuple exists because $\langle\mathcal{A}, a\rangle \cong \langle\mathcal{B}, b\rangle$ and $\Phi_e(z_{e,s}) = a$.

Now suppose that $t + 1$ is the first stage after $s + 1$ at which condition 3 holds, and let ε be as in that condition. Then $\mathcal{B} \vDash \neg((\Gamma_t \wedge \theta_t^{\varepsilon})[h_{e,t}])$. On the other hand, $\mathcal{B} \vDash \Gamma_t[h_{e,t}]$. Thus $\mathcal{B} \vDash (\Gamma_t \wedge \theta_t^{1-\varepsilon})[h_{e,t}]$. Since $h_{e,t}$ is an extension of $h_{e,s+1}$, $\mathcal{B} \vDash (\Gamma_t \wedge \theta_t^{1-\varepsilon})[h_{e,s+1}]$. But then by 2 above, $\mathcal{A} \vDash (\Gamma_t \wedge \theta_t^{1-\varepsilon})[\Phi_e \upharpoonright r_e + 1]$. But by part b of condition 3, $\mathcal{B} \vDash \neg((\Gamma_t \wedge \theta_t^{1-\varepsilon})[g_t \upharpoonright r_e + 1])$. By Proposition 6.21(5), there exists a u and an ε such that $\mathcal{A} \vDash \neg(\theta_u^{\varepsilon}[\Phi_e])$ and $\mathcal{B} \vDash \neg(\theta_u^{1-\varepsilon}[g_t \upharpoonright r_e + 1])$. But then θ_u^{ε} must be in Γ_{u+1}, so that $\mathcal{A} \nvDash \Gamma_{u+1}$, contrary to our assumption.

So condition 3 holds only finitely often. Say it never holds after stage $s_4 \geq s_3$. Since condition 2 holds infinitely often, there are infinitely many e-expansionary stages. Thus, since R_e is never satisfied, Φ_e is a computable isomorphism from C to $\mathcal{A}$. Furthermore, $h_e = \lim_s h_{e,s}$ is well-defined, and in fact $h_e(x) = h_{e,s}(x)$ for the least $s > s_4$ for which $h_{e,s}(x)$ is defined. Since $\mathcal{B} \vDash \Gamma_s[h_{e,s}]$ for all $s > s_4$, h_e is a computable isomorphism from C to $\mathcal{B}$.

Thus $h_e \circ \Phi_e^{-1}$ is a computable isomorphism from $\mathcal{A}$ to $\mathcal{B}$. But if we let $z = \lim_s z_{e,s}$, then $h_e \circ \Phi_e^{-1}(a) = h_e(z) = b$. Thus in fact $h_e \circ \Phi_e^{-1}$ is the desired computable isomorphism from $\langle\mathcal{A}, a\rangle$ to $\langle\mathcal{B}, b\rangle$.

Finally, suppose for the sake of a contradiction that every e is active only finitely often. It is not hard to see that at any e-expansionary stage, one of conditions 1, 2, or 3 must hold. Thus, if there are infinitely many e-expansionary stages then R_e is eventually permanently satisfied.

As we have mentioned, if s is a stage such that, for each $i < e$ and each $t \geq s$, R_i is not active at stage t and $r_{e,t} = r_{e,s}$, then for all $t \geq s$, $g_t \upharpoonright r_{e,s} + 1 = g_s \upharpoonright r_{e,s} + 1$ and $\mathcal{B} \vDash \Gamma_t[g_t \upharpoonright r_{e,s} + 1]$. So the fact that each e is active only finitely often implies that $g = \lim_t g_t$ exists and is an isomorphism from C to $\mathcal{B}$.

Thus C is isomorphic, but not computably isomorphic, to $\mathcal{A}$, contradicting the computable categoricity of $\mathcal{A}$. □

Thus 1-decidability suffices to guarantee the persistence of computable categoricity. We will see in the next section that, without such an assumption, computable categoricity need not be persistent. Moreover, the equivalence of computable categoricity with having a Scott family established by Goncharov under the assumption of 2-decidability does not hold for all 1-decidable structures (Theorem 7.19).

6.2 Nonpersistence of Computable Categoricity

We now see that the addition of even a single constant for any element of a computably categorical structure can change its dimension.

Theorem 6.22 (Cholak, Goncharov, Khoussainov and Shore [1999]) *For each $k \in \omega$ there is a computably categorical $\mathcal{A}$ such that the expansion $\mathcal{A}'$ of $\mathcal{A}$ gotten by adding on a constant naming any element of $\mathcal{A}$ has dimension exactly k.*

Idea of Proof (for $k = 2$). We first construct a (uniformly) computably enumerable family of distinct pairs of sets $S = \{f(i) | i \in \omega\} = \{(A_i, B_i) | i \in \omega\}$ which is *symmetric*, i.e. for every $i \in \omega$ there is a $j \in \omega$ such that $f(i) = (A_i, B_i) = (B_j, A_j)$. In addition to the computable enumeration f, there is one other natural computable enumeration of this family, $\tilde{f}$ defined by $\tilde{f}(i) = (B_i, A_i)$. This family S is constructed (by a $0''$ type priority argument) to have dimension 2 in the sense that there is no computable function g such that $f = \tilde{f}g$ but, for every one-one computable enumeration h of the family, there is a computable function g such that $f = hg$ or $\tilde{f} = hg$. The two enumerations of this family are then coded symmetrically into a graph so that the whole structure is computably categorical. If one adds on a constant, however, it distinguishes between the two coded enumerations and so one has a structure of dimension 2. □

For $k > 2$, one can generalize the notion of symmetric family to one-one enumerations f of families S of k-tuples of sets. The combinatorial details become fairly complicated. A simpler approach to a proof of the general theorem is provided in the next section as a corollary to some results on degree spectra.

7. Degree Spectra of Relations

Another important topic in computable model theory that turns out to be closely connected to computable categoricity is that of the dependence of the computability properties of relations not included in the language of a given structure on its presentation. For example, in "standard" presentations of $\langle \mathbb{N}, \leq \rangle$ the successor function is computable but it is not computable in every presentation (Proposition 6.1). Similarly, standard presentations of the algebraically closed field $\tilde{\mathbb{Q}}$ of characteristic 0 and infinite transcendence degree make the relation of algebraic dependence computable but not all presentations do. (Indeed, if algebraic dependence is computable in both of two isomorphic computable algebraically closed fields then they are computably isomorphic. However, Corollary 6.12 says that if they have infinite transcendence degree their dimension is infinite.) On the other hand, these particular relations are easily seen to always be co-computably enumerable and computably enumerable respectively. Others remain computable or computably enumerable in every presentation. Such relations were singled out and studied by Ash and Nerode [1981].

Definition 7.1 *(Ash and Nerode)*
If $R \subseteq A^n$ is an n-ary relation on a structure $\mathcal{A}$, R is *intrinsically computable (computably enumerable)* if $f[R]$ is computable (computably enumerable) for every isomorphism $f : \mathcal{A} \to \mathcal{B}$.

Example 7.2 $\langle \mathbb{N}, \leq \rangle$: Successor is not intrinsically computable.

Example 7.3 $\langle \mathbb{N}, s \rangle$: Every computable relation is intrinsically computable.

We know that the two structures discussed above, $\langle \mathbb{N}, \leq \rangle$ and $\widetilde{\mathbb{Q}}$, are not computably categorical while similar ones (such as $\langle \mathbb{N}, s \rangle$ and algebraically closed extension of $\mathbb{Q}$ of finite transcendence degree) are computably categorical and in each of them this phenomena (of a relation being computable in one presentation and not in another) does not arise. One might naturally ask if computable categoricity guarantees that a relation computable in one presentation is computable in all. The answer is both yes and no. If we restrict our attention to relations that are definable or even *invariant* under all automorphisms the answer is yes.

Proposition 7.4 *If a structure $\mathcal{A}$ is computably categorical then every definable relation R (or one invariant under automorphisms) on $\mathcal{A}$ that is computable in any presentation of $\mathcal{A}$ is intrinsically computable, i.e. computable in every presentation of $\mathcal{A}$.*

Proof. Suppose $\mathcal{A}$ is computably categorical, $R^{\mathcal{A}}$ is computable, and g is an isomorphism from $\mathcal{A}$ to $\mathcal{B}$. We wish to show that $g[R^{\mathcal{A}}]$ is computable. As $\mathcal{A}$ is computably categorical, there is a computable isomorphism $f : \mathcal{A} \to \mathcal{B}$. $R^{\mathcal{A}}$ and $\overline{R}^{\mathcal{A}}$ are computable and so their images under f are computably enumerable and complementary and hence computable. As R is invariant under automorphisms, in particular under $g^{-1}f$, $f[R^{\mathcal{A}}] = g[R^{\mathcal{A}}]$ and so $g[R^{\mathcal{A}}]$ is also computable. □

So for computably categorical structures the effectiveness of definable properties is independent of the presentation. If we ask instead that every computable relation on $\mathcal{A}$ (definable or not) be intrinsically computable, the answer to our question is no. Computable categoricity does not suffice to guarantee that every computable relation is intrinsically computable. (See Example 7.7 below.) Instead we are led to a stronger notion.

Definition 7.5 $\mathcal{A}$ is *computably stable* if every isomorphism $f : \mathcal{A} \to \mathcal{B}$ is computable.

Example 7.6 $\langle \mathbb{N}, s \rangle$ is computably stable. Indeed, every isomorphism between two presentations is uniquely determined by the computable procedure of sending the least element in one presentation to the least one in the other and then proceeding by recursion as in Example 6.5.

Example 7.7 $\langle \mathbb{Q}, \leq \rangle$ is computably categorical but not computably stable. In fact, given any two presentations of $\langle \mathbb{Q}, \leq \rangle$, the usual back and forth argument shows that there are continuum many isomorphisms between them. Moreover, the usual back and forth argument can be run in each of countably many intervals to, for example, construct an automorphism taking a computable subset (such as $\mathbb{Z}$) to a noncomputable one (any set consisting of one element from each interval $(x, x+1)$ for $x \in \mathbb{Z}$).

Proposition 7.8 (Ash and Nerode [1981]) *$\mathcal{A}$ is computably stable if and only if every computable relation on $\mathcal{A}$ is intrinsically computable.*

Proof. As every isomorphism between presentations of $\mathcal{A}$ is computable, the argument of Proposition 7.4 shows that the image of any computable relation R on $\mathcal{A}$ under any isomorphism is computable. For the other (if) direction, consider $\mathcal{A}$ as a structure on the set $\mathbb{N}$ and the relation R giving, in $\mathcal{A}$, the successor function on $\mathbb{N}$. If $f : \mathcal{A} \to \mathcal{B}$ is an isomorphism and $R^{\mathcal{B}} = f[R^{\mathcal{A}}]$ is computable then the construction of Example 6.5 computes f. □

One can, in fact, give a more informative characterization of computable stability like that provided for computable categoricity in terms of Scott families by Theorem 6.16. In place of a sequence of formulas each of which determines a sequence of elements of the given structure $\mathcal{A}$ up to automorphisms, one needs a sequence of formulas that uniquely define the elements of $\mathcal{A}$. On the other hand, we now only need the 1-decidability of $\mathcal{A}$ for the characterization.

Theorem 7.9 (Ash and Nerode [1981], Goncharov [1975]) *If $\mathcal{A}$ is 1-decidable then $\mathcal{A}$ is computably stable if and only if there are constants $\overline{c} \in A$ and a computable sequence $\phi_i(\overline{c}, x)$ of existential formulas such that for each i there is a unique $a \in A$ satisfying ϕ_i and each $a \in A$ satisfies some ϕ_i.*

Proof sketch. It is easy to see that the existence of a family as described insures that every isomorphism $f : \mathcal{A} \to \mathcal{B}$ is computable as, once the image of the constants $\overline{c}$ are fixed, f must send the unique solution of each $\phi_i(\overline{c}, x)$ in $\mathcal{A}$ to the solution of the same formula in $\mathcal{B}$. The proof of the other direction (only if) of this theorem involves a finite injury priority argument. One attempts to build a $\mathcal{B}$ isomorphic to the given $\mathcal{A}$ by a Δ_2 isomorphism but not by any computable isomorphism. The least failure of this diagonalization requirement occurs only because the elements on which we might diagonalize are uniquely defined from those fixed by higher priority requirements. These already fixed elements are the constants $\bar{c}$ required. The portions of the diagram of $\mathcal{B}$ to which we have committed ourselves at various stages of the construction provide the desired formulas ϕ_i when various extra parameters are replaced by existentially quantified variables. □

More generally, we would like to know when a specified computable (or computably enumerable) relation is intrinsically computable or computably enumerable. An examination of the two examples considered above, $\langle \mathbb{N}, \leq \rangle$ and $\widetilde{\mathbb{Q}}$, gives us a clue as to when a relation is intrinsically c.e. The relation $P(x, y)$ on $\mathbb{N}$ saying that y is not the immediate successor of x is definable in the structure $\langle \mathbb{N}, \leq \rangle$ by the existential formula $\exists z((x < z < y) \vee (y < z < x) \vee (x = y))$ and so is computably enumerable in any presentation of $\langle \mathbb{N}, \leq \rangle$. The binary relation $D(x, y)$ saying that x and y are algebraically dependent is equivalent to the disjunction of an infinite computable list of existential formulas ϕ_n each asserting (in the language of fields) that there is a nonzero polynomial of degree n in x and y which equals 0. Any such relation is again clearly computably enumerable in any presentation of a field. To enumerate the dependent pairs, one simply dovetails the searches for witnesses for each of the existential formulas ϕ_n. These phenomena suggest a definition.

Definition 7.10 A relation $R(x_1, \ldots, x_n)$ on a structure $\mathcal{A}$ is *formally computably enumerable* if it is equivalent to a disjunction $\bigvee \phi_i(x_1, \ldots, x_n)$ of a computable sequence of existential formulas ϕ_i with free variables $x_1, \ldots, x_n$. R is *formally computable* if both R and $\overline{R}$ are formally computably enumerable.

Clearly, any formally computable (computably enumerable) relation is intrinsically computable (computably enumerable). Ash and Nerode [1981] prove that, under mild decidability conditions, this condition is also necessary.

Theorem 7.11 (Ash and Nerode [1981]) *If $R \subseteq A^n$ and $\langle \mathcal{A}, R \rangle$ is 1-decidable, then R is intrinsically computably enumerable if and only if it is formally computably enumerable. R is intrinsically computable if and only if it is formally computable.*

Actually, the 1-decidability of $\langle \mathcal{A}, R \rangle$ is a bit stronger than what Ash and Nerode need. They only need to be able to decide for each $\bar{c}$ in A and each existential $\phi(\bar{c}, \bar{x})$ if there is an $\bar{a} \notin R$ such that $\mathcal{A} \models \phi(\bar{c}, \bar{a})$. However, some conditions are necessary as Goncharov [1980a] and Manasse [1982] have constructed examples of intrinsically c.e. relations which are not formally c.e. There has been a lot of work, primarily by Ash, Ash and Knight and their students generalizing these results (under stronger decidability conditions) to syntactic characterizations of relations being intrinsically Σ_α or Δ_α for all levels α of the hyperarithmetic hierarchy. They also provide similar generalizations of the notions and results on computable categoricity and stability to higher levels of the hierarchy of computable infinitary formulas. These papers include Ash [1986], [1986a], [1987]; Ash and Knight [1990], [1994], [1995], Barker [1988] and Chisholm [1990a]. Related results when the notions are relativised to the degree of noncomputable models can be found in Ash, Knight, Manasse and Slaman [1989], Ash, Knight and Slaman [1993] and Chisholm [1990]. Here the results are proven by forcing arguments and the extra decidability hypotheses are not needed.

Faced with a computable (or c.e.) relation R on $\mathcal{A}$ which is not intrinsically computable (or c.e.), what can we say about its image under isomorphisms? In particular, how complicated can $f[R]$ be for a (computable) relation R on $\mathcal{A}$ and an arbitrary isomorphism $f : \mathcal{A} \to \mathcal{B}$ (with $\mathcal{B}$ computable, of course). An approach to this question is suggested by the following definition.

Definition 7.12 If $R \subseteq A^n$ is an n-ary relation on $\mathcal{A}$, the *degree spectrum of* R, $DgSp(R)$, is $\{\deg_T(f[R]) \mid f : \mathcal{A} \to \mathcal{B} \text{ is an isomorphism}\}$.

There are a number of results giving conditions under which the degree spectrum of a computable relation consists of precisely some particular standard class of degrees such as all the degrees, the c.e. degrees, etc. We concentrate on the issue of finding instances where the spectrum is finite and the connections between this issue and the dimension of the given structure. The first results of this sort are due to Harizanov. Here is one example.

Theorem 7.13 (Harizanov [1993]) *There is an $\mathcal{A}$ and an R on $\mathcal{A}$ such $\mathcal{A}$ has exactly two computable presentations and $DgSp(R) = \{\mathbf{0}, \mathbf{c}\}$ with* c *noncomputable and Δ_2^0.*

The next problem (that remained open for some time) was whether Δ_2^0 could be replaced by c.e. in this result or, more generally, what is possible for intrinsically c.e. relations especially for structures of finite dimension. Goncharov has announced a solution, based on work with Khoussainov, constructing a structure $\mathcal{A}$ of dimension 2 with a relation R on $\mathcal{A}$ with degree spectrum consisting of $\mathbf{0}$ and a nonzero c.e. c. He constructs families of c.e. sets and codes them into a structure. Khoussainov and Shore have independently directly constructed directed graphs of each finite dimension n with relations having various degree spectra. Moreover, these structures can be simply modified to provide examples of ones for each n which are computably categorical but when expanded by a constant have dimension n. We first state the main result for dimension 2.

Theorem 7.14 (Khoussainov and Shore [1998]) *There is a rigid directed graph $\mathcal{A}$ (i.e. one with no nontrivial automorphisms) of dimension 2 and a subset R of A such that $DgSp(R) = \{\mathbf{0}, \mathbf{c}\}$ with* c *noncomputable and c.e. Moreover, the relation $P = \{(x, y) | x \in R^{\mathcal{A}_0} \wedge y \in R^{\mathcal{A}_1} \wedge$ there is an isomorphism from $\mathcal{A}_0$ to $\mathcal{A}_1$ which extends the map $x \mapsto y\}$ is computable.*

We sketch the proof of this theorem in § 9. For now we give some generalizations and corollaries.

Theorem 7.15 (Khoussainov and Shore [1998]) *For any computable partially ordered set $\mathcal{D}$ there is a rigid directed graph $\mathcal{A}$ of dimension the cardinality of $\mathcal{D}$*

and a subset R of $\mathcal{A}$ such that $DgSp(R) \cong \mathcal{D}$. (The ordering on $DgSp(R)$ is given by Turing reducibility.) Indeed, we can also guarantee that $R^{\mathcal{B}}$ is c.e. for every computable presentation $\mathcal{B}$ of $\mathcal{A}$ and that, if $\mathcal{D}$ has a least element, then the least element in $DgSp(R)$ is 0. Moreover, there is a uniformly computable sequence $\mathcal{A}_i$ of representatives of the computable isomorphism types of $\mathcal{A}$ such that the relation $P = \{(x, y) | x \in R^{\mathcal{A}_i} \wedge y \in R^{\mathcal{A}_j} \wedge$ there is an isomorphism from $\mathcal{A}_i$ to $\mathcal{A}_j$ which extends the map $x \mapsto y\}$ is computable.

Corollary 7.16 *For each natural number $k \geq 2$ there exists a computably categorical structure $\mathcal{B}$ whose expansion by finitely many constants has exactly k many computable isomorphism types.*

Proof. Take the structure $\mathcal{A}$ given by Theorem 7.15 for the partial order consisting of k many incomparable elements. Let $\mathcal{A}_i$, $1 \leq i \leq k$ be the computable representatives of the computable isomorphism types of $\mathcal{A}$. So, in particular the sets $R^{\mathcal{A}_i}$ are Turing incomparable. We use the computability of P to paste the $\mathcal{A}_i$ together to produce a $\mathcal{B}$ as required. More precisely, $\mathcal{B}$ consists of the disjoint union of the $\mathcal{A}_i$ and the edges of $\mathcal{B}$ are the ones in each $\mathcal{A}_i$. In addition, $\mathcal{B}$ has an extra binary predicate defined by the relation P in the theorem and an equivalence relation E whose equivalence classes are the $\mathcal{A}_i$.

Clearly $\mathcal{B}$ is a computable structure. Now let $\mathcal{B}'$ be any computable presentation of $\mathcal{B}$. Let $\mathcal{A}'_1$ and $\mathcal{A}'_2$ be two equivalence classes in $\mathcal{B}'$. These two substructures of $\mathcal{B}'$ considered as graphs are isomorphic to $\mathcal{A}$. Hence $\mathcal{A}'_1$ is computably isomorphic to one of $\mathcal{A}_1, \ldots, \mathcal{A}_k$. Without loss of generality suppose that $\mathcal{A}'_1$ is computably isomorphic to $\mathcal{A}_1$ via a computable function $f_1 : \mathcal{A}_1 \to \mathcal{A}'_1$. If $\mathcal{A}'_2$ were computably isomorphic to $\mathcal{A}_1$ via a computable function $f_2 : \mathcal{A}_1 \to \mathcal{A}'_2$, then we would be able to decide $R^{\mathcal{A}_1}$ in $\mathcal{A}_1$ as follows: x in $\mathcal{A}_1$ belongs to $R^{\mathcal{A}_1}$ if and only if $(f_1(x), f_2(x)) \in P$. Hence all the structures $\mathcal{A}'_1, \ldots, \mathcal{A}'_k$ are pairwise noncomputably isomorphic and so represent all the computable isomorphism types of $\mathcal{A}$, i.e. are computably isomorphic to $\mathcal{A}_1, \ldots, \mathcal{A}_k$ (in some order). Hence $\mathcal{B}'$ is clearly computably isomorphic to $\mathcal{B}$ and so $\mathcal{B}$ is computably categorical.

Now let a be any element from $\mathcal{A}_1$. Consider the expanded structures $\mathcal{B}_i$ consisting of $\mathcal{B}$ with the new constant interpreted as a_i, the image of a in $\mathcal{A}_i$. It is clear that all the $\mathcal{B}_i$ are isomorphic but not computably so. Thus the dimension of $\mathcal{B}_i$ is at least k. On the other hand, as $\mathcal{A}$ is rigid there are no choices other than the a_i as the interpretation of a in $\mathcal{B}$. Thus, by the computable categoricity of $\mathcal{B}$, any structure isomorphic to say $\mathcal{B}_1$ must be computably isomorphic to one of the $\mathcal{B}_i$ and so the dimension of these structures is precisely k as required. □

Corollary 7.17 (Khoussainov and Shore [1998]) *There exists a computably categorical structure without a Scott family.*

Proof. If structure of previous corollary had a Scott family it would remain computably categorical when constants were added. □

A similar construction provides an example showing that even if the structure is persistently computably categorical it need not have a Scott family.

Theorem 7.18 (Khoussainov and Shore [1998]) *There exists a structure without a Scott family such that every expansion of the structure by a finite number of constants is computably categorical.*

Kudinov independently proved more.

Theorem 7.19 (Kudinov [1996]) *There is a computably categorical* 1 *-decidable structure $\mathcal{A}$ with no Scott family.*

Proof sketch. Kudinov slightly modifies a family of computable enumerations constructed by Selivanov [1976] and then codes the family as a unary algebra in such a way as to produce a computably categorical structure with a decidable existential theory but no Scott family. □

Of course, this Theorem shows that the assumption of 2-decidability was necessary in Goncharov's characterization (Theorem 6.16) of computably categorical structures as ones with Scott families. By Millar's result on persistence (Theorem 6.20), Kudinov's structure is persistently computably categorical and so is also a witness to Theorem 7.18.

A very natural question is whether every c.e. degree can be realized (with 0) as a degree spectrum. Hirschfeldt has recently answered this question by adapting and extending the methods presented here.

Theorem 7.20 (Hirschfeldt [1999]) *For every c.e. degree* $\mathbf{c}$ *there is an* $\mathcal{A}$ *and a relation* R *on* $\mathcal{A}$ *such that* $DgSp(R) = \{\mathbf{0}, \mathbf{c}\}$. *Indeed* $\mathbf{c}$ *can be replaced by any uniformly c.e. array of c.e. degrees.*

Hirschfeldt's construction precisely controls the degree spectrum of the relation R but does not control dimension of $\mathcal{A}$. Thus the following question is still open.

Question 7.21 (Goncharov and Khoussainov [1997]) *Which* n*-tuples of c.e. degrees can be realized as the degree spectrum of a relation on a structure of dimension* n*?*

If we move beyond the c.e. degrees there are a few results by Harizanov on possible degree spectra but not much is known. However, we should point out that several natural strengthenings of these results can ruled out by classical descriptive set theoretic results.

Remark 7.22 For a given relation R on a computable structure $\mathcal{A}$, the set $\{R^{\mathcal{B}} \mid \mathcal{B}$ is a computable presentation of $\mathcal{A}\}$ is Σ^1_1 in R. Thus, there are countable partial orderings that cannot be realized in the c.e. degrees as the degree spectrum of any relation R on any computable structure $\mathcal{A}$. (Just consider one that is too complicated to be Σ^1_1.) Similarly, such a partial ordering with least element cannot be realized anywhere in the Turing degrees as the degree spectrum of a computable relation R on a computable structure $\mathcal{A}$. Nor can it be true that any finite set of degrees can be realized as the degree spectrum of any relation R on a computable structure $\mathcal{A}$. Indeed, any degree spectrum containing both a hyperarithmetic degree and a nonhyperarithmetic degree is uncountable as any Σ_1^1 set with a nonhyperarithmetic member is uncountable.

8. Algebraic Examples

In § 6 we saw several examples of theories whose models all have dimension 1 or ω and algebraic conditions characterizing the models in each class. The theories of this sort considered there were linear orderings, Boolean algebras and algebraically closed fields. We cite two more.

Theorem 8.1 *A real closed field of finite transcendence degree over* $\mathbb{Q}$ *is computably stable. One of infinite transcendence degree has dimension* ω.

Proof. If a real closed field $\mathcal{A}$ has finite transcendence degree over $\mathbb{Q}$ and $f : \mathcal{A} \to \mathcal{B}$ is an isomorphism, let $a_1, \ldots, a_n$ be a transcendence basis for $\mathcal{A}$ over $\mathbb{Q}$ and $b_1, \ldots, b_n$ be their image in $\mathcal{B}$. Calculate $f(a)$ for any element a of A by first finding an equation over $\mathbb{Q}[a_1, \ldots, a_n]$ satisfied by a. Find all its solutions and the place of a among these solutions in the order of $\mathcal{A}$. Now, $f(a)$ must be the solution of the same equation over $\mathbb{Q}[b_1, \ldots, b_n]$ which lies in the same place among all the solutions in $\mathcal{B}$ listed in order. Thus f is computable. On the other hand, if $\mathcal{A}$ is of infinite transcendence degree then by Theorem 6.11, it has dimension ω. (Note that as the theory of real closed fields has effective quantifier elimination, every computable model is decidable. Moreover, the prime model of $Th(\mathcal{A}, \overline{c})$ for any finite list $\overline{c}$ of elements of $\mathcal{A}$ is of finite transcendence degree and so not $\mathcal{A}$ itself.) □

Theorem 8.2 (Goncharov [1981]) *If* $\mathcal{A}$ *is an abelian group then it has dimension* 1 *or* ω.

The proof of this result is particularly interesting because it relies on important sufficient condition for a structure to have dimension ω.

Theorem 8.3 (Goncharov [1982]) *If there is a* Δ_2^0 *isomorphism between* $\mathcal{A}$ *and* $\mathcal{B}$ *but no computable one then* $\mathcal{A}$ *has dimension* ω.

On the other hand, the results described in § 6.2 and § 7, as well as many earlier papers, supply examples of structures of dimension n for each $n \in \omega$. Indeed, our results supply examples of structures of dimension n whose presentations are characterized by the Turing degree of a specific relation on the structure. Moreover, representatives of the n many computable isomorphism types of these structures can be pasted together to produce a single computably categorical structure such that an expansion by constants yields a structure of dimension n. Of course, when there are characterization theorems that show that the dimension must be 1 or ω such constructions are not possible. On the other hand, for many familiar algebraic theories for which we cannot provide such a dichotomy and characterization, it is possible to construct examples of models not only of each finite dimension but also ones exhibiting the additional properties enjoyed by the examples in § 7.

Theorem 8.4 (Goncharov [1980a], [1981]; Goncharov and Dobrotun [1989], Goncharov, Molokov and Romanovski [1989]; Kudinov (personal communication); Hirschfeldt, Khoussainov, Slinko and Shore [1999]) *For each of the following theories and each $n \in \omega$, there is a model $\mathcal{A}$ with a subset R such that the dimension of $\mathcal{A}$ is n and the degree spectrum of R consists of n different c.e. degrees. Moreover, for each n there is a model $\mathcal{A}$ which is computably categorical but some expansion by constants has dimension n: graphs, lattices, partial orders, nilpotent groups, rings (with zero divisors) and integral domains. In each case the subset R can be taken to be a substructure of the appropriate type.*

The results on the existence of models of each of these theories of each finite dimension are due to various people (most to Goncharov and his coauthors, the one for integral domains is due to Kudinov). They were typically first proved by codings of families of c.e. sets. For graphs, the results involving degree spectra and extensions by constants are due to Khoussainov and Shore and are described in § 7. (Actually, the original paper used directed graphs but an examination of the construction shows that it is possible to use undirected graphs instead.) All the other ones involving degree spectra and extensions by constants have been proven by Hirschfeldt, Khoussainov, Slinko and Shore [1999].

Although direct constructions are sometimes possible, these results can all be derived from the results on graphs by finding a sufficiently effective coding of graphs into models of each theory. The idea is that, if the coding is sufficiently effective, all the computability properties involved carry over. Thus all these theories are not only undecidable but the codings (of say graphs) needed to prove that they are universal (i.e. code all of predicate logic) are highly effective. (In addition to simple codings of the domain and edge relation on the initial graph, an important issue is the effective reversibility of the coding. That is, one wants the model coding a given graph to effectively determine the original graph.) On the other hand, the theories discussed in § 6 whose models are all either computably categorical or of dimension ω are decidable and have strong structure theorems that are used in the proofs. We expect that there are natural theories that are neither "so decidable" as those of § 6 nor "so undecidable" as the ones in Theorem 8.4. In particular, we suggest the theory of fields as a good test case as it is undecidable but the proofs of undecidability (that we know) interpret $\mathbb{N}$ in a rather specific way rather than arbitrary structures.

9. The Basic Theorem on Degree Spectra

In this section we sketch the proof of the Theorem 7.14, the case $n = 2$ of our main theorem on degree spectra.

Theorem 7.14 (Khoussainov and Shore [1998]) *There is a rigid directed graph $\mathcal{A}$ of dimension 2 with computable (but not computably isomorphic) presentations $\mathcal{A}_0$ and $\mathcal{A}_1$ and a subset R of A such that $DgSp(R) = \{0, \mathbf{c}\}$ with* c *noncomputable and c.e. Moreover, the relation $P = \{(x, y) | x \in R^{\mathcal{A}_0} \wedge y \in R^{\mathcal{A}_1} \wedge$ there is an isomorphism from $\mathcal{A}_0$ to $\mathcal{A}_1$ which extends the map $x \mapsto y\}$ is computable.*

Proof sketch. Our directed graph $\mathcal{A}$ will consist of disjoint components $[B_i]$ all of one special type. The graph we denote by $[B]$ is uniquely determined by the set $B \subseteq \{n | n \geq 5\}$. It consists of one 3-cycle and one n-cycle for each $n \in B$. In addition, there is one element of the 3-cycle, called the *top* of the graph, from which there is an edge to one element of each n-cycle for $n \in B$. This element of the n-cycle is called the *coding location for n*. For convenience, we also denote $[\{n\}]$ by $[n]$. We build up our graph using two operations, + and $\cdot$. The sum $[A] + [B]$ of two graphs is simply their disjoint union. The product, $[A] \cdot [B]$, of two graphs of our special form is gotten by taking disjoint copies of $[A]$ and $[B - A]$ and identifying the top elements (and the associated 3-cycles) in each of the two graphs. For example $[5] \cdot [6] \cong [\{5, 6\}]$ and $[\{5, 6, 7\}] \cdot [\{6, 7, 8, 9\}] \cong [\{5, 6, 7, 8, 9\}]$. Note that $[A] \cdot [B] \cong [B] \cdot [A]$.

Our plan is to construct our graph $\mathcal{A} = [B_0] + [B_1] + \cdots + [B_n] + \cdots$ together with enumerations of the sets B_i so that $B_i - B_j \neq \emptyset$ for $i \neq j$ (and indeed we guarantee that $B_{i,s} - B_{j,s} \neq \emptyset$ for every s and $i \neq j$). So clearly

- $\mathcal{A}$ is rigid.

The required R will be a subset of the coding points in $\mathcal{A}$. We enumerate two presentations $\mathcal{A}_0$ and $\mathcal{A}_1$ of $\mathcal{A}$ as $\mathcal{A}_{0,s}$ and $\mathcal{A}_{1,s}$ each isomorphic to $[B_{0,s}] + [B_{1,s}] + \cdots$ and the interpretations R_i (as $R_{i,s}$) of R in $\mathcal{A}_i$ so that

- R_0 is computably enumerable but not computable: As the construction proceeds we enumerate the elements x of R_0 so as to make the set enumerated noncomputable by a standard diagonalization procedure.
- R_1 is computable: As we enumerate a number x into R_0, we make sure that the corresponding element y of $\mathcal{A}_1$ is a new large number. Thus R_1 is enumerated in increasing order.
- $P = \{\langle x, y\rangle | x \in R_0 \wedge y \in R_1 \wedge (\exists f : \mathcal{A}_0 \cong \mathcal{A}_1)(f(x) = y)\}$ is computable: By the procedure alluded to above for choosing the $y \in R_1$ corresponding to a given $x \in R_0$, the pairs $\langle x, y\rangle \in P$ are enumerated in increasing order.
- $\mathcal{A}_0 \not\cong_c \mathcal{A}_1$: This is guaranteed by the previous requirements that R_0 is computable but R_1 is not. By the rigidity of $\mathcal{A}$, there is only one isomorphism from $\mathcal{A}_0$ to $\mathcal{A}_1$ and it must take R_0 to R_1. If it were computable it would preserve the computability of the interpretation of R.
- Every computable presentation $\mathcal{G}_j$ of $\mathcal{A}$ is computably isomorphic to $\mathcal{A}_0$ or $\mathcal{A}_1$:

Our plan here is to define maps $r_{j,s}$ so that at every stage s of the construction at which it still looks as if $\mathcal{G}_j$ might be isomorphic to $\mathcal{A}$, $r_{j,s}$ is a monomorphism from $\mathcal{G}_{j,s}$ into $\mathcal{A}_{i,s}$ (for $i = 0$ or 1) and that, at the end of stage s, if we cannot extend the current map $r_{j,s}$ then we switch so that $r_{j,s+1}$ is a monomorphism from $\mathcal{G}_{j,s+1}$ into $\mathcal{A}_{1-i,s+1}$. If, after some stage t, we never switch our potential isomorphism then $\cup\{r_{j,s} | t \leq s\}$ is, in fact, the desired computable isomorphism from $\mathcal{G}_j$ to $\mathcal{A}_i$. On the other hand, if we switch infinitely often we guarantee that there is a *special component* S_j of $\mathcal{G}_j$ which is not a component of $\mathcal{A}$ and so $\mathcal{G}_j$ is not isomorphic to $\mathcal{A}$.

The crucial idea needed for the construction is how to diagonalize to make R_0 noncomputable while its isomorphic image R_1 remains computable and also maintain control over the potential isomorphisms between $\mathcal{G}_j$ and $\mathcal{A}_i$. The diagonalization procedure is based on two symmetric operations $\mathbf{L}$ (*left*) and $\mathbf{R}$ (*right*) on sequences of graphs $[B_i]$.

Definition 9.1 $\mathbf{L}([B_1], \ldots, [B_n])$ is the graph

$$[B_1] \cdot [B_2] + \cdots + [B_{n-1}] \cdot [B_n] + [B_n] \cdot [B_1].$$

$\mathbf{R}([B_1], \ldots, [B_n])$ is the graph

$$[B_1] \cdot [B_n] + [B_1] \cdot [B_2] + \ldots + +[B_{n-1}] \cdot [B_n].$$

We apply the $\mathbf{L}$ operation, for example, to a graph $\mathcal{G}$ whose components include the $[B_i]$ by removing all the $[B_i]$ and inserting $\mathbf{L}([B_1], \ldots, [B_n])$. We also adopt the convention that the elements of the component $[B_i]$ are the same ones in the corresponding subgraph in the component $[B_i] \cdot [B_{i+1}]$ of $\mathbf{L}([B_1], \ldots, [B_n])$ while those elements in the new graph corresponding to ones in $[B_{i+1}]$ of the original graph are new elements in $[B_i] \cdot [B_{i+1}]$ (with 1 for $n+1$ when $i = n$). This convention is important for establishing computability properties of the graphs being constructed.

We will apply an $\mathbf{R}$ operation in the construction (to $\mathcal{A}_1$) only when we also apply an $\mathbf{L}$ one (to $\mathcal{A}_0$). We also have the corresponding convention that the elements of the component $[B_{i-1}]$ are the same ones in the corresponding graph in the component $[B_{i-1}] \cdot [B_i]$ of $\mathbf{R}([B_1], \ldots, [B_n])$ while those elements in the new graph corresponding to ones in $[B_i]$ of the original graph are new elements in $[B_{i-1}] \cdot [B_i]$ (with 0 for n when $i = 1$).

The following lemma is immediate from the definitions.

Lemma 9.2 *For any sequence* $[B_1], \ldots, [B_n]$ *of graphs,* $\mathbf{L}([B_1], \ldots, [B_n])$ *and* $\mathbf{R}([B_1], \ldots, [B_n])$ *are isomorphic and extend* $[B_1] + \cdots + [B_n]$. *Moreover, if* $\mathcal{G}$ *has the* $[B_i]$ *as components then replacing their sum with* $\mathbf{L}([B_1], \ldots, [B_n])$ *or* $\mathbf{R}([B_1], \ldots, [B_n])$ *produces two isomorphic graphs each extending* $\mathcal{G}$. □

The plan for diagonalization is now easily described. To make sure that $R_0 \neq \phi_e$, we choose numbers a_e, b_e and c_e and insert copies of $[a_e]$, $[b_e]$ and $[c_e]$ into $\mathcal{A}_0$ and $\mathcal{A}_1$. For definiteness, say that x_e is the (number which is) the coding location for a_e in these graphs. We now wait for $\phi_e(x_e)$ to converge to 0. If it never does we do nothing and so win as x_e is not in R_0. If $\phi_e(x_e)$ converges to 0 at stage s, we replace the components $[a_e]$, $[b_e]$ and $[c_e]$ in $\mathcal{A}_0$ and $\mathcal{A}_1$ by $\mathbf{L}([b_e], [a_e], [c_e])$ and $\mathbf{R}([b_e], [a_e], [c_e])$, respectively; put x_e into R_0 and its image in $\mathcal{A}_1$ into R_1. The crucial point here is that, by our conventions, the image of x_e (as an element of $\mathbf{L}([b_e], [a_e], [c_e])$ in $\mathcal{A}_0$) in $\mathbf{R}([b_e], [a_e], [c_e])$ and so in $\mathcal{A}_1$ is a new large number. Thus we diagonalize for R_0 but keep R_1 computable.

The remaining issue is how to simultaneously satisfy the requirements that, if isomorphic to $\mathcal{A}$, $\mathcal{G}_j$ is computably isomorphic to $\mathcal{A}_0$ or $\mathcal{A}_1$. Consider the requirement for a single $\mathcal{G}$. Following the idea described above, we choose a *special component* $[S]$ of $\mathcal{G}$ and make its image in the $\mathcal{A}_i$ participate in infinitely many of the left and right operations done for diagonalizations. We have some definition of expansionary stage that measures the extent of a possible isomorphism between $\mathcal{G}$ and $\mathcal{A}$. If there are only finitely many such expansionary stages then $\mathcal{G}$ is not isomorphic to $\mathcal{A}$ and no other actions are necessary. So suppose there are infinitely many expansionary stages.

At each expansionary stage s we have a monomorphism r_s from $\mathcal{G}$ into $\mathcal{A}_i$ and components $[S_{i,s}]$ of $\mathcal{A}_i$ (for $i = 0$ or 1) corresponding to the special component $[S]$ of $\mathcal{G}$. If we wish to diagonalize at a coding location x_e in the range of r, we wait for the next expansionary stage s and perform $\mathbf{L}$ and $\mathbf{R}$ operations in $\mathcal{A}_0$ and $\mathcal{A}_1$, respectively, on the sequence $[b_e], [a_e], [c_e], [P_e], [S_{i,s}], [Q_e]$. Here P_e and Q_e are either numbers chosen in advance for the requirement for $\mathcal{G}$ or sets that have participated in one of these two locations in a previous operation for $\mathcal{G}$. In any case, all of these components are in the range of r_s when we perform the operations. Suppose r_s mapped $\mathcal{G}$ into $\mathcal{A}_i$. The crucial point is that when we next get an expansionary stage at t and it is possible to extend r_s so as to keep $[S]$ mapped into $[S_{i,s}]$ then it is possible to extend r_s to be a map of $\mathcal{G}$ into $\mathcal{A}_i$ at t. The key idea here is that each component in the original sequence can "grow into" only one of two components in the final one, itself or the one immediately to its left (or right depending on whether $i = 0$ or 1). Thus if the image of one of the components remains fixed then we can see (in the reverse order of the operation performed) that each component in turn remains fixed as it has no other place to go. In this case, we extend r_s to r_t still mapping $\mathcal{G}$ into $\mathcal{A}_i$. If it is not possible to keep $[S]$ mapped into $[S_{i,s}]$ then we change r so as to define r_t as a map from $\mathcal{G}$ into $\mathcal{A}_{1-i}$. This also means that $[S_{i,t}]$ is not the same component in $\mathcal{A}_i$ as was $[S_{i,s}]$ (or we could have kept it fixed). (Actually it is the component that had been $[P_e]$ or $[Q_e]$ depending on the specifics of the situation.) We now guarantee never to use the old $[S_{i,s}]$ component in any future operation.

The ultimate consequence of such a procedure is that, if we change the range of r_t infinitely often, $[S]$ becomes infinite in $\mathcal{G}$ but each component $[S_{i,s}]$ that is a potential image of $[S]$ in $\mathcal{A}_i$ is involved in only finitely many operations and so is itself finite. Thus, in this case, $\mathcal{G}$ is not isomorphic to $\mathcal{A}$. On the other hand, if $\mathcal{G}$ actually is isomorphic to $\mathcal{A}$, we keep extending r_s from some stage on while never changing the $\mathcal{A}_i$ to which it maps $\mathcal{G}$. In this case, we arrange the definition of r_s so that if it eventually maps onto $\mathcal{A}_i$ and so determines the required computable isomorphism from $\mathcal{G}$ to $\mathcal{A}_i$.

We have, of course, omitted some of the combinatorics (particularly the way in which we extend the domain of r) even in this case of one $\mathcal{G}$ requirement. The full construction consists of using a module of this sort for each requirement on a typical $0''$ priority tree. Of course, the precise actions for a diagonalization requirement at a node α (e.g. which special components have to go into the sequence on which the operations are performed and in what order they go on this list) depend on the outcome of nodes β of higher priority contained in α which are devoted to various $\mathcal{G}_j$. (The choices here are whether there are infinitely many expansionary stages or not and if so whether the range of r_s is fixed as i from some point on or we change it infinitely often.) The details can be found in Khoussainov and Shore [1998].

We take this opportunity to point out two corrections that should be made to the details of the general construction found in Khoussainov and Shore [1998]. The first is that whenever one applies (or considers the application of) an operation for a node β to a sequence of the form

$$B^k_{n+1}, X^k, C^k_{n+1}, B^k_n, [S]^k_{\beta_n,t+1}, C^k_n, \ldots D^k_1, B^k_1, [S]^k_{\beta_1,t+1}, C^k_1$$

one should instead use its *β-transform* which is defined to be the sequence $B^k_{i_1}$, $[S]^k_{\beta_{i_1},t+1}, C^k_{i_1}, B^k_{i_2}, [S]^k_{\beta_{i_2},t+1}, C^k_{i_2}, \ldots, B^k_{i_m}, [S]^k_{\beta_{i_m},t+1}, C^k_{i_m}, B^k_{n+1}, X^k, C^k_{n+1}, B^k_{j_1}$, $[S]^k_{\beta_{j_1},t+1}, C^k_{j_1}, B^k_{j_2}, [S]^k_{\beta_{j_2},t+1}, C^k_{j_2}, \ldots, B^k_{j_r}, [S]^k_{\beta_{j_r},t+1}, C^k_{j_r}$ where $i_1, i_2, \ldots, i_m$ list, in order, the i such that the designated isomorphism for the $\beta_i = \beta \restriction (3i+1)$ such that $\beta(3i+1) \neq w$ at $t+1$ is $r^0_{\beta_i,t+1}$ and $j_1, j_2, \ldots, j_r$ list, in order, the j such that the designated isomorphism for the $\beta_j = \beta \restriction (3j+1)$ such that $\beta(3j+1) \neq w$ is $r^1_{\beta_{i,t+1}}$.

The second concerns the marking of numbers with the symbols $\square^\beta_w$. No numbers should be marked in Case 1 of the construction. As a result, condition 1 in Subcase 2.1 should be weakened by not requiring that the image of $r^k_{\beta_i,t+1}$ has nonempty intersection with B^k_i if the designated isomorphism for β_i is $r^0_{\beta_i,t+1}$ or with C^k_i if the designated isomorphism for β_i is $r^1_{\beta_i,t+1}$. Instead, the marking take place at the end of each stage of the construction as follows:

At the end of stage $t+1$ we do some additional cancellation and marking. Suppose $t+1$ is a γ recovery stage. If there are any uncancelled components isomorphic to $[b_{\beta,\gamma,u}]$ or $[c_{\beta,\gamma,u}]$ for $u \leq t$ (and $\beta \supseteq \gamma$) which, necessarily, have not participated in any operation, we cancel them and appoint new ones $[b_{\beta,\gamma,t+1}]$ or $[c_{\beta,\gamma,t+1}]$, respectively. We now mark all of the following with $\square^\gamma_w$ if they are not already so marked:

1. Any cancelled component.
2. Any component associated with a node β to the left of γ.
3. Any component of the form $[q_{\beta,t+1}]$ with $\beta \supseteq \gamma$.
4. Any components of the form $[b_{\beta,t+1}]$, $[c_{\beta,t+1}]$ or $[p_{\beta,t+1}]$ for $\beta \supseteq \gamma$.

5. Any components of the form $[b_{\beta,\beta_i,t+1}]$ or $[c_{\beta,\beta_i,t+1}]$ for $\beta \supseteq \beta_i \supset \gamma$.
6. Any components of the form $[b_{\beta,\gamma,t+1}]$ if $\beta \supseteq \gamma$ and the designated isomorphism for γ is $r^1_{\gamma,t+1}$or of the form $[c_{\beta,t+1}]$ if $\beta \supseteq \gamma$ and the designated isomorphism for γ is $r^0_{\gamma,t+1}$.
7. If, at $t+1$, we performed an operation on the β-transform of a sequence $B^0_{n+1}, X^0, C^0_{n+1}, B^0_n, [S]^0_{\beta_n}, C^0_n, \dots D^0_1, B^0_1, [S]^0_{\beta_1,t+1}, C^0_1$ (and so $\gamma \subseteq \beta$), then, for $k = 0, 1$, we mark B^k_i or C^k_i if $\beta_i \subseteq \gamma$ and it has previously participated in an operation; we mark B^k_i if $\beta_i \subseteq \gamma$ and the designated isomorphism for β_i is $r^1_{\beta_i,t+1}$; we mark C^k_i if $\beta_i \subseteq \gamma$ and the designated isomorphism for β_i is $r^0_{\beta_i,t+1}$; and we mark both B^k_i and C^k_i for $\beta_i = \gamma$.

The changes needed in the verifications to take advantage of these corrections are straightforward. A complete corrected version of the paper can be found at http://math.cornell.edu/~shore/.□

10. Bibliography

Ash, C. J. [1986], Recursive labeling systems and stability of recursive structures in hyperarithmetical degrees, *Trans. Amer. Math. Soc.* **298**, 497-514.

Ash, C. J. [1986a], Stability of recursive structures in arithmetical degrees, *Ann. Pure Appl. Logic* **32**, 113–135.

Ash, C. J. [1987], Categoricity in hyperarithmetical degrees, *Ann. Pure Appl. Logic* **34**, 1–14.

Ash, C. J., Cholak, P. and Knight, J. F. [1997], Permitting, forcing, and copying of a given recursive relation, *Ann. Pure Appl. Logic* **86**, 219–236.

Ash, C. J. and Knight, J. F. [1990], Pairs of recursive structures, *Ann. Pure Appl. Logic* **46**, 211–234.

Ash, C. J. and Knight, J. F. [1994], A completeness theorem for certain classes of recursively infinitary formulas, *Math. Logic Quart.* **40**, 173–181.

Ash, C. J. and Knight, J. F. [1995], Possible degrees in recursive copies, *Ann. Pure Appl. Logic* **75**, 215–221.

Ash, C. J. and Knight, J. F. [1997], Possible degrees in recursive copies II, *Ann. Pure Appl. Logic* **87**, 151–165.

Ash, C. J. and Knight, J. F. [1999], *Computable Structures and the Hyperarithmetical Hierarchy*, in preparation.

Ash, C., Knight, J., Manasse, M. and Slaman, T. [1989], Generic copies of countable structures, *Ann. Pure Appl. Logic* **42**, 195–205.

Ash, C. J., Knight, J. F. and Slaman, T. A. [1993], Relatively recursive expansions. II, *Fund. Math.* **142**, 147–161.

Ash, C. J. and Nerode, A. [1981], Intrinsically recursive relations, in *Aspects of Effective Algebra* (Clayton, 1979), Upside Down A Book Co., Yarra Glen, Vic., 26–41.

Baldwin, J. T. [1973], α_T is finite for $\aleph_1$-categorical T, *Trans. A.M.S.* **181**,

37-51.

Baldwin, J. T. and Lachlan, A. H. [1971], On strongly minimal sets, *J. Symbolic Logic* **36**, 79–96.

Barker, E. J. [1988], Intrinsically Σ^0_α-relations, *Ann. Pure Appl. Logic* **39**, 105–130.

Chang, C. C. and Keisler, H. J. [1990], *Model Theory*, Third edition, Studies in Logic and the Foundations of Mathematics **73**, North-Holland, Amsterdam-New York.

Chisholm, J. [1990], Effective model theory vs. recursive model theory, *Journal of Symb. Logic* **55**, 1168-1191.

Chisholm, J. [1990a], The complexity of intrinsically r.e. subsets of existentially decidable models, *Journal of Symb. Logic* **55**, 1213-1232.

Cholak, P., Goncharov S. S., Khoussainov, B. and Shore R. A. [1999], Computably Categorical Structures and Expansions by Constants, *Journal of Symb. Logic*, to appear.

Ershov, Yu. L. [1977], *Teoriya Numeratsii* (Russian) [*Theory of Numerations*], Matematicheskaya Logika i Osnovaniya Matematiki. [Monographs in Mathematical Logic and Foundations of Mathematics] "Nauka", Moscow.

Ershov, Yu. L. [1980], *Problemy Razreshimosti i Konstruktivnye Modeli* (Russian)[*Decision Problems and Constructivizable Models*], Matematicheskaya Logika i Osnovaniya Matematiki [Monographs in Mathematical Logic and Foundations of Mathematics] "Nauka", Moscow.

Ershov, Yu. L. and Goncharov, S. S. eds. [1986], *Logicheskaya Tetrad'* (Russian) [Logic Notebook], Nereshennye voprosy matematicheskoĭ logiki. [Unsolved problems in mathematical logic] Akad. Nauk SSSR Sibirsk. Otdel., Inst. Mat., Novosibirsk, 1986.

Ershov, Yu. L. and Goncharov, S. S. [1998], Elementary theories and their constructive models, in Ershov et al. [1998].

Ershov, Yu. L., Goncharov, S. S., Nerode, A. and Remmel, J. B. eds., Marek, V. W. assoc. ed. [1998], *Handbook of Recursive Mathematics*, Elsevier, New York.

Goncharov, S. S. [1973], Constructivizability of superatomic Boolean algebras (Russian), *Algebra i Logika* **12**, 31-40; translation in *Algebra and Logic* **12** (1973), 17–22.

Goncharov, S. S. [1975], Autostability, and computable families of constructivizations (Russian), *Algebra i Logika* **14**, 647–680; translation in *Algebra and Logic* **14** (1975), 392–409.

Goncharov, S. S. [1978], Constructive models of $\aleph_1$-categorical theories (Russian), *Mat. Zametki (Akad Nauk SSR)* **23**, 885–888; translation in *Math. Notes (Acad. Sci. USSR)* **23** (1978), 486–487.

Goncharov, S. S. [1978a], Strong constructivizability of homogeneous models (Russian), *Algebra i Logika* **17**, 363–388, 490; translation in *Algebra and Logic* **17** (1978), 247–263.

Goncharov, S. S. [1980], Autostability of models and abelian groups (Russian),

Algebra i Logika **19**, 23–44, 132; translation in *Algebra and Logic* **19** (1980), 13–27.

Goncharov, S. S. [1980a], The problem of the number of nonautoequivalent constructivizations (Russian), *Algebra i Logika* **19**, 621–639, 745; translation in *Algebra and Logic* **19** (1980), 401–414.

Goncharov, S. S. [1981], Groups with a finite number of constructivizations (Russian), *Dokl. Akad. Nauk SSSR* **256**, 269–272; translation in *Sov. Math. Dokl.* **23** (1981), 58-61.

Goncharov, S. S. [1982], Limit equivalent constructivizations (Russian), in *Mathematical Logic and the Theory of Algorithms*, Trudy Inst. Mat. **2**, "Nauka" Sibirsk. Otdel., Novosibirsk, 4–12.

Goncharov, S. S. [1997], *Countable Boolean Algebras and Decidability*, Consultants Bureau, New York.

Goncharov, S. S. [1998], Autostable models and their algorithmic dimension, in Ershov et al. [1998].

Goncharov, S. S. and Dzgoev, V. D. [1980], Autostability of models (Russian), *Algebra i Logika* **19**, 45–58; translation in *Algebra and Logic* **19** (1980), 28–37.

Goncharov, S. S. and Drobotun, B. N. [1989], The algorithmic dimension of nilpotent groups (Russian), *Sibirsk. Mat. Zh.* **30**, 52–60, 225; translation in *Siberian Math. J.* **30** (1989), 210–217.

Goncharov, S. S. and Khoussainov, B. [1997], On the spectrum of the degrees of decidable relations (Russian), *Dokl. Akad. Nauk* **352**, 301–303; translation in *Dokl. Math.* **55** (1997), 55-57.

Goncharov, S. S., Molokov, A. V. and Romanovski, N. S. [1989], Nilpotent groups of finite algorithmic dimension (Russian), *Sibirsk. Mat. Zh.* **30**, 82–88; translation in *Siberian Math. J.* **30** (1989), 63–68.

Goncharov, S. S. and Nurtazin, A. T. [1973], Constructive models of complete decidable theories (Russian), *Algebra i Logika* **12**, 125–142, 243; translation in *Algebra and Logic* **12** (1973), 67–77.

Griffor, E. ed. [1999], *Handbook of Computability Theory*, North-Holland.

Harizanov, V. S. [1993], The possible Turing degree of the nonzero member in a two element degree spectrum, *Ann. Pure Appl. Logic* **60**, 1–30.

Harizanov, V. S. [1998], Pure computable model theory, in Ershov et al. [1998].

Harizanov, V. S. [2000], *Computable Model Theory*, in preparation.

Harrington, L. [1974], Recursively presentable prime models, *J. Symbolic Logic* **39**, 305–309.

Herwig, B., Lempp, S. and Ziegler, M. [1999], *Proc. Amer. Math. Soc.*, to appear.

Hirschfeldt, D. [1999], Ph. D. Thesis, Cornell University.

Hirschfeldt, D., Khoussainov, B., Slinko, A. and Shore R. A. [1999], Degree spectra and computable categoricity in algebraic structures, in preparation.

Hodges, W. [1993], *Model Theory*, Encyclopedia of Mathematics and its Applications **42**, Cambridge University Press, Cambridge, England.

Kalantari, I. [1998], A bibliography of recursive algebra and model theory, in

Ershov et al. [1998].

Khisamiev, N. G. [1974], Strongly constructive models of a decidable theory (Russian), *Izv. Akad. Nauk Kazah. SSR Ser. Fiz.-Mat.*, 83–84.

Khoussainov, B., Nies, A. and Shore, R. A. [1997], Computable models of theories with few models, *Notre Dame J. Formal Logic* **38**, 165–178.

Khoussainov, B. and Shore, R. [1998], Computable Isomorphisms, Degree Spectra of Relations, and Scott Families, *Ann. of Pure and Applied Logic*, to appear.

Kudaibergenov, K. Ž. [1979], A theory with two strongly constructible models (Russian), *Algebra i Logika* **18**, 176–185, 253; translation in *Algebra and Logic* **18** (1979), 111–117.

Kudaibergenov, K. Ž. [1980], Constructivizable models of undecidable theories (Russian), *Sibirsk. Mat. Zh.* **21**, 155–158, 192.

Kudinov, O. V. [1996], An autostable 1-decidable model without a computable Scott family of $\exists$-formulas (Russian), *Algebra i Logika* **35**, 458–467, 498; translation in *Algebra and Logic* **35** (1996), 255–260.

LaRoche, P. [1977], Recursively represented Boolean algebras, *Notices Amer. Math. Soc.* **24**, A-552.

Manasse, M. [1982], *Techniques and Counterexamples in Almost Categorical Recursive Model Theory*, Ph. D. Thesis, University of Wisconsin, Madison.

Metakides, G. and Nerode, A. [1979], Effective content of field theory, *Ann. Math. Logic* **17**, 289–320.

Millar, T. [1978], Foundations of recursive model theory, *Ann. Math. Logic* **13**, 45–72.

Millar, T. [1979], Complete, decidable theory with two decidable models, *J. Symbolic Logic* **44**, 307–312.

Millar, T. [1983], Omitting types, type spectrums, and decidability, *J. Symbolic Logic* **48**, 171–181.

Millar, T. [1986], Recursive categoricity and persistence, *J. Symbolic Logic* **51**, 430–434.

Millar, T. [1999], Abstract recursive model theory, in Griffor [1999].

Morley, M. [1976], Decidable models, *Israel J. Math.* **25**, 233–240.

Nurtazin, A. T. [1974], Strong and weak constructivizations and computable families (Russian), *Algebra i Logika* **13**, 311–323, 364; translation in *Algebra and Logic* **13** (1974), 177–184.

Remmel, J. B. [1981], Recursive isomorphism types of recursive Boolean algebras, *J. Symbolic Logic* **46**, 572–594.

Remmel, J. B. [1981a], Recursively categorical linear orderings, *Proc. Amer. Math. Soc.* **83**, 387–391.

Robinson, R. M. [1950], An essentially undecidable axiom system, *Proc. Intl. Cong. Math.*, Cambridge 1950, v. 1, 729-730.

Sacks, G. E. [1963], On the degrees less than $0'$, *Ann. of Math. (2)* **77**, 211–231.

Selivanov, V. L. [1976], Numerations of families of general recursive functions, *Algebra and Logic* **15**, 205-226, 1976; translation in *Algebra and Logic* **15** (1976), 128–141.

Shoenfield, J. R. [1971], *Degrees of Unsolvability*, Mathematical Studies **2**, North-Holland, Amsterdam.

Tennenbaum, S. [1959], Non-Archimedian models of arithmetic, *Notices Amer. Math. Soc.* **6**, 270.

Vaught, R. L. [1961], Denumerable models of complete theories, in *Infinitistic Methods, Proceedings of the Symposium on the Foundations of Mathematics, Warsaw 2-9, September 1959*, Pegamon Press, Oxford, 301-321.

A Survey on Canonical Bases in Simple Theories

Byunghan Kim *

1 Introduction

The class of *simple* first order theories has recently been intensively studied. Simple theories were introduced by Shelah ([24]), who attempted to find a strictly broader class than the class of stable theories, in which forking is still reasonably well-behaved. After some years of neglect, the present author proved, in his doctoral thesis, that *forking* satisfies almost all the basic properties which hold in stable theories ([15],[16]). In fact, a large amount of the machinery of stability theory, invented by Shelah, is valid in the broader class of simple theories. All stable theories are simple, but there are also simple unstable theories, such as the theory of the random graph. Many of the theories of particular algebraic structures which have been studied recently (pseudofinite fields ([11],[12], [13],[4]), algebraically closed fields with a generic automorphism ([3]), smoothly approximable structures ([14], [5],[10]) turn out to be simple unstable, too.

Since a general survey of simple theories written by the author and A. Pillay appeared in the Bulletin of Symbolic Logic ([18]), here we shall focus on the notion of canonical base and Zilber's theorem in the context of simple theories. Roughly speaking, Zilber's theorem (Theorem 4.4) on a strongly minimal set says that any nontrivial ω-categorical, strongly minimal set is essentially a vector space over a finite field. Zilber used the notion of canonical base (for stable theories) significantly in the proof of this theorem. In the context of simple theories, it was conjectured that the underlying structure of any nontrivial ω-categorical, minimal set (a set of SU-rank 1) is again a vector space over a finite field. Recently Bradd Hart, Anand Pillay and the present author found the right notion of canonical base for simple theories ([7]). Using this notion, a result aiming for the conjecture was obtained (Theorem 4.7). However the full conjecture was recently refuted by Hrushovski. In section

*Supported by NSF grant DMS-9803425

2, we shall introduce the notion of canonical base in stable theories, as a generalization of the notion of "code" (or base) of a definable set. In section 3, we shall describe the extended notion of canonical base in simple context and give a sketch of a proof of the existence of canonical bases in simple theories. In section 4, in order to understand Zilber's theorem more precisely, we shall define a (combinatorial) "geometry", and discuss related problems in an ω-categorical supersimple theory.

We assume that the reader has some familiarity with the basics of forking in stable and simple theories, as developed in [25], [16] and [19]. However, we briefly recall from [16] and [19] some basic facts and definitions we need. A (partial) type p *divides* over a set A, if there are an L-formula $\varphi(\bar{x}, \bar{y})$ and a set of tuples $\{\bar{c}_i | i < \omega\}$ such that $p \vdash \varphi(\bar{x}, \bar{c}_0)$, $tp(\bar{c}_i/A) = tp(\bar{c}_0/A)$ for all $i < \omega$, and $\{\varphi(\bar{x}, \bar{c}_i) | i < \omega\}$ is k-inconsistent for some $k \in \omega$ (i.e. any subset of size k is inconsistent). A type p *forks* over A, if p implies some finite disjunction of formulas, each of which divides over A. A first order complete theory T is said to be *simple* if, for any type $p \in S(B)$, p does not fork over some subset A of B with $|A| \leq |T|$. In a simple theory, the notion of forking coincides with the notion of dividing. The theory T is called *supersimple* if, for any type $p \in S(B)$, p does not fork over some finite subset A of B. We also recall that T is *unstable* if there are a formula $\psi(\bar{x}, \bar{y})$ and tuples $\bar{b}_i, \bar{c}_i$ $(i < \omega)$ such that $\models \psi(\bar{b}_i, \bar{c}_j)$ if and only if $i \leq j$. A theory T is said to be *stable* if T is not unstable, and *superstable* if T is stable and supersimple.

In [16] and [24], it is shown that, for simple T, nonforking has the following properties. (i) *Extension* : for any $p \in S(A)$ and $A \subseteq B$, p has a *nonforking extension* q in $S(B)$, namely q is an extension of p which does not fork over A. (ii) *Symmetry*: $tp(\bar{b}/A\bar{c})$ does not fork over A if and only if $tp(\bar{c}/A\bar{b})$ does not fork over A. (iii) *Transitivity*: if $A \subseteq B \subseteq C$ and $p \in S(C)$, then p does not fork over A if and only if p does not fork over B and the restriction of p to B does not fork over A. Hence nonforking equips an arbitrary simple theory with a nice notion of independence, namely A is independent from B over C if for each $\bar{a} \in A$, $tp(\bar{a}/B \cup C)$ does not fork over C. Recall that in a stable theory T, every complete type over a model, and in fact every strong type, has a unique nonforking extension to any superset of the domain. The analogue of this in a simple theory is the so-called *Independence Theorem.*

Fact 1.1 *(The Independence Theorem over a model) Assume that T is simple. Let B and C be independent over a model M $(M \subseteq B, C)$. If $tp(\bar{d}/M) = tp(\bar{e}/M)$ and both $tp(\bar{d}/B)$ and $tp(\bar{e}/C)$ do not fork over M, then $tp(\bar{d}/B) \cup tp(\bar{e}/C)$ does not fork over M .*

Tuples $\bar{a}, \bar{b}$ are said to have the same *Lascar strong type over A* ($Lstp(\bar{a}/A) = Lstp(\bar{b}/A)$) if there are models $M_1, ..., M_k$, each of which contains A, and tuples $\bar{a} = \bar{a}_0, \bar{a}_1, ..., \bar{a}_k = \bar{b}$ such that $tp(\bar{a}_{i-1}/M_i) = tp(\bar{a}_i/M_i)$ for $1 \leq i \leq k$. In [19], the following is also shown.

Fact 1.2 *(The Independence Theorem for Lascar strong types) Assume that T is simple. Let B and C be independent over A ($A \subseteq B, C$). If $Lstp(\bar{d}/A) = Lstp(\bar{e}/A)$ and both $tp(\bar{d}/B)$ and $tp(\bar{e}/C)$ do not fork over A, then there is $\bar{a}$ such that $tp(\bar{e}/C) \cup tp(\bar{d}/B) \subseteq tp(\bar{a}/B \cup C)$, $tp(\bar{a}/B \cup C)$ does not fork over A, and $Lstp(\bar{a}/A) = Lstp(\bar{d}/A)$.*

Now let us mention two important ranks: $D(-, \psi, k)$ rank, and SU rank. Given a formula $\psi(x, y)$ in L and a positive number $k \in \omega$, the crucial inductive clause in the definition of $D(-, \psi, k)$ rank is: if $p(x)$ is a partial type, then $D(p(x), \psi(x, y), k) \geq n + 1$ if there is some infinite set $\{b_i : i < \omega\}$ such that $\{\psi(x, b_i) : i < \omega\}$ is k-inconsistent, and $D(p(x) \cup \{\psi(x, b_i)\}, \psi, k) \geq n$ for all i. The main clause for SU rank is: if $p(x)$ is a complete type, then $SU(p) \geq \alpha + 1$ if p has a forking extension $q(x)$ (also a complete type) such that $SU(q) \geq \alpha$. For stable T, this rank is the same as U rank. $D(-, \psi, k)$ rank and SU rank reflect forking in simple theories and supersimple theories respectively.

The notation here is fairly standard. T is a complete theory with no finite models in a first order language L. Types, denoted by p, q, are n-types and possibly partial. We work in a $\bar{\kappa}$-saturated model $\bar{M}$ (sometimes in $\bar{M}^{eq}$), for some big cardinal $\bar{\kappa}$ as usual. Tuples $\bar{a}, \bar{b}, \bar{c}... \in \bar{M}$ are finite. Sometimes we simply write $a, b, c, ..$ for finite tuples. Sets $A, B, C...$ are subsets of $\bar{M}$, and models are elementary submodels of $\bar{M}$, the cardinalities of all of which are strictly less than $\bar{\kappa}$. A type-definable set is a solution set in $\bar{M}^n$ of a partial n-type over some set. Algebraic closure and definable closure are denoted by $acl(-)$ and $dcl(-)$, respectively. When we work in $\bar{M}^{eq}$, we use the same notation for each closure.

2 Canonical bases in stable theories

In this section we shall discuss the notion of canonical base, which is a quite subtle, but powerful notion in stability theory. Although canonical bases were introduced by Shelah, the following theorem of André Weil ([26, p.19]) yields an analogue in algebraically closed fields. Let $\bar{K}$ be an algebraically closed field, and let V be a Zariski closed set in $\bar{K}$. Then a subfield k is said to be *a field of definition* of V, if $I(V) = \{f \in \bar{K}[\bar{x}] : f(\bar{a}) = 0 \text{ for all } \bar{a} \in V\}$ is generated by polynomials in $k[\bar{x}]$.

Theorem 2.1 *Let $\bar{K}$ be an algebraically closed field, and let V be Zariski closed in $\bar{K}$. Then V has a unique smallest perfect field of definition k.*

Moreover, any automorphism of $\bar{K}$ fixes V setwise if and only if it fixes k pointwise.

The field of definition k of V is a model theoretic definable closure of a finite tuple $\bar{c}$, which is a smallest perfect field containing $\bar{c}$. Hence any automorphism of $\bar{K}$ fixes V setwise if and only if it fixes $\bar{c}$ pointwise. It follows that algebraically closed fields have elimination of imaginaries ([23]). (See Definition 2.3 below.)

Theorem 2.2 *Let $\bar{K}$ be an algebraically closed field, and let V be a constructible set in $\bar{K}$. Then there is a finite tuple $\bar{d}$ in $\bar{K}$ such that any automorphism of $\bar{K}$ fixes V setwise if and only if it fixes $\bar{d}$ pointwise. We call $\bar{d}$ a* code *for V.*

For example, if a curve V in an algebraically closed field K (saturated) is defined by $y = c_0 + c_1x + ... + c_nx^n$, where $\{c_0, ..., c_n\}$ is algebraically independent over the prime field, then $(c_0, ..., c_n)$ is a code for V.

Now the issue is, given a definable (or type-definable) set X, to find a set B such that any automorphism of $\bar{M}$ fixes X setwise if and only if it fixes B pointwise.

First, we define a *base* for a definable set.

Definition 2.3 *Let T be arbitrary, and let X be a definable set in $\bar{M}^n$. We say a tuple $\bar{c}$ is a* base *(or* code*) of X if for any automorphism f of $\bar{M}$, f fixes X setwise if and only if f fixes $\bar{c}$ pointwise.*

We note that if $\bar{c}$ is a base for a definable set X, then X is definable over $\bar{c}$. Moreover if X is defined over A, then $\bar{c}$ is contained in $dcl(A)$.

We say T has *elimination of imaginaries* if every definable set has a base.

Fact 2.4 *Suppose that there are two terms (having no variables) c, d in L such that $T \models c \neq d$. Then the following are equivalent.*

1. *T has elimination of imaginaries.*
2. *For each n, and for every $\emptyset$-definable equivalence relation $E(\bar{x}, \bar{y})$ on $\bar{M}^n$, there is a definable function $f : \bar{M}^n \to \bar{M}^m$ such that $E(\bar{a}, \bar{b})$ if and only if $f(\bar{a}) = f(\bar{b})$.*

In fact, for any theory T, the associate T^{eq} (the theory of M^{eq}) has elimination of imaginaries: Let X be defined by a formula $\varphi(\bar{x}, \bar{a})$, where $\varphi(\bar{x}, \bar{y})$ is an L-formula. Consider the equivalence relation $E(\bar{y}, \bar{y}')$ defined by $\forall \bar{x}(\varphi(\bar{x}, \bar{y}) \leftrightarrow \varphi(\bar{x}, \bar{y}'))$. It is easy to see that $\bar{a}/E$ is a base for X in $\bar{M}^{eq}$.

Second, we introduce the notion of base for a complete type over $\bar{M}$.

Definition 2.5 *Let $\widehat{p} \in S(\bar{M})$. We say a set B is a* base *of $\widehat{p}$ if for any automorphism f of $\bar{M}$, $f(\widehat{p}) = \widehat{p}$ if and only if f fixes B pointwise.*

In general, even in $\bar{M}^{eq}$, $\widehat{p}$ does not need to have a base (e.g. any nonalgebraic type in a random graph has no base). But if T is stable, then every complete type $\widehat{p}$ over $\bar{M}$ has a base in $\bar{M}^{eq}$: As $\widehat{p}$ is definable, for every L-formula $\varphi(\bar{x}, \bar{y})$, there is an $\bar{M}$-formula $\psi_\varphi(\bar{y})$ (which we call the φ-definition of $\widehat{p}$) such that, for $\bar{b} \in \bar{M}$, $\varphi(\bar{x}, \bar{b}) \in \widehat{p}$ if and only if $\models \psi_\varphi(\bar{b})$. It is easy to check that the union of codes for $\psi_\varphi(\bar{y})$ (ranging over $\varphi(\bar{x}, \bar{y}) \in L$) is a base of $\widehat{p}$. Hence, for stable T, if T has elimination of imaginaries, then every complete type over $\bar{M}$ has a base in $\bar{M}$. The converse is not true in general, but true for a stable theory of fields.

Why are we interested in the notion of base? One of the examples in which it plays an important role is the proof of Zilber's theorem on ω-categorical strongly minimal sets. A major consequence of the theorem is that a totally categorical structure is not finitely axiomatizable. These will be discussed in section 4.

Finally, we shall investigate the notion of *canonical base* of a stationary type in a stable theory. Until the end of this section, we are working in $\bar{M}^{eq}$.

Definition 2.6 *Suppose that T is stable. Let $p \in S(A)$ be stationary (i.e. p has a unique nonforking extension over any superset of A.) Let $\widehat{p}$ be a unique nonforking extension of p to $\bar{M}$, and let c be a base for $\widehat{p}$. We call $dcl(c)$ the* canonical base *for p, denoted by $Cb(p)$.*

We note that $Cb(p)$ is a base for $\widehat{p}$, and moreover if d is a base for $\widehat{p}$, then $Cb(p) = dcl(d)$. For a stationary type p, since $\widehat{p}$ is definable over the domain set of p, $Cb(p)$ is the definable closure of the set of codes for the φ-definitions of p, as $\varphi(\bar{x}, \bar{y})$ ranges over L.

Canonical bases have the following properties. Let $p \in S(A)$ be stationary.

1. $Cb(p) \subseteq dcl(A)$, and p does not fork over $Cb(p)$. Moreover the restriction of p to $Cb(p)$ is stationary.

2. For any automorphism f, $f(p)$ is parallel to p (namely they have a common nonforking extension) if and only if f fixes $Cb(p)$ pointwise.

3. If $q \in S(B)$ is parallel to p, then $Cb(p) \subseteq acl(B)$. If in addition q is stationary, then $Cb(p) \subseteq dcl(B)$, hence $Cb(p) = Cb(q)$.

Hence, we can think of $Cb(p)$ as a unique smallest definably closed subset of $dcl(A)$ such that p does not fork over $Cb(p)$, and the restriction of p to $Cb(p)$ is stationary.

Example 2.7 *Let us examine canonical bases in the transparent, but insightful theory of an equivalence relation $E(x, y)$ having infinitely many infinite classes. It is easy to see that for any $b \notin A$ with bEc for some $c \in A$, $Cb(tp(b/A)) = dcl(b/E)$.*

Naturally, we can ask what the right notion of canonical base is in a simple theory. Since definability of types plays a key role in the proof of their existence in a stable theory, a good generalization might seem problematic. It can be done, however, as shown by Bradd Hart, Anand Pillay and the author ([7]). We shall discuss this in the next section.

3 Canonical bases in simple theories

We begin with an example.

Example 3.1 *In Example 2.7, let us assume that there is one more binary relation $R(x, y)$ such that $R(x, y)$ implies $E(x, y)$, and R defines a random graph inside each E-class. This theory is simple unstable. Now choose a point $b \in \bar{M}$. Suppose that $\{\hat{A}, \hat{B}\}$ partitions $\{x \in \bar{M} : xEb\}$, and that there is an automorphism f of $\bar{M}$ with $f(\hat{A}) = \hat{B}$. Let $\widehat{p}$ be a type over $\bar{M}$ determined by the set of formulas $\{xEb, xRc$ for $c \in \hat{A}$, $\neg xRd$ for $d \in \hat{B}$, and $x \neq a$ for all $a \in \bar{M}\}$. Then b/E is no longer a base for $\widehat{p}$ in the sense of definition 2.5. In fact there is no base for $\widehat{p}$ in $\bar{M}^{eq}$.*

In Example 3.1, although b/E is not a base for $\widehat{p}$ in the sense of definition 2.5, still it has a connection with $\widehat{p}$. Namely, $\{b/E\}$ is a smallest subset of $\bar{M}^{eq}$ contained in any $dcl(A)(\subseteq \bar{M}^{eq})$ such that $\widehat{p}$ does not fork over A. Now let $\wp = \{tp(c/A) | A \subseteq \bar{M}, cEb, c \notin A$ and dEb for some $d \in A\}$. Then it is easy to see that for any automorphism f, $f(\wp) = \wp$ iff $f(b/E) = b/E$. But obviously there are $p, q \in \wp$ such that even $p \cup q$ is inconsistent. However for any $p, q \in \wp$, there is $r \in \wp$ such that p and r have a common nonforking extension, and r and q have a common nonforking extension.

The idea of finding canonical bases in simple theories is quite similar to the previous example. Namely, we want to find a code for the transitive closure of parallelism.

Definition 3.2 *Let T be simple.*

1. *A complete type $p(x) \in S(A)$ is an* amalgamation base *if the Independence Theorem holds for $p(x)$. That is, whenever B, C are supersets of A which are independent over A, and $q_1(x)$, $q_2(x)$ are nonforking extensions of $p(x)$ over B, C, respectively, then q_1 and q_2 have a common nonforking extension $r(x) \in S(B \cup C)$.*

2. *For an amalgamation base p, define* $\wp_p = \{r | r$ *is an amalgamation base and there are amalgamation bases* $p_0, \ldots, p_n$ *such that* $p = p_0$, $r = p_n$, *and* p_i, p_{i+1} *have a common nonforking extension, for* $i < n\}$.

3. *For an amalgamation base p, we say a set* $B \subseteq \bar{M}^{eq}$ *is a canonical base for p, if every automorphism fixes* $\wp_p$ *setwise if and only if it fixes B pointwise.*

For simple T, by the Independence Theorem over a model, every complete type over a model is an amalgamation base. Similarly we can think of every Lascar strong type as an amalgamation base. Hence if Lascar strong type is strong type, then every type over an algebraically closed set in $\bar{M}^{eq}$ is an amalgamation base.

For an arbitrary T, we say that T has *elimination of hyperimaginaries* on finite tuples if for any type-definable equivalence relation $r(x, y)$ over $\emptyset$ on $\bar{M}^n$, and for any complete n-type $p \in S(\emptyset)$, there are $\emptyset$-definable equivalence relations $E_i(x, y)$ $(i \in I)$ such that

$$p(x) \wedge p(y) \models r(x, y) \leftrightarrow \bigwedge_i E_i(x, y).$$

Similarly, we can think of elimination of hyperimaginaries for infinite tuples. We now state the following important theorem on the existence of canonical bases in simple theories.

Theorem 3.3 *Let T be simple. Moreover suppose that T has elimination of hyperimaginaries. Let* $p \in S(A)$ *be an amalgamation base. Then there is a canonical base* $B \subseteq \bar{M}^{eq}$.

If we work with hyperimaginaries as in [7], we do not need to assume elimination of hyperimaginaries, and may choose B as a single hyperimaginary element. Canonical bases in simple theories share important properties of canonical bases in stable theories (see Proposition 3.4). In fact, for stable T, the two notions of canonical bases coincide, and also the notions of stationary type and amalgamation base are the same. In particular canonical bases exist in $\bar{M}^{eq}$.

There is a well known example of Poizat which does not have elimination of hyperimaginaries (see [18]). We summarize several important results on elimination of hyperimaginaries. Obviously elimination of hyperimaginaries implies that Lascar strong type is strong type whenever equality of Lascar strong type is type-definable (e.g. in simple theories ([17])). Moreover, for simple T, the converse of 3.3 is true. Namely, if every amalgamation base has a canonical base in $\bar{M}^{eq}$, then T has elimination of hyperimaginaries ([21]). Hence if T is stable, then T has elimination of hyperimaginaries (see also [22]). The same (for finite tuples) is true for any small theory ([17]).

Recently Buechler, Pillay and Wagner proved that every supersimple theory has elimination of hyperimaginaries ([1]). It is still open whether any simple first order theory has elimination of hyperimaginaries.

Proposition 3.4 *Let T be simple, and p an amalgamation base. Let $Cb(p) \subseteq \bar{M}^{eq}$ be a canonical base for p. Then the following properties hold.*

1. *$Cb(p) \subseteq dcl(A)$, and p does not fork over $Cb(p)$. Moreover the restriction of p to $Cb(p)$ is an amalgamation base.*

2. *If $q \in S(B)$ is a complete type, and has a common nonforking extension with p, then $Cb(p) \subseteq acl(B)$. If in addition q is an amalgamation base, then $Cb(p) \subseteq dcl(B)$, hence $dcl(Cb(p)) = dcl(Cb(q))$.*

Hence, we can think of $dcl(Cb(p))$ as the unique smallest definably closed subset of $dcl(A)$ such that p does not fork over $Cb(p)$, and the restriction of p to $Cb(p)$ is an amalgamation base.

Now let us describe the proof of Theorem 3.3. We use the same notation as 3.3. Let $q(y) = tp(a)$. We define relations inside q (i.e. on the set of realizations of q). We denote realizations of q by $b, c, \ldots$ Write $R_0(b, c)$ if $p(x, b)$ and $p(x, c)$ have a common nonforking extension. Let $R(-,-)$ be the transitive closure of $R_0(-,-)$, so R is an equivalence relation. The main difficulty lies in showing that R is a type-definable equivalence relation in q. Then, as T has elimination of hyperimaginaries, there are $\emptyset$-definable equivalence relations E_i such that, in q, R is equivalent to the conjunction of the E_i. Now the canonical base of $p(x, a)$ is $B = \bigcup_i \{a/E_i\}$. We shall now sketch a proof of the type-definability of R: It is not difficult to see that $R_0(-,-)$ is type-definable, using a $D(-,\psi,k)$-rank argument. Now enumerate all ranks $D(-,\psi_i(x,y),k_i)$, where $\psi_i(x,y) \in L, k_i \in \omega$, as $\{D_i | i \in |L|\}$. Define n_i inductively by

$$n_i = max\{D_i(b/a) : R_0(a,b) \text{ and } D_j(b/a) = n_j \text{ for } j < i\}.$$

Note that $n_i \leq D_i(q)$, hence the maximum exists. We define a type-definable relation $R_1(-,-)$ by

$$R_1(x,y) \text{ iff } R_0(x,y) \wedge \bigwedge D_i(y/x) \geq n_i.$$

Then the following can be shown.

Proposition 3.5 *The following are equivalent.*

1. *$R(b, c)$.*

2. *For some d, $R_1(b, d)$ and $R_1(c, d)$.*

3. *For some d such that* $\underset{c}{d \mathop{\downarrow} b}$ *and* $\underset{b}{d \mathop{\downarrow} c}$, *both* $R_1(b,d)$ *and* $R_1(c,d)$ *hold.*

Clearly $3 \Rightarrow 2 \Rightarrow 1$. Now $1 \Rightarrow 3$ can be shown by an induction argument, using the following two observations:

First, using the properties of amalgamation bases, one can see that if $\underset{c}{d \mathop{\downarrow} b}$, $R_0(d,c)$ and $R_0(b,c)$, then $R_0(d,b)$. Second, if $R_1(c,d)$, $\underset{c}{d \mathop{\downarrow} b}$ and $R_0(b,d)$, then $R_1(b,d)$, since $n_i \geq D_i(d/b) \geq D_i(d/bc) = D_i(d/c) \geq n_i$.

4 ω-categorical supersimple theories

One of the aims of developing the notion of canonical base is to extend Zilber's theorem (Theorem 4.4) to the context of ω-categorical supersimple theories. In order to understand the meaning of Zilber's theorem, we need the following series of definitions. The complete account of most material discussed in this section can be found in chapter 2 of Pillay's book [20].

Definition 4.1 *Let* X *be a set. If an operation* $cl : \mathcal{P}(X) \to \mathcal{P}(X)$ *satisfies the following properties, then we say that* (X, cl) *is a pregeometry.*

1. *For* $A \subseteq X$, $A \subseteq cl(A)$.

2. *For* $A \subseteq X$, $cl(cl(A)) = cl(A)$.

3. *For* $A \subseteq X$ *and* $a, b \in X$, *if* $a \in cl(A \cup b) \setminus cl(A)$, *then* $b \in cl(A \cup a)$.

4. *If* $A \subseteq X$ *and* $a \in cl(A)$, *then there is finite* $A_0 \subseteq A$ *such that* $a \in A_0$.

We say a pregeometry (X, cl) is a *geometry* if $cl(\emptyset) = \emptyset$ and $cl(\{a\}) = \{a\}$ for all $a \in X$. If (X, cl) is a pregeometry, then we can associate a canonical geometry $(\hat{X}, \hat{cl})$: $\hat{X} = X_0/\sim$ where $X_0 = X \setminus cl(\emptyset)$, and $\sim$ is an equivalence relation on X given by $a \sim b$ iff $cl(\{a\}) = cl(\{b\})$. For $\{a_i/\sim |i \in I\} \subseteq \hat{X}$, $\hat{cl}(\{a_i/\sim |i \in I\}) = \{b/\sim |b \in cl(\{a_i : i \in I\})\}$.

Let (X, cl) be a pregeometry. For $A, B \subseteq X$, we say that A is *independent over* B if $a \notin cl((A \setminus \{a\}) \cup B)$ for any $a \in A$. We say that $A_0 \subseteq A$ is a *basis* for A over B if $A \subseteq cl(A_0 \cup B)$ and A_0 is independent over B. Then any two bases for A over B have the same cardinality, denoted by $dim(A/B)$. For $A, B, C \subseteq X$, if $dim(A'/C) = dim(A'/B \cup C)$ for any finite $A' \subseteq A$, then we say that A is *independent from* B *over* C.

Definition 4.2 *Let* (X, cl) *be a pregeometry.*

1. *(X, cl) is trivial if, for any $A \subseteq X$, $cl(A) = \bigcup\{cl(\{a\}) : a \in A\}$.*

2. *(X, cl) is locally modular if for any closed A, B with $A \cap B \neq \emptyset$, A is independent from B over $A \cap B$, or equivalently if $dim(A) + dim(B) = dim(A \cup B) + dim(A \cap B)$.*

If a pregeometry is trivial or locally modular, so is its associated canonical geometry.

Example 4.3 1. *Let X be any set, and $cl(A) = A$ for any $A \subseteq X$. Then (X, cl) is a trivial geometry.*

2. *Let F be a division ring, and V an infinite dimensional vector space over F. For $A \subseteq V$, let $cl(A)$ be a linear F-subspace of V spanned by A. Then (V, cl) is a nontrivial locally modular pregeometry. The associated geometry is called projective geometry over F.*
   Again, let F be a division ring, and V an infinite dimensional vector space over F. For $A \subseteq V$, let $afcl(A)$ be a smallest F-affine subspace of V containing A. (An affine subspace of V is of the form $a + W$ where $a \in V$ and W is a linear F-subspace of V.) Then $(V, afcl)$ is a nontrivial locally modular geometry.

3. *Let K be an algebraically closed field of infinite transcendence degree. Then (K, acl) (where acl is the field-theoretic algebraic closure) is a pregeometry which is neither trivial nor locally modular.*

Let X be the solution set of a strongly minimal formula (i.e. every definable subset of X is either finite or cofinite), or of a complete type over A of SU-rank 1. (We call X a strongly minimal set, or a set of SU-rank 1, respectively.) It is clear that (X, cl) is a pregeometry, where $cl(-) = acl(- \cup A) \cap X$. Moreover, for $B, C, D \subseteq X$, B is independent from C over D (in the pregeometry) if and only if B is independent from C over $D \cup A$ (in forking sense), if and only if B is independent from C over D in T_A.

All examples in 4.3 are strongly minimal. In the 1970s, it was conjectured by Zilber that the above are only examples, i.e. strongly minimal pregeometries are either locally modular, or interpret an algebraically closed field. (Note that a trivial pregeometry is locally modular.) But Hrushovski constructed a strongly minimal set whose geometry is not locally modular and which does not even interpret a group ([8]). However, Zilber proved that his conjecture is true for an ω-categorical theory ([27],[28],[29]). In that case (infinite) fields do not arise, and pregeometries are locally modular.

Theorem 4.4 *Let X be strongly minimal and ω-categorical. Then the pregeometry of X is locally modular.*

Zilber's theorem implies that a nontrivial geometry in an ω-categorical strongly minimal set is isomorphic to an affine or projective geometry over a finite field. The same is true for an ω-categorical set of U-rank 1. Later, Cherlin, Harrington and Lachlan studied more generally ω-categorical, ω-stable structures in [6]. There, Cherlin proved that these structures must have finite U-rank; he also gave an alternative proof of Theorem 4.4, using the classification of finite simple groups. From this, Zilber derived the non-finite axiomatizability of ω-categorical ω-stable structures. The point is that canonical bases play an important role in all of these proofs.

There is a notion of *modularity* which is closely related to that of *local modularity*. Some authors call it 1-*basedness*. For the following definition, we work in $\bar{M}^{eq}$.

Definition 4.5 *Let T be simple. Suppose that T has elimination of hyperimaginaries (e.g. supersimple theories). Let X be a type-definable set over some parameter A. We say that X is* modular *if for every $B, C \subseteq acl(X \cup A)$, B is independent from C over $acl(B \cup A) \cap acl(C \cup A)$. We say that T is* modular *if $\bar{M}$ is , i.e. for any $B, C (\subseteq \bar{M})$, B is independent from C over $acl(B) \cap acl(C)$.*

Remark 4.6 *In 4.5, since T has elimination of hyperimaginaries, every type over an algebraically closed set (in $\bar{M}^{eq}$) is an amalgamation base, as we noted before. Also, in that case, it is easy to see that a type-definable set X over A is modular if and only if for every $a \in acl(X \cup A)$ and for every set B, $Cb(tp(a/acl(B \cup A))) \subseteq acl(aA)$.*

From Buechler's work on modular theories ([2]) it follows that ω-categorical, ω-stable structures are modular.

In [14], Kantor, Liebeck and Macpherson investigated smoothly approximable structures, the class of which is a natural proper superclass containing the class of ω-categorical, ω-stable structures. A *smoothly approximable structure* is a countable relational ω-categorical structure M which is the union of an increasing chain of finite "homogeneous substructures" of M. By a "homogeneous substructure" of M we mean a subset A of M, such that for any finite tuples a, b chosen from A, a has the same type as b in M if and only if there is an automorphism of M which fixes A setwise and takes a to b. The typical example of a smoothly approximable structure (which is not ω-stable) is an infinite-dimensional vector space over a finite field, equipped with a non-degenerate bilinear form. Smoothly approximable structures are not finitely axiomatizable. Hrushovski and Cherlin's work implies that those structures are modular and have finite SU-rank ([10], [5]). It follows that the geometry of a set of SU rank 1 in a smoothly approximable structure is locally modular, and hence either trivial, or isomorphic to an affine or projective geometry over a finite field.

The next canonical proper superclass of the class of smoothly approximable structures is the class of ω-categorical, supersimple structures. (It is well-known that an ω-categorical, superstable structure is ω-stable.) A random graph is an example of an ω-categorical, supersimple structure, which is not smoothly approximable. But it has a trivial geometry. A nontrivial example is an infinite-dimensional vector space over a finite field, equipped with a binary relation defining a random graph.

Using the notion of canonical base, the following is proved in [7].

Theorem 4.7 *Let T be supersimple and ω-categorical. Then T is modular if and only if T has finite SU rank and all sets of SU rank* 1 *are modular.*

It is natural to ask whether the class of ω-categorical, supersimple structures possesses the important properties of the previously discussed two subclasses, such as modularity of T, local modularity of the geometry of a set of SU-rank 1, and the finite rank property. Namely, it was once asked whether, in Theorem 4.7, supersimplicity and ω-categoricity of T simply implies modularity of T. But even if we assume that T is supersimple, ω-categorical and modular, Theorem 4.7 is not strong enough to imply local modularity of an SU rank 1 set in a model of T.

Recently, Hrushovski constructed an ω-categorical SU-rank 1 structure whose geometry is not locally modular ([9]) by generalizing his construction method used for refuting various conjectures in stability theory, including Zilber's conjecture on strongly minimal sets mentioned above. This implies that the structure is not modular, and it also implies that the geometry of the structure is neither trivial, nor isomorphic to an affine or projective geometry over a finite field. However, the question whether every ω-categorical, supersimple structure has finite SU rank is still open.

References

[1] S. Buechler, A. Pillay and F. O. Wagner, 'Superstrong is strong enough', preprint (1998).

[2] S. Buechler, 'Locally modular theories of finite rank', *Annals of Pure and Applied Logic* 30 (1986), 83-95.

[3] Z. Chatzidakis and E. Hrushovski, 'Model theory of difference fields', to appear in *Transactions of American Mathematical Society.*

[4] Z. Chatzidakis, L. van den Dries, and A. J. Macintyre, 'Definable sets over finite fields', *J. reine angew. Math.* 427 (1992), 107-135.

[5] G. Cherlin, 'Large finite structures with few types', to appear in *Proceedings of 1996 NATO ASI meeting at the Fields Institute.*

[6] G. Cherlin, L. Harrington and A. H. Lachlan, 'ω-categorical, ω-stable structures', *Annals of Pure and Applied Logic* 28 (1985), 103-135.

[7] B. Hart, B. Kim, and A. Pillay, 'Coordinatization and canonical bases in simple theories', preprint (1997).

[8] E. Hrushovski, 'A new strongly minimal set', *Annals of Pure and Applied Logic* 62 (1993), 147-166.

[9] E. Hrushovski, 'Simplicity and the Lascar group', preprint (1997).

[10] E. Hrushovski, 'Finite structures with few types', in *Finite and Infinite Combinatorics in Sets and Logic*, NATO ASI Series C: Math. Phys. Sci. 411, Kluwer, Dordrecht, 1993, 175-187.

[11] E, Hrushovski, 'Pseudofinite fields and related structures', preprint (1992).

[12] E. Hrushovski and A. Pillay, 'Groups definable in local fields and pseudofinite fields', *Israel Journal of Math.* 85 (1994), 203-262.

[13] E. Hrushovski and A. Pillay, 'Definable subgroups of algebraic groups over finite fields', *J. reine angew. Math.* 462 (1995), 69-91.

[14] W. M. Kantor, M. W. Liebeck, and H. D. Macpherson, 'ω-categorical structures smoothly approximated by finite structures', *Proceedings of London Math. Soc.* (3) 59 (1989), 439-463.

[15] B. Kim, *Simple first order theories*, Ph. D. Thesis, Univ. of Notre Dame, 1996.

[16] B. Kim, 'Forking in simple unstable theories', *Journal of London Math. Soc.* 57 (1998), 257–267.

[17] B. Kim, 'A note on Lascar strong types in simple theories', to appear in *Journal of Symbolic Logic.*

[18] B. Kim and A. Pillay, 'From stability to simplicity', *Bulletin of Symbolic Logic* (4) 1 (1998), 17-36.

[19] B. Kim and A. Pillay, 'Simple theories', *Annals of Pure and Applied Logic* 88 (1997) 149-164.

[20] A. Pillay, *Geometric stability theory*, Oxford University Press, Oxford, 1996.

[21] A. Pillay and D. Lascar, *Hyperimaginaries and automorphism groups*, preprint (1998).

[22] A. Pillay and B. Poizat, 'Pas d'imaginaires dans l'infini!', *Journal of Symbolic Logic* 52 (1987), 400-403.

[23] B. Poizat, *Groupes stables*, Nur Al-Mantiq Wal-Ma'rifah, Villeurbanne, 1987.

[24] S. Shelah, 'Simple unstable theories', *Annals of Math. Logic* 19 (1980), 177-203.

[25] S. Shelah, *Classification theory and the number of non-isomorphic models*, 2nd edition, North-Holland, Amsterdam, 1990.

[26] A. Weil, *Foundations of algebraic geometry*, AMS Colloquium Publications 29, AMS, New York, 1946.

[27] B. I. Zilber, 'Strongly minimal countably categorical theories', (in Russian) *Sibirsk Matematika Zhurnal* 21, No. 2 (1980) 98-112.

[28] B. I. Zilber, 'Strongly minimal countably categorical theories II', (in Russian) *Sibirsk Matematika Zhurnal* 25, No. 3 (1980) 71-88.

[29] B. I. Zilber, 'Strongly minimal countably categorical theories IIi', (in Russian) *Sibirsk Matematika Zhurnal* 25, No. 4 (1984) 63-77.

True Approximations and Models of Arithmetic

Julia F. Knight
University of Notre Dame

0 Introduction

All structures here have recursive language and recursive universe–we could take the universe to be ω or a finite initial segment. A structure $\mathcal{A}$ is identified with its atomic diagram $D(\mathcal{A})$. Thus, the Turing degree of $\mathcal{A}$ is the degree of $D(\mathcal{A})$. Non-standard models of PA are among the most self-aware of all mathematical structures, but they do not know that they are non-standard. This results in some interesting properties, both model-theoretic and recursion-theoretic.

The self-awareness properties are related to the definability of satisfaction, for formulas of bounded complexity, coupled with a coding of definable sets in divisibility types. For an arbitrary structure $\mathcal{A}$, $T_n(\mathcal{A}) = \text{Th}(\mathcal{A}) \cap \Sigma_n$ is $\Sigma^0_n(\mathcal{A})$, uniformly in $\mathcal{A}$ and n, for $n \geq 1$. Thanks to the self-awareness properties, we have Feferman's observation that if $\mathcal{A}$ is a non-standard model of PA, then $T_n(\mathcal{A}) \leq_T \mathcal{A}$ for all n.

The fact that a non-standard model of PA does not know that it is non-standard means that we can carry out certain constructions and operations internally (i.e., definably), which would be impossible externally. In particular, we can determine a total function on the model by giving the value at 0 and saying how the value at $x+1$ is determined from the value at x – never mind that the order type of the model is $\omega + (\omega^* + \omega) \cdot \eta$.

A remarkable result of Matijasevich implies that satisfaction of a formula with only bounded quantifiers can be determined effectively from the atomic diagram [Mat], [Ma]. For a standard model, this is clear, since the searches involved are finite. For a non-standard model, checking something which begins $\forall x < b\ \exists y < c \ldots$, for *infinite* elements b, c, looks more complicated, from the outside.

Using Matijasevich's Theorem, Solovay [So$_2$] proved the following result, on a model's ability to find indices for the fragments of its theory.

Approximation Lemma: *For a non-standard model $\mathcal{A}$ of PA, there exist* *functions t_n, $\Delta^0_n(\mathcal{A})$ uniformly in n, such that $\lim_{s\to\infty} t_n(s)$ is an index for $T_n(\mathcal{A})$ as a set recursive in $\mathcal{A}$, and for all $s, t_n(s)$ is an index for a subset of $T_n(\mathcal{A})$.*

For an arbitrary structure $\mathcal{A}$ and tuple $\overline{a}$ in $\mathcal{A}$, the Σ_n type of $\overline{a}$, $T_n(\mathcal{A},\overline{a})$, is $\Sigma^0_n(\mathcal{A})$, uniformly in $\overline{a}$ and n. Using the same reasoning as in Solovay's Approximation Lemma, we can show that for a non-standard model $\mathcal{A}$ of PA, there are functions $t_n(\overline{a}, s)$, $\Delta^0_n(\mathcal{A})$ uniformly in n, such that for all tuples $\overline{a}$ in $\mathcal{A}$, $\lim_{s\to\infty} t_n(\overline{a}, s)$ is an index for $T_n(\mathcal{A},\overline{a})$ as a set recursive in $\mathcal{A}$, and for all $s, t_n(\overline{a}, s)$ is an index for a subset of $T_n(\mathcal{A},\overline{a})$.

For most structures $\mathcal{A}$, $\{\deg(\mathcal{B}) : \mathcal{B} \cong \mathcal{A}\}$ is closed upwards. For models of PA, this was proved by Solovay and by Marker [Mar]; the general case is in [K_1]. Since the standard model is recursive, it has copies of all degrees. Solovay [So$_1$] characterized the degrees of non-standard models of TA, and Marker [Mac-Mar] simplified the characterization. Later, Solovay [So$_2$] characterized the degrees of models of other completions of PA.

It turns out to be easier to construct a model, controlling the degree, if we specify more than just the theory. Solovay characterized the degrees of non-standard models of a particular completion of PA by characterizing the degrees of models with a given "Scott set". The result for non-standard models of TA, after Marker's simplification, says that the degrees of the models having Scott set $\mathcal{S}$ are the degrees of "enumerations" of $\mathcal{S}$. Hence, the degrees of non-standard models of TA are the degrees of enumerations of the appropriate Scott sets.

Solovay did not publish this work. The model construction half of the result for TA follows from a result in [K$_2$]. The other half becomes obvious after Marker's simplification. The model construction half of the result for other completions of PA follows from a result in [K$_3$]. The other half, the Approximation Lemma, is not obvious, and no proof has been published.

The aim of the present paper is to give a readable account of these results. The results in [K$_2$] and [K$_3$], on models which "represent" a given Scott set, apply to arbitrary theories, not just completions of PA. The model construction results here have the same general nature, but they are simpler. The theorem from [K$_2$] is replaced here by a weaker result. The proof in [K$_2$] is a transfinitely nested priority construction, described in terms of Harrington's "workers". Here this is replaced by two finite injury constructions. The theorem from [K$_3$] is strengthened here, in a minor way. Both the proof in [K$_3$] and the one here are transfinitely nested priority constructions, but the one here should be more readable. Finally, for completeness, with Solovay's kind permission, we include a proof of the Approximation Lemma.

Priority constructions with more than three levels are never simple. For this reason, it seems worthwhile pointing out the abstract properties which make a given construction work, in case the same properties appear again. In [A$_1$], there is a "metatheorem" for arbitrarily nested priority constructions. This metatheorem has a number of applications [A$_1$], [A$_2$], [B], [A – K$_1$], [H], and [D]. However, neither it nor any of the related metatheorems in [A – K$_2$], [A – K$_3$], and [K$_4$] applies to the setting in [K$_3$]. There is a metatheorem in [K$_3$], but it is more complicated than necessary, and in addition, there is a minor, but confusing mistake (on p. 225, line 15, ρ and ρ' are supposed to have *even* length, not odd). The present paper contains a new, simpler metatheorem.

In §1, we recall the definition of a Scott set and say what it means for a structure $\mathcal{A}$ to represent a Scott set. In §2, there is an organized collection of results on structures representing a given Scott set, with proofs of all but one. In §3, there are proofs of Solovay's results, using the results from §2. The new metatheorem is in §4. In §5, this metatheorem is used to prove the result which was left unproved in §2. In §6, we indicate what little progress has been made in answering certain natural questions.

1 Definitions and basic results

A *Scott set* is a set $\mathcal{S} \subseteq P(\omega)$ such that

(1) if $X \in \mathcal{S}$ and $Y \leq_T X$, then $Y \in \mathcal{S}$,

(2) if $X, Y \in \mathcal{S}$, then $X \oplus Y \in \mathcal{S}$,

(3) if $T \subseteq 2^{<\omega}$ is an infinite tree in $\mathcal{S}$, then T has a path p in $\mathcal{S}$; equivalently, if A is a consistent set of sentences in $\mathcal{S}$, then A extends to a complete theory in $\mathcal{S}$.

Scott sets are naturally associated with completions of PA [Sc]. For a theory T in the language of PA and a set $x \subseteq \omega$, X is *representable* with respect to T if there is a formula φ such that for $n \in X$, $T \vdash \varphi(S^{(n)}(0))$, and for $n \notin X$, $T \vdash \neg\varphi(S^{(n)}(0))$. The collection of sets representable with respect to T is denoted by Rep(T). For example, Rep(TA) is the collection of arithmetical sets.

Theorem 1.1 (Scott) *For a countable collection $\mathcal{S} \subseteq P(\omega)$, the following are equivalent:*

(1) there is a completion T of PA such that $\mathrm{Rep}(T) = \mathcal{S}$,

(2) $\mathcal{S}$ is a Scott set.

For a non-standard model $\mathcal{A}$ of PA, $SS(\mathcal{A})$ is the collection of sets of the form $d_a = \{k \in \omega : \mathcal{A} \models p_k \mid a\}$, for $a \in \mathcal{A}$. We call $SS(\mathcal{A})$ the *Scott set* of $\mathcal{A}$ (the *standard system* of $\mathcal{A}$ is another name).

Theorem 1.2 (Scott) *For a non-standard model $\mathcal{A}$ of PA, $SS(\mathcal{A})$ is a Scott set.*

For any *countable* Scott set $\mathcal{S}$, there exists a non-standard model $\mathcal{A}$ of PA such that $SS(\mathcal{A}) = \mathcal{S}$. The same is true for a Scott set of cardinality $\aleph_1$ [N]. Without CH, it is an open question whether this holds for Scott sets of cardinality greater than $\aleph_1$.

We shall define what it means for an arbitrary structure $\mathcal{A}$ to *represent* a Scott set $\mathcal{S}$. We need some terminology to describe fragments of types. A Σ_n *formula* is a formula in prenex normal form with n blocks of quantifiers, starting with $\exists$, followed by an open formula. A B_n *formula* is a Boolean combination of finitary Σ_n formulas. A complete B_n *type* is the set of B_n formulas in some complete type; i.e., it is the set of all B_n formulas true of some tuple in some structure. A structure $\mathcal{A}$ *represents* a Scott set $\mathcal{S}$ if for all complete B_n types $\Gamma(\overline{u}, x)$ and all tuples $\overline{c}$ in $\mathcal{A}$, $\Gamma(\overline{c}, x)$ is realized in $\mathcal{A} \Leftrightarrow \Gamma \in \mathcal{S}$ and $\Gamma(\overline{c}, x) \cup D^c(\mathcal{A})$ is consistent.

The definition above is as in [M]. The definition in [K$_2$] said only that the complete B_n types realized in $\mathcal{A}$ are the ones in $\mathcal{S}$ which are consistent with Th($\mathcal{A}$)–there was no homogeneity condition. Thus, the definition here is stronger, in general, although the difference does not show up in models of arithmetic. If $\mathcal{A}$ is a non-standard model of PA, then $SS(\mathcal{A})$ is the *unique* Scott set represented by $\mathcal{A}$. Some structures, such as the standard model of arithmetic, do not represent any Scott set. Other structures, such as $(Q, <)$ or an algebraically closed field of infinite transcendance degree, represent all Scott sets.

For a complete theory T and a countable Scott set $\mathcal{S}$, T has a model representing $\mathcal{S}$ if and only if $T_n = T \bigcap \Sigma_n \in \mathcal{S}$, for all n. We then say that $\mathcal{S}$ is *appropriate* for T. If T is a completion of PA, the Scott sets appropriate for T are those which contain the elements of Rep(T). To give conditions under which a given theory has a model representing a given Scott set, and having a specified degree, we must consider not just the elements of the Scott set, but also "enumerations" of the Scott set.

For a countable collection $\mathcal{S} \subseteq P(\omega)$, an *enumeration* is a relation $R \subseteq \omega \times \omega$ such that $\mathcal{S} = \{R_n : n \in \omega\}$, where $R_n = \{k : (n, k) \in R\}$. An *R-index* for a set $X \in \mathcal{S}$ is a number n such that $R_n = X$. If $\mathcal{A}$ is a non-standard model of PA with universe ω, then $R = \{(a, n) : \mathcal{A} \models p_n \mid a\}$ is an enumeration of $SS(\mathcal{A})$, called the *canonical* enumeration. The following is well known.

Proposition 1.3. *Let $\mathcal{A}$ be a non-standard model of PA with universe ω. If R is the canonical enumeration of $SS(\mathcal{A})$, then $R \leq_T \mathcal{A}$.*

Proof: To determine whether $\mathcal{A} \models p_n \mid a$, using $D(\mathcal{A})$, we remember the division algorithm and search for a sentence in $D(\mathcal{A})$ of the form $p_n b + S^{(r)}(0) = a$ for some $r < p_n$.

For a countable Scott set $\mathcal{S}$, an *effective* enumeration is an enumeration R equipped with functions f, g, and h which witness the fact that $\mathcal{S}$ is a Scott set:

(1) if $\varphi_e^{R_i} = \chi_X$, then $f(i, e)$ is an R-index for X,

(2) $g(i, j)$ is an R-index for $R_i \oplus R_j$, and

(3) if R_i is (the set of Gödel numbers for elements of) an infinite tree $T \subseteq 2^{<\omega}$, then $h(i)$ is an R-index for a set X such that χ_X is a path through T.

An effective enumeration is said to be *recursive* relative to X if the enumeration and the three functions are all recursive in X. The next result says that an arbitrary enumeration can be replaced by an effective one.

Theorem 1.4 (Marker) *Let $\mathcal{S}$ be a countable Scott set. If $\mathcal{S}$ has an enumeration recursive in X, then it has an effective enumeration which is recursive relative to X.*

The proof of this result in [Mac-Mar] uses the self-awareness properties of non-standard models of PA. The idea is to choose a completion T of PA in $\mathcal{S}$, construct a model $\mathcal{A}$ of T such that $SS(\mathcal{A}) = \mathcal{S}$ and $D^c(\mathcal{A}) \leq_T X$, and let this model provide the effective enumeration. No direct proof is known, and the reader is urged to look for one.

2 Results on structures representing a Scott set

This section includes an organized collection of results on structures representing a given Scott set. There are three basic theorems, followed by combinations. The first basic theorem is related to a result of Lerman and Schmerl [Le-Sch], and also to a result of Goncharov and Peretyatkin [G],[P]. It gives some indication how the notion of a structure representing a Scott set is used.

Theorem 2.1. *Let $\mathcal{S}$ be a countable Scott set and let $\mathcal{A}$ be a structure representing $\mathcal{S}$. Suppose $\mathcal{S}$ has an enumeration $R \leq_T X$ and $Q = \{(i, \overline{a}) : R_i$ is the B_1 type of $\overline{a}\}$ is $\Sigma_2^0(X)$. Then there exists $\mathcal{B} \cong \mathcal{A}$ such that $\mathcal{B} \leq_T X$.*

Proof: This is a finite injury construction. We may suppose that R is equipped with functions making it an effective enumeration of $\mathcal{S}$. By Theorem 1.4, there is an effective enumeration recursive in X. Moreover, using $\Delta_2^0(X)$, we can pass effectively from an index in one enumeration to an index in the other. We shall use the three facts below.

Fact 1: There is a function $g_1 \leq_T X$ such that for all tuples $\overline{a}$ from $\mathcal{A}$, $\lim_{s\to\infty} g_1(\overline{a}, s)$ is some i with $(i, \overline{a}) \in Q$.

Fact 2: There is a function $g_2 \leq_T X$ such that if i is an index for the B_1 type $\Gamma(\overline{u})$ realized by $\overline{c}$ and $\gamma(\overline{u}, x)$ is an existential formula with $\exists x\, \gamma(\overline{u}, x) \in \Gamma(\overline{u})$, then $g_2(i, \gamma)$ is an index for a B_1 type $\Gamma^*(\overline{u}, x) \supseteq \Gamma(\overline{u}) \cup \{\gamma(\overline{u}, x)\}$, where $\Gamma^*(\overline{c}, x)$ is realized by some $a \in \mathcal{A}$.

Proof of Fact 2: We form a recursive list of all existential formulas $\varphi(\overline{u}, x)$ in variables $\overline{u}, x$, with the given $\gamma(\overline{u}, x)$ first. We put $\gamma(\overline{u}, x)$ into Γ^*. Then, proceeding down the list, we add $\varphi(\overline{u}, x)$ if it is consistent with what we have so far; i.e., letting $\psi(\overline{u}, x)$ be the conjunction of the existential formulas already in Γ^*, we add $\varphi(\overline{u}, x)$ to Γ^* provided that $\exists x[\psi(\overline{u}, x) \& \varphi(\overline{u}, x)] \in \Gamma(\overline{u})$. Finally, we let $\Gamma^*(\overline{u}, x)$ be the B_1 type generated by the existential formulas that we have included, together with the negations of the ones that we have omitted. Clearly, $\Gamma^*(\overline{u}, x) \leq_T \Gamma(\overline{u})$. Since our enumeration of $\mathcal{S}$ is effective, using X, we can pass effectively from an index for Γ to an index for Γ^*. The B_1 type Γ^* has the feature that it is consistent with *all* completions of Γ. Since $\mathcal{A}$ represents $\mathcal{S}$, $\Gamma^*(\overline{c}, x)$ is realized.

Fact 3: There is a function $g_3 \leq_T X$ such that if i is an index for a B_1 type $\Gamma(\overline{u}, x)$ realized by $\overline{c}$ and some a, then $\lim_{s\to\infty} g_3(i, \overline{c}, s)$ is such an a.

Let B be an infinite recursive set. Let $(a_n)_{n\in\omega}$, $(b_n)_{n\in\omega}$, be recursive lists of the elements of $\mathcal{A}$ and B, respectively. Fix a recursive list of the atomic sentences with constants from B. We determine a function F from B to $\mathcal{A}$ and we enumerate $D(\mathcal{B})$ so that F will be an isomorphism from $\mathcal{B}$ onto $\mathcal{A}$. There are obvious requirements,

R_{2r}: put a_r into $\mathrm{ran}(F)$

R_{2r+1}: put b_r into $\mathrm{dom}(F)$.

At stage s, we have determined a finite set $\delta_s \subseteq D(\mathcal{B})$. We have tentatively determined a chain $(p_0, \ldots, p_n)$ of partial functions from B to $\mathcal{A}$. We suppose that for $k \leq n$, p_k maps $\overline{b}_k$ to $\overline{a}_k$, where $\overline{b}_0 = \overline{a}_0 = \emptyset$; if $k = 2r + 1$, then $\overline{a}_k = \overline{a}_{k-1}, a_r$; and if $k = 2r+2$, then $\overline{b}_k = \overline{b}_{k-1}, b_r$. We also have a tentatively determined sequence of (indices for) B_1 types Γ_k for $0 \leq k \leq n$, and, if n is odd, also for $k = n+1$, such that the following conditions hold:

(1) for $k \leq n$, and for $k = n+1$ if k is odd, $\Gamma_k(\overline{b}_k) \cup \delta_s$ is consistent; i.e., $\Gamma_k(\overline{b}_k)$ contains the existential sentence which is obtained from the conjunction of δ_s by quantifying out the constants *not* in $\overline{b}_k$,

(2) Γ_0 is the B_1 type of $\emptyset$,

(3) for odd $k \leq n$, Γ_k seems to be the B_1 type realized by $\overline{a}_k$–according to g_1 with s,

(4) for even k with $0 < k \leq n+1$, Γ_k is sure to be realized by $\overline{a}_{k-1}$ and *some* element, provided that Γ_{k-1} is realized by $\overline{a}_{k-1}$,

(5) for even $k = 2r+2 < n$, p_k maps b_r to an element c–given by g_3 with s–such that $\overline{a}_{2r}, a_r, c$ seems to realize Γ_{2r+2}.

At stage $s+1$, δ_{s+1} will be the result of adding to δ_s either the next atomic sentence α, or its negation. For the sequence $(p_0, \ldots, p_n)$ we do one of the following:

(a) extend the sequence, adding some p_{n+1},

(b) for some $k \leq n$, drop p_i for all $k \leq i \leq n$, having determined that at least one of the guesses behind p_k was not stable by stage s,

(c) leave the sequence as is, having determined that at least one of the guesses needed for p_{n+1} did not become stable by stage s.

Let δ_{s+1} be the result of adding to δ_s either α or $\neg\alpha$, so that for the greatest possible $n' \leq n$, $\Gamma_k(\overline{b}_k) \cup \delta_{s+1}$ is consistent for all $k \leq n'$ and for $k = n'+1$ if n' is odd. Let $n'' \leq n'$ be greatest such that for all $k \leq n''$, the stage s guesses behind p_k match the stage $s+1$ guesses.

Case 1: Suppose $n'' < n$.

In this case, we drop p_i for $n'' < i \leq n$, keeping $\Gamma_{n''+1}$ if n'' is odd. This is action (a).

Case 2: Suppose $n'' = n$.

In this case, we attempt to determine p_{n+1}, as follows:

Subcase 2(i): Suppose $n = 2r$.

Let Γ_{n+1} be the type which $\overline{a}_n, a_r$ seems to realize, according to g_1 with $s+1$. We look for a constant d, one appearing in δ_{s+1} or the first new one, such that $\Gamma_{n+1}(\overline{b}_n, d) \cup \delta_{s+1}$ is consistent. If we find d , then we let p_{n+1}

map $\overline{b}_n, d$ to $\overline{a}_n, a_r$. Also, using g_2, we determine $\Gamma_{n+2} \supset \Gamma_{n+1}$, such that $\Gamma_{n+2}(\overline{b}_n, d, b_r)$ is consistent with δ_{s+1}. This is action (a). If we do not find d, then we leave p_{n+1} undefined. This is action (c).

Subcase 2(ii): Suppose $n = 2r + 1$.

Say we have determined Γ_n and Γ_{n+1} such that if Γ_n is realized by $\overline{a}_n$, then Γ_{n+1} is realized by $\overline{a}_n, c$, for some c, and $\Gamma_{n+1}(\overline{b}_n, b_r) \cup \delta_{s+1}$ is consistent. Using g_3 with $s+1$, we obtain c which seems to serve, and we let p_{n+1} map $\overline{b}_n, b_r$ to $\overline{a}_n, c$. This is action (a).

All the requirements are eventually satisfied. Suppose we have $(p_0, \ldots, p_n)$ taking care of the first n requirements, $n = 2r$. When the guess at the type Γ_{n+1} of $\overline{a}_{n+1}$ becomes stable, then we will have p_{n+1} taking care of R_{2r}. We will also have a type Γ_{n+2} which is sure to be realized by $\overline{a}_{n+1}$ and some c. When our guess at such a c becomes stable, then we will have p_{n+2} taking care of R_{2r+1}. This completes the proof.

The hypotheses of Theorem 2.1 may have seemed obscure. However, the next two results give conditions under which these hypotheses are satisfied.

Theorem 2.2 *Let T be a complete theory. Suppose $R \leq_T X$ is an enumeration of a Scott set $\mathcal{S}$, and t is a $\Delta^0_3(X)$ function such that for all n, $t(n)$ is an R-index for $T_n = T \cap \Sigma_n$. Then T has a model $\mathcal{A}$, representing $\mathcal{S}$, such that $Q = \{(i, \overline{a}) : R_i$ is the complete B_1 type of $\overline{a}\}$ is $\Sigma^0_2(X)$.*

Note: The hypotheses imply that $\mathcal{S}$ is appropriate for T.

Proof: This is another finite injury construction. Let A be an infinite recursive set of constants, for the universe of $\mathcal{A}$. The set inherits from ω the usual recursive ordering. Fix a recursive list of triples $(k, j, \overline{c})$, where $k, j \in \omega$ and $\overline{c}$ is a tuple of constants from A, such that if $(k, j, \overline{c})$ is n^{th} on the list, then $k \leq n$ and $\overline{c}$ is included among the first n constants. There are two requirements for each n:

A_n: Assign a complete B_{n+1} type to a tuple including the first n constants from A.

Wi_n: For the nth triple $(k, j, \overline{c})$ on our list, if R_j is a complete B_k type, $R_j(\overline{c}, x)$ is witnessed or else is inconsistent with the complete B_{n+1} type assigned to the first n constants.

Approximation of T: Let t^* be a $\Delta^0_2(X)$ function such that for $n \geq 2$, $\lim_{s \to \infty} t^*(n, s) = t(n)$.
We suppose that for all n and s, there is some complete theory T' such that for all $n \leq s$, $t^*(n+2, s)$ is an R-index for $T' \cap \Sigma_{n+2}$.

At stage s, we have a finite sequence $t_2, i_0, \overline{a}_0, \ldots, t_{r+2}, i_r, \overline{a}_r$, for some $r \leq s$. The sequence represents an attempt to satisfy requirements A_n and Wi_n for $n \leq r$ so that the following hold:

(1) $t_{n+2} = t^*(n+2, s)$,

(2) R_{i_n} is a complete type in which the Σ_{n+2} sentences (with no constants from A) are just those of R_{t_n+2},

(3) $\overline{a}_n$ is a sequence of constants, including the first n elements of A,

(4) for the nth triple $k, j, \overline{c})$, if R_j is a complete B_k type, then either $R_j(\overline{c}, a) \subseteq R_{i_n}(\overline{a}_n)$ for some $a \in \overline{a}_n$, or else $R_{i_n}(\overline{a}_n) \cup R_j(\overline{c}, x)$ is inconsistent,

(5) $R_{i_n}(\overline{a}_n) \cap B_{n+1} \subseteq R_{i_{n+1}}(\overline{a}_{n+1})$.

At stage $s+1$, either we extend the sequence, or, for some $n \leq r$, we change t_{n+2}, i_n, and $\overline{a}_n$, and drop any later terms. In either case, we maintain the B_1 type assigned to $\overline{a}_r$.

Case 1: For all $n \leq r$, $t^*(n+2, s+1) = t^*(n+2, s)$.

In this case, we extend the sequence, attempting requirements A_{r+1} and Wi_{r+1}. Let $t_{r+3} = t^*(r+3, s+1)$. Using $\Delta^0_2(X)$, we determine i_{r+1} and $\overline{a}_{r+1}$ such that $\overline{a}_{r+1} \supseteq \overline{a}_r$, the first $r+1$ constants are included in $\overline{a}_{r+1}$, $R_{i_{r+1}}$ is a complete type with exactly the same Σ_{n+3} sentences as $R_{t_{r+3}}$, $R_{i_r}(\overline{a}_r) \cap B_{r+1} \subseteq R_{i_{r+1}}(\overline{a}_{r+1})$, and for the $(r+1)^{\text{st}}$ triple $(k, j, \overline{c})$, if R_j is a complete B_k type, then either $R_j(\overline{c}, a) \subseteq R_{i_{r+1}}(\overline{a}_{r+1})$ for some $a \in \overline{a}_{r+1}$, or else $R_{i_{r+1}}(\overline{a}_{r+1}) \cup R_j(\overline{c}, x)$ is inconsistent.

Case 2: For some first $n \leq r$, $t^*(n+2, s+1) \neq t^*(n+2, s)$.

In this case, we replace t_{n+2}, i_n, and $\overline{a}_n$, by t, i and $\overline{a}$ chosen as follows. We let $t = t^*(n+1, s+1)$. Note that $R_{i_{n-1}}(\overline{a}_{n-1}) \cap B_n \subseteq R_{i_r}(\overline{a}_r)$, where R_{i_r} has the same Σ_{n+2} sentences as $R_{t_{n+2}}$ and R_t has the same Σ_{n+1} sentences. We take i and $\overline{a}$ such that $\overline{a} \supseteq \overline{a}_r$, R_i is a complete type with the same Σ_{n+2} sentences as R_t, $R_i(\overline{a}) \supseteq R_{i_r}(\overline{a}_r) \cap B_n$, and for the nth triple $(k, j, \overline{c})$, either $R_j(\overline{c}, a) \subseteq R_i(\overline{a})$ for some $a \in \overline{a}$, or else $R_i(\overline{a}) \cup R_j(\overline{c}, x)$ is inconsistent.

We have said how to determine the finite sequence for each stage s. For each n, there is some stage after which the first $3n$ terms in the sequence remain fixed. Therefore, the limit of the sequence of finite sequences exists. It is an infinite sequence $t_2, i_0, \overline{a}_0, t_3, i_1, \overline{a}_1, t_4, i_2, \overline{a}_2, \ldots$ such that $\bigcup_n R_{t_{n+2}} = T$ and $\bigcup_n R_{i_n}(\overline{a}_n) \cap B_{n+1} = D^c(\mathcal{A})$, where $\mathcal{A}$ is a model of T representing $\mathcal{S}$. Note that the complete B_1 type realized by $\overline{c}$ is R_i if and only if R_i is equal to the appropriate restriction of the type assigned to some $\overline{a} \supseteq \overline{c}$ at some stage in the construction. Therefore, $Q = \{(i, \overline{a}) : R_i$ is the complete B_1 type realized by $\overline{a}\}$ is $\Sigma^0_2(X)$.

The next result is a strengthening of Theorem 2.2, with a weaker hypothesis.

Theorem 2.3 *Let T be a complete theory. Suppose $R \leq_T X$ is an enumeration of a Scott set $\mathcal{S}$, with functions t_n, $\Delta^0_n(X)$ uniformly in n, such that for each n, $\lim_{s\to\infty} t_n(s)$ is an R-index for $T_n = T \cap \Sigma_n$, and for all $s, t_n(s)$ is an index for a subset of T_n. Then T has a model $\mathcal{A}$, representing $\mathcal{S}$ such that $Q = \{(i, \overline{a}) : R_i$ is the B_1 type of $\overline{a}\}$ is $\Sigma^0_2(X)$.*

The proof of Theorem 2.3 requires a transfinitely nested priority construction. It will be postponed until §5.

Theorem 2.1 and Theorem 2.2 combine to give the following.

Theorem 2.4 *Let T be a complete theory. Suppose $R \leq_T X$ is an enumeration of a Scott set $\mathcal{S}$, and t is a $\Delta^0_3(X)$ function such that for all n, $t(n)$ is an R-index for $T_n = T \cap \Sigma_n$. Then T has a model $\mathcal{B}$, representing $\mathcal{S}$, such that $\mathcal{B} \leq_T X$.*

Proof: By Theorem 2.2, T has a model $\mathcal{A}$, representing $\mathcal{S}$, such that $Q = \{(i, \overline{a}) : R_i$ is the B_1 type realized by $\overline{a}\}$ is $\Sigma^0_2(X)$. By Theorem 2.1, there is a model $\mathcal{B} \cong \mathcal{A}$ such that $\mathcal{B} \leq_T X$.

Note: The proof of Theorem 2.4, from Theorems 2.1 and 2.2, is based on two finite injury priority constructions.

Theorem 2.1 and 2.3 combine to give the result below. The statement is the same as in $[K_3]$, although the definition of what it means for $\mathcal{A}$ to represent $\mathcal{S}$ has changed.

Theorem 2.5 *Let T be a complete theory. Suppose $R \leq_T X$ is an enumeration of a Scott set $\mathcal{S}$, with functions $t_n(s)$, $\Delta^0_n(X)$ uniformly in n, such that for each n, $\lim_{s\to\infty} t_n(s)$ is an R-index for $T_n = T \cap \Sigma_n$, and for all s, $t_n(s)$ is an index for a subset of T_n. Then T has a model $\mathcal{B}$, representing $\mathcal{S}$, such that $\mathcal{B} \leq_T X$.*

Proof: By Theorem 2.3, T has a model $\mathcal{A}$, representing $\mathcal{S}$, such that $Q = \{(i, \overline{a}) : R_i$ is the B_1 type realized by $\overline{a}\}$ is $\Sigma^0_2(X)$. By Theorem 2.1, there exists $\mathcal{B} \cong \mathcal{A}$ such that $\mathcal{B} \leq_T X$.

In the next section, Theorems 2.4 and 2.5 will be applied to Solovay's results.

3 Models of arithmetic

Solovay [So₁] characterized the degrees of non-standard models of TA representing a given Scott set. The standard model does not represent any Scott set. Recall that the Scott sets appropriate for TA are the ones which contain $TA \cap \Sigma_n$ for all n; equivalently, they are the Scott sets which contain all the arithmetical sets.

Theorem 3.1 (Solovay) *For any Scott set $\mathcal{S}$ containing the arithmetical sets, and any set X, the following are equivalent:*

(1) there is a non-standard model $\mathcal{A}$ of TA such that $\mathcal{SS}(\mathcal{A}) = \mathcal{S}$ and $\mathcal{A} \leq_T X$,

(2) there is an effective enumeration of $\mathcal{S}$ recursive in X.

Using Theorem 2.1, Marker [Mar-Mac] simplified the characterization, replacing the effective enumeration by an arbitrary enumeration. The set of degrees of models and the set of degrees of enumerations are both closed upward. Thus, the characterization may be stated as follows.

Theorem 3.2 (Solovay, Marker) *For any Scott set $\mathcal{S}$ containing the arithmetical sets, the degrees of models of TA which represent $\mathcal{S}$ are the degrees of enumerations of $\mathcal{S}$.*

Corollary 3.3 (Solovay, Marker) *The degrees of non-standard models of TA are the degrees of enumerations of Scott sets which contain the arithmetical sets.*

Proof of Theorem 3.2: First, suppose $\mathcal{A}$ is a non-standard model of TA such that $\mathcal{S} = \mathcal{SS}(\mathcal{A})$. We may assume that $\mathcal{A}$ has universe ω. By Proposition 1.3, the canonical enumeration of $\mathcal{SS}(\mathcal{A})$ is recursive in $\mathcal{A}$. Now, suppose $\mathcal{S}$ is a Scott set containing the arithmetical sets and R is an enumeration of $\mathcal{S}$. Let $t(n)$ be the first R-index of $TA \cap \Sigma_n$. The function t is $\Delta^0_3(R)$. To see why this is so, note that $TA \cap \Sigma_{n+1}$ is recursive in $(TA \cap \Sigma_n)'$, uniformly in n. Using $\Delta^0_3(R)$, we can pass effectively from an index for $TA \cap \Sigma_n$ to one for $(TA \cap \Sigma_n)$, and from that to one for $(TA \cap \Sigma_{n+1})$. Applying Theorem 2.4, we get a non-standard model $\mathcal{A}$ of TA such that $\mathcal{SS}(\mathcal{A}) = \mathcal{S}$ and $\mathcal{A} \leq_T R$.

For an arbitrary completion T of PA, Solovay [So₂] characterized the degrees of models of T representing a given Scott set as follows.

Theorem 3.4 (Solovay) *Let T be a completion of PA, and let $\mathcal{S}$ be a Scott set appropriate for T. Then for all sets X, the following are equivalent:*

(1) there is a (non-standard) model $\mathcal{A}$ of T such that $\mathcal{SS}(\mathcal{A}) = \mathcal{S}$ and $\mathcal{A} \leq_T X$,

(2) *there is an enumeration* R *of* S, *recursive in* X, *equipped with functions* t_n, $\Delta^0_n(X)$ *uniformly in* n, *such that for all* n, $\lim_{s\to\infty} t_n(s)$ *is an* R*-index for* $T_n = T \cap \Sigma_n$, *and for all* s, $t_n(s)$ *is an* R*-index for a subset of* T_n.

Corollary 3.5 (Solovay) *If* T *is a completion of PA, the degrees of (non-standard) models of* T *are the degrees of sets* X *with an enumeration* $R \leq_T X$ *of a Scott set appropriate for* T, *equipped with functions* t_n $\Delta^0_n(X)$ *uniformly in* n, *such that for all* n, $\lim_{s\to\infty} t_n(s)$ *is an* R*-index for* $T_n = T \cap \Sigma_n$, *and for all* s, $t_n(s)$ *is an* R*-index for a subset of* T_n.

Proof of Theorem 3.4: First, suppose that $R \leq_T X$ is an enumeration of S, equipped with functions t_n, $\Delta^0_n(X)$ uniformly in n, such that for all n, $\lim_{s\to\infty} t_n(s)$ is an R-index for T_n, and for all s, $t_n(s)$ is an R-index for a subset of T_n. By Theorem 2.5, T has a model $\mathcal{A}$, representing S, such that $\mathcal{A} \leq_T X$.

We have proved the model construction half of Theorem 3.4 using Theorem 2.5. Theorem 2.5 was based on Theorem 2.3, which we will prove in §5. For the other half of Theorem 3.4, we need Solovay's Approximation Lemma. This is based on the magical result below [Mat], [Ma].

Lemma 3.6 (Matijasevich) *For a formula* δ *with only bounded quantifiers, we can find an existential formula* α *and a universal formula* β *such that both* α *and* β *are equivalent to* δ *over PA.*

Here is the Approximation Lemma.

Lemma 3.7 *Let* $\mathcal{A}$ *be a non-standard model of PA with universe* ω, *and let* R *be the canonical enumeration of* $SS(\mathcal{A})$. *Then there are functions* t_n, $\Delta^0_n(\mathcal{A})$ *uniformly in* n, *such that* $\lim_{s\to\infty} t_n(s)$ *is an* R*-index for* $T_n(\mathcal{A})$, *and for* $r < s$, $R_{t_n(r)} \subseteq R_{t_n(s)}$.

Proof: We have the usual type ω ordering the universe of $\mathcal{A}$. We also have a natural ordering definable in $\mathcal{A}$. To avoid confusion, in this proof, we write $<_\omega$ for the former. Let $\mathrm{Sat}_n(x)$ be the usual Σ_n formula defining truth for Σ_n sentences–this formula has the feature that if k is the Gödel number of a Σ_n sentence φ, then $\mathrm{PA} \vdash \mathrm{Sat}_n(S^k(0)) \leftrightarrow \varphi$. Let $\varphi_n(u)$ say that $\forall x < u[p_x \mid u \rightarrow \mathrm{Sat}_n(x)]$.

Claim 1: We can find a Σ_n formula $\varphi^*_n(u)$ which is equivalent to $\varphi_n(u)$ over PA.

Proof of Claim 1: First, by minor rearrangement and bringing bounded quantifiers across unbounded ones we obtain a formula with n alternating blocks of quantifiers, starting with $\exists$, followed by a formula β which has only bounded quantifiers. Then, using Lemma 3.6, we replace β by a formula which is either Σ_1 or Π_1, matching the last block of unbounded quantifiers.

Claim 1 is important because we can enumerate $D^c(\mathcal{A}) \cap \Sigma_n$ using a procedure which is $\Delta_n^0(\mathcal{A})$, uniformly in n. If $\mathcal{A} \models \varphi_n^*(b)$, then $\varphi_n^*(b)$ will eventually appear in our enumeration.

Claim 2: For each n, there is some index b for $T_n(\mathcal{A})$ such that $\mathcal{A} \models \varphi_n^*(b)$.

Proof of Claim 2: Let $\psi(u, v) = \forall x < u[\mathrm{Sat}_n(x) \leftrightarrow p_x \mid v]$. Since $\exists v\psi(u, v)$ holds for all standard elements, it must hold for some non-standard element a. If $\mathcal{A} \models \psi(a, b)$, then b is the desired index.

Now we can define the functions t_n. For $n = 1$, t_1 is constant, with value equal to the $<_\omega$-first R-index for $T_1(\mathcal{A})$. Suppose $n \geq 2$. Let $t_n(0)$ be an R-index for $\emptyset$. Supposing that $t_n(s) = a$, we determine $t_n(s + 1)$, using $\Delta_n^0(\mathcal{A})$, as follows. We carry out $s + 1$ steps in the enumeration of $D^c(\mathcal{A}) \cap \Sigma_n$, and search for $b <_\omega s + 1$ such that

(1) for some standard x, $\mathcal{A} \models p_x \mid b$ and $p_x \nmid a$,

(2) for all standard x, $\mathcal{A} \models p_x \mid a \rightarrow p_x \mid b$,

(3) $\varphi_n^*(b)$ has appeared in $D^c(\mathcal{A}) \cap \Sigma_n$.

If we find b satisfying (1)-(3), then $t_n(s + 1)$ is the $<_\omega$-first; otherwise, $t_n(s + 1) = a$.

Remark: We need $n \geq 2$ to check (1) and (2) using $\Delta_n^0(\mathcal{A})$.

We argue that t_n has the desired features. For all s, $t_n(s)$ is an R-index for a subset of $T_n(\mathcal{A})$. This clearly holds for $s = 0$. By (3), it holds for $s + 1$ if $t_n(s + 1) \neq t_n(s)$. By (1) and (2), if $t_n(s + 1) \neq t_n(s)$, then the set with index $t_n(s + 1)$ is strictly larger than the one with index $t_n(s)$. Thus, if $t_n(s)$ is an index for $T_n(\mathcal{A})$, then for $s' > s$, $t_n(s') = t_n(s)$. It is enough to show that there is such an s.

By claim 2, there is an index b for $T_n(\mathcal{A})$ such that $\mathcal{A} \models \varphi_n^*(b)$. Take the $<_\omega$-first such b, and suppose that $\varphi_n^*(b)$ has appeared within s steps of enumerating $D^c(\mathcal{A}) \cap \Sigma_n$. Assuming that $t_n(s)$ is not an index for $T_n(\mathcal{A})$, $t_n(s + 1)$ is either b or some $b' <_\omega b$ such that $b' \neq t_n(r)$ for all $r \leq s$. Thus, within finitely many steps, we arrive at an index for $T_n(\mathcal{A})$.

Lemma 3.7 is all we need for Theorem 3.4. However, the same proof yields the following.

Proposition 3.8 *Let $\mathcal{A}$ be a non-standard model of PA with universe ω, and let R be the canonical enumeration of $SS(\mathcal{A})$. Then there are functions $t_n(\overline{a}, s)$, $\Delta_n^0(\mathcal{A})$ uniformly in n, such that for all tuples $\overline{a}$ in $\mathcal{A}$, $\lim_{s\to\infty} t_n(\overline{a}, s)$ is an R-index for $T_n(\mathcal{A}, \overline{a})$, and for all s, $R_{t_n(\overline{a},s)}$ is an index for a subset of $T_n(\mathcal{A}, \overline{a})$.*

4 Metatheorem

This section gives a new metatheorem, which we shall use to prove Theorem 2.3. In Theorem 2.3, we must enumerate a $\Sigma_2^0(X)$ set Q, satisfying requirements like those in Theorem 2.2 with information given by a sequence of functions t_n, where t_n is $\Delta_n^0(X)$, uniformly in n, for $n \geq 2$. For simplicity, we first state the metatheorem for constructions which enumerate an *r.e.* set, satisfying requirements with information given by functions which are Δ_n^0 uniformly in n, for $n \geq 1$. We then relativize to $\Delta_2^0(X)$.

We describe an abstract setting similar to that in [A_1]. An *alternating sequence* on L and U is a sequence $\sigma = u_0\ell_0u_1\ell_1 \ldots$ with $u_k \in U$ and $\ell_k \in L$. An *alternating tree* on U and L is a non-empty set P of finite alternating sequences, closed under initial segments. A *path* through P is an infinite sequence whose finite initial segments are all in P. We shall consider trees P with the property that every element extends to a path. If σ is an element of P of even length, then we may write $\ell(\sigma)$ for the last term.

An *enumeration function* on L is a function E mapping L to the set of finite subsets of ω. If $\pi = u_0\ell_0u_1\ell_1 \ldots$ is a path through P (or other infinite sequence with some terms in L), we let $E(\pi) = \bigcup_n E(\ell_n)$. An *instruction function* on P is a function q defined on elements of P of even length (ending in L), such that if $q(\sigma) = u$, then $\sigma u \in P$. A *run* of (P, q) is a path π through P in which the terms from U are chosen by q. We suppose that U and L are r.e. sets, P is an r.e. alternating tree on U and L, and E is a partial recursive enumeration function on L. For $\sigma \in P$ of length $2n$, $q(\sigma)$ will be the limit of a special Δ_{n+1}^0 approximation.

We indicate roughly what these objects are supposed to represent. Elements of U represent possible pieces of information (not necessarily true), while elements of L represent possible steps in a construction (as seen from the top level). Elements of P represent possible partial constructions in which the first few requirements are satisfied based on the first few pieces of information given. The instruction function gives true information.

In [A_1] the object of the construction is to produce a run π of (P, q) such that $E(\pi)$ is r.e. The object here is the same. The conditions in the metatheorem from [A_1] involved further binary relations defined on L. Here we have binary relations $\leq_n$ for $n < \omega$. These are defined on $L \cup U$, not just on L. As in [A_1], we suppose that the relations $\leq_n$ are uniformly r.e. The metatheorem in [A_1] has two versions, depending on whether the top level is a successor ordinal or a limit ordinal.

The version for successor ordinals α is for an "α-system" with an instruction function which is Δ_α^0. The version for limit ordinals α is for a "special (α_n) system" with a "special (α_n)" instruction function, where $(\alpha_n)_{n\in\omega}$ is a recursive increasing sequence of ordinals with limit α. The nth piece of information – the value of the instruction function on a sequence of length $2n$–is $\Delta^0_{\alpha_n}$, uniformly in n.

We are about to define $(n+1)$ *approximation system* and $(n+1)$ *approximated instruction function* . These resemble the special $(n+1)$ system and special $(n+1)$ instruction function. In particular, the first piece of information is obtained from Δ^0_1, the second is obtained from Δ^0_2, and so on. There are differences. In a run $u_0\ell_0u_1\ell_1\ldots$ of the special $(n+1)$ system and special $(n+1)$ instruction function, u_n is computed *directly* using Δ^0_{n+1}, and $\ell_n \leq_{n+1} \ell_{n+1}$. By contrast, in a run $u_0\ell_0u_1\ell_1\ldots$ of the $(n+1)$ approximation system and $(n+1)$ approximated instruction function, u_n is computed as the limit of a Δ^0_{n+1} function, and $\ell_n \leq_n \ell_{n+1}$.

Let L, U, P, E, and $\leq_n$ be as described above. For $u, v \in U$, we may write $u \lesseqgtr_n v$ to mean that $u \leq_n v$ and $v \leq_n u$. We call $\mathcal{A} = (L, U, P, E, \leq_{n\, n<\omega})$ an $(n+1)$ *approximation system* if the following conditions hold.

(1) $\leq_n$ is transitive and reflexive for all n,

(2) $\leq_n$ implies $\leq_m$ for all $m < n$,

(3) for all $\ell, \ell' \in L$, $\ell \leq_0 \ell'$ implies $E(\ell) \subseteq E(\ell')$,

(4) if $u_0\ell_0u_1\ell_1\ldots$ in P, then for all n, $u_n \leq_{n+1} \ell_n$ and $u_n \lesseqgtr_{n+1} u_{n+1}$,

(5) If $\sigma u \in P$, where length$(\sigma) = 2n$, there exists ℓ such that $\sigma u\ell \in P$ and $\ell(\sigma) \leq_{n-1} \ell$; moreover, if $u \leq_{n_0} \ell^1 \leq_{n_1} \ldots \leq_{n_{k-1}} \ell^k$ and $\ell(\sigma) \leq_{n-1} \ell^1$, where $n+1 \geq n_0 > \ldots > n_k \geq 0$, and $n-1 \leq n_1$, then we may take ℓ such that $\ell^i \leq_{n_i} \ell$ for $i = 1, \ldots, k$ (for $n = 0$, we omit the parts of the condition which refer to $\ell(\sigma)$).

Suppose q is an instruction function on P. We say that q is $(n+1)$ *approximated* if there are functions q_n, Δ^0_{n+1} uniformly in n, such that for σ in P of length $2n$,

(1) $q(\sigma) = \lim_{s\to\infty} q_n(\sigma, s)$,

(2) if $q_n(\sigma, s) = u$, then $\sigma u \in P$, and

(3) $q_{n-1}(\sigma, s) \geq_{n+1} q_{n-1}(\sigma, s+1)$.

Here is the new metatheorem.

Theorem 4.1 *Let $\mathcal{A} = (L, U, P, E, \leq_{n\, n\in\omega})$ be an $(n+1)$ approximation system, and let q be an $(n+1)$ approximated instruction function for P. Then (P, q) has a run π such that $E(\pi)$ is r.e. Moreover, $\pi \mid (2n)$ is Δ^0_{n+1}, uniformly in n.*

Proof: We may suppose that for $\sigma = u_0\ell_0u_1\ell_1\ldots$ in P, $\ell_n \leq_n \ell_{n+1}$ (using Condition (5) and replacing P by a smaller tree, if necessary). The outline of the proof is as follows. We define a family of trees P^n for $n \geq 1$, r.e. uniformly in n, together with instruction functions p_n which are Δ^0_n uniformly in n. The definitions will be such that the first $2n-2$ terms in a run of (P^n, p_n) form a partial run of (P, q). In addition, a Δ^0_n run π^n of (P^n, p_n) yields (in a uniform way) a Δ^0_{n+1} run π^{n+1} of (P^{n+1}, p_{n+1}) such that $\pi^n \mid (2n-2) = \pi^{n+1} \mid (2n-2)$ and $E(\pi^n) = E(\pi^{n+1})$. Then from a Δ^0_1 run π^1 of (P^1, p_1), we obtain runs π^n of (P^n, p_n), Δ^0_n uniformly in n, and a run π of (P, q) such that $\pi \mid (2n-2) = \pi^n \mid (2n-2)$ and $E(\pi) = E(\pi^1)$. Since $E(\pi^1)$ is r.e., so is $E(\pi)$, and π is the desired run of (P, q).

A *picture* is a pair $c = (\sigma u; \tau)$, where $\sigma u \in P$, and $\tau = n_0\ell^1n_1\ldots n_{k-1}\ell^k$ is as in the hypothesis of Condition (5); i.e., if length$(\sigma) = 2n$, then $u \leq_{n_0} \ell^1 \leq_{n_1} \ldots \leq_{n_{k-1}} \ell^k$ and $\ell(\sigma) \leq_{n-1} \ell^1$, where $n \geq n_0 > \ldots > n_{k-1} > 0$ and $n_1 \geq n-1$, with modifications if $n = 0$. We allow $\tau = \emptyset$. Let $\ell(c)$ be the last term of τ if $\tau \neq \emptyset$ and the last term of σ if $\tau = \emptyset$. We may write $u(c)$ for u, and $\tau(c)$ for τ. If $\tau(c) = \emptyset$, then we may identify c with $u(c)$. We say that ℓ *completes* c if ℓ is as in the conclusion of Condition (5); i.e., $\sigma u\ell \in P$, $\ell(\sigma) \leq_{n-1} \ell$, $\ell^i \leq_{n_i} \ell$ for $i = 1, \ldots, k-1$, and $\ell^k \leq_0 \ell$. Let C be the set of all pictures, and let C^n consist of the pictures $c = (\sigma u; \tau)$ in which the numbers $n_0, \ldots, n_{k-1}$ occurring in τ, if any, are all $\geq n$.

The elements of P^n will be finite sequences $\sigma = c_0\ell_0c_1\ell_1\ldots$, with $c_k \in C^n$, $\ell_k \in L$. If length$(\sigma) \leq 2n$, then σ is in P^n if and only if σ is in P (after identification of c_k with $u(c_k)$). In general, an element of P^n is a sequence $\sigma = c_0\ell_0c_1\ell_1\ldots$ satisfying the following conditions:

(1) $\sigma \mid (2n)$ is in P, and for $k \geq n$, $c_k = (\sigma_k u_k; \tau_k)$, where $\sigma_k \supseteq \sigma \mid (2n-2)$,

(2) for $k \geq n$, $\ell(c_k) = \ell_{k-1}$ and $u(c_k) \leq_n u(c_{k-1})$ (after identification).

(3) for $k \geq n$, ℓ_k completes c_k and $\ell_{k-1} \leq_{n-1} \ell_k$.

We define $p_n(\sigma)$ simultaneously with σ^+, where σ^+ approximates a partial run of (P^{n+1}, p_{n+1}) in which the even terms come from σ. If σ is an element of P^n of even length $\leq 2n-2$, then $\sigma^+ = \sigma$. If length$(\sigma) < 2n-2$, then $p_n(\sigma) = q(\sigma)$, and if length$(\sigma) = 2n-2$, then $p_n(\sigma) = q_{n-1}(\sigma, n)$. The definition continues by induction. We say how to determine σ^+ and $p_n(\sigma)$ for σ in P^n of length $2s \geq 2n$, given a Δ^0_n index for the restriction of p_n to sequences of even length $< 2s$ and a Δ^0_{n+1} index for the restriction of p_{n+1} to sequences of even length $\leq 2s$.

Approximations: Since p_{n+1} (and its restrictions) are Δ^0_{n+1}, there are natural Δ^0_n approximations p^*_{n+1}. We may suppose that for ρ of even length $< 2n-2$, $p^*_{n+1}(\rho, t) = q(\rho)$; for ρ of length $2n-2$, $p^*_{n+1}(\rho, t) = q_n(\rho, t)$, and always $p^*_{n+1}(\rho, t)$ has some value d such that $\rho d \in P^{n+1}$.

Let $\sigma_k = \sigma \mid (2k)$, for $k \leq s$. We say that $\sigma \in P^n$ *follows* p_n if the terms in C^n are chosen by p_n. First, consider the case where σ does *not* follow p_n. If k is greatest $< s$ such that either $k = n-1$ or σ_k follows p_n, then $\sigma^+ = \sigma_k^+$, and $p_n(\sigma)$ is the first c we find such that $\sigma c \in P^n$. Now, consider the case where σ follows p_n. Then σ^+ will be an element of P^{n+1} beginning with σ_{n-1}, and the next term, if any, will be $q_{n-1}(\sigma_{n-1}, s+1)$. We maintain the following conditions:

(1) (a) $\ell(\sigma^+) \leq_{n-2} \ell(\sigma)$ if length$(\sigma^+) = 2n-2$,

(b) $\ell(\sigma^+) \leq_{n-1} \ell(\sigma)$ if length$(\sigma^+) = 2n$,

(c) $\ell(\sigma^+) \leq_n \ell(\sigma)$ if length$(\sigma^+) > 2n$,

(2) $q_{n-1}(\sigma_{n-1}, s) \leq_n \ell(\sigma)$.

We are about to give case-by-case definitions. Assuming that (1) holds for proper initial segments, and (2) holds for σ, we define σ^+ and $p_n(\sigma)$ so as to guarantee (1) for σ and (2) for extensions of σ of length $2s+2$ (partial runs of (P^n, p_n)).

Case I: Suppose $q_{n-1}(\sigma_{n-1}, s+1) = u \neq q_{n-1}(\sigma_{n-1}, s)$.

Then $\sigma^+ = \sigma_{n-1}$, and $p_n(\sigma) = (\sigma^+ u; n\ell(\sigma))$. Since $\sigma \in P^n$, $\ell(\sigma^+) \leq_{n-2} \ell(\sigma)$. By the properties of q_{n-1}, $u \leq_n q_{n-1}(\sigma_{n-1}, s)$, and since (2) holds for σ, $q_n(\sigma_{n-1}, s) \leq_n \ell(\sigma)$. Therefore, (1) holds for σ, and $p_n(\sigma) \in C^n$. If ℓ completes $p_n(\sigma)$ and $\ell(\sigma) \leq_{n-1} \ell$, then $\sigma^+ u\ell \in P^{n+1}$. Since $u \leq_n \ell$, (2) holds for $\sigma p_n(\sigma)$.

In the remaining cases, our first approximation to σ^+ is $\rho = \sigma_{s-1}^+ p_{n+1}^*(\sigma_{s-1}^+, s)\ell(\sigma)$, where this is an element of P^{n+1}. Let $\rho_k = \rho \mid (2k)$.

Case II: Suppose we are not in Case I, and $p_{n+1}^*(\rho_n, s+1) = u_n \neq p_{n+1}^*(\rho_n, s)$.

Then $\sigma^+ = \rho_n$, and $p_n(\sigma) = (\sigma^+ u_n; n\ell(\sigma))$. Then $\ell(\sigma^+) \leq_{n-1} \ell(\sigma)$, so, (1) holds for σ. If $u_{n-1} = q_{n-1}(\sigma_{n-1}, s+1) = q_{n-1}(\sigma_{n-1}, s)$, then since (2) holds for σ, $u_{n-1} \leq_n \ell(\sigma)$. Since $u_n \leq_n u_{n-1}$, we have $u_n \leq_n \ell(\sigma)$, so $p_n(\sigma) \in C^n$. If ℓ completes $p_n(\sigma)$ and $\ell(\sigma) \leq_{n-1} \ell$, then $\sigma^+ u_n \ell \in P^{n+1}$. Then $u_{n-1} \leq_n u_n \leq_n \ell$, so (2) holds for $\sigma p_n(\sigma)\ell$.

Case III: Suppose we are not in Case I or II, and for some $n < k < \text{length}(\rho)$, $p_{n+1}^*(\rho_k, s+1) = d \neq p_{n+1}^*(\rho_k, s)$.

Then $\sigma^+ = \rho_k$, for the first such k. If $p_{n+1}^*(\rho_k, s+1) = d = (vu; \tau)$, then $p_n(\sigma) = (vu; \tau n\ell(\sigma))$. Since $\rho \in P^{n+1}, \ell(d) = \ell(\rho_k) \leq_n \ell(\rho) = \ell(\sigma)$. Therefore, (1) holds for σ and $p_n(\sigma) \in C^n$. If ℓ completes $p_n(\sigma)$ and $\ell(\sigma) \leq_{n-1} \ell$, then ℓ completes d and $\ell(\rho_k) \leq_n \ell$, so $\rho_k d\ell \in P^{n+1}$. Then by the same reasoning as in Case II, (2) holds for $\sigma p_n(\sigma)\ell$.

Case IV: Suppose we are not in Case I, II, or III.

Then $\sigma^+ = \rho$. If $p^*_{n+1}(\rho, s+1) = d = (\nu u; \tau)$, then $p_n(\sigma) = (\nu u; \tau n\ell(\sigma))$. Since $\ell(\rho) = \ell(\sigma)$, (1) holds for σ. Since $\rho d \in P^{n+1}$, $\ell(d) = \ell(d)$. Therefore, $p_n(\sigma) \in C^n$. If ℓ completes $p_n(\sigma)$, then it completes d, and $\ell(\sigma) \leq_n \ell$, so $\rho d\ell \in P^{n+1}$. Then, again by the same reasoning as in Case II, (2) holds for $\sigma p_n(\sigma)\ell$.

We have defined the trees P^n and the instruction functions p_n.

Lemma 4.2 *For any Δ^0_n run π of (P^n, p_n), there is a Δ^0_{n+1} run π^* of (P^{n+1}, p_{n+1}) such that $\pi^* \mid (2n-2) = \pi \mid (2n-2)$, and the terms of π^* which are in L form a subsequence of π. Moreover, from a Δ^0_n index for π, we can find a Δ^0_{n+1} index for π^*.*

Proof: Let π be a Δ^0_n run of (P^n, p_n), and let $\pi_k = \pi \mid (2k)$. We define a Δ^0_{n+1} sequence of numbers $s(k)$, where length$(\pi^+_{s(k)}) = 2k$ and for all $t > s$, $\pi^+_t \supseteq \pi^+_{s(k)}$. Let $s(0) = 0$. Suppose that we have determined $s(k)$. For all sufficiently large t, $p^*_{n+1}(\pi^+_{s(k)}, t)$ is constant, with value $p_{n+1}(\pi^+_{s(k)})$. Let $s(k+1)$ be the first $s > s(k)$ such that for all $t \geq s$, $p^*_{n+1}(\pi^+_{s(k)}, t)$ is constant. Now, letting $\pi^* \mid (2k) = \pi^+_{s(k)}$, we obtain the desired run π^* of (P^{n+1}, p_{n+1}).

Suppose that for each n, π^n is a run of (P^n, p_n). We call $(\pi^n)_{1 \leq n < \omega}$ a *coherent family* if for each n, π^n and π^{n+1} are related as π and π^* are in Lemma 4.2.

Lemma 4.3 *There is a coherent family $(\pi^n)_{1 \leq n < \omega}$, where π^n is a run of (P^n, p_n), Δ^0_n uniformly in n.*

Proof Let π^1 be a recursive run of (P^1, p_1). Given π^n, a Δ^0_n run of (P^n, p_n), let π^{n+1} be the Δ^0_{n+1} run of (P^{n+1}, p_{n+1}) obtained from π^n as in Lemma 4.2.

Now, we are ready to complete the proof of Theorem 4.1. Take a coherent family $(\pi^n)_{1 \leq n < \omega}$ as in Lemma 4.3, and let π be the run of (P, q) such that $\pi \mid (2n-2) = \pi^n \mid (2n-2)$ for all n. Then $\pi \mid (2n-2)$ is Δ^0_n uniformly in n. Since all π^n preserve at least $\leq_0$, $E(\pi) = E(\pi^1)$. Since π^1 is recursive, $E(\pi^1)$ is r.e., so $E(\pi)$ is r.e.

To prove Theorem 2.3, we need the relativization of Theorem 4.1 which is stated below. Let $\mathcal{A} = (L, U, P, E, \leq_{n\, 1 \leq n < \omega})$, where L, U, and P are $\Sigma^0_2(X)$, $\leq_n$ is $\Sigma^0_2(X)$, uniformly in n, and E is partial $\Delta^0_2(X)$. We call $\mathcal{A}$ an $(n+1)$ *approximating system relative to* $\Delta^0_2(X)$ if the following conditions hold:

(1) $\leq_n$ is transitive and reflexive,

(2) $\leq_n$ implies $\leq_m$ for $m < n$,

(3) for $\ell, \ell' \in L$, $\ell \leq_1 \ell'$ implies $E(\ell) \subseteq E(\ell')$,

(4) for $u_0\ell_0u_1\ell_1\ldots$ in P, $u_n \lneqq_{n+2} u_{n+1}$ and $u_n \leq_{n+2} \ell_n$,

(5) for $\sigma u \in P$ with σ of length $2n$, there exists ℓ such that $\sigma u\ell \in P$ and $\ell(\sigma) \leq_n \ell$. Moreover, if $u \leq_{n_0} \ell^1 \leq_{n_1} \cdots \leq_{n_{k-1}} \ell^k$ and $\ell(\sigma) \leq_n \ell^1$, where $n+2 \geq n_0 > \ldots n_k \geq 1$ and $n \leq n_1$, then we may take ℓ such that $\ell^i \leq_{n_i} \ell$ for $i = 1, \ldots, k$. (For $n = 0$, we drop the parts of the condition which refer to $\ell(\sigma)$.)

An instruction function q is $(n+1)$ *approximated relative to* $\Delta^0_2(X)$ if there are functions q_n, $\Delta^0_{n+2}(X)$, uniformly in n, such that for $\sigma \in P$ of length $2n$,

(1) $q(\sigma) = \lim_{s\to\infty} q_n(\sigma, s)$,

(2) if $q_n(\sigma, s) = u$, then $\sigma u \in P$, and

(3) $q_n(\sigma, s) \geq_{n+2} q_n(\sigma, s+1)$.

Here is the relativized version of Theorem 4.1.

Corollary 4.4 *Suppose $\mathcal{A} = (L, U, P, E, \leq_{n\ 1\leq n<\omega})$ is an $(n+1)$ approximating system relative to $\Delta^0_2(X)$. Suppose q is an instruction function for P which is $(n+1)$ approximated relative to $\Delta^0_2(X)$. Then (P, q) has a run π such that $E(\pi)$ is $\Sigma^0_2(X)$. Moreover, $\pi \mid (2n)$ is $\Delta^0_{n+2}(X)$, uniformly in n.*

5 Application

In this section, we apply Corollary 4.4 to prove the following result, which was stated in §2.

Theorem 2.3 *Let T be a complete theory. Suppose $R \leq_T X$ is an enumeration of a Scott set $\mathcal{S}$, with functions t_n, $\Delta^0_n(X)$ uniformly in n, such that for each n, $\lim_{s\to\infty} t_n(s)$ is an R-index for $T_n = T \cap \Sigma_n$, and for all s, $t_n(s)$ is an index for a subset of $T_n = T \cap \Sigma_n$. Then T has a model $\mathcal{A}$, representing $\mathcal{S}$, such that $Q = \{(i, \overline{a}) : R_i$ is the B_1 type of $\overline{a}\}$ is $\Sigma^0_2(X)$.*

Proof: To apply Corollary 4.4, we define $(L, U, P, E, \leq_{n\ 1\leq n<\omega})$, forming an $(n+1)$ approximating system relative to $\Delta^0_2(X)$, together with an instruction function q which is $(n+1)$ approximated relative to $\Delta^0_2(X)$. We consider theories, types, etc. in the language of T. Let A be an infinite recursive set of new constants. Let L consist of the pairs $\ell = (i, \overline{a})$ where R_i is a complete type in variables corresponding to $\overline{a}$. For $\ell = (i, \overline{a})$ in L, let $E(\ell)$ consist of the pairs $(j, \overline{c})$, where $j < \text{length}(\overline{a})$, $\overline{c} \subseteq \overline{a}$, and R_j is the restriction of R_i to a B_1 type in variables corresponding to $\overline{c}$. For $\ell = (i, \overline{a})$ and $\ell' = (i', \overline{a}')$ in L, and $n \geq 1$, let $\ell \leq_n \ell'$ if $\overline{a} \subseteq \overline{a}'$ and $R_i(\overline{a}) \cap B_n \subseteq R_i'(\overline{a}')$. Let $\ell \subseteq \ell'$ if $\overline{a} \subseteq \overline{a}'$ and $R_i(\overline{a}) \subseteq R_i'(\overline{a}')$.

Let $U = \bigcup_n U_n$, where $U_n = \{u : R_u$ is the Σ_n of part of a complete theory$\}$. We define $\leq_n$ on U such that $u \leq_n u'$ if $R_{u'} \cap \Sigma_n \subseteq R_u$. Finally, we define $\leq_n$ for $u \in U$ and $\ell = (i, \overline{a})$ in L such that $u \leq_n \ell$ if the Σ_n *sentences* in R_i are among the sentences in R_u.

Remarks: (1) If $\ell \leq_n m$ and $k < n$, then there exists $\ell' \supseteq \ell$ such that $m \leq_k \ell'$.

(2) If $u \leq_n m$ and $k < n$, then there exists $\ell = (i, \overline{a})$ such that $m \leq_k \ell$ and the Σ_n sentences of R_i are precisely those of R_u.

(3) $\leq_n$ is transitive on $L \cup U$.

We define P so that for each path $\pi = u_0 \ell_0 u_1 \ell_1 \ldots$, $\bigcup_n R_{u_n} = T(\pi)$ is a complete theory, and the ℓ_n's carry out a construction like that in Theorem 2.2, yielding a model $\mathcal{A}_\pi$ of $T(\pi)$ such that $\mathcal{A}_\pi$ represents $\mathcal{S}$. We define q so that for a run $\pi = u_0 \ell_0 u_1 \ell_1 \ldots$ of (P, q), $\bigcup_n R_{u_n} = T$. Let $\lim_{s \to \infty} t_n(s) = t_n^*$. We may assume that $t_{n+1}(0) \lesssim_n t_n^*$. Even before defining P precisely, we can describe $q(\sigma)$ and $q_n(\sigma, s)$, for $\sigma \in P$ of even length.

To begin with, let $q(\emptyset) = t_2^*$ and let $q_0(\emptyset, s) = t_2^*$ for all s. Now, consider a non-empty sequence $\sigma = u_0 \ell_0 \ldots u_{n-1} \ell_{n-1}$. Assuming that $u_k = t_{k+2}^*$ for all $k < n$, let $q(\sigma) = t_{n+2}^*$, and let $q_n(\sigma, s) = t_{n+2}(s)$, for all s. If $u_k \neq t_{k+1}^*$ for some $k < n$, then $q(\sigma)$ and $q_n(\sigma, s)$ are defined trivially. Let $q(\sigma)$ be the first u such that R_u is a complete Σ_{n+2} theory with $u_{n-1} \lesssim_{n+1} u$, and let $q_n(\sigma, s) = q(\sigma)$, for all s.

As in the proof of Theorem 2.2, we form a recursive list of triples $(k, j, \overline{c})$, for $\overline{c}$ a tuple of constants and $k, j \in \omega$, such that if $(k, j, \overline{c})$ is the nth triple, then $k \leq n$ and the elements of $\overline{c}$ are among the first n constants. For $\sigma = u_0 \ell_0 u_1 \ell_1 \ldots$, with $u_n \in U_{n+2}$ and $\ell_n \in L$, let $\sigma \in P$ if and only if the following conditions hold:

(1) $u_n \lesssim_{n+2} u_{n+1}$,

(2) $\ell_n \leq_{n+1} \ell_{n+1}$,

(3) if $\ell_n = (i_n, \overline{a}_n)$, then R_{u_n} is the set of Σ_{n+2} sentences of R_{i_n}, $\overline{a}_n$ includes the first n constants, and for the nth triple $(k, j, \overline{c})$, if R_j is a complete B_k type, then either there exists $a \in \overline{a}_n$ such that $R_j(\overline{c}, a) \subseteq R_{i_n}(\overline{a}_n)$ or else $R_j(\overline{c}, x) \cup R_{i_n}(\overline{a}_n)$ is inconsistent.

Lemma 5.1 *Suppose $\pi = u_0\ell_0u_1\ell_1\ldots$ is a path through P, where $\ell_n = (i_n, \overline{a}_n)$. Then $T(\pi) = \bigcup_n R_{u_n}$ is a complete theory, and $\bigcup_n(R_{i_n}(\overline{a}_n) \cap B_{n+1})$ is the complete diagram of a model $\mathcal{A}_\pi$ of $T(\pi)$ such that $\mathcal{A}_\pi$ represents $\mathcal{S}$. Moreover, $E(\pi) = \bigcup_n E(\ell_n)$ is the set of pairs $(j, \overline{c})$ such that R_j is the complete B_1 type realized by $\overline{c}$ in $\mathcal{A}_\pi$.*

Proof of Lemma: From the properties of P, it is clear that there is a complete theory $T(\pi)$ such that $R_{u_n} = T(\pi) \cap \Sigma_{n+2}$ for all n. Since R_{u_n} is the set of Σ_{n+2} sentences in R_{i_n}, $R_{i_n}(\overline{a}_n) \cap B_{n+1}$ is consistent with R_{u_n}, so it is consistent with $T(\pi)$. Since $\ell_n \leq_{n+1} \ell_{n+1}$ for all n, $\bigcup_n(R_{i_n}(\overline{a}_n) \cap B_{n+1})$ is consistent. If $k \leq n$ and $\overline{c} \subseteq \overline{a}_n$, then a B_k type which is witnessed in $R_{i_n}(\overline{a}_n)$ is also witnessed in $\bigcup_n(R_{i_n}(\overline{a}_n) \cap B_{n+1})$, and one which is inconsistent with $R_{i_n}(\overline{a}_n)$ is also inconsistent with $\bigcup_n(R_{i_n}(\overline{a}_n) \cap B_{n+1})$. It follows that $\bigcup_n(R_{i_n}(\overline{a}_n) \cap B_{n+1})$ is the complete diagram of a model $\mathcal{A}_\pi$ of $T(\pi)$, where $\mathcal{A}_\pi$ represents $\mathcal{S}$. From the definition of E, it is clear that $E(\pi) = \{(j, \overline{c}) :$ R_j is the B_1 type realized by $\overline{c}\}$.

We must show that $(L, U, P, E, \leq_{n\, 1\leq n<\omega})$ is an $(n+1)$ approximatingsystem relative to $\Delta^0_2(X)$. Conditions (1), (2), (3), and (4) are clear. Condition (5) is guaranteed by the next lemma.

Lemma 5.2 *Let $\sigma u \in P$, where* length$(\sigma) = 2n$. *Then there exists ℓ such that $\sigma u \ell \in P$. Moreover, if $u \leq_{n_0} \ell^1 \leq_{n_1} \ldots \leq_{n_{k-1}} \ell^k$ and $\ell(\sigma) \leq_n \ell^1$, where $n + 2 \geq n_0 > \ldots > n_k \geq 1$ and $n \leq n_1$, then we may take ℓ such that $\ell^i \leq_{n_i} \ell$ for $i = 1, \ldots, k$.*

Proof of Lemma: Say $\ell(\sigma) = (i, \overline{a})$. Since $\sigma u \in P$, the Σ_{n+2} part of R_u matches the set of Σ_{n+2} sentences of R_i. Then $R_i(\overline{a}) \cap B_{n+1}$ is consistent with any theory whose Σ_{n+2} part is R_u. Therefore, there exists $\ell = (j, \overline{c}) \in L$ such that $R_j(\overline{a}') \supseteq R_i(\overline{a}) \cap B_{n+1}$ and the Σ_{n+2} sentences of R_j are those of R_u. We may suppose, extending if necessary, that ℓ takes care of witnessing the nth triple. Then $\sigma u \ell \in P$.

Now, suppose $u \leq_{n_0} \ell^1 \leq_{n_1} \cdots \leq_{n_{k-1}} \ell^k$ and $\ell(\sigma) \leq_n \ell^1$, where $n + 2 \geq n_0 > \ldots > n_k \geq 1$ and $n \leq n_1$. Using Remark (1) and working our way back from k, we obtain $\mathfrak{m} \supseteq \ell^1$ such that $\ell^i \leq_{n_i} \mathfrak{m}$ for $i = 2, \ldots, k$. By Remark (2), since $u \leq_{n_0} \mathfrak{m}$, there exists $\ell = (j, \overline{c})$ such that $\mathfrak{m} \leq_{n_1} \ell$ and the Σ_{n+2} sentences of R_j are those of R_u. Since $n \leq n_1$, we have $\ell(\sigma) \leq_n \ell$. Extending if necessary, we may suppose that ℓ takes care of witnessing the nth triple. Then $\sigma u \ell \in P$.

We have shown that $\mathcal{A} = (L, U, P, E, \leq_{n\, 1\leq n})$ is an $(n+1)$ approximating system relative to $\Delta_2^0(X)$. Moreover, the instruction function q is $(n+1)$ approximated relative to $\Delta_2^0(X)$ by the functions q_n described above. We are in a position to apply Corollary 4.4. We obtain a run $\pi = u_0\ell_0u_1\ell_1\ldots$ of (P,q) such that $E(\pi)$ is $\Sigma_2^0(X)$. If $\ell_n = (i_n, \overline{a}_n)$, then by Lemma 5.1 together with the definition of q, $\bigcup_n (R_{i_n}(\overline{a}_n) \cap B_n)$ is the complete diagram of a model $\mathcal{A}_\pi$ of the given theory T such that $\mathcal{A}_\pi$ represents $\mathcal{S}$, and $E(\pi) = \bigcup_n E(\ell_n) = \{(j,\overline{c}) : R_j \text{ is the } B_1 \text{ type realized by } \overline{c}\}$. This completes the proof of Theorem 2.3.

6 Questions

There are natural questions about degrees of non-standard models of TA, some of them posed before Solovay's results.

Question 1 (Lerman [Le]): Is there a non-standard model $\mathcal{A}$ of TA with no enumeration of just the arithmetical sets recursive in $\mathcal{A}$?

Question 2 (Abramson-Knight): Is there a non-standard model $\mathcal{A}$ of TA such that $\deg(\mathcal{A})$ is a minimal upper bound for the arithmetical degrees?

Lachlan and Soare [L-S] gave an affirmative answer to Question 1, using Corollary 3.3. They produced a "generic" enumeration E of a Scott set J which includes the collection $\mathcal{S}$ of arithmetical sets, and they proved that there is no enumeration of $\mathcal{S}$ recursive in E. By Corollary 3.3, there is a non-standard model of TA recursive in E.

McAllister [M] extended the result of [L-S], showing that for a large class of completions T of PA, there is a model $\mathcal{A}$ of T with no enumeration of Rep(T) recursive in $\mathcal{A}$. McAllister did not use Corollary 3.3 or Corollary 3.5. Instead, he gave a direct construction of a "generic" model representing an appropriate Scott set. McAllister's methods do not handle all completions of PA, and this leads to an open question.

Question 3 (Detlefsen): Is there a completion T of PA such that forall models $\mathcal{A}$ of T, there is an enumeration of Rep(T) recursive in $\mathcal{A}$?

Using his construction of generic models representing a given Scott set, McAllister [M] also proved the following.

Theorem 6.1 (McAllister). *For a complete theory T and a Scott set $\mathcal{S}$ appropriate for T, the following are equivalent:*

(1) for all models $\mathcal{A}$ of T representing $\mathcal{S}$, $X \leq_T \mathcal{A}$,

(2) χ_X is enumeration reducible to some complete Σ_1 type Γ such that Γ is consistent with T and $\Gamma \in \mathcal{S}$.

Question 2 is still open. We may generalize the question to arbitrary completions T of PA .

Question 4: If T is a completion of PA, is there a (non-standard) model $\mathcal{A}$ of T such that $\deg(\mathcal{A})$ is a minimal upper bound of $\{\deg(T_n) : n \in \omega\}$?

There is no completion T of PA for which the answer to Question 4 is known. We may vary Question 2 further.

Question 5: Is there a non-standard model of TA whose degree is minimal among degrees of such models? Similarly, if T is some other completion of PA, does $\{\deg(\mathcal{A}) : \mathcal{A}$ is a model of $T\}$ have a minimal element?

References

[A_1] Ash, C. J. " Recursive labelling systems and stability of recursive structures in hyperarithmetical degrees", *Trans. Amer. Math. Soc.*, vol. 298 (1986), pp. 497–514. Corrections: Ibid vol. 300 (1988), p. 851.

[A_2] Ash, C. J. "Categoricity in hyperarithmetical degrees", *Annals of Pure and Applied Logic*, vol. 34 (1987), pp. 1–14.

[$A - K_1$] Ash, C. J., and Knight, J. F. "Pairs of recursive structures", *Annals of Pure and Applied Logic*, vol. 46 (1990), pp. 211–234.

[$A - K_2$] Ash, C. J., and Knight, J. F. "Mixed systems", *J. Symb. Logic*, vol. 59 (1994), pp. 1383–1399.

[$A - K_3$] Ash, C. J., and Knight, J. F. "Ramified systems", *Annals of Pure and Applied Logic*, vol. 70 (1994), pp. 205–221.

[B] Barker, E. "Intrinsically Σ^0_α relations", *Annals of Pure and Appl. Logic*, vol. 39 (1988), pp. 105–130.

[D] Davey, K. J. "Inseparability in recursive copies", *Annals of Pure and Applied Logic*, vol. 68 (1994), pp. 1–52.

[F] Feferman, S. "Arithmetical definable models of formalizable arithmetic", *Notices of Amer. Math. Soc.*, vol. 5 (1958), p. 679.

[G] Goncharov, S. S. " Strong constructivizability of homogeneous models", *Algebra i Logika*, vol. 17 (1978).

[H] Hurlburt, K. "Sufficiency conditions for theories with recursive models", *Annals of Pure and Applied Logic*, vol. 55 (1992), pp. 305–320.

[K_1] Knight, J. F. "Degrees coded in jumps of orderings", *J. Symb. Logic*, vol. 51 (1986), pp. 1034–1042.

[K_2] Knight, J. F. " Degrees of models with prescribed Scott set", *Classification: Proc. of Joint U.S.-Israel Workship*, ed. by Baldwin, 1987, pp. 182–191.

[K_3] Knight, J. F. " Requirement systems", *J. Symb. Logic*, vol. 60 (1995), pp. 222–245.

[K_4] Knight, J. F. "Coding a family of sets", to appear in *Annals of Pure and Applied Logic.*

[L – S] Lachlan, A. H., and Soare, R . I. "Models of arithmetic and upper bounds of arithmetic sets", *J. Symb. Logic*, vol. 59 (1994), pp. 977–983.

[Le] Lerman, M. "Upper bounds for the arithmetical degrees", *Annals of Pure and Applied Logic*, vol. 29 (1985), pp. 225–254.

[Le – Sch] Lerman, M., and Schmerl, J. "Theories with recursive models", *J. Symb. Logic*, vol. 44 (1979), pp. 59–76.

[M] McAllister, A. M. *Computability in Structures Representing a Scott Set*, Ph.D. thesis, University of Notre Dame, 1997.

[Ma] Manin, Y. I. *A Course in Mathematical Logic*, Springer-Verlag, 1977.

[Mac – Mar] Macintyre, A., and Marker, D. "Degrees of recursively saturated models", *Trans. of the Amer. Math. Soc.*, vol. 282 (1984), pp. 539–554.

[Mar] Marker, D. "Degrees of models of true arithmetic", in *Proc. of the Herbrand Symp.*, ed. by Stern, North-Holland, 1982, pp. 233–242.

[Mat] Matijasevich, Yu. "Enumerable sets are Diophantine", *Doklady Akademii Nauka SSSR*, vol. 191 (1970), pp. 272–282 (Russian).

[N] Nadel, M. "On a problem of MacDowell and Specker", J. Symb. Logic, vol. 45 (1980), pp. 612–622.

[P] Peretyat'kin, M. G. "Criterion for strong constructivizability of a homogeneous model", *Algebra i Logika*, vol. 17 (1978).

[Sc] Scott, D. "Algebras of sets binumberable in complete extensions of arithmetic", Recursive Function Theory, (ed. by Dekker), Amer. Math. Soc., 1962, pp. 117–22.

[So_1] Solovay, R. preprint circulated in 1982.

[So_2] Solovay, R. personal correspondence, 1991.

On the Topological Stability Conjecture

Ludomir Newelski*
Mathematical Institute of Wroclaw University
Mathematical Institute of the Polish Academy of Sciences

0 Introduction

Throughout we assume T is a complete theory in a countable language L and we work within a monster model $C = C^{eq}$ of T. In this survey paper we will sketch some ideas leading to the topological stability conjecture. Also we will show how this conjecture is related to stable model theory.

One of the central open problems in model theory is Vaught's conjecture, saying that if T has *few* (that is, $< 2^{\aleph_0}$) countable models, then T has countably many of them. Thus far, this conjecture was proved for ω-stable theories [SHM], for superstable theories of finite rank [Bu2] and in some other cases [Ne10](see also [Ls] for more information on Vaught's conjecture).

Vaught's conjecture refers to countable models and it became a yardstick with which we measure the level of our understanding of them. In fact, for a model theorist Vaught's conjecture is interesting mainly because it leads to various structural theorems on countable models. Usually such theorems say that if T has few countable models, then these models can be described so that it becomes possible to count them. So usually (and also in this paper) we assume T has $< 2^{\aleph_0}$ countable models, or at least that T is small (meaning that $S_n(\emptyset)$ is countable for all n).

In [Ne2] I proposed the following approach to the problem of describing a countable model M^* of T. It may happen that we know the structure of a definable piece $Q = \Phi(M^*)$ of M^*. Then describing M^* reduces to showing how M^* "envelopes" Q. In other words, we want to describe models in the class

$$K_Q = \{M \models T : M \text{ is countable and } \Phi(M) = Q\}.$$

*Research supported by KBN grant 2 P03A 006 09.
Mailing address: Mathematical Institute, Wrocław University, pl.Grunwaldzki 2/4, 50-384 WROCŁAW, Poland.

This is related to classification over a predicate (see e.g. [HHM]), however in our approach we do not regard the set Q as being fixed pointwise.

Any model $M \in K_Q$ is a result of choosing countably many elements $a_n, n < \omega$, of C so that $M = \{a_n, n < \omega\} \cup Q$ satisfies the Tarski-Vaught test. Considering this procedure leads to the notion of pseudotype over Q, which is important for investigating K_Q.

One of the first problems we encountered here was to prove that assuming few countable models, there are countably many "good" Q-pseudotypes. The topological stability conjecture asserts that all good Q-pseudotypes are "topologically stable", which in turn implies that there are countably many of them. We use lower case letters $a, b, c, \ldots$ to denote finite tuples of elements of C.

We shall use some basic notions from descriptive set theory (cf e.g. [Ke]). Let X be a Polish topological space, that is a separable complete metric space. A subset of X is called meager if it is a countable union of nowhere dense subsets of X. The σ-field of subsets of X generated by the open subsets of X is called the field of Borel subsets of X. We say that $A \subseteq X$ has the Baire property if for some open $O \subseteq X$ and meager $N \subseteq X$, we have $A = O \Delta X$. All Borel (and even all analytic) subsets of X have the Baire property. A topological group, whose underlying topological space is Polish, is called a Polish group.

1 Pseudotypes

Throughout, Φ is a countable disjunction of some types over $\emptyset$, M^* is a countable model of T and $Q = \Phi(M^*)$. Sometimes we will consider a special case, where Φ is a single formula. We are interested in describing models in the class $K_Q = \{M \prec C : M \text{ is countable and } \Phi(M) = Q\}$, up to isomorphism.

If $M = \{a_n, n < \omega\}$, $N = \{b_n, n < \omega\}$ are countable models of T and for every n, $tp(a_{<n}) = tp(b_{<n})$ (that is, $f(a_{<n}) = b_{<n}$ for some $f \in Aut(C)$), then M and N are isomorphic.

Let $Aut(Q) = \{f \in Aut(C) : f[Q] = Q\}$ and for $A \subseteq C$ let $Aut(Q/A) = \{f \in Aut(Q) : f\lceil A = id_A\}$. If $M = \{a_n, n < \omega\}$, $N = \{b_n, n < \omega\} \in K_Q$ and for every n there is $f \in Aut(Q)$ with $f(a_{<n}) = b_{<n}$, then also $M \cong N$. So we introduce the notion of a pseudotype as follows.

For $p, q \in S(Q)$ we write $p \sim q$ iff $f(p) = q$ for some $f \in Aut(Q)$. So $\sim$ is an equivalence relation on $S(Q)$. W call $p/\sim$ (the $\sim$-class of p) a Q-pseudotype (or just a pseudotype) of p. In other words, we say that a, b have the same Q-pseudotype if $f(a) = b$ for some $f \in Aut(Q)$. We see that in the context of K_Q pseudotypes play the role of types.

For $A \subseteq M^*$ we define a localized version $\sim_A$ of $\sim$ (on $S(QA)$) by: $p \sim_A q$

iff $f(p) = q$ for some $f \in Aut(Q/A)$, and we call $p/{\sim_A}$ a Q-pseudotype of p over A.

Also, let $\mathcal{B}_Q(A) = \{p \in S(QA) : p \text{ is not realized in any } N \in K_Q \text{ containing } A\}$. It is easy to see [Ne2], that $\mathcal{B}_Q(A)$ is a meager F_σ-subset of $S(QA)$, and is closed under $\sim_A$. We call the types in $\mathcal{B}_Q(A)$ bad (since they do not contribute to building models in K_Q) and the types in $S(QA) \setminus \mathcal{B}_Q(A)$ good (over A) (we apply a similar terminology to pseudotypes).

The group $Aut(Q/A)$, restricted to QA, is a Polish group acting in a continuous way on $S(QA)$,which is a Polish space. Pseudotypes over A are just the orbits of this action. This shows that $\sim_A$ is an analytic equivalence relation.

To analyze pseudotypes we use a local Scott analysis of C. Let A be a countable subset of C. We define equivalence relations $x \equiv_A y(\alpha)$ on tuples of elements of C by recursion on $\alpha \in Ord$.

(1) $a \equiv_A b(0)$ if $a \equiv b$ (that is, $tp(a) = tp(b)$).

(2) For limit δ, $a \equiv_A b(\delta)$ if $a \equiv_A b(\alpha)$ for every $\alpha < \delta$.

(3) $a \equiv_A b(\alpha+1)$ if for every $c \subseteq A$ there are $d, e \subseteq A$ with $ac \equiv_A bd(\alpha)$ and $ae \equiv_A bc(\alpha)$.

This differs from the original Scott approach in that we start from the level of elementary equivalence instead of atomic equivalence. If $p, q \in S(A)$ then we write $p \equiv_A q(\alpha)$ if $a \equiv_A b(\alpha)$ holds for some (equivalently: every) $a \models p$ and $b \models q$. We denote the equivalence relation $x \equiv_A y(\alpha)$ on tuples of elements of C and on $S(A)$ by $E_A(\alpha)$.

When $c \subseteq C$, then $a \equiv_A b(\alpha, c)$ means $ac \equiv_A bc(\alpha)$, and we extend this definition to $S(A)$ in the case where $c \subseteq A$. We denote the equivalence relation $x \equiv_A y(\alpha, c)$ by $E_A(\alpha, c)$.

We define the Scott height $SH(a/A)$ as the minimal α such that for $b, c \subseteq A$, $ab \equiv_A ac(\alpha)$ implies $ab \equiv_A ac(\alpha+1)$. The Scott height $SH(A)$ is the minimal α such that for $a, b \subseteq A$, $a \equiv_A b(\alpha)$ implies $a \equiv_A b(\alpha+1)$. For $p \in S(A)$, $SH(p)$ is $SH(a/A)$ for any $a \models p$.

So $SH(A)$ and $SH(a/A)$ are countable and for $\alpha < \omega_1$, $E_A(\alpha)$ is Borel on $S(A)$. As in the Scott isomorphism theorem, for $p \in S(Q)$, $p/{\sim} = p/E_Q(\gamma+2)$, where $\gamma = SH(p)$ [Ne2,Ne3]. This shows that pseudotypes are Borel and have the Baire property.

One of the first problems related to K_Q was to prove that if T has few countable models, then there are countably many good Q-pseudotypes [Ne2]. This problem is still open. However since pseudotypes are Borel, I tried to analyze them topologically.

Definition 1.1 *Assume A is a finite subset of $M \in K_Q$. We say that $p \in S(QA)$ is Q-isolated over A if $p/{\sim_A}$ is not meager. "Q-isolated" means Q-*

isolated over $\emptyset$. Similarly a Q-isolated pseudotype is just a non-meager pseudotype. We say that M is Q-atomic (over A) if for every $a \subseteq M$, $tp(a/QA)$ is Q-isolated over A.

Notice that all Q-isolated pseudotypes are good, and there are at most countably many of them. So the problem of counting good pseudotypes would be solved if all of them were Q-isolated. Unfortunately it is not so, although Q-isolation yields a nice description of $\aleph_0$-categoricity of the class K_Q.

Proposition 1.2 ([Ne2,Ne3]) *(1) If $I(T, \aleph_0) < 2^{\aleph_0}$, then Q-isolated types are dense in $S_n(Q)$.*
(2) If Q-isolated types are dense in $S_n(Q)$ for every $n > 0$, then there is a Q-atomic model $M \in K_Q$. Moreover, any two Q-atomic models are isomorphic.
(3) The following conditions are equivalent:
(a) all models in K_Q are isomorphic,
(b) all models in K_Q are Q-atomic,
(c) all good types in $S_n(Q), n < \omega$, are Q-isolated.
(4) If $M \in K_Q$ is Q-atomic, then $SH(M) \leq SH(Q) + 1$. If $p \in S(Q)$ is Q-isolated, then $SH(p) \leq SH(Q) + 1$.

In [Ne3] we give the following characterization of Q-isolation.

Proposition 1.3 *Assume $p \in S(Q)$ and $\alpha^* = SH(Q)$. The following conditions are equivalent.*
(1) p is Q-isolated.
(2) For every $a \subseteq Q$, $p/\sim_a$ is not nowhere dense.
(3) For every $a \subseteq Q$, $p/E_Q(\alpha^* + 1, a)$ is not nowhere dense.
(4) $p/E_Q(\alpha^* + 3)$ is not meager.

Question 1.4 *Can we replace $\alpha^* + 3$ in proposition 1.3(4) by $\alpha^* + 1$?*

2 Relative Q-isolation and topological stability

Let us recall the argument showing that there are countably many Q-isolated pseudotypes. Assume $p \in S(Q)$. If p is Q-isolated, then $cl(p/\sim)$ (the topological closure of $p/\sim$) has non-empty interior and $p/\sim$ is co-meager in $cl(p/\sim)$. It follows that there are countably many Q-isolated pseudotypes.

We know that there are also good pseudotypes, which are not Q-isolated (unless all models in K_Q are isomorphic). To count them, we may use topological arguments similar to the above one. For this purpose, the following notion of relative Q-isolation may be useful.

Suppose $X \subseteq S(Q)$ is closed. We say that X is Q-invariant [over $A \subseteq Q$] if X is invariant under the action of $Aut(Q)$ on $S(Q)$ [$Aut(Q/A)$ on $S(A)$, respectively]. For instance, for any $p \in S(Q)$, the set $cl(p/\sim)$ is Q-invariant.

Definition 2.1 *Assume $X \subseteq S(Q)$ is closed and Q-invariant [over a finite $A \subseteq Q$]. We say that $p \in S(Q)$ is Q-isolated in X [over A] if $p/\sim$ [$p/\sim_A$, respectively] is a relatively non-meager subset of X.*

As in Proposition 1.2 we have

Proposition 2.2 ([Ne3]) *If $I(T, \aleph_0) < 2^{\aleph_0}$ and $X \subseteq S(Q)$ is closed, Q-invariant and contains a good type, then X contains a type Q-isolated in X.*

It is easy to see that if $X \subseteq S(Q)$ is closed, Q-invariant, $p \in X$ and $\beta > SH(Q)$, then $p/E(\beta) \subseteq X$ [Ne3,Lemma 1.1]. Also, a generalized version of Proposition 1.3 holds.

Proposition 2.3 *Assume $X \subseteq S(Q)$ is closed, Q-invariant, $p \in X$ and $\alpha^* = SH(Q)$. Then the following conditions are equivalent.*
(1) p is Q-isolated in X.
(2) For every $a \subseteq Q$, $p/\sim_a$ is not nowhere dense in X.
(3) For every $a \subseteq Q$, $p/E_Q(\alpha^* + 1, a)$ is not nowhere dense in X.
(4) $p/E_Q(\alpha^* + 3)$ is not meager in X.

It follows that if p is Q-isolated in X, then $SH(p) \leq \alpha^* + 1$, where $\alpha^* = SH(Q)$. Since every Borel equivalence relation with $< 2^{\aleph_0}$ classes has countably many classes, we see that if $I(T, \aleph_0) < 2^{\aleph_0}$, then there are countably many pseudotypes of the form $p/\sim$, where $p \in S(Q)$ is Q-isolated in X over some $a \subseteq Q$.

Definition 2.4 ([Ne3]) *We say that $p \in S(Q)$ is topologically stable (τ-stable, for short), τ-based on $a \subseteq Q$, if any of the following equivalent conditions holds.*
(1) p is Q-isolated in X over A for some closed set $X \subseteq S(Q)$, which is Q-invariant over A.
(2) $p/\sim_a$ is co-meager in $cl(p/\sim_a)$.
(3) For every $b \subseteq Q$ with $a \subseteq b$ we have $p/\sim_b$ is not nowhere dense in $p/\sim_a$.
In this case we say also that the pseudotype $p/\sim$ is τ-stable, based on a.

We know that $I(T, \aleph_0) < 2^{\aleph_0}$ implies that there are countably many τ-stable good pseudotypes. So the folowing conjecture would solve the problem of counting good pseudotypes.

Conjecture 2.5 (topological stability conjecture [Ne3]) *If $I(T, \aleph_0) < 2^{\aleph_0}$, then every good pseudotype is τ-stable.*

We also call Conjecture 2.5 the τ-stability conjecture. I must admit that this conjecture neither implies nor is implied directly by Vaught's conjecture. Still it seems a reasonable approach to count pseudotypes. Also, attempts to prove the τ-stability conjecture led me to some interesting results, presented in the next sections. Whereas Vaught's conjecture implies that under the few models assumption there are countably many good pseudotypes, Conjecture 2.5 says more: not only are there countably many of them, but they are also τ-stable.

So the τ-stability conjecture should be easier to refute than Vaught's conjecture. However, as time passed I was becoming more and more confident of it. In a sequence of papers I managed to prove the τ-stability conjecture in many cases (see the next section of this paper). Also, one could argue for this conjecture as follows.

Suppose Conjecture 2.5 fails. Then there is a good $p \in S(Q)$ and an increasing sequence of finite tuples $a_n \subseteq Q, n < \omega$ such that for every n, $p/{\sim_{a_{n+1}}}$ is nowhere dense in $p/{\sim_{a_n}}$.

It is rather hard to find examples of such a sequence of "vanishing orbits", and thus far in such examples T has $2^{\aleph_0}$ countable models. So it seems that the existence of such a sequence means that T is rather complicated. Also it is my feeling that the theories with few countable models really have strong regularity properties, usually revealed in the proofs of Vaught's conjecture for them.

In some cases however we are able to count good pseudotypes under the few models assumption without deciding the τ-stability conjecture for them. This is a puzzling situation. Specifically, suppose now $\Phi(x)$ is just a single formula (or a countable disjunction of formulas). Then we can consider Q as a structure in its own right (in the case where Φ is a countable disjunction of formulas, this is a many-sorted structure), and $Th(Q)$ does not depend on the choice of M^*, so we denote it by $T\lceil\Phi$.

It may happen that Vaught's conjecture is proved for $T\lceil\Phi$ and for some expansions of $T\lceil\Phi$ by constants. This enables us to count good pseudotypes in $S(Q)$.

To be more precise, let c be a new constant and $L' = L\cup\{c\}$. Let $p \in S(Q)$ be good. Say, $p = tp(a/Q)$ for some $a \in N \in K_Q$. Interpreting c as a makes N into an L'-structure N'. Then Q_p denotes the set Q with the structure induced from N'. Let $T_p = Th(Q_p)$. Clearly, T_p and Q_p do not depend on the choice of a. So T_p is an expansion of $T\lceil\Phi$.

Lemma 2.6 ([Ne3]) *(1) For good $p, q \in S(Q)$, $p \sim q$ iff $Q_p \cong Q_q$.*
(2) If $p \equiv q(\omega)$, then $Q_p \equiv Q_q$, hence $T_p = T_q$.
(3) If $I(T, \aleph_0) < 2^{\aleph_0}$ then for every good $p \in S(Q)$, $I(T_p, \aleph_0) < 2^{\aleph_0}$.

If $T\lceil\Phi$ is ω-stable or superstable of finite rank, then so is T_p for every good $p \in S(Q)$, so by [SHM,Bu2], Vaught's conjecture holds for T_p. Assume $I(T,\aleph_0) < 2^{\aleph_0}$. There are countably many classes of $E_Q(\omega)$ on $S(Q) \setminus \mathcal{B}_Q(\emptyset)$, each of them is a union of some number of pseudotypes. By lemma 2.6, pseudotypes in $p/E_Q(\omega)$ (for good $p \in S(Q)$) correspond to some countable models of T_p (expansions of Q). Since $I(T_p,\aleph_0) = \aleph_0$, we get that there are countably many good pseudotypes in this case.

This argument does not decide if the pseudotypes in question are τ-stable. In order to decide this one must look into the proof of Vaught's conjecture for the theories $T\lceil\Phi$ and T_p. We will do so in the next section.

3 Some special cases of the topological stability conjecture

In this section we shall examine some special cases of the topological stability conjecture, considered mainly in [Ne7]. The simpler Q and the more stable T is, the easier it should be to verify this conjecture.

Case 0. T is ω-stable.

By [Ne3, Remark 2.9], for any $p \in S(Q)$, if $p/\sim$ is countable then for some finite $a \subseteq Q$, $p/\sim_a$ is a singleton. Hence p is τ-stable, τ-based on a. This shows that the τ-stability conjecture is true for ω-stable T.

Next we shall consider cases where Q is simple.

Case 1. $\Phi(x)$ is a formula over $\emptyset$ (or a disjunction of such formulas) and $Q = \Phi(M^*)$ is minimal as a model of $T\lceil\Phi$.

In this case by [Ne3, Theorem 4.3], for good $p, q \in S(Q)$, Q_p is a prime model of T_p and $p \equiv_Q q(\omega)$ implies $p \sim q$, hence assuming few models we get that there are countably many good pseudotypes. If additionally T is stable, then [Ne3, Theorem 4.11] shows that in this case every good type in $S(Q)$ is τ-stable (the proof involves canonical bases and Cantor-Bendixson's rank). Without stability assumption the τ-stability conjecture is open here.

From now on in this section we shall assume that T is stable and small and $\Phi(x)$ is a countable disjunction of formulas over $\emptyset$.

Case 2. $T\lceil\Phi$ is $\aleph_0$-categorical.

In this case the τ-stability conjecture translates into a statement about expansions of some $\aleph_0$-categorical theories. Specifically, for any $a \subseteq C$ consider $T(a)\lceil\Phi$, the theory obtained from T by naming a and restricting to Φ. Equivalently this theory is an expansion of $Th(Q)$ by naming countably many constants from $Cb(tp(a/Q))$, where $Cb(tp(a/Q))$ is the canonical base of $tp(a/Q)$ [Ne7, Lemma 1.2]. If $a \subseteq Q$, then clearly by Ryll-Nardzewski's theorem, this theory is still $\aleph_0$-categorical. But if a lies outside Q, then expanding $T\lceil\Phi$ by naming an infinite set of constants from $Cb(tp(a/Q))$ might well destroy its $\aleph_0$-catgoricity. It turns out that this is closely related to the

τ-stability conjecture.

Theorem 3.1 ([Ne7,Theorem 2.5]) *Assume T is stable and $T\lceil\Phi$ is ω-categorical. Then the following conditions are equivalent.*
(1) Every good $p \in S(Q)$ is τ-stable.
(2) Every good $p \in S(Q)$ is τ-stable based on $\emptyset$.
(3) For every $a \subseteq C$, $T(a)\lceil\Phi$ is $\aleph_0$-categorical.

Hence for stable T and $T\lceil\Phi$ $\aleph_0$-categorical the τ-stability conjecture is equivalent to the following conjecture.

Conjecture 3.2 ([Ne7, Conjecture 1.1]) *Assume T is stable, $I(T, \aleph_0) < 2^{\aleph_0}$ and $a \subseteq C$. If $T\lceil\Phi$ is $\aleph_0$-categorical then $T(a)\lceil\Phi$ is $\aleph_0$-categorical.*

Also notice that if the τ-stability conjecture is true here, then a stronger statement holds: every good $p \in S(Q)$ is τ-stable, based on $\emptyset$.

The proof of Theorem 3.1 relies on the analysis of pseudotypes in the case, where Q is atomic or saturated (as a model of $T\lceil\Phi$). We shall say more about this later. The next theorem partially confirms Conjecture 3.2.

Theorem 3.3 ([Ne7, Theorem 1.3]) *Assume T is small, stable, $T\lceil\Phi$ is $\aleph_0$-categorical, $\aleph_0$-stable and $a \subseteq \Phi(C)$. Then $T(a)\lceil\Phi$ is $\aleph_0$-categorical.*

The proof of this theorem uses Morley rank and leads in [Ne7] to several new properties of types of finite weight in a small stable theory. In general, Conjecture 3.2 remains open. It seems hard to attack since there are only a few examples (of Hrushovski) of ω-categorical, stable, but not ω-stable structures.

Case 3. Q is atomic or saturated (as a model of $T\lceil\Phi$).

In this case to examine the status of the τ-stability conjecture we describe τ-stable types. Recall from [Sh, IV, Definition 2.1] that $p \in S(A)$ is $F^s_{\aleph_0}$-isolated if for some finite $a \subseteq A$, $p\lceil a \vdash p$. A is $F^s_{\aleph_0}$-atomic over B if for every $a \subseteq A$, $tp(a/B)$ is $F^s_{\aleph_0}$-isolated. A topological analysis of pseudotypes yields the following proposition.

Proposition 3.4 ([Ne7, Corollary 2.4]) *Assume T is stable, $a \subseteq Q$ and $p = tp(c/Q)$ is good.*
(1) If Q is atomic then p is τ-stable τ-based on a iff Q is atomic over ac (equivalently: over $aCb(p)$).
(2) If Q is saturated, then p is τ-stable τ-based on a iff Q is $F^s_{\aleph_0}$-atomic over $aCb(p)$.

Hence for atomic or saturated Q , the τ-stability conjecture may be restated as follows.

Conjecture 3.5 *Assume T is stable, $I(T,\aleph_0) < 2^{\aleph_0}$ and $p = tp(c/Q)$ is good.*
(1) If Q is atomic then Q is atomic over ac for some $a \subseteq Q$.
(2) If Q is saturated then Q is $F^s_{\aleph_0}$-atomic over $aCb(p)$ for some $a \subseteq Q$.

When T is superstable with few countable models, then this conjecture follows from the vanishing multiplicities conjecture, proved by myself in [Ne4]. Predrag Tanovic proved (unpublished) a stronger version of Conjecture 3.5(1) for superstable T, for any atomic countable set A in place of a definable set Q, and for arbitrary type $p \in S(A)$.

In [Ne7] I proved partially Conjecture 3.5 in the case where $T\lceil\Phi$ is ω-stable (then the assumption $I(T,\aleph_0) < 2^{\aleph_0}$ is weakened to smallness, and in 3.5(1) we get $a = \emptyset$) and when $T\lceil\Phi$ is superstable of finite rank (this relies on [Bu2]). Hence we have the following theorem.

Theorem 3.6 ([Ne7, Theorems 2.6, 2.8]) *Assume T is small, stable and Q is atomic or saturated.*
(1) If $T\lceil\Phi$ is ω-stable then every good $p \in S(Q)$ is τ-stable τ-based on $\emptyset$. Moreover there are countably many good pseudotypes.
(2) If $T\lceil\Phi$ is superstable of finite rank, then every good $p \in S(Q)$ is τ-stable (not necessarily τ-based on $\emptyset$).

As mentioned in Section 2, when $T\lceil\Phi$ is ω-stable and $I(T,\aleph_0) < 2^{\aleph_0}$, then by [SHM] there are countably many good pseudotypes. Now we shall examine this case.

Case 4. $T\lceil\Phi$ is ω-stable.

Basically, proving the τ-stability conjecture in this case consists in carefully reading the proof of Vaught's conjecture for ω-stable theories. This is because counting the orbits $p/\sim$, $p \in S(Q)$ reduces to counting the models of T_p. Sometimes this requires some new facts about small stable theories (like in [Ne7, Section 1]). The strongest result I have here is the following theorem.

Theorem 3.7 ([Ne8, Theorem 2.9]) *Assume T is small, stable, $T\lceil\Phi$ is ω-stable and (either 1-based or bounded or of finite rank), with $I(T\lceil\Phi,\aleph_0) < 2^{\aleph_0}$. Then for every good $p \in S(Q)$, $p/\sim$ is open in $cl(p/\sim) \cap (S(Q) \setminus \mathcal{B}_Q(\emptyset))$. In particular, p is τ-stable τ-based on $\emptyset$. Moreover, there are countably many good pseudotypes.*

I could not remove from this theorem the assumption that $T\lceil\Phi$ is 1-based, bounded or of finite rank. Neither could I prove the τ-stability conjecture in Case 4. It should be mentioned here that in the above cases often we prove that every good type is τ-stable τ-based on $\emptyset$.

4 On multiplicities of types

In section 3 we saw that in some special cases the τ-stability conjecture leads to interesting questions. The most interesting ideas occurred in the case when $acl(Q)$ is homogeneous and T is superstable. The τ-stability conjecture led in this case to the conjecture of vanishing multiplicities (formulated in [Ne2], proved in [Ne4]) and the discovery of several new notions in stable model theory. These were the notions of meager forking, meager type, m-independence, $\mathcal{M}$-rank, traces of types and $*$-finite tuples.

The main motivation for investigating models in K_Q is Vaught's conjecture. Since Vaught's conjecture was proved for ω-stable theories, the next natural object of interest became the superstable ones. Then the easiest case to consider was when Q is atomic. Actually the method for counting pseudotypes for atomic Q works for a more general case where $acl(Q)$ is homogeneous (we work in $C = C^{eq}$).

So in this section (unless otherwise stated) we assume that T is small superstable and $acl(Q)$ is homogeneous. In this case pseudotypes are closely related to sets of stationarizations of some types, which are a special case of traces of types [Ne6]. We shall recall this notion now.

Assume A is a finite set and $s(x)$ is a (possibly incomplete) type over C. We define the trace of s over A as the set $Tr_A(s) = \{tp(a/acl(A)) : a \models s\} = \{r(x) \in S(acl(A)) : r(x) \cup s(x) \text{ is consistent}\}$.

We see that $Tr_A(s)$ is a closed subset of $S(acl(A))$. $Tr_A(a/C)$ abbreviates $Tr_A(tp(a/C))$ and $Tr(s)$ denotes $Tr_\emptyset(s)$. For any $p \in S(A)$ let $St(p)$ denote the set of non-forking extensions of p over C (that is, stationarizations of p over C). This set is homeomorphic to the set $Tr_A(p)$ (which is the set of stationarizations of p over A) and also to $Tr_B(p)$, whenever $B \subseteq A$ and p does not fork over B.

Suppose $A \subseteq C$, $p \in S(A)$ and $q \in S(C)$ is a non-forking extension of p. This means that $St(q)$ is a closed subset of $St(p)$. $Aut(C/C)$ acts continuously and transitively on $St(q)$. This implies that

($*$) either $St(q)$ is nowhere dense in $St(p)$ or $St(q)$ is open in $St(p)$.

Equivalently, using the homeomorphism between $St(p)$ and $Tr_A(p)$, we have that $Tr_A(q)$ is either nowhere dense or open in $Tr_A(p)$. If the latter case holds, we say that q is an m-free extension of p (m stands for "multiplicatively").

Now suppose $p \in S(Q)$ does not fork over a finite $A \subseteq Q$. Let $\pi : S(C) \to S(Q)$ be restriction and let $X_A(p) = \pi(St(p\lceil A))$. The next lemma explains why we are interested in $X_A(p)$.

Lemma 4.1 *If $acl(Q)$ is homogeneous, then $X_A(p) = cl(p/\sim_A)$.*

Proof. This is proved essentially in [Ne3, Proposition 2.7]. Here we give an explicit proof. Cleary $X_A(p)$ is closed and contains $p/{\sim_A}$. So we must show that $p/{\sim_A}$ is dense in $X_A(p)$. If $E(x,y) \in FE(A)$, $\alpha = a/E$ for some $a \models p$ and β is an E-class containing a realization of $p\lceil A$, then $\alpha, \beta \in acl(Q)$ and $\alpha \equiv \beta(A)$, so by homogeneity there is $f \in Aut(Q/A)$ with $f(\alpha) = \beta$. Clearly $\beta = f(a)/E$, showing that $p/{\sim_A}$ is dense in $X_A(p)$.

The next proposition is essentially [Ne3, Proposition 2.7]. It follows easily from Lemma 4.1.

Proposition 4.2 *Assume $p \in S(Q)$ does not fork over a finite $A \subseteq Q$ and $acl(Q)$ is homogeneous. Then the following conditions are equivalent.*
(1) p is τ-stable, τ-based on A.
(2) For every finite $B \subseteq Q$ containing A, $p\lceil B$ is an m-free extension of $p\lceil A$.

Proof. (1) $\rightarrow$ (2). Since π is continuous, Lemma 4.1 gives that $St(p\lceil B)$ is not nowhere dense in $St(p\lceil A)$. (*) gives that $St(p\lceil B)$ is open in $St(p\lceil A)$. For (2) $\rightarrow$ (1) use the fact that $\pi\lceil St(p\lceil A)$ is open (this is the open mapping theorem of Lascar-Poizat).

Now suppose $p \in S(Q)$ is not τ-stable, and does not fork over a finite set $A \subseteq Q$. Hence there is an increasing sequence of finite sets $A_n \subseteq Q$, $n < \omega$, with $A = A_0$ such that for every n, $p\lceil A_{n+1}$ is not an m-free extension of $p\lceil A_n$, that is $Tr_A(p\lceil A_{n+1})$ is nowhere dense in $Tr_A(p\lceil A_n)$. Hence we obtain a sequence of sets of stationarizations $St(p\lceil A_n), n < \omega$, such that $St(p\lceil A_{n+1})$ "vanishes" in $St(p\lceil A_n)$.

Also, the existence of such a sequence is equivalent to p not being τ-stable. So the τ-stability conjecture leads to the investigation of sets of stationarizations of types. If $I(T,\aleph_0) < 2^{\aleph_0}$, then these sets are not arbitrary.

Theorem 4.3 ([Ne1]) *If $I(T,\aleph_0) < 2^{\aleph_0}$ and $p \in S(A)$ is weakly minimal, then either p is isolated (that is, $Tr_A(p)$ is open in $S(acl(A))$) or p has finite multiplicity (that is, $Tr_A(p)$ is finite).*

This theorem (conjectured earlier by J.Saffe) was decisive in the proof of Vaught's conjecture for weakly minimal theories [Bu1],[Ne1]. Theorem 4.3 implies that if $I(T,\aleph_0) < 2^{\aleph_0}$ and p is weakly minimal, then in (*) we can say that either $Tr_A(q)$ is open in $Tr_A(p)$ or $Tr_A(q)$ is finite. In particular we see that in this case

(**) there is no increasing sequence of finite sets $A_n, n < \omega$, and non-forking extensions $p_n \in S(A_n), n < \omega$ of p with $Tr_A(p_{n+1})$ nowhere dense in $Tr_A(p_n)$ for all n.

In fact, by Theorem 4.3 in the weakly minimal case the length of such a sequence is bounded by 2. In [Ne2, Conjecture 3.5] I conjectured that (**) is true for any superstable theory with few countable models and any $p \in S(A)$ (where A is finite), and called this the vanishing multiplicities conjecture. I proved this conjecture in [Ne4] and a stronger version in [Ne6]

Theorem 4.4 **([Ne4,Ne6])** *If T is superstable and $I(T, \aleph_0) < 2^{\aleph_0}$ then for every finite A and $p \in S(A)$, (**) holds. Moreover, for every such p there is a natural number n such that the longest sequence in (**) has length n, and $n \leq U(p) + 1$.*

As an application we get the τ-stability conjecture in the case where T is superstable and $acl(Q)$ is homogeneous (in fact, also in a more general case, where Q is atomic over finitely many Morley sequences).When Φ is a countable disjunction of formulas, $T\lceil\Phi$ is ω-stable or superstable of finite rank and T is superstable, then by the proof of Vaught's conjecture in the corresponding cases [SHM,Bu2], Q is atomic over a countable skeleton built-up from Morley sequences. Theorem 4.4 also implies the τ-stability conjecture in this case.

The first natural case, where the τ-stability conjecture is open, is the case where T is stable and Q is atomic. It is possible that the τ-stability conjecture could also lead in this case to a new insight into the structure of stable models.

To prove Theorem 4.4 I developed in [Ne4] and subsequent papers [Ne5, Ne6, Ne9, Ne11] a new technique for investigating traces of types. The starting point here was the notion of meager forking.

Suppose A is a set of parameters and $\varphi(x)$ is a formula over C. The open mapping theorem yields that if $\varphi(x)$ does not fork over A, then $Tr_A(\varphi)$ has non-empty interior in $S(acl(A))$. More generally, for any $p(x) \in S(A)$, if $p(x) \cup \{\varphi(x)\}$ does not fork over A, then $Tr_A(\varphi) \cap Tr_A(p)$ has non-empty interior in $Tr_A(p)$.

However, if $\varphi(x)$ forks over A, then there are two possibilities: either $Tr_A(\varphi) \cap Tr_A(p)$ has non-empty interior in $Tr_A(p)$ or $Tr_A(\varphi) \cap Tr_A(p)$ is nowhere dense (equivalently: meager) in $Tr_A(p)$. If for every formula $\varphi(x)$ forking over A the latter case holds, then we say that forking is meager on p.

If T is small stable, we say that a regular type r is meager if there is a regular isolated type $p \in S(A)$ (for some finite set A) such that p is non-orthogonal to r and forking is meager on p. (The definition of a meager type may be given also in an arbitrary stable theory, but then it is more complicated; it involves the notion of a p-formula [Ne4]). As an example notice that any properly weakly minimal non-trivial type is meager.

In [Ne4] I proved that any meager type is locally modular. In [Ne5] I gave a procedure for producing meager types in a small superstable theory. This procedure involves an ordinal-valued $\mathcal{M}$-rank measuring the size of the set of stationarizations of a type p (introduced in [Ne2]). The notion of an m-free extension leads to the notion of m-independence [Ne9]. In [Ne11] it is proved that in a small stable T m-independence is the strongest natural notion of independence strengthening forking independence.

m-independence, $\mathcal{M}$-rank, meager types and so-called $*$-finite tuples [Ne9] are tools for investigating multiplicities of types. They are parallel in many ways to forking independence, U-rank, regular types and imaginaries in classical stability theory [Sh]. These new notions enable us to prove Vaught's conjecture in some new cases [Ne10]. Also they lead to several new questions and conjectures (for instance the $\mathcal{M}$-gap conjecture [Ne6]).

The τ-stability conjecture is still open. I think that it may lead to several interesting ideas. It may be regarded as a part of a more general program of developing a model theory for K_Q. I think it would be desirable to develop a model theory for K_Q even under some facilitating assumptions. Some results in this direction are given in [Ne8].

References

[Bu1] S. Buechler, *Classification of small weakly minimal sets I*, in: **Classification theory, proceedings, Chicago 1985** (J. T. Baldwin, ed.), Springer 1987, 32-71.

[Bu2] S. Buechler Vaught's conjecture for superstable theories of finite rank, preprint 1993.

[HHM] W. Hodges, I. M. Hodkinson, D. Macpherson, *Omega-categoricity, relative categoricity and coordinatization*, Ann. Pure Appl. Logic 46 (1990), 169-199.

[Ke] A. Kechris, **Classical Descriptive Set Theory**, Springer, New York 1995.

[Ls] D. Lascar, *Why some people are excited by Vaught's conjecture*, J. Symb. Logic 50 (1985), 973-983.

[Ne1] L. Newelski, *A proof of Saffe's conjecture*, Fund. Math. 134 (1990), 143-155.

[Ne2] L. Newelski, *A model and its subset*, J. Symb. Logic 57 (1992), 644-658.

[Ne3] L. Newelski, *Scott analysis of pseudo-types*, J. Symb. Logic 58 (1993), 648-663.

[Ne4] L. Newelski, *Meager forking*, Ann. Pure Appl. Logic 70 (1994), 141-175.

[Ne5] L. Newelski, *$\mathcal{M}$-rank and meager types*, Fund. Math. 146 (1995), 121-139.

[Ne6] L. Newelski, *$\mathcal{M}$-rank and meager groups*, Fund. Math. 150 (1996), 149-171.

[Ne7] L. Newelski, *On atomic or saturated sets*, J. Symb. Logic 61 (1996), 318-333.

[Ne8] L. Newelski, *On the prime model property*, Proc. AMS 124 (1996), 2519-2525.

[Ne9] L. Newelski, *$\mathcal{M}$-gap conjecture and m-normal theories*, Israel J. Math., to appear.

[Ne10] L. Newelski, *Vaught's conjecture for some meager groups*, Israel J. Math., submitted.

[Ne11] L. Newelski, *Flat Morley sequences*, J. Symb. Logic, submitted.

[Sh] S. Shelah, **Classification theory**, 2nd edition, North Holland, 1990.

[SHM] S. Shelah, L. Harrington, M. Makkai, *A proof of Vaught's conjecture for ω-stable theories*, Israel J. Math. 49 (1984), 259-280.

A Mahlo-Universe of Effective Domains with Totality

Dag Normann

Abstract

We construct a typed hierarchy of effective algebraic domains with totality of height the first recursively Mahlo ordinal. The hierarchy is based on the empty type and the domains for singleton, boolean values and natural numbers, and it is closed under dependent sums and pro-ducts of effectivly parameterised families of types, and under universes closed under a very general universe operator.

1 Introduction

Given a type theory, there are several ways to produce a semantics for the theory. An important distinction will be between intuitionistic and classical semantics. It is an understandable view that a classical semantics for an intuitionistic type theory is of little value. On the other hand, if one construct some natural, classical semantics for a type-theory, by interpreting the type operators described in the theory, one gets an impression of the classical strength of these operators, even inside a constructive environment. In [14] we constructed a hierarchy of domains with totality and density having the complexity of Kleene-recursion in the functional ${}^3E$. Taking the hereditarily effective version of this hierarchy, the complexity will be that of the first non-recursive ordinal ω_1^{CK}. In a sense this is the minimal complexity of effective dependent sums and products, and corresponds to the minimal model of KP-set theory.
If we introduce the W-type constructor, the corresponding minimal classical model will be of complexity the first recursively inaccessible ordinal. This is essentially established in Normann [13]. There are corresponding results of equivalent proof-theoretical strength between certain extensions of KP-theory and certain intuitionistic type theories, see Griffor and Rathjen [5] and Setzer [17]. Type theory with one universe, but without the W-operator corresponds to KP-theory with one admissible, while type theory with one universe *and* the W-operator corresponds to KP-theory with one recursively inaccessible

ordinal. The correspondence between results on proof-theoretical strength and on the complexity of the minimal models can not be a coincidence. With this paper we will extend this pattern one step.

When we use the term *minimal model* this is not accurate. It is of course possible to construct semantics of these type theories of semicomputable complexity; we introduce formal interpretations of every type proved to exist and every object proved to be of that type. Since the set of proofs is computable, the model will be semicomputable. What we mean by minimal here, is that we view the constructors used in the theory, we then make classical interpretations of these constructors inside the category of effective domains with totality, and finally consider the minimal hierarchies closed under these constructors.

In this paper we will go the other way around. In the previous estimates of complexity we have somehow viewed transfinite computations as transfinite propositions, and we have used standard (and not so standard) *propositions as types* translations to express termination and value of transfinite computations in the typed hierarchy. Here we will consider computations relative to the functional known as *the superjump* defined by Gandy [4], and see what sort of type constructors will be required in order to express these computations as properties of the typed hierarchy. A. Setzer on one hand, and E.R. Griffor and M. Rathjen on the other, has given a proof-theoretical analysis (unpublished) of Mahloness, and their results and the results of this paper supplement each other.

Domains

The domains in this paper will be *algebraic domains* or *Scott-Ershov-domains* as defined in Stoltenberg-Hansen, Lindström and Griffor [18]. We view a domain element as an ideal in a partially ordered set of finitary compacts. Thus we use the symbol "$\subseteq$" for the ordering relation on any domain, including the domain of continuous functions. We will assume that the reader is familiar with the theory of domains as presented in e.g. [18].

Acknowledgements

My work on this started when I visited the University of Uppsala in September '95. At several occations I had discussions with Ulrich Berger on a related problem, representing computations in the superjump in a typed hierarchy of domains with dense totality. Many of the ideas used in this paper grew from the discussions with him. The workshop in Uppsala in April '96 was most stimulating, as was discussions with Griffor and Setzer earlier on. My final inspiration for working out this approach came through Rathjen's visit to Oslo in October '96. It is still my belief that the original approach via uniformly dense universes can be carried out, and a proof along these lines will give a much more fine structured analysis of the computations relative to the superjump.

I am grateful to the referee for pointing out ways to improve the exposition and for discovering misprints, mis-spellings and inaccuracies in the original version.

2 The Mahlo-universe

In this section we will introduce a typed hierarchy of effective domains with a notion of totality restricted to the effective elements. We first introduce the underlying domains via a standard fix-point construction. Following the format of the papers Kristiansen and Normann [8, 9] and Normann [14, 13, 11] we will construct one domain T, being the domain of *type descriptions*, and an interpretation map $I(t)$ interpreting each $t \in T$ as a domain. See also Berger [1, 2, 3] or Waagbø [19, 20] for a discussion.
I will be a parameterisation in the sense of Palmgren and Stoltenberg-Hansen [16]. As a part of our construction, we will give interpretations of universes. Here we will follow ideas from Berger [1, 2, 3], though we will not be bothered with problems of density.

Definition 1

a) Let $\mathbb{N}$ be the natural numbers, let $\mathbb{B}$ be the set of Boolean values $\{tt, ff\}$ and let O, N and B be extra atomic formal objects.
Let o, Π, Σ and U be syntactical entities used in the denotation of compacts.
We will define the (for the sake of this paper) universal domain V via defining the set of compacts V_0 with standard preorderings as follows:

1. $\bot$, O, N, B, all natural numbers and all boolean values are compacts in V_0 with the flat ordering.
2. If $u_1, \ldots, u_n, v_1, \ldots, v_n$ are compacts in V_0 with the standard consistency requirements for function-spaces, then $\{(u_1, v_1), \ldots, (u_n, v_n)\}$ is a compact in V_0.
These compacts are ordered like compacts in any function domain.
3. If u and v are compacts in V_0, then (u, v), $\Pi(u, v)$, $\Sigma(u, v)$ and $o(u, v)$ are compacts in V_0.
Such compacts with different prefixes are incomparable.
Two compacts with the same prefix will be ordered as for ordered pairs.

b) By recursion we define a subset T_0 of V_0.
Simultaneously, we define a subset $I_0(t)$ of V_0 for each $t \in T_0$.
All these subsets will be closed under least upper bounds of pairs bounded in V_0, and they will contain $\bot$. Moreover, if $t \sqsubseteq t'$, then $I_0(t) \subseteq I_0(t')$. We will leave the verification of these facts to the reader.

1. $\bot \in T_0$ and $I_0(\bot)) = \{\bot\}$.
2. $O \in T_0$ and $I_0(O) = \{\bot\}$.
3. $N \in T_0$ and $I_0(N) = \mathbb{N}_\bot$.
4. $B \in T_0$ and $I_0(B) = \mathbb{B}_\bot$.
5. If $t \in T_0$ and $F = \{(u_1, s_1), \ldots, (u_n, s_1)\}$ is a compact with $u_1, \ldots, u_n \in I_0(t)$ and $s_1, \ldots, s_n \in T_0$, then $\Sigma(t, F)$ and $\Pi(t, F)$ are in T_0.
$I_0(\Sigma(t, F)) = \{(u, v) \mid u \in I_0(t) \wedge v \in I_0(\sqcup\{s_i \mid u_i \sqsubseteq v\})\}$.
$I_0(\Pi(t, F)) = \{\{(w_1, v_1), \ldots, (w_m, v_m)\} \mid \forall j \leq m[w_j \in I_0(t) \wedge v_j \in I_0(\sqcup\{s_i \mid u_i \sqsubseteq w_j\})]\}$.
6. If $\Phi = \{(s_1, t_1), \ldots, (s_n, t_n)\}$ is a compact with all $s_i, t_j \in T_0$, then $(U, \Phi) \in T_0$.
$I_0(U, \Phi)$ is defined recursively, partly copying the definition of T_0 itself.
Simultaneously we will define a map $\rho : I_0(U, \Phi) \rightarrow T_0$, independently of Φ:
   - O, B and N are in $I_0(U, \Phi)$ with $\rho(O) = O$ etc.
   - If $t \in I_0(U, \Phi)$ and F is a compact function mapping $I_0(\rho(t))$ to $I_0(U, \Phi)$, we let $\Pi(t, F)$ and $\Sigma(t, F)$ be in $I_0(U, \Phi)$.
   We let $\rho(\Pi(t, F)) = \Pi(\rho(t), \rho \circ F)$, and likewise for Σ.
   - If $\Psi \sqsubseteq \Phi$ is compact and $s \in I_0(U, \Psi)$, then $o(\Psi, s) \in I_0(U, \Phi)$ with $\rho(o(\Psi, s)) = \Psi(\rho(s))$.

Remark 1 T_0 represents finite syntactic descriptions of types, i.e. the elements t contain information about how the interpretation $I_0(t)$ is constructed. If $t = (U, \Phi)$ the interpretation of t will be a universe. This universe will consist of a domain of syntactic forms. The syntactic form will then say if the type in question is a base type, a dependent sum, a dependent product or if it is obtained by application of the operator Φ of which we take the closure. This is the point of the symbol o, $o(\Psi, t)$ denotes the type obtained by applying Ψ to the type denoted by t.

There are technical reasons for accepting notations for $\Psi(\rho(t))$ for all $\Psi \subseteq \Phi$, we just need those subelements.

The set of compacts has an effective enumeration.

In order to turn T_0 into a domain and I_0 into a parameterisation, we will

have to take the ideal completion. In this paper we will only be dealing with the effective elements of these domains, i.e. with ideals that correspond to recursive enumerations of (indices for) compacts.
This is reflected in the following definition:

Definition 2 *Let T be the set of effective ideals of compacts in T_0, ordered by inclusion.*
For each $t \in T$, let $I_0(t) = \bigcup\{I_0(t_0) \mid t_0 \in t\}$.
Let $I(t)$ be the set of effective ideals in $I_0(t)$

We are now ready to define the hierarchy of well-formed types and the set of total objects in each well-formed type:

Definition 3 By induction on the countable ordinal α we define $T_\alpha \subseteq T$ and the total elements $\bar{I}(t)$ for $t \in T_\alpha$ as follows:

O, B and N are in T_α for all α with the obvious set of total elements.

If α is a limit ordinal, then $T_\alpha = \bigcup_{\beta<\alpha} T_\beta$.

If $\alpha = \beta + 1$ we let

$\Sigma(t, F) \in T_\alpha$ if $t \in T_\beta$ and $F(x) \in T_\beta$ for all $x \in \bar{I}(t)$.

$\Pi(t, F) \in T_\alpha$ if $t \in T_\beta$ and $F(x) \in T_\beta$ for all $x \in \bar{I}(t)$.

In both cases the total objects are defined in the obvious way.
If $\Phi : T \to T$ we define $I_\alpha(U, \Phi)$ as the union of the inductively defined sets $I_{\gamma,\alpha}(U, \Phi)$ for $\gamma < \alpha$ as follows:

We perform the same inductive definition as for T_α with one restriction and one suplement.
The restriction is that we do not add any element s to $I_{\gamma+1,\alpha}(U, \Phi)$ unless $\rho(s) \in T_\beta$.
The suplement is that if $\Psi \subseteq \Phi$, $s \in I_{\gamma,\alpha}(U, \Psi)$ and $\Psi(t) \in T_\beta$ for all $t \in T_\beta$ with $t \subseteq \rho(s)$, then $o(\Psi, s) \in I_{\gamma',\alpha}(U, \Phi)$ for all γ' with $\gamma < \gamma' < \alpha$.

We then let $(U, \Phi) \in T_\alpha$ if $I_\alpha(U, \Phi)$ is closed under Π and Σ and if $\Phi(t) \in T_\beta$ whenever $t \subseteq \rho(s)$, $t \in T_\beta$ and $s \in I_\alpha(U, \Phi)$. In this case $I_\alpha(U, \Phi) = \bar{I}(U, \Phi)$.
Finally we let

$$\bar{T} = \bigcup_{\alpha<\omega_1} T_\alpha$$

Remark 2 The intuition is that we add a well-formed universe to our hierarchy when we have an operator that maps well-formed types to well-formed types; at least when the operator is restricted to its own closure. The definition of totality of $o(\Psi, s)$ is as it is in order to ensure certain regularity properties discussed in the next section.

Our main result will be that this hierarchy will have the first recursively Mahlo ordinal ρ_0 as its closure ordinal. In our hierarchy there is no way to 'tell in advance' when we may form a universe from an operator, we can only do so when it is established that we have a closure of the operator. This aspect is in our view the essence of the Mahlo property, and any axiomatised version of Mahlo type theory must take this into account.

Lemma 1 $T_{\rho_0+1} = T_{\rho_0}$

Proof
We will prove this by induction on the rank of $s \in T_{\rho_0+1}$.
If s denotes a base type, the lemma is trivial, and if s denotes a dependent sum or product the lemma follows since ρ_0 is admissible.
If $s = (U, \Phi)$ where $\Phi : T \to T$ we consider the ordinal function $\hat{\Phi} : \rho_0 \to \rho_0$ defined by

$$\hat{\Phi}(\gamma) = \mu\beta[\forall s \in I_{\gamma,\rho_0}(U, \Phi)(\rho(s) \in T_\beta)].$$

Because the local universes are closed under dependent sums and products, their closure ordinals will be admissible. By the Mahlo-property then, α, the least admissible ordinal closed under $\hat{\Phi}$,will be less than ρ_0.
α will be the rank of s, so $s \in T_{\rho_0}$.

Remark 3 We might add the W-operator as one of our basic operators inside each universe. The closure ordinals of each universe would then be recursively inaccessible, without this changing the argument.

3 Digression

The hierarchy investigated in this paper is an extension of the single-valued version of the hierarchy from Normann [11]. Waagbø [19, 20] uses a similar hierarchy to give aninterpretation of basic intuitionistic type theory.

In both these papers natural equivalence relations on the set of well-formed type expressions and on the total elements of equivalent types are constructed. These equivalence relations will correspond to objects having the same extentional interpretation. In this digression we will indicate how these equivalence relations and their characterisations can be extended to the universes. We will not need this result for the characterisation of the complexity. Thus we do not give all the details. A more general treatment of the kind of results established here will be given in the forthcomming Normann [15].

In this digression it is of no importance if we consider the full hierarchy or restrict ourselves to the effective elements. All arguments will hold in both cases.

We let D be the domain where each interpretation $I(t)$ will be a subdomain of D. We will use the notation D_0, $I_0(t)$, T_0 etc. for the set of compacts in each domain.

Lemma 2 *There is a partial map $\nu : D_0 \to T_0$ such that for all $t \in T$ and $\delta \in D_0$:*

$$\text{If } \delta \in I_0(t) \text{ then } \nu(\delta) \subseteq t \text{ and } \delta \in I_0(\nu(\delta)).$$

Proof
The proof of Normann [11] can be used in all cases except for the new universe operator, and the extension to that case is simple by recursion on the possible elements of $I(U, \Phi)$, we simply collect the amount of Φ used in the formation of the compact.

Corollary 1 *For each t_1 and t_2 in T we have that*

$$I_0(t_1 \cap t_2) = I_0(t_1) \cap I_0(t_2).$$

Definition 4 Let X and Y be domains. We say that X is a *full subdomain* of Y if X is a subdomain of Y and for all $x \in X$ and $y \in Y$, if $y \subseteq x$ in the sense of Y, then $y \in X$.

Lemma 3 *Let $\Phi \subseteq \Phi_1$. Then $I(U, \Phi)$ is a full subdomain of $I(U, \Phi_1)$.*

The proof is trivial

Definition 5 **a)** If $s \subseteq t \in T$ and $x \in I(s)$ we let x^t be the minimal extension of x to an element of $I(t)$

b) If $s \subseteq t \in T$ and $y \in I(t)$, we let y_s be the restriction of y to an element of $I(s)$.

Theorem 1 **a)** *If $t \in \bar{T}$ and $t \subseteq t_1$, then*

i) $t_1 \in \bar{T}$.

ii) *If $x \in \bar{I}(t)$, then $x^{t_1} \in \bar{I}(t_1)$.*

iii) *If $y \in \bar{I}(t_1)$, then $y_t \in \bar{I}(t)$.*

b) *If t_1 and t_2 are in $\bar{T}$ and $\{t_1, t_2\}$ is bounded, then $t_1 \cap t_2 \in \bar{T}$.*

c) *If t_1 and t_2 are as above, x_1 and x_2 are total in $I(t_1)$ and $I(t_2)$ resp. and $\{x_1, x_2\}$ is bounded, then $x_1 \cap x_2$ is total in $I(t_1 \cap t_2)$.*

Proof
We prove the theorem by simultanous induction on the rank of t and the rank of the bound on t_1 and t_2. Sufficient methods are given in Normann [11] or in Waagbø [19, 20] in all cases except the formation of universes, so let us consider this case.

We first sketch the proof of a). Let $\Phi \subseteq \Phi_1$.
We prove ii) and iii) by recursion on the formation of the universes. As a consequence of iii) we obtain that the construction of $\bar{I}(U, \Phi_1)$ will be closed, so $(U, \Phi_1) \in \bar{T}$. Thus i) will hold.

In the proof of ii) and iii), the only new case is the use of the operator. For ii) this case is trivial, since $I(U, \Phi)$ is a full subdomain of $I(U, \Phi_1)$ and we actually prove that every total object in $I(U, \Phi)$ also is total as an element of $I(U, \Phi_1)$.
In order to prove iii) we need the following
Claim
Let Ψ_1 and Ψ_2 be bounded and let $s_1 \in I_\alpha(U, \Psi_1)$ and $s_2 \in I_\alpha(U, \Psi_2)$. Assume that $\{s_1, s_2\}$ is bounded. Then

$$s_1 \cap s_2 \in I_\alpha(U, \Psi_1 \cap \Psi_2)$$

Proof of claim:
We prove this for $I_{\gamma,\alpha}(U, \Psi)$ by induction on γ uniformly for all Ψ, with $\alpha = \beta + 1$.
For all cases but the use of the operator, we can use the methods from Normann [11].
So, let

$$s_1 = o(\Psi_1', t_1) \in I_{\gamma+1,\alpha}(U, \Psi_1)$$

$$s_2 = o(\Psi_2', t_2) \in I_{\gamma+1,\alpha}(U, \Psi_2)$$

where $\{\Psi_1, \Psi_2\}$ is bounded and $\{s_1, s_2\}$ is bounded.
Then the t's and the Ψ''s are bounded and we may use the induction hypothesis to obtain

$$t_1 \cap t_2 \in I_{\gamma,\alpha}(U, \Psi_1' \cap \Psi_2')$$

and thus that $\rho(t_1 \cap t_2) \in T_\beta$.
By assumption $\Psi_1'(\rho(t_1 \cap t_2)) \in T_\beta$ and $\Psi_2'(\rho(t_1 \cap t_2)) \in T_\beta$, so $(\Psi_1' \cap \Psi_2')(\rho(t_1 \cap t_2)) \in T_\beta$. It follows that

$$s_1 \cap s_2 = o(\Psi_1' \cap \Psi_2', t_1 \cap t_2) \in I_{\gamma+1,\alpha}(U, \Psi_1 \cap \Psi_2)$$

This ends the proof of the claim.

We now prove iii) in this case.

Let $t_1 = (U, \Psi_1)$, $t = (U, \Psi)$ and $y = o(\Psi_1, x)$, where $\Psi_1 \subseteq \Phi_1$.
Let $\Psi = \Phi \cap \Psi_1$. By the *claim* Ψ is total and $o(\Psi, x_{(U,\Psi)}) \subseteq y_t$. By the induction hypothesis, $o(\Psi, x_{(U,\Psi)})$ is total, so $y_t \in \bar{I}(U, \Phi_1)$. This ends the proof of a).

The proofs of b) and c) follow the same pattern and are omitted. All cases except the universe formation are as in Normann [11] and the remaining case is mainly taken care of by the *claim*. This ends the proof of the theorem.

Definition 6 We define two binary relations $\sim$ on $\bar{T}$ and $\approx$ on $\Sigma(t \in \bar{T})\bar{I}(t)$ as follows:

$O \sim O$ etc. for atomic elements of $\bar{T}$.

$(N, 17) \approx (N, 17)$ etc. for atomic total elements of atomic types.

$\Sigma(s, F) \sim \Sigma(t, G)$ if $s \sim t$ and for all $x \in \bar{I}(s)$ and all $y \in \bar{I}(t)$, if $(s, x) \approx (t, y)$, then $F(x) \sim G(y)$.
In this case $(\Sigma(s, F), (x, u)) \approx (\Sigma(t, G), (y, v))$ if $(s, x) \approx (t, y)$ and $(F(x), u) \approx (G(y), v)$.

$\Pi(s, F) \sim \Pi(t, G)$ if $s \sim t$ and for all $x \in \bar{I}(s)$ and all $y \in \bar{I}(t)$, if $(s, x) \approx (t, y)$, then $F(x) \sim G(y)$.
In this case $(\Pi(s, F), f) \approx (\Pi(t, G), g)$ if $(F(x), f(x)) \approx (G(y), g(y))$ whenever $(s, x) \approx (t, y)$.

In the case (U, Φ) we first define a relation $\hat{\approx}$ between elements of $\bar{I}(U, \Phi_1)$ and elements of $\bar{I}(U, \Phi_2)$ for arbitrary Φ_1 and Φ_2 (omitting some indices on $\hat{\approx}$) as follows:
For base types we define $\hat{\approx}$ as the identity.
We let $\Sigma(s_1, F_1) \hat{\approx} \Sigma(s_2, F_2)$ if $s_1 \hat{\approx} s_2$ and whenever x_1 and x_2 are total in $I(\rho(s_1))$ and $I(\rho(s_2))$ resp., and $(\rho(s_1), x_1) \approx (\rho(s_2), x_2)$, then $F_1(x_1) \hat{\approx} F_2(x_2)$.
We define $\hat{\approx}$ for Π-types in the analogue way.
$o(\Psi_1, s_1) \hat{\approx} o(\Psi_2, s_2)$ if $s_1 \hat{\approx} s_2$ and for all $t_1 \subseteq \rho(s_1)$ and $t_2 \subseteq \rho(s_2)$, if $t_1 \sim t_2$, then $\Psi_1(t_1) \sim \Psi_2(t_2)$.

We then let $(U, \Phi_1) \sim (U, \Phi_2)$ if for all s_1, s_2, t_1 and t_2, if $s_1 \hat{\approx} s_1$, $t_1 \subseteq \rho(s_1)$, $t_2 \subseteq \rho(s_2)$ and $t_1 \sim t_2$, then $\Phi_1(t_1) \sim \Phi_2(t_2)$.

Remark 4 The idea is to define the obvious notion of extentional equality by recursion on the inductive definition of $\bar{T}$. We will prove that these relations indeed are equivalence relations. A consequence will be that every object (function, parameterisation) will be extentional.

Theorem 2 a) *Given* $s_1, s_2 \in \bar{T}$ *we have*

$$s_1 \sim s_2 \Leftrightarrow s_1 \cap s_2 \in \bar{T}$$

b) *Given* $s_1, s_2 \in \bar{T}$, $x_1 \in \bar{I}(s_1)$ *and* $x_2 \in \bar{I}(s_2)$ *we have*

$$(s_1, x_1) \approx (s_2, x_2) \Leftrightarrow s_1 \sim s_2 \wedge x_1 \cap x_2 \in \bar{I}(s_1 \cap s_2)$$

c) $\sim$ *and* $\approx$ *are equivalence relations.*

d) *If* $t_1 \sim t_2$ *then* $\mathrm{rank}(t_1) = \mathrm{rank}(t_2)$.

Proof
We prove a) and b) by induction. c) follows as in Normann [11] and d) is proved by a simple induction.
The cases *Base type*, Σ-*type* and Π-*type* are handled as in Normann [11]. In order to handle the last case we need the following
Claim
Let $s_1 = (U, \Phi_1)$ and $s_2 = (U, \Phi_2)$ be total, and let $\hat{\approx}$ be the relation between elements of $\bar{I}(s_1)$ and $\bar{I}(s_2)$ given in the definition.
Then for each $t_1 \in I_{\gamma,\alpha}(s_1)$ and $t_2 \in I_{\gamma,\alpha}(s_2)$ we have

$$t_1 \hat{\approx} t_2 \Leftrightarrow t_1 \cap t_2 \in I_{\gamma,\alpha}(s_1 \cap s_2)$$

Proof of claim
We use induction on γ. The basic types, Σ-types and Π-types are handled as in [11].
So let $t_1 = o(\Psi_i, t_1') \in I_{\gamma+1,\alpha}(U, \Phi_1)$ and $t_2 = o(\Psi_2, t_2') \in I_{\gamma+1,\alpha}(U, \Phi_2)$.
First assume that $t_1 \hat{\approx} t_2$. Then $t_1' \hat{\approx} t_2'$ and by the induction hypothesis, $t_1' \cap t_2' \in I_{\gamma,\alpha}(\Phi_1 \cap \Phi_2)$. It follows that $t_1' \cap t_2' \in I_{\gamma,\alpha}(\Psi_1 \cap \Psi_2)$.
Let $t \subseteq \rho(t_1' \cap t_2')$ with $t \in T_\beta$ ($\alpha = \beta + 1$). Then $\Psi_1(t) \sim \Psi_2(t)$. It follows that $(\Psi_1 \cap \Psi_2)(t) \in T_\beta$, and as a consequence we obtain

$$o(\Psi_1 \cap \Psi_2, t_1' \cap t_2') \in I_{\gamma+1,\alpha}(U, \Phi_1 \cap \Phi_2).$$

Conversly, assume that

$$o(\Psi_1 \cap \Psi_2, t_1' \cap t_2') \in I_{\gamma+1,\alpha}(U, \Phi_1 \cap \Phi_2).$$

Then $t_1' \cap t_2' \in I_{\gamma,\alpha}(U, \Phi_1 \cap \Phi_2)$, so by the induction hypothesis $t_1' \hat{\approx} t_2'$.
Let $s_1' \subseteq \rho(t_1')$ and $s_2' \subseteq \rho(t_2')$ be such that $s_1' \sim s_2'$.
Then $s_1' \cap s_2' \in T_\beta$ and $\rho(t_1' \cap t_2') \in T_\beta$. These two objects are consistent, so by Theorem 1 we have that

$$s' = s_1' \cap s_2' \cap \rho(t_1' \cap t_2') \in T_\beta.$$

By assumption, $(\Psi_1 \cap \Psi_2)(s') \in T_\beta$, since $o(\Psi_1 \cap \Psi_2, t_1' \cap t_2')$ is total.
Clearly

$$(\Psi_1 \cap \Psi_2)(s') \subseteq \Psi_1(s_1') \cap \Psi_2(s_2') \in T_\beta$$

so $\Psi_1(s_1') \sim \Psi_2(t_2')$.
This shows that $o(\Psi_1, t_1') \approx o(\Psi_2, t_2')$, and the claim is proved.

We now prove a). b) will be a direct consequence of a) and the claim.

1. Let $(U, \Phi_1) \sim (U, \Phi_2)$.
   Let $t_1 \in \bar{I}(U, \Phi_1)$ and $t_2 \in \bar{I}(U, \Phi_2)$ with $t_1 \approx t_2$.
   By the claim $t_1 \cap t_2 \in \bar{I}(U, \Phi_1 \cap \Phi_2)$.
   We then argue as in the case $o(\Psi, t)$ in the proof of the claim and see that if $t \subseteq \rho(t_1 \cap t_2)$ with $t \in T_\beta$ we have that $(\Phi_1 \cap \Phi_2)(t) \in T_\beta$. This establishes $\Rightarrow$.

2. Let $(U, \Phi_1 \cap \Phi_2)$ be total.
   Let $t_1 \approx t_2$, and let $t_1' \subseteq \rho(t_1)$ and $t_2' \subseteq \rho(t_2)$ with $t_1' \sim t_2'$.
   Then, as for the corresponding case in the proof of the claim,

   $$(\Phi_1 \cap \Phi_2)(t_1' \cap t_2') \in T_\beta$$

   This establishes $\Leftarrow$.

End of proof.

4 Simulation

In section 6 we will be simulating computations in the functional S, *The Superjump*, which will be defined in section 5. This will involve the simulation of natural numbers obtained as the result of transfinite computations.

There are of course various formats that we could choose for the simulation of such computations. In Normann [14, 12] we have used the so called representations. The representation technique will require that the total objects in each type is dense, a property that we certainly do not have.

Here we will use the same method of simulation that was introduced in the unpublished [10] and reused in [13].

Definition 7 **a)** Let $t \in \bar{T}$ and let $\nu : I(t) \to \mathrm{N}_\perp$ be continuous.
We say that (t, ν) *simulates* the number n if $\bar{I}(t) \neq \emptyset$, ν is total and constant n on $\bar{I}(t)$.

b) If $t = (U, \Phi) \in \bar{T}$ and $s \in \bar{I}(t)$, we say that (s, ν) *t-simulates* n if $(\rho(s), \nu)$ simulates n.

We will in reality be working with operations on simulations, but technically we can only deal with operations on T. In this section we will see that any operator on the set of simulations can be translated to an operator on T. First we translate a simulation to an element of T.

Definition 8 a) If $t_1 = t_{tt}$ and $t_2 = t_{ff}$ are two elements in T, we let $t_1 \oplus t_2 = \Sigma(B, t_x)$

b) We let $N_0 = B$, and recursively we let $N_{k+1} = N_k \oplus B$ be elements of T.

c) Let $t \in T$ and let $\nu : I(t) \to \mathbb{N}_\bot$ be continuous.
We let $[t, \nu]$ be $\Pi(t, F)$, where $F(x) = N_{\nu(x)}$.

Lemma 4 *There is a continuous map C (for collapse) such that for any $t \in T$*

i) *$C(t)$ is of the form $[t_1, \nu]$.*

ii) *If $t = [t_1, \nu]$, then $C(t) = t$.*

Proof
If t is not a product, we just select some code for a simulation, e.g. the product of the constant B over N.
If t is a product $\Pi(t_1, F)$, we construct the parameterisation G that for each $x \in I(t_1)$ is $F(x)$ if $F(x) = N_k$ for some k, and is B if $F(x)$ is inconsistent with all N_k, undefined otherwise.

Lemma 5 *Let C be as above.*
If $t \in \bar{T}$, then $C(t) \in \bar{T}$.

The proof is left for the reader.

Definition 9 A *pre-simulation* will be a pair (t, ν), where $t \in \bar{T}$ and $\nu : I(t) \to \mathbb{N}_\bot$ is total.

It is clear that there is a bicontinuous 1-1 correspondence between pre-simulations and products of total parameterisations with values of the form N_k. From now on we will work with pre-simulations and simulations, but consider all operators constructed as operators on T, composing with C for objects t that do not directly correspond to pre-simulations.

Our next step will be to transfer all pre-simulations to simulations, not altering the simulated value in case the input already is a simulation.

Lemma 6 *There is a continuous total map ϕ from the set of pre-simulations to the set of simulations such that if (t, ν) simulates n, then $\phi(t, \nu)$ simulates n.*

Proof
Let (t, ν) be given. For x and y in $I(t)$, let
$B(x, y) = B$ if $\nu(x) = \nu(y) \in \mathbb{N}$
$B(x, y) = O$ if $\nu(x) \in \mathbb{N}$, $\nu(y) \in \mathbb{N}$, but $\nu(x) \neq \nu(y)$.
$B(x, y) = \bot_T$ otherwise.
Let $C = \Pi(x \in I(t))\Pi(y \in I(t))B(x, y)$, and let t_1 be a code for

$$(I(t) \times C) \oplus ((I(t) \times C) \to \mathbb{N}_\bot))$$

Let $\nu_1(left(z_1, z_2)) = \nu(z_1)$ and let $\nu_1(right(z)) = 0$.
It is easy to se that the left hand side contains total elements if and only if the right hand side does not. The left hand side contains total elements if and only if both C and $I(t)$ does so, and this is the case exactly when $\bar{I}(t) \neq \emptyset$ and ν is total and constant on $\bar{I}(t)$.
This ends the proof of the lemma. As a consequence of this lemma we will

operate on simulations and produce new simulations, blowing the operators up to operators from the set of total pre-simulations to the set of simulations by composing with ϕ. In the sequel we will do so without further explanation on what is really going on.

We may extend the notion of simulating a number to simulation of a function:

Definition 10 Let $f : \mathbb{N} \to \mathbb{N}$ be a function. A *simulation* of f will be a family $\{(t_n, \nu_n)\}_{n \in \mathbb{N}}$ where (t_n, ν_n) is a simulation of $f(n)$ for each n.

Remark 5 A simulation of a function will be a sequence of simulation of numbers. We have seen how these individual simulations can be represented by elements in $\bar{T}$. Thus a simulation of a function can be seen as a dependent family from $\bar{T}$. Such dependent families can again be represented as elemens in $\bar{T}$ using dependent products. In this respect we can say that elements of $\bar{T}$ may simulate a function on $\mathbb{N}$.

We will show that the functions f that can be simulated by objects in $\bar{T}$ in this respect will be exactly the functions that appears in Gödel's L before the first recursively Mahlo ordinal ρ_0. The easy inclusion is given by Lemma 1.

5 The Superjump

Our method for proving the main theorem will be by simulating computations in the type 3 functional S known as the superjump. Kleene [7] gave a general definition of computations $\{e\}(\vec{\phi}) \approx n$ where e is a natural number, and $\vec{\phi}$ is a sequence of total functionals of pure finite type. The definition is given as an inductive definition with 9 clauses, S1 - S9, and for each clause we give an index coding the clause, the signature of the arguments and the indices of the immediate subcomputations. Below we will give the definition without specifying the construction of each index. For a more complete definition, we refer to the original paper [7] or to any textbook on the subject. We omit clause S5, covering primitive recursion, since this clause can be reduced to the 8 other clauses. Here x will denote a natural number

Definition 11 The relation $\{e\}(\vec{\phi}) = n$ is inductively defined as follows:

S1 $\{e\}(x, \vec{\phi}) = x + 1$

S2 $\{e\}(\vec{\phi}) = q$

S3 $\{e\}(x, \vec{\phi}) = x$

S4 $\{e\}(\vec{\phi}) = \{e_1\}(\{e_2\}(\vec{\phi}), \vec{\phi})$

S6 $\{e\}(\vec{\phi}) = \{e_1\}(\tau(\vec{\phi}))$ where τ is a permutation.

S7 $\{e\}(x, f, \vec{\phi}) = f(x)$

S8 $\{e\}(\psi, \vec{\phi}) = \psi(\lambda\xi\{e_1\}(\xi, \psi, \vec{\phi}))$

S9 $\{e\}(x, \vec{\phi}, \vec{\psi}) = \{x\}(\vec{\phi})$

Remark 6 Non-terminating computations will be introduced via S9. In S4 we will assume that $\{e_2\}(\vec{\phi})$ terminates and gives a value y, and then that $\{e_1\}(y, \vec{\phi})$ terminates. In S8, ψ must be of type $k + 2$ and ξ will range over the total objects of type k. We will assume that $\{e_1\}(\xi, \psi, \vec{\phi})$ will terminate for all ξ.

Gandy [4] introduced the type three functional S defined as follows:

Definition 12 Let F be a total functional of type 2, and let e be a natural number.
$S(F, e) = 0$ if $\{e\}(F) = 0$, otherwise $S(e, F) = 1$.

We call S *The Superjump* since it is a jump operator for arbitrary functionals of type 2.

Harrington [6] showed that a function f is computable in S if and only if $f \in L_{\rho_0}$ where ρ_0 is the first recursively Mahlo ordinal. We will show that any function f that is computable in S can be simulated by an element of $\bar{T}$, and thereby show that every $f \in L_{\rho_0} \cap \mathbb{N}^{\mathbb{N}}$ can be simulated. The other way around follows from Lemma 1.

6 Simulating computations

The aim

In this section we will use the notation $\{e\}^S(f_1, \ldots, f_n)$, or simply the notation $\{e\}^S(\vec{f})$ for a computation relative to S. Here each f_i will either be a function or a number, which sort will be clear from the index.
Our notation for computations relative to a type two functional F will be similar, $\{d\}^F(\vec{n})$.

Uniformly in each index e for a computation $\{e\}^S(\vec{f})$ we will construct a continuous operator $\Phi_e : T^n \to T$ such that if $\vec{t} \in \bar{T}^n$ are simulations of $\vec{f}$, and $\{e\}^S(\vec{f}) = m$, then $\Phi_e(\vec{t}) \in \bar{T}$ and $\Phi_e(\vec{t})$ is a simulation of m.

We will define the operators using the fix-point theorem for domains, but we will simultanously give the induction steps needed in order to prove that our construction works.

We will use the following notational conventions: When t is an element of T representing a simulation of a number, we let the simulation be the pair (t', ν) where this notation will commute with the use of indices. Similarily, the operators Φ_e will be split into two operators ϕ_e and μ_e giving the object and the function of the simulation $\Phi_e(\vec{t})$.
We will only use this notation when it simplifies our construction.

Basic computations

There are four clauses giving the basic computations.
$\{e\}^S(x, \vec{f}) = x + 1$. If (s, ν) is a simulation for x, we use $(s, \nu + 1)$ (with the obvious meaning of $\nu + 1$) as a simulation of $x + 1$.
The cases S2 and S3 are even more trivial.
In the case S7, we let (s, ν) be a simulation of x, and $\{(s_i, \nu_i)\}_{i \in \mathbb{N}}$ be a simulation of f. Then $\phi = \Sigma(s, s_{\nu(x)})$ and $\mu(x, y) = \nu_{\nu(x)}(y)$ will be a simulation of $f(x)$.
It is trivial to show that our constructions of simulation for basic computations work.

Composition, permutation, enumeration

The three cases S4, S6 and S9 are fairly simple.

S4 $\{e\}(\vec{f}) = \{e_1\}(\{e_2\}(\vec{f}), \vec{f})$.
We assume that we have constructed Φ_{e_1} and Φ_{e_2}
We might then simply use the composition $\Phi_{e_1}(\Phi_{e_2}(\vec{t}), \vec{t})$, but it will be an advantage to code in the subsimulations more explicitly.

Let $c(n)$ be some canonical simulation of the number n. We then let

$$\phi_e(\vec{t}) = \Sigma(x \in I(\phi_{e_2}(\vec{t})))\phi_{e_1}(c(\mu_{e_2}(\vec{t})(x)), \vec{t})$$

and we let

$$\mu_e(\vec{t})(x, y) = \mu_{e_1}(c(\mu_{e_2}(\vec{t})(x)), \vec{t})(y)$$

S6 $\{e\}(\vec{f}) = \{e_1\}(\tau(\vec{f}))$. Let

$$\Phi_e(\vec{t}) = \Phi_{e_1}(\tau(\vec{t}))$$

S9 $\{e\}(e', \vec{f}, \vec{g}) = \{e'\}(\vec{f})$. Let

$$\phi_e(t, \vec{t}, \vec{r}) = \Sigma(x \in I(t'))\phi_{\nu(x)}(\vec{t})$$

$$\mu_e(x, y) = \mu_{\nu(x)}(\vec{t})(y).$$

Application

$\{e\}^S(d, \vec{f}) = S(d, \lambda g\{e_1\}^S(g, \vec{f}))$
We let $F(g) = \{e_1\}^S(g, \vec{f})$ in this section.
Recall that $S(d, F) = 0$ if $\{d\}(F) = 0$, while $S(d, F) = 1$ otherwise.
We will do the final proof by induction on the length of the computations in S, so we may assume that F is total, and consequently deduce from the induction hypothesis that if $t \in \bar{T}$ and $\vec{t}$ are simulations of $\vec{f}$, then

$$\Phi(t) = \Phi_{e_1}(t, \vec{t})$$

will also be an element of $\bar{T}$. Here Φ will depend continuously on the choice of $\vec{t}$. We then have $(U, \Phi) \in \bar{T}$.

Our first step will be to construct simulations s of $\{d\}^F(\vec{n})$ inside $\bar{I}(U, \Phi)$ uniformly in simulations for $\vec{n}$ in the same universe. By this we will mean that $\rho(s)$ is a simulation in $\bar{T}$.
For each d being an index accepting a type two functional and k numbers as inputs, we construct a continuous function $\Psi_d : (I(U, \Phi))^k \to I(U, \Phi)$ transforming a simulation $\vec{s}$ of $\vec{n}$ to a simulation of $\{d\}^F(\vec{n})$ whenever the latter terminates.
S7 does not apply here, and all cases exept S8 is handled exactly as in the major construction.
S8: $\{d\}^F(\vec{n}) = F(\lambda m\{d_1\}^F(m, \vec{n}))$
By the induction hypothesis it is trivial to construct a simulation s for $g = \lambda m\{d_1\}(m, \vec{n})$, i.e. $\rho(s)$ is a simulation of g in $\bar{T}$.

By the grand induction hypothesis and the construction of Φ, $\Phi(\rho(s))$ will be a simulation of $F(g)$. We then use $o(\Phi, s)$ as a simulation of
$F(g) = \{d\}(F, \vec{n})$ in $\bar{I}(U, \Phi)$.
This ends our construction.

The second step will be to use this to construct a type that contains total elements if and only if $\{d\}(F) = 0$. The idea is to take any element x in $\bar{I}(U, \Phi)$ and ask if x is a simulation of $\{d\}(F)$ constructed as in the first step. The reason why this will work is that $\bar{I}(U, \Phi)$ is inductively defined, and by recursion on this induction we can compare the object with the ones used to simulate the results of computations.

Lemma 7 *There is a continuous function σ defined on sequences of the form $(d, \vec{n}, s)$ where $s \in \bar{I}(U, \Phi)$ such that $\sigma(d, \vec{n}, s)$ is a simulation of 1 if s can be extended to a simulation of $\{d\}^F(\vec{n})$ obtained from a simulation in $\bar{I}(U, \Phi)$ for $\vec{n}$, and $\sigma(d, \vec{n}, s)$ is a simulation of 0 otherwise.*

Proof
We define σ by the 7 cases corresponding to S1-4, S6, S8 and S9, but the proof that σ fullfills the lemma will be by induction on the rank of s in $\bar{I}(U, \Phi)$.
We will not give all the details. In section 4 we established how the use of quantifiers or boolean combinations can be transfered to continuous operations on simulations. Thus when we in the construction below 'check' something, we mean that we construct a simulation of the truth value of the statement.

We now give a scetch of the argument, noticing that the constructions will be by recursion on the inductive definition of $\bar{I}(U, \Phi)$.
S1: $\{d\}^F(x, \vec{n}) = x + 1$
We simply have to check if s is a simulation of $x + 1$.
S2 and S3 are equally simple.
S4: $\{d\}^F(\vec{n}) = \{d_1\}^F(\{d_2\}^F(\vec{n}), \vec{n})$
In this case, s' has to be a sum $\Sigma(s_1', F)$ where s_1 is a simulation of $\{d_2\}^F(\vec{n})$ and for each total x, $F(x)$ is a simulation of the appropriate $\{d_1\}^F(m, \vec{n})$. This can be checked.
In the case S8, $\{d\}^F(\vec{n}) = F(\lambda m\{d_1\}^F(m, \vec{n}))$, we must have that
$s = o(\Phi', s_1)$ for some $\Phi' \subseteq \Phi$, where s_1 can be extended to a simulation of $\lambda m\{d_1\}^F(m, \vec{n})$. By the induction hypothesis, this can be checked.
The other cases are also easy, we can first check if s is locally of the correct form, and then use the induction hypothesis to check if the subtypes are the simulations we want them to be.
This ends the proof.

The third stage will be to use this to construct a simulation of

$$\{e\}^S(d, \vec{f}) = S(d, \lambda g\{e_1\}^S(g, \vec{f}))$$

in the case S8. But we simply have to ask if there is any element s of $\bar{I}(U, \Phi)$ such that $\sigma(d, \cdot, s)$ is a simulation of 1 and s is a simulation of 0. If this is the case, the simulation we construct should be a simulation of 0, otherwise it should be a simulation of 1. We have already developed standard techniques to do this inside $\bar{T}$.

The Main Theorem

We are now ready to state our main result:

Theorem 3 *We use the notation from the paper.*
The first recursively Mahlo ordinal ρ_0 is the least ordinal α such that $T_{\alpha+1} = T_\alpha$.

Proof
By lemma 1, ρ_0 is an upper bound for this least α.
On the other hand it is easy to see that $T_\beta \in L_{\omega+\beta+1}$ for all β, and if f has a simulation in T_β, then $f \in L_{\omega+\beta+\omega}$.
If f is computable in S, then by our main construction, f can be simulated in $\bar{T}$. By Harrington [6] ρ_0 is the least α such that $f \in L_\alpha$ whenever f is computable in S. The theorem follows.

Remark 7 In this proof we have used Harrington's result because it was available. The advantage was that we would never have to worry if the operators Φ considered in the construction actually defined universes. If we were to use some other system for the first recursively Mahlo, say viewing S as a monotone partial operator defined on any d and partial F containing at least enough information to determine $S(d, F)$, we would have to be more careful.

The use of Harrington's result indicates that we might form a Mahlo-hierarchy in an impredicative way; it is sufficient for our purpose to construct universes from operators that will remain total at the end of the construction.

References

[1] Berger, U. *Density theorems for the domains-with-totality semantics for dependent types*, to appear in the proceedings from the workshop *Domains 2*, TU-Braunschweig May 1996.

[2] Berger, U. *Continuous functionals of dependent and transfinite types*, Habilitationsschrift, München 1997.

[3] Berger, U. *Continuous functionals of dependent and transfinite types*, This volume.

[4] Gandy, R. O. *General recursive functionals of finite type and hierarchies of functions*, A paper given at the Symposium on Mathematical Logic held at the University of Clermond-Ferrand, June 1962.

[5] Griffor, E.R. and Rathjen, M. *The Strength of some Martin Löf Type Theories*, Arch. Math. Logic 33, 347-385 (1994).

[6] Harrington, L. *The superjump and the first recursively Mahlo ordinal*, in Fenstad and Hinman (eds.) Generalized Recursion Theory, North-Holland (1974) 43-52.

[7] Kleene, S.C. *Recursion in functionals and quantifiers of finite types I*, T.A.M.S. 91, (1959) 1-52.

[8] Kristiansen, L. and Normann, D *Semantics for some type constructors of type theory*, in Behara, Fritsch and Lintz (eds.) Symposia Gaussiana, Conf A, Walter de Gruyter & Co. (1995), 201-224.

[9] Kristiansen, L. and Normann, D. *Total objects in inductively defined types*, Archive for Mathematical Logic 36 (1997) 405-436.

[10] Normann, D. *Wellfounded and non-wellfounded types of continuous functionals*, Oslo Preprint Series in Mathematics No 6 (1992).

[11] Normann, D. *A hierarchy of domains with totality, but without density*, in Cooper, Slaman and Wainer (eds.) Computability, Enumerability, Unsolvability, Cambridge University Press (1996) 233-257.

[12] Normann, D. *Representation theorems for transfinite computability and definability*, Oslo Preprint Series in Mathematics No. 20 (1996).

[13] Normann, D. *Hereditarily effective typestreams* , Archive for Mathematical Logic 36 (1997) 219 - 225.

[14] Normann, D. *Closing the gap between the continuous functions and recursion in* ${}^3E$, Archive for Mathematical Logic 36 (1997), 269-287.

[15] Normann, D. *Categories of domains with totality*, Oslo Preprint Series in Mathematics No. 4 (1997).

[16] Palmgren, E. and Stoltenberg-Hansen, V. *Domain interpretations of Martin-Löf's partial type theory*, Annals of Pure and Applied Logic 48 (1990) 135-196

[17] Setzer, A. *Proof theoretical strength of Martin-Löf Type Theory with W-type and one Universe*,Thesis, Ludwig-Maximilians-Universität München (1993).

[18] Stoltenberg-Hansen, V., Lindström, I. and Griffor, E.R. *Mathematical Theory of Domains*, Cambridge University Press (1994).

[19] Waagbø, G. *Domains-with-totaloty semantics for intuitionistic type theory*, Disertation, Oslo 1997.

[20] Waagbø, G. *Denotational semantics for Intuitionistic Type Theory using a Hierarchy of Domains with Totality*, To appear in Archives for Mathematical Logic.

Logic and Decision Making

David E. Over
University of Sunderland

Abstract

There are largely separate psychological literatures on logical reasoning and on decision making. This division has limited psychological theories in both areas, and particularly held up the study of ordinary deductive reasoning. People do not ordinarily reason from a restricted set of arbitrary assumptions taken, in effect, to be certainly true. Much more often, they try to perform inferences from all their relevant beliefs, holding few of these with absolute confidence, or they perform inferences from statements made to them, and treat few of these as absolutely reliable. They recognise that their premises have some degree of probability or uncertainty, and they consequently have more or less confidence in them. This affects their confidence in, and probability judgements about, their conclusions. People also face the problem of getting evidence to assess how probable or uncertain their premises are, and this calls for further probability judgements, as well as utility judgements and decision making. Much better understanding of people's deductive reasoning in realistic contexts will be achieved by integrating research on it with research on decision making and probability and utility judgments.

Introduction

There has been extensive psychological research on people's ability at deductive reasoning (Evans, Newstead, Byrne, 1993), but this field has been largely separate from psychological research on human probability judgements and decision making (Baron, 1994). There have been some attempts recently to integrate these fields (Evans and Over, 1996a), but the division between them has limited psychological theories in both, and has been especially harmful to the study of ordinary logical reasoning.

The standard approach in psychology to the study of deductive reasoning has been to ask participants in experiments to assume the truth of arbitrary premises outside of the context of any realistic problem. The participants have then been asked to indicate whether or not some given conclusion validly follows from these premises, or alternatively expected to generate their own valid conclusions. Until recently, not enough attention was paid to how the participants interpreted the nature of the experimental task or the premises themselves. One possibility which has been almost totally ignored is that participants may naturally view an experimental task as one in which they are to rely, at least to some extent, on judgements of probability or uncertainty, or to make a decision combining probability and utility judgements. I shall focus on this possibility in this paper, and aim to show that we cannot correctly describe people's deductive ability without taking into account their probability judgements and their decision making.

Suppose we hear a weather forecaster predict that there will be rain if the wind changes direction. When we notice later that the wind has changed direction, we can perform the valid inference Modus Ponens to infer that there will be rain, using the forecaster's conditional statement as the major premise. But clearly to make an important decision on this basis, such as whether to irrigate a plot of land, we cannot simply assume that the forecaster's statement is true, nor take it as certain. This prediction is more or less probably true, and can also be said to be uncertain to some degree, where its degree of uncertainty is 1 minus its probability. Given some uncertainty in the prediction, how do we come to have some degree of confidence in the conclusion of Modus Ponens? How do we make a judgement about the probability that it will rain? Such questions have been almost totally ignored by the psychologists who have studied human deductive reasoning, and the main psychological theories of this reasoning give us no answers to them. The result is that these theories have limited application to people's actual reasoning in the real world, and tell us nothing about how this reasoning is integrated with their decision making. Clearly, people's finite cognitive abilities alone imply that they will have some tendency to be illogical and incoherent, especially in experiments presenting abstract problems distant from ordinary affairs. There is evidence, however, that people's performance improves when they are given realistic problems requiring both valid inference and good probability judgements or decision making. (See Manktelow and Over, 1993, for essays on the debate in psychology and philosophy about human rationality, and Evans and Over, 1996a, for an attempt to resolve it.)

Uncertain premises

It is a commonplace observation that we have to live with a great deal of uncertainty in the real world. Psychological research, however, has mainly concerned inference from premises which are supposed to be assumed true, though we often have to perform valid inferences in ordinary affairs from statements or beliefs we do not have absolute confidence in, as they are not certainly true. Ordinary people can hardly be expected to know the technical distinction, often made in philosophical logic, between a valid inference and a sound inference, i.e. a valid one with true premises. (As Politzer and Braine, 1991, point out, more psychologists themselves should take account of this distinction.) The possibility has always to be considered that participants in an experiment are refusing to accept a valid conclusion because they lack confidence in the premises. In technical terms, they are unsure that the inference is sound because they find the premises too uncertain, i.e. with a probability too far from 1.

Consider experiments on so-called belief bias. (These are reviewed by Evans et al., 1993, and their significance discu ssed in Evans and Over, 1996a, 1997a.) People have some tendency not to accept obviously unbelievable conclusions of valid inferences, even if they have been asked to assume the premises. That is, they tend not to endorse quite so many valid unbelievable conclusions as valid believable conclusions. This tendency is even present for very easy valid inferences, a fact that some psychologists have found hard to explain in their theories without ad hoc devices (as in Oakhill, Johnson-Laird, Garnham, 1989). But clearly, if people find a valid conclusion unbelievable, they have grounds for being suspicious of a premise or the conjunction of the premises. The fact that many people may express this suspicion by claiming that the conclusion is not valid does not necessary mean that they have no concept of logical validity. It may only mean that they do not use the word 'valid' in precisely the same way a logician or cognitive psychologist does. Indeed, there is evidence that people do have some concept of what it is to make assumptions and to perform valid inferences from these, even when they are uncertain. Validity itself has a big effect when conclusions are unbelievable. That is, people do endorse far more valid unbelievable conclusions than invalid unbelievable conclusions.

Belief bias can be dramatically reduced by experimental instructions which particularly stress that the premises are to be assumed true and only conclusions endorsed which necessarily follow from these premises (Newstead, Pollard, Evans, Allen, 1992, Experiment 5). This supports the view that people have some deductive competence, and some concept of what it is to make assumptions and perform valid inferences from them. (See Evans and Over, 1997a, for a general argument that people have a modest degree of deductive competence.) However, much more is needed than this ability in ordinary affairs, where so little is certain, and there is indeed evidence that people are highly sensitive to uncertainty in the premises of a valid inference. They take account of uncertainty in their beliefs as premises (George, 1995), and will indicate low confidence in the valid conclusion of even such a basic inference as Modus Ponens, when given the major and minor premises as statements in a dialogue (Stevenson and Over, 1995, Experiment 2). Suggested uncertainty in a possible dialogue is also the crucial factor, in my view, in cases like the following.

If John goes fishing, he will have a fish supper.
John goes fishing.
Therefore, John will have a fish supper.

If John goes fishing, he will have a fish supper.
If John catches a fish, he will have a fish supper.
John goes fishing.
Therefore, John will have a fish supper.

Almost all participants in an experiment will endorse the conclusion of the first inference: Modus Ponens is accepted by as high a proportion of people as one can get in any psychological experiment on reasoning. But Modus Ponens will be suppressed in the second inference: far few people hold that the conclusion validly follows there. Byrne (1989) was the first to run experiments of this type, but she explained the results in terms of the peculiarities of one version of mental models theory (Johnson-Laird and Byrne, 1991). My view is quite different. Think of a natural circumstance in which one speaker would assert the second conditional after another speaker had asserted the first. The pragmatic point of the second conditional assertion would normally be to cast doubt on the first premise. (See Grice, 1989, on such pragmatic suggestions in conversations.) Support for this view comes from experiments in which the degree of doubt in the first, major premise is manipulated, by varying the conditional probability that John is lucky given that he goes fishing. This systematically affects the extent to which participants hold that the conclusion of Modus Ponens validly follows and also their degree of confidence in it (Stevenson and Over, 1995).

More evidence comes from experiments using examples like this:

Professor of Medicine: If Bill has typhoid he will make a good recovery.
First year Student: If Bill has malaria he will make a good recovery.
Bill has typhoid.

Now the second conditional premise above has little effect on people's willingness to perform Modus Ponens and be confident in its conclusion. But there is a significant effect if the Profession of Medicine asserts the second conditional premise after the student asserts the first (Over and Stevenson,

1996). People thus take sensible account in their reasoning of the reliability of the speaker who asserts a premise.

It could again be countered that people are making mistakes in these experiments. They should really distinguish sharply between an invalid argument and one which valid but unsound or probably unsound, i.e. with clearly false or uncertain premises. We can all agree that this should be so, but that does not mean that people are seriously illogical in ordinary affairs. Of course, people are not going to be perfect in the way they respond to uncertainty in their reasoning, whatever the source of the uncertainty, but as yet we know little about how they do respond to it. This is simply the result of the fact that experiments on reasoning have almost always required the participants to make inferences from assumptions, and not from their actual beliefs or statements in a realistic dialogue.

Those psychologists who restrict themselves to questions about valid inference from arbitrary assumptions perhaps think they can avoid difficult questions about the normative standard for correct response. Moving beyond the examples they generally focus on, they would not find it so easy to presuppose that this standard is given by elementary extensional logic. De Finetti (1972, Introduction) famously argued that this logic is limited by covering only '...the ideal case of absolute certainty.' His Bayesian interpretation of probability theory has been called a logic of partial beliefs, expressed in subjective probability judgements. This includes the special case in which the premises are certain, e.g. when they are the axioms of number theory. But the more general normative position is clear for the Bayesian. The probability of the conclusion of a valid inference, whether or not its premises are certain, should not be strictly less than the probability of the conjunction of its premises. Equivalently, the degree of uncertainty of the conclusion should not strictly exceed the sum of the uncertainties of the premises. People's actual probability judgements do not, to repeat, always conform to probability theory, not even to the principle that the probability of a proposition cannot be strictly higher than what it logically implies, for they will sometimes judge a conjunction to be more probable than one of its conjuncts (Tversky and Kahneman, 1983). However, this happens only in special cases, and has never been found to hold when people actually infer the conjunct from the conjunction.

In Bayesian probability theory, the precise probability of the conclusion of an inference with uncertain premises is given by the theorem on total probability, where C is the conclusion and P is the conjunction of the premises:

$$(\text{TT})\ \text{Prob(C)} = \text{Prob(P)Prob(C/P)} + \text{Prob(notP)Prob(C/notP)}$$

In ordinary affairs, TT is often impossible to conform to. For a valid inference, Prob(C/P) is of course 1, and people's modest deductive competence may enable them to grasp this, certainly if the inference is a basic as Modus Ponens. But we may have no idea how to work out Prob(P), as the premises may not be independent of each other, and we may even be unable to make probability judgements about the individual premises. If we do have confidence in, say, the conditional statement of an authority like a professor of medicine, we may find it hard to express this judgement in numerical terms. Last but not least, Prob(C/notP) may be impossible to make a non-arbitrary judgement about. One could try to avoid such limitations on TT with an alternative normative theory, such as that of Shafer (1976), in which the left-hand side of TT could be assigned a higher value than the right-hand side. A range of normative views on uncertain reasoning have been proposed, and some of these have found applications in the study of expert systems and artificial intelligence (Pearl and Shafer, 1990). But George (1997) points out that most of these are psychologically implausible, if they are suggested as descriptive accounts of people's ordinary judgements of probability and uncertainty.

The descriptive support theory of Tversky and Koehler (1994) would allow that right-hand side of TT to be greater than the left-hand side. People may give too low a value for the valid conclusion of an argument with uncertain premises because they do not reflect that the conclusion could still be true even if some of the premises are false. They may only grasp this possibility when it is made explicit in some way, as in the right-hand side of TT. The problem is that support theory was originally intended to explain people's general probability judgements and decision making, and no attempt has yet been made to apply it specifically to the study of valid reasoning from uncertain premises. Much more important, however, has been the failure of most psychologists studying valid reasoning to pay any attention to uncertainty in the premises.

The psychology of the conditional

The greatest limitation in current psychological theories of logical inference is their failure to account adequately for ordinary conditionals in natural language, especially when these are more or less probable or uncertain. Let us consider first some influential proposals in psychology about mental models, using the conditional:

If this match is struck, then it will light

According to Johnson-Laird and Byrne (1991), people will construct what they call a mental model for this conditional that has the following form.

[s] l

...

The first row above, as in a truth table, explicitly represents a state of affairs in which the match is struck and it lights. The square brackets are said by Johnson-Laird and Byrne to indicate that the striking of the match is exhaustively represented - there is no model in which the match is struck but does not light - and the dots to represent implicit models in which the match is not struck. Given as well the minor premise of Modus Ponens, stating that the match is struck, people are held to remove the dots, leaving them with only the explicit model, and in that the conclusion follows, i.e. the match lights.

The general use of square brackets in Johnson-Laird and Byrne (1991) is unclear in some respects (Hodges, 1993), but their specific use, and that of the dots, in simple conditional inferences makes them appear no more than notational variants of the conditional itself. In effect, Johnson-Laird and Byrne propose a kind of Modus Ponens rule, allowing us to infer the existence of a state of affairs in which both the minor premise and the conclusion are true. At this point, in other words, the proposal gives really us little more than a mental version of Modus Ponens. Johnson-Laird, Byrne, and Schaeken (1994) introduce three levels of mental modelling that people can go through, with accompanying so-called mental footnotes doing the work of the square brackets and dots. In these levels, the mental models people are said to construct are not what mathematical logicians call models. One way to look at what is supposed to happen in the mind, according to Johnson-Laird et al., is that an ordinary conditional is finally represented by the disjunctive normal form for the truth-functional, material conditional. Another way to look at this mental process is that, at the most explicit level, the three rows of the truth table which make the material conditional true are displayed. Either way, an ordinary conditional of the form 'if p then q' is finally made logically equivalent to 'notp or q'. The notorious problems of this equivalence are compounded in this proposal. Take a case of strengthening the antecedent:

If this match is struck, then it will light.
If this match is immersed in water and struck, then it will light.

This inference is, of course, valid for the material conditional but intuitively invalid for the ordinary conditional, a fact that the psychology of reasoning ought to explain (Cummins, 1997). Strengthening the antecedent is just one of a host of valid inferences for the material conditional which are intuitively invalid for the ordinary conditional. In spite of these problems, some philosophers and logicians have argued that the ordinary indicative conditional is the material conditional, but that pragmatic factors make many valid inferences for it appear invalid. (See Edgington, 1995, and Woods,

1997, for a survey of these arguments.) Johnson-Laird has a view of this general type. (See especially Johnson-Laird, 1995). Yet he does not model any such pragmatic factors. If he is right about how people understand ordinary conditionals and perform inferences from them, strengthening the antecedent should appear obviously valid to everyone in any pragmatic context (Over and Evans, 1997a). People should find it easy to add a term i for the immersion of the match to the first row of the above mental model, where the match is exhaustively represented as being struck and also as lighting. Johnson-Laird has yet to give an account of deduction from uncertain premises. His theory is about deduction from assumptions, and not from beliefs or statements which are judged to have some probability less than 1. With his account of the ordinary conditional, he would have to hold that its probability should be judged to be the probability of the material conditional, but this is just intuitively wrong. Suppose the match is so thoroughly soaked in water that no one wants to waste time striking it, making it obviously highly probable that it will not be struck. Johnson-Laird's theory implies in this case that people with fully explicit mental models will judge it highly probable that, if the match is struck, then it will light. A material conditional is highly probable if the negation of its antecedent is highly probable, but there is no evidence that people will make such apparently absurd probability judgements about the ordinary conditional, even if they are encouraged to construct what Johnson-Laird considers fully explicit mental models for it.

Another influential psychological theory of the conditional is that of Braine and O'Brien (1991). They did not restrict, for the limited processing powers of human beings, the rows of truth tables to yield mental models, but rather adapted natural deduction rules to give them what they think of as a mental logic for reasoning. They put constraints on the rule of conditional proof to try to avoid the problems of treating the ordinary conditional as the truth-functional conditional. But these constraints have a factual component, so that for them validity is determined partly by matters of fact and not purely by logical form. About the above example of strengthening the antecedent, for instance, they remark casually that it is invalid in their system unless the match happens to be of a special type that lights when it is wet (p. 190). They basically confuse specifying valid inference for the conditional in natural language with a psychological process account of how people find out whether such a conditional is in fact true (Over and Evans, 1997b). They also fail to give any logical semantics for their conditional: they do not specify in which states of affairs it is true and in which it is false. Partly for this reason, it is obscure what it would mean for their conditional to have a degree of probability or uncertainty (Evans and Over, 1997b).

In ordinary affairs, people would often express, in probability judgements, more or less confidence that, if the match is struck, it will light, depending, say, on whether they thought it likely that the match had got wet. For Braine and O'Brien the probability of this conditional is not that of a material conditional asserting that the match will not be struck or it will light. This is as much we know about probability and their conditional. They do not tell us how we can make a probability judgement about their conditional, nor how such a judgement would affect, for example, our confidence in the conclusion of Modus Ponens with this conditional as a major premise.

Stalnaker (1968, 1984) developed a proper intensional logic for the ordinary conditional with an analysis in terms of possible worlds, which are more or less similar to the actual world. A conditional is true in the actual world, by this analysis, if its consequent is true in the most similar possible world (to the actual world) in which its antecedent is true. Strengthening the antecedent is not valid for this intensional conditional: it can turn out that a match lights in the most similar world in which it is struck, but does not light in the most similar one in which it is soaked in water and struck. This analysis has been given a natural deduction formulation, and Rips (1994) has suggested basing a mental logic on that. But Stalnaker's theory is itself highly controversial, and there is a serious debate about the relation between it and probability judgements (Edgington, 1995; Woods, 1997). As is clear from this debate, there should be some close relation between the ordinary conditional and conditional probability judgements, and there is some empirical evidence that people do make this connection, though much more work needs to be done on this (Stevenson and Over, 1995). Note also that strengthening the antecedent is a monotonic aspect of extensional logic, and some researchers in expert systems and AI have developed default and nonmonotonic logics for reasoning in the presence of uncertainty. Some of these logics can be related to a Stalnaker-type analysis of the ordinary conditional (Lehman and Magidor, 1992; Brewka, Dix, Konolige, 1997). However, there has been little psychological investigation of whether such logics can help us to understanding ordinary reasoning from uncertain premises.

Integrating a psychological theory of conditional reasoning with one of uncertainty and decision making will require much more than the relatively simple modifications of extensional logic that are now on offer from most psychologists. Nevertheless, there is already a lively research area which should tell us something important about how people try to find out whether a conditional is more or less probable or uncertain, and we now move on to this.

Investigating conditionals

We know little as yet about how people reason from uncertain premises in general and uncertain conditional premises in particular. Of course, the rationality of people's inferences depends not just on how they do this, but also on whether they make good judgements about which conditional premises are more or less acceptable, or more or less probable or uncertain, in the first place. The question is whether people investigate conditional hypotheses in a rational way. Now this has been the topic of one of the most widely used experimental paradigms in the study of reasoning: the Wason selection task (Wason, 1968).

Participants in the original version of the selection task, which we shall call an abstract indicative task, were presented with an indicative conditional of the form 'if p then q', like the following.

If there an A on one side of a card, there is a 4 on the other side.

They were told that this claim was about four cards in front of them with only one side visible. There was a p card, a notp card, a q card, and a notq card, so in our example these cards might respectively show A, T, 4, 7. The participants were also told that their object was to select just those cards that, when turned over, might reveal whether the conditional was true or false. Given that the claim is supposed to be about only the four cards displayed, the correct answers are the p card and the notq card, or in our example, the A card and the 7 card. Most participants in an abstract indicative task do not choose the p card and the notq card. Most select the p card alone (Evans, et al., 1993). They thus appeared irrational to the early experimenters on this type of task, who took it as an example of how Modus Tollens is not always endorsed as a valid inference. Their presupposition was that participants should assume that the conditional is true, and then use it as the major premise of Modus Tollens to infer validly that there will be notp on the other side of the notq card. The experimenters presupposed further that the conditional would be true without qualification if q was on the other side of the p card and notp on the other side of the notq card, false without qualification if notq was on the other side of the p card or p on the other side of the notq card, and that no other card was at all relevant. However, most ordinary conditionals are about more items than four and cannot usually be found simply true or false by examining only two of these under ideal, or nearly ideal, conditions. In a more realistic case, a sample of items almost always has to be taken, yielding some degree of probability or uncertainty for the conditional.

Another basic point is that realistic cases are not all of the same logical type. Sometimes the question is not about the truth, or probable truth, of an indicative conditional to do with some matter of fact, such as a relation between letters and numbers, but rather about whether someone has violated a deontic rule for guiding or regulating human action. Psychologists took some time to recognise that there are deontic selection tasks, and that these are logically different from both abstract and realistic indicative selection tasks. Cheng and Holyoak (1985) took an important step in this direction by extending the work of Johnson-Laird, Legrenzi, and Legrenzi (1972) on the following conditional.

If a letter is sealed, it must have a 20 cent stamp on it.

This is a deontic conditional, containing the deontic modal 'must' and expressing an obligation. It states a proper rule for guiding behaviour, and does not purport, as an indicative conditional would, to describe some supposed matter of fact. The object of a deontic selection task is find out if someone has violated a deontic rule, which is just given as a true or actual regulation. Participants were presented with the above deontic conditional and four envelopes: one showing it was sealed, one that was unsealed, one that it had 20 cent stamp on it, and one showing that it had a lower value stamp on it. Their task was to act as postal workers and to select just those cards that might show that the deontic conditional had been violated by someone. Those who were familiar with postal rules or regulations like this one tended to choose the sealed envelope, i.e. the equivalent of the p card, and the envelope with the low value stamp on it, i.e. the equivalent of the notq card. Even participants who were unfamiliar with such regulations made these choices if they were told that the point of the rule was to increase profit from personal mail. Note again how this deontic task differs from an indicative task. In the later, a card with p on one side and notq on the other would falsify an indicative conditional, but here a sealed envelope with a low value stamp on it does not falsify the deontic conditional, if that even makes sense. What it really indicates is that someone has violated a regulation which is presupposed to be a true or proper one.

The first psychologists to study deontic tasks did mistakenly think of them as having the same logical form as indicative tasks. What they thought they needed to explain was how tasks of the same logical form could elicit such different responses from the participants. They concluded that people do not have a mental logic that they apply to all problems of the same logical form, but rather have pragmatic schemes (Cheng and Holyoak, 1985) or specialised so-called Darwinian algorithms (Cosmides, 1989) for finding the right answer to questions with a certain kind of factual content. However, deontic tasks can only be fully understood as problems in decision making, calling for probability and utility judgements. Showing why this is so, in the next section, will help us to return to indicative tasks to see how they too, in spite of their different logical form, require probability and utility judgements.

Decision making in deontic tasks

Rips (1994) has suggested that there could be a mental deontic logic that people apply in deontic selection tasks. But existing deontic logics specify valid inference forms for propositions containing deontic modal connectives, particularly ones for 'it is obligatory that' and 'it is permissible that', and the concept of someone's committing a violation of a deontic rule does not figure in them. Many logics of this type have been proposed, and perhaps the least controversial theorem in any of them is that a proposition is obligatory if it is a logical truth (Chellas, 1980). Theorems like this can tell us nothing about how to find violators in problems like deontic selection tasks, which are much more like exercises in decision making than in deontic logic. For these tasks, one must use the concept of expected utility (Manktelow and Over, 1991; Over and Manktelow, 1993). In our example, one should turn over the sealed envelope - the p 'card' - and the envelope with the low value stamp on it - the notq 'card' - because only these choices give one the chance of finding a violator of the deontic postal rule. That outcome is of value if one has the job of catching violators, wishes to punish them to prevent future violations, or just has a dislike of letting people get away with breaking rules. Sometimes the correct solution, i.e. the best choice by decision theoretic standards, is not the selection of the equivalent of the p and notq cards. Consider this conditional permission statement made by a mother to her son:

If you tidy your room, you may go out to play.

Participants in a deontic selection task can be cued to the mother's perspective on this deontic conditional, and then they select the card showing that the room has not been tidied - the notp case - and the card showing the son has gone out to play - the q case (Manktelow and Over, 1991). Clearly, neither extensional nor deontic logic implies that these are the cards the mother should select. It takes a decision theoretic analysis to rationalise these choices as the ones which give her a chance of getting a benefit, by finding whether there is a case in which the room has not been tidied but her son has gone out to play.

Deontic selection tasks work equally well if the rule is prudential in nature, e.g. 'if you clean up spilt blood, you must wear rubber gloves' (Manktelow and Over, 1990). Here one gets utility by uncovering violators who may be spreading dangerous diseases. First of all the rule makes sense as deontic regulation. It will be asserted because cleaning up spilt blood and wearing rubber gloves has higher expected utility than cleaning up spilt blood and not wearing rubber gloves. Suppose there are cards indicating on one side whether someone has or has not cleaned up spilt blood, and on the other side, whether someone has or has not worn rubber gloves. Turning over a card with 'has cleaned up spilt blood' or 'has not used rubber gloves' could reveal a case of unsafe practice, while not turning it over could only save some small amount of time and energy. Turing over a card with 'has not cleaned up spilt blood' or 'has worn rubber gloves' is simply a waste of time and energy.

Kirby (1994) showed how to express these points intuitively using concepts from signal detection theory. Let us call turning over a card and finding a violator a 'hit', and turning it over and not finding one a 'false alarm'. Not turning over a card recording a violation on the other side is then a 'miss', and not turning it over when it does not do that a 'correct rejection'. Let Prob(hit) be the probability that one will make a hit by turning over a card, Util(hit) be the utility of doing that, and so on for the other possibilities. When for any card

Prob(hit)Util(hit) + Prob(false alarm)Util(false alarm)

is greater than

Prob(miss)Util(miss) + Prob(correct rejection)Util(correct rejection)

one should decide to turn the card over. Participants in deontic selection tasks do tend to conform this decision rule, when it is to their benefit to make hits, avoid misses, and not to waste effort on useless cards.

Consider the following immigration rule (from Cheng and Holyoak, 1985) about cards recording, on one side whether a traveller is entering the country or in transit, and on the other, the inoculations that traveller has.

If 'entering' is on one side, 'cholera' must be on the other side.

Manktelow, Sutherland, and Over (1995), in a new type of selection task, told participants that cholera is common is tropical countries, and presented them with many cards, indicating which country the traveller is coming from on one side and the inoculations on the other. The participants had a greater tendency to choose cards with, say, 'entering from India' on them than 'entering from Iceland'. A simple application of detachment for the immigration rule, the equivalent of Modus Ponens for deontic conditionals, should have led them to choose all 'entering' cards. But they, or at least a significant number of them, conformed to the above decision procedure instead. Notice that it is not that the probability of a hit is low for a card with 'entering from Iceland' on it, but rather the (expected) utility of such a hit is low. Even if Icelandic violators enter the country, they are unlikely to be carrying cholera, and so no benefit would compensate for the trouble of catching and stopping them.

Deontic reasoning is technically nonmonotonic and indeed easily affected by extra information that alters probabilities or utilities. Other experiments also illustrate this (Manktelow, Kilpatrick, and Over, 1996). Participants will rank, on a scale from 1 to 10, the serious of a violation of a rule, indicating, say, whether breaking a speed limit is aggravated by poor visibility or mitigated by the urgent need to take someone to hospital. To go with this, an inference tasks like the following can be devised.

If a car driver travels above 30 mph in a built-up area, s/he is liable for a fine.

A car driver travels above 30 mph in a built up area and is stopped by the police.

What follows?

In different manipulations of this task, information is added about aggravating or mitigating circumstances. Participants are asked to choose one of ten conclusions, ranging from in strength from a statement the driver 'must be fined' to 'may be fined' and on to 'must not be fined'. The strength of the conclusion parallels the serious of the violation, being highly sensitive to the presence or absence of the extra information about aggravating or mitigating circumstances.

Cognitive psychology does not yet have much to offer as a theory of such nonmonotonic deontic reasoning, and yet it is remarkable how well people do, by obvious intuitive standards, in all these experiments on deontic conditionals. The strong evidence linking reasoning with these conditionals to probability and utility has yet to be developed into anything like a proper account of their logic and semantics, or of judgements of the seriousness of their violations. (See Manktelow and Over, 1995, for an attempt to make a start on this project.) An analogous problem deeply affects the psychological study of ordinary indicative conditionals. The uncertainty of these conditionals should affect, and does affect, the inferences that people perform from them. But psychology has made little progress in understanding how people represent the content of these conditionals or their degree of uncertainty.

Some progress should be made by studying more realistic non-deontic, indicative selection tasks. The original abstract task presented participants with an unrealistic indicative conditional, to do with numbers and letters, in a most unusual context: the instructions stipulated that this conditional was a claim only about the four cards. In spite of these instructions, participants might apply procedures to these tasks that are much more appropriate in real-world settings. Consider as a realistic example the conditional hypothesis used to state Hempel's paradox (Evans and Over, 1996a; Nickerson, 1996) in the philosophy of science:

(H) If a bird is a raven, then it is black.

Let us suppose that this hypothesis H does not seem very probable to us: it has some fairly high degree of uncertainty. We will not then have a great deal of confidence in the conclusion of a Modus Ponens inference with H as the major premise. However, we can try to get evidence which may confirm H, i.e. make H more probable or less uncertain, so that we can have more confidence in using H as the major premise of Modus Ponens. Alternatively, our evidence may disconfirm H, and then perhaps we can forget about using

H as a premise in inferences.

How then should we gather evidence about H? The equivalent of choosing the p card when investigating this conditional is to examine a raven to see whether it is black, and the equivalent of choosing the notq card is to examine a non-black thing to see whether it is a raven. But in this natural realistic context, it is clearly wrong to do the latter. The set of non-black things is huge and heterogeneous, and by trying to search it, we are most unlikely to find a counterexample to H even if one exists (cf. Kirby, 1994). The most efficient way to investigate H is by observing ravens alone, and then one should be able to tell whether it is probably true or probably false. A realistic indicative selection task is a good experimental representation of making decisions which may confirm or disconfirm an actual conditional hypothesis. This is indeed decision making, and so is to that extent like a deontic task, but a realistic indicative task has the point of investigating the probable truth of a conditional, and not of finding possible violations of a deontic rule which is just given as true.

Epistemic utility

The psychologists who first developed the selection task presupposed a Popperian philosophy of science (Popper, 1959), according to which it was only legitimate to try to falsify a hypothesis. Trying to confirm it - to get evidence that it is probably true - was supposed to be a mistake, but this puritanical position has been heavily criticised from a Bayesian point of view (Howson and Urbach, 1993). Taking confirmation into account, we can perhaps achieve a deeper understanding of what participants are doing in indicative selection tasks, particularly ones with a realistic content in which a sample of items has to made to find out whether a conditional is probably true or probably false. Oaksford and Chater (1994) gave a kind of Bayesian analysis of a wide range of selection tasks. Their analysis raises many questions, but these could lead to better psychological theories of the ordinary conditional.

Consider a realistic conditional, like H above about ravens. As already noted, we may begin by judging this hypothesis H to have a rather low prior probability. But imagine that a great deal of data about individual birds has been put on cards, on one side of which the species of an individual bird is given and on the other side its colour. Now we can gather evidence about H by turning over some of these cards. For instance, we might turn over a p card - indicating that some raven has been examined - to find a q on the other side - indicating that that raven was black. Call this a pq piece of evidence, which, in fact, is not the same as turning over a q card to find p on the other side, a qp piece of evidence (Over and Jessop, in press). Bayes's theorem from probability theory tells us how to calculate Prob(H/pq), i.e. the posterior probability of the conditional hypothesis H given this possible outcome evidence. We can do this provided we can make a judgement about Prob(H), i.e. the prior probability of the conditional, and Prob(pq/H) and Prob(pq/notH), i.e. the probability, respectively, of the evidence given the conditional and of the evidence given the negation of the conditional. Since p is visible on the p card, Prob(pq/H) is the probability that a q is on the other of this card given the conditional, and Prob(pq/notH) the probability that q is there given its negation. Participants should grasp that Prob(pq/H) = 1 if they have any logical ability. Oaksford and Chater made a number of other assumptions that help them predict what people will do in a selection task. One is that, in ordinary affairs, the p and q sets tend to be relatively small, as the set of ravens is relative to the set of all physical things, or even to the set of all birds. Another assumption they make is that notH is represented by people as a state of affairs in which q is independent of p. With these and other assumptions, Oaksford and Chater were able to calculate the posterior probability of H that would result from turning over any of the cards and thereby getting a particular piece of evidence.

The participants in a standard selection task do not turn over any cards, but they might have expectations about what they would find in the way of confirmation if they did. For example, they may expect the p card to provide more confirmation or disconfirmation than any other card. For after turning it over, one gets either a pq or a pnotq outcome, and by Bayes's theorem, the former piece of evidence highly confirms the conditional hypothesis and the latter even more highly disconfirms it. The participants may rationally judge that no other card would be so valuable, no matter what is on the other side of it. Now to express such a judgment formally, we need some normative measure of the expected epistemic utility of a choice, i.e. of how far one expects turning over a card to confirm or disconfirm H. For this purpose, Oaksford and Chater used expected uncertainty reduction. We have identified above the uncertainty of a proposition with 1 minus its probability, but Oaksford and Chater are thinking of the degree of uncertainty of someone's state of mind. They measure the initial degree of this uncertainty with the following formula from information theory.

(1) - Prob(H)logProb(H) - Prob(notH)logProb(notH)

The highest level of this uncertainty for us is clearly when we judge H as likely to be true as false, i.e. Prob(H) = 0.5. At this level, we would have no confidence in using H as the major premise of Modus Ponens: H would be too improbable to give us any confidence in the conclusion. But from this level, this uncertainty would decrease after a p card was turned over to reveal some evidence in the form, say, of a pq outcome. The new level of this uncertainty is given by the following.

(2) - Prob(H/pq)logProb(H/pq) - Prob(notH/pq)logProb(notH/pq)

Subtracting (1) from (2) is a measure of uncertainty reduction. In this instance, the result will be negative, of course, but reversing the sign gives Oaksford and Chater a measure of information gain in, say, bits. The information gain from a pnotq outcome can be derived in the same way, and then the expected information gain can be defined for the p card.

Oaksford and Chater demonstrated that, when the p and q set sizes are small, the p card has the greatest expected information gain, followed by the q card, the notq card, and finally the notp card. In this Bayesian analysis, the participants have a good reason for not selecting the not very informative notq card. More than this, Oaksford and Chater tried to explain the frequency with which all the cards are chosen in standard indicative selection tasks, as p is chosen most frequently in these, followed by q, then notq, and finally notp. However, they make a large number of assumptions to do this. Some of these can certainly be questioned, and there are some experimental results that arguably go against their full analysis (Evans and Over, 1996a,b; Laming, 1996; Almog and Sloman, 1996; Oaksford and Chater, 1996; Green, Over, Pyne, 1997).

There is also the normative question of whether we should always value reducing uncertainty as our epistemic goal. Uncertainty can increase after we get some evidence in some cases. For instance, the conditional hypothesis may strike us initially so implausible that we begin by assigning it some low probability, taking Prob(H) as 0.1 and putting us in a low state of uncertainty about it. Then we may get some strong evidence for it and Prob(H) becomes 0.5 for us. Now by Oaksford and Chater's definitions we have, in effect, paid an epistemic cost by gaining uncertainty and losing information. But we would actually be incoherent if we did not think our evidence had given us an epistemic benefit by taking us closer to the truth. Suppose we go on to gather more evidence and end up concluding that Prob(H) = 0.9. By Oaksford and Chater's definitions, we are in the same state of uncertainty as when we initially felt Prob(H) = 0.1. Yet it is surely wrong to say that we have made no epistemic progress after doing so much to confirm H.

There are other technical measures of epistemic utility that do not have these unfortunate implications. Oaksford and Chater (1996) modified their analysis by introducing one of these that gives the same expected value as their first measure. However, they continued simply to presuppose that so many bits of information are always of equal value to people, and this is an aspect of their approach that is not fully Bayesian. Bits of information, however defined, will sometimes have more subjective value to people and sometimes less, depending on their general epistemic utilities and how these are related to other utilities. No one is willing pay any amount of money to get any number of bits of information about any hypothesis. As we all know, a strong argument has to made for any grant award, no matter how small! Even for a subjectively interesting hypothesis, people may have a greater preference for finding disconfirmation of it if they believe it to be probably false than if they believe it to be probably true (cf. Evans and Over, 1996a,b). Individual people will assign, in effect, different utilities to making correct or incorrect assertions about some matter. They may, for example, derive more or less satisfaction from being right or wrong about ravens, depending on such facts as whether they are professional ornithologists or not.

Oaksford and Chater held that their particular analysis is one of 'optimal data selection selection', but Klauer (1997) has pointed out that it is not 'optimal' in at least one statistical sense of this word. Instead of the utilities of investigating a hypothesis, he refers to certain losses one might suffer when doing this. There is the cost of asserting H when H is false and that of asserting notH when H is true, and there are the costs, in time, energy, or money, of experimenting or gathering data. From the expected costs, one can calculate the overall risk of some procedure for investigating H. Klauer argued that any such procedure that is properly optimal should minimise the overall risk, or at least be asymptotically optimal by becoming in the long run as good as any procedure that does minimise the overall risk, but Oaksford and Chater's procedure does not satisfy this criterion (Chernoff, 1972).

Perhaps that criterion is not Oaksford and Chater's. They rather vaguely wrote of people's search for data as optimally adapted to the environment, and so they may have had in mind optimality as defined by evolutionary theory. But they have no argument that maximising expected uncertainty reduction would be adaptive in the strict biological sense. Relating that construct to reproductive success under primitive conditions is certainly not trivial. It is also difficult to believe that ordinary people could efficiently use Oaksford and Chater's equations, even if much of this work were done implicitly in some connectionist system in the brain. What is at least required is a plausible and easy to apply heuristic that has some justification.

Consider the likelihood ratio, which appears in one form of Bayes's theorem and is another possible normative measure of epistemic utility, in the sense of diagnosticity (Evans and Over, 1996a). An example of this ratio is Prob(pq/H)/Prob(pq/notH), which indicates how far a pq piece of evidence discriminates between H and notH, the diagnosticity of pq. Where H is still the ravens hypothesis, pq is rb, an observation of raven that is black. To find rb highly diagnostic, i.e. a good piece of evidence that H is true, we do not even have to assign very precise numbers to Prob(rb/H) and Prob(rb/notH), or do much more than simply noting that the former is much higher than the latter. Using such differences in probability to estimate the value of evidence is an easy heuristic that is highly correlated with diagnosticity as measured by the likelihood ratio (Slowiaczek, Klayman, Sherman, Skov, 1992). We can also be helped by our knowledge of the environment. For example, we might know that birds of a species are of the same colour, or at most males and females differ, and then for us Prob(rb/notH) will be especially low. (See Nisbett, Krantz, Jepson, Kunda, 1983, on common knowledge about species.)

Causal conditionals

Bayesian, or more generally decision theoretic, analyses of problems like the selection task are most likely to get support when these are realistic cases about an environment with an intuitive structure. There are grounds for thinking that many causal conditionals may be of this type (Green and Over, 1997; Over and Jessop, in press). There are theoretical problems about these conditionals, such as whether they strictly assert the existence of a causal relation or just pragmatically convey that. But they are certainly justified by appealing to a supposed causal relation, and our degree of confidence in any of them depends on our belief about the strength of this relation. Compare the following instances of Modus Ponens (from Cummins, Lubart, Alksnis, Rist, 1991).

If my finger is cut, then it bleeds.
My finger is cut.
Therefore, it bleeds.

If I eat candy often, then I have cavities.
I eat candy often.
Therefore, I have cavities.

People tend to have more confidence in the conclusion of the first inference above than in the conclusion of the second. This difference is correlated with the number of disabling conditions people can list for the conditionals: they can think of few reasons why a cut finger would not bleed but many why eating candy often might not lead to cavities (Cummins et al., 1991). Knowledge of such disabling conditions affects how probable or uncertain we judge the major premise to be, with a resulting influence on our confidence in the conclusion. We may have high confidence in an instance of a causal conditional with few disabling conditions, and so be happy to use it as a major premise of Modus Ponens. We may rightly take a different attitude to an instance of a causal conditional with many disabling conditions.

There are also good grounds, normative and descriptive, for holding that the strength of a claimed causal relation depends on how much greater Prob(e/c) is than Prob(e/notc), where e is a statement of the effect and c of the supposed cause (Cheng and Novick, 1992). We have relatively low confidence in what is conveyed by the second major premise above because we judge Prob(I get cavities/I eat candy often) to be not much higher than

Prob(I get cavities/I do not eat candy often). This judgement is the result of the many disabling conditions we can think of, such as going to the dentist often, and other correlated factors, such as the number of people we know who do not get cavities though they have a taste for candy or sweets. Another important fact about people's causal judgement is that they tend to show more interest in cases in which the suppose cause is followed or not followed by the effect, ce or cnote sequences, than in cases in which the lack of the cause is followed or not followed by the effect, notce or notcnote sequences. At first, any such tendency might seem irrational, as knowledge of both types of case may appear necessary for estimating the difference between Prob(e/c) and Prob(e/notc). However, Anderson and Sheu (1995) pointed out how to justify the tendency on Bayesian grounds, given that c and e are uncommon states of affairs, as they tend to be in most realistic examples.

The point can perhaps best be made using a causal selection task (Green and Over, 1997; Over and Jessop, in press). Suppose we are interested in the following causal conditional about a well in a village we are visiting, conveying that a causal relation exists between drinking from the well and getting cholera.

(R) If we drink from the well, we will get cholera.

Assume as a first type of example that few people drink from the well and few get cholera. Then whether or not the well drinking causes cholera, i.e. whether or not R holds, most people who do not drink from the well will not have cholera, and most people who do not have cholera will not drink from the well. In the more abstract terms of a corresponding selection task, most notp cards will have notq on the other side, and most notq cards will have notp on the other side, whether R holds or not. We would expect to find mostly notpnotq and notqnotp sequences, and Prob(notpnotq/R) is little different from Prob(notpnotq/notR), and similarly for Prob(notqnotp/R) and Prob(notqnotp/notR). Thus doing the equivalent of turning over notp cards and notq cards, examining non-well drinkers or non-cholera victims, will have little epistemic utility and tend to waste our time. On the other hand, Prob(pq/R) will be significantly greater than Prob(pq/notR), and Prob(qp/R) greater than Prob(qp/notR). Doing the equivalent of turning over p cards and q cards, examining well drinkers and cholera victims, will consequently have high epistemic utility for us, greatly affecting our confidence in the conditional.

There is a complete switch in a second type of example, when most people drink from the well and most get cholera. Here we expect most well drinkers to have cholera, and most cholera sufferers to be well drinkers, whether or not R holds. In other words, we will expect, in this new example, pq and qp sequences, while Prob(pq/R) and Prob(pq/notR), and Prob(qp/R) and Prob(qp/notR), will be little different from each other. What has come to have high epistemic utility for us is the equivalent of turning over notp cards and notq cards, examining non-well drinkers and non-cholera victims. There are now significant differences between Prob(notpnotq/R) and Prob(notpnotq/notR), and between Prob(notqnotp/R) and Prob(notqnotp/notR). Therefore, the best decision has become to select notp cards and notq cards.

Participants in causal selection tasks conform to these points, by tending to select p cards and q cards in the first type of example, but notp cards and notq cards in the second type of example (Green and Over, 1998). Of course, a full decision theoretic analysis of investigating a causal hypothesis should take account of its judged importance, e.g. whether it concerns a serious disease like cholera or a minor one, and of the costs of searching for evidence. People also show themselves to be sensitive to the manipulation of these factors (Green and Over, 1997). Causal tasks are then one of the best places to study how people make decisions to assess the probability or uncertainty of ordinary conditionals of a significant type. After they have made this assessment, their response to these conditionals as the premises of valid inferences can also be profitably studied.

Conclusion

Psychological research on deductive reasoning has almost totally concentrated on inference from premises that are just to be assumed true. However, most ordinary deductive reasoning has to take account of some degree of uncertainty in the premises. The result is that little is known about how people respond to such uncertainty, or about how they assess its degree in the first place. Realistic selection tasks are relevant to the latter question, but this has not been recognised by most psychologists. Gaining a better account of ordinary deductive reasoning calls for much more than the modifications of extensional logic found in current theories of mental models and mental logic. It requires, to begin with, the integration of research on valid inference with that on decision making and probability and utility judgements. More specifically, it is necessary to achieve a much better understanding of how people assess and perform inferences from ordinary conditionals: indicative, deontic, causal, and others. Intensional and nonmontonic logics are potentially highly relevant to this goal, and that is another reason why psychologists who are studying ordinary reasoning will have to revise their assumptions, or at least find some of them uncertain.

Acknowledgement

I should like to thank Wilfred Hodges, Jeff Paris, and Rosemary Stevenson for helpful comments on a draft of this paper.

References

[1] Almog, A. and Sloman, S.A. (1996). Is deontic reasoning special? Psychological Review, 103, 374–380.

[2] Anderson, J.R. and Sheu, C.-F. (1995). Causal inferences as perceptual judgments. Memory and Cognition, 23, 510–524.

[3] Baron, J. (1994). Thinking and deciding (2nd ed.). Cambridge: Cambridge University Press.

[4] Braine, M.D.S. and O'Brien, D.P. (1991). A theory of If: A lexical entry, reasoning program, and pragmatic principles. Psychological Review, 98, 182–203.

[5] Brewka, G., Dix, J. and Konolige, K. (1997). Nonmonotonic reasoning: An overview. Stanford: CSLI Publications.

[6] Byrne, R.M.J. (1989). Suppressing valid inferences with conditionals.

[7] Cheng, P.W. and Holyoak, K.J. (1985). Pragmatic reasoning schemas. Cognitive Psychology, 17, 391–416.

[8] Cheng, P.W. and Novick, L.R. (1992). Covariation in natural causal induction. Psychological Review, 99, 365–382.

[9] Cosmides, L. (1989). The logic of social exchange: Has natural selection shaped how humans reason? Studies in Wason's selection task. Cognition, 31, 187–276.

[10] Chellas, B.F. (1980). Modal logic: An introduction. Cambridge: Cambridge Cambridge University Press.

[11] Chernoff, H. (1972). Sequential analysis and optimal design. Philadelphia: Society for Industrial and Applied Mathematics.

[12] Cummins, D.D. (1997). Rationality: Biological, psychological, and normative theories. Current Psychology of Cognition, 16, 78–86.

[13] Cummins, D.D., Lubart, T., Alknis, O. and Rist, R. (1991). Conditional reasoning and causation. Memory and Cognition, 19, 274–282.

[14] De Finetti, B. (1972). Probability, induction, and statistics. New York: John Wiley.

[15] Edgington, D. (1995). On conditionals. Mind, 104, 235–329. Evans, J.St.B.T., Newstead, S.E., Byrne, R.M.J. (1993). Human reasoning: The psychology of deduction. Hove: Erlbaum.

[16] Evans, J.St.B.T. and Over, D.E. (1996a). Rationality and reasoning. Hove: Psychology Press.

[17] Evans, J.St.B.T. and Over, D.E. (1996b). Rationality in the selection task: Epistemic utility and uncertainty reduction. Psychological Review, 103, 356–363.

[18] Evans, J.St.B.T. and Over, D.E. (1997a). Rationality in reasoning: The problem of deductive competence. Current Psychology of Cognition, 16, 3–38.

[19] Evans, J.St.B.T. and Over, D.E. (1997b). Reply to Barrouillet and Howson. Current Psychology of Cognition, 16, 399–405.

[20] George, C. (1995). The endorsement of the premises: Assumption-based or belief-based reasoning. British Journal of Psychology, 86, 93–113.

[21] George, C. (1997). Reasoning from uncertain premises. Thinking and Reasoning, 3, 161–189.

[22] Green, D.W., Over, D.E. and Pyne, R.A. (1997). Probability and Choice in the selection task. Thinking and Reasoning, 3, 209–236.

[23] Green, D.W. and Over, D.E. (1997). Causal inference, contingency tables, and the selection task. Current Psychology of Cognition, 16, 459–487.

[24] Green, D.W. and Over, D.E. (1998). Decision theoretic effects in testing a causal conditional. Unpublished manuscript, University College London.

[25] Grice, H.P. (1989). Studies in the way of words. Cambridge, Mass.: Harvard University Press.

[26] Hodges, W. (1993). The logical content of theories of deduction. Behavioral and Brain Sciences, 16, 353–354.

[27] Howson, C. and Urbach, P.M. (1993). Scientific reasoning: The Bayesian approach. (Second Edition). Chicago: Open Court.

[28] Johnson-Laird, P.N. (1995). Inference and mental models. In S.E. Newstead and J.St.B.T. Evans, Perspectives on thinking and reasoning: Essays in honour of Peter Wason. Hove: Erlbaum.

[29] Johnson-Laird, P.N. and Byrne, R.M.J. (1991). Deduction. Hove: Erlbaum.

[30] Johnson-Laird, P.N., Byrne, R.M.J. and Schaeken, W. (1994). Why models rather than rules give a better account of propositional reasoning: A reply to Bonatti and to O'Brien, Braine, Yang. Psychological Review, 101, 734–739.

[31] Johnson-Laird, P.N., Legrenzi, P. and Legrenzi, M.S. (1972). Reasoning and a sense of reality. British Journal of Psychology, 63, 395–400.

[32] Kirby, K.N. (1994). Probabilities and utilities of fictional outcomes in Wason's four card selection task. Cognition, 51, 1–28.

[33] Klauer, K.C. (1997). On the normative justification for information gain in Wason's selection task. Unpublished manuscript, University of Bonn.

[34] Laming, D. (1996). On the Analysis of irrational data selection: A critique of Oaksford and Chater (1994). Psychological Review, 103, 364–373.

[35] Lehman, D. and Magidor, M. (1992). What does a conditional knowledge base entail? Artificial Intelligence, 55, 1–60.

[36] Manktelow, K.I., Kilpatrick, S.A. and Over, D.E. (1996). Pragmatic effects on deontic inference. Presented at The Third International Conference on Thinking, University College London, 29-31 August.

[37] Manktelow, K.I. and Over, D.E. (1990). Deontic thought and the selection task. In K.J. Gilhooly, M. Keane, R.H. Logie and G. Erdos (Eds), Lines of Thinking, Vol. 1, Chichester: Wily.

[38] Manktelow, K.I. and Over, D.E. (1991). Social roles and utilities in reasoning with deontic conditionals. Cognition, 39, 85–105.

[39] Manktelow, K.I. and Over, D.E. (Eds). (1993). Rationality. London: Routledge.

[40] Manktelow, K.I. and Over, D.E. (1995). Deontic reasoning. In S.E. Newstead and J.St.B.T. Evans (Eds), Perspectives on thinking and reasoning. Hove: Erlbaum.

[41] Manktelow, K.I., Sutherland and E.J., Over, D.E. (1995). Probabilistic factors in deontic reasoning. Thinking and Reasoning, 1, 201–220.

[42] Newstead, S.E., Pollard, P., Evans, J.St.B.T. and Allen, J. (1992). The source of belief bias in syllogistic reasoning. Cognition, 45, 257–284.

[43] Nickerson, R.S. (1996). Hempel's paradox and Wason's selection task: Logical and psychological puzzles of confirmation. Thinking and Reasoning, 2, 1–32.

[44] Nisbett, R.E., Krantz, D.H., Jepson, D.H. and Kunda, Z. (1983). The use of statistical heuristics in everyday inductive reasoning. Psychological Review, 90, 339–363.

[45] Oaksford, M. and Chater, N. (1994). A rational analysis of the selection task as optimal data selection. Psychological Review, 101, 608–631.

[46] Oaksford, M. and Chater, N. (1996). Rational explanation of the selection task. Psychological Review, 103, 381–391.

[47] Oakhill, J., Johnson-Laird, P.N. and Garnham, A. (1989). Believability and syllogistic reasoning. Cognition, 31, 117–140.

[48] Over, D.E. and Evans, St.B.T. (1997a). Two cheers for deductive competence. Current Psychology of Cognition, 16, 255–278.

[49] Over, D.E. and Evans, St.B.T. (1997b). The nature of mental logic: A reply to O'Brien. Current Psychology of Cognition, 16, 823–830.

[50] Over, D.E. and Jessop, A. (in press). Rational analysis of causal conditionals and the selection task. In M. Oaksford and N. Chater (Eds), Rational models of cognition. Oxford: Oxford University Press.

[51] Over, D.E. and Manktelow, K.I. (1993). Rationality, utility and deontic reasoning. In K.I. Manktelow and D.E. Over (Eds), Rationality. London: Routledge.

[52] Over, D.E. and Stevenson, R.J. (1996). Uncertainty, reasoning, and decision making. Presented at The Third International Conference on Thinking, University College London, 29–31 August.

[53] Politzer, G. and Braine, M.D.S. (1991). Responses to inconsistent premises cannot count as suppression of valid inferences. Cognition, 38, 103–108.

[54] Popper, K.R. (1959). The logic of scientific discovery. London: Hutchinson.

[55] Rips, L.J. (1994). The psychology of proof. Cambridge, Mass: MIT Press.

[56] Rumain, B., Connell, J. and Braine, M.D.S. (1983). Conversational comprehension processes are responsible for reasoning fallacies in children as well as adults. Developmental Psychology. 19, 471–481.

[57] Shafer, G. (1976). A mathematical theory of evidence. Princeton: Princeton University Press.

[58] Shafer, G. and Pearle, J. (Eds) (1990). Readings in uncertain reasoning. San Mateo, CA: Morgan Kaufman.

[59] Slowiaczek, L.M., Klayman, J., Sherman, S.J. and Skov, R.B. (1992). Information selection and use in hypothesis testing: What is a good question, and what is a good answer? Memory and Cognition, 20, 392–405.

[60] Stalnaker, R. (1968). A theory of conditionals. In N. Rescher (Ed), Studies in logical theory. Oxford: Blackwell.

[61] Stalnaker, R. (1884). Inquiry. Cambridge, Mass: MIT Press. Stevenson, R.J. and Over, D.E. (1995). Deduction from uncertain premises. Quarterly Journal of Experimental Psychology, 48A, 613–643.

[62] Tversky, A. and Kahneman, D. (1983). Extensional vs intuitive reasoning: The conjunction fallacy in probability judgement. Psychological Review, 90, 293–315.

[63] Tversky, A. and Koehler, D.J. (1994). Support theory: A nonextensional representation of subjective probability. Psychological Review, 101, 547–567.

[64] Wason, P.C. (1968). Reasoning about a rule. Quarterly Journal of Experimental Psychology, 20, 273–281.

[65] Woods, M. (1997). Conditionals. Oxford: Clarendon Press.

The Sheaf of Locally Definable Scalars Over a Ring

Mike Prest
Department of Mathematics
University of Manchester
Manchester M13 9PL
UK

Abstract

We introduce some geometric aspects of the model theory of modules and show that for many rings this geometry is rather wild in that it embeds every affine variety.

1 Two topologies

Let R be any ring and let $\text{Mod} - R$ denote the category of right R-modules. To this category is associated a pair of topological spaces which live on a common underlying set. The set is the set of isomorphism classes of indecomposable pure-injective (=algebraically compact) right R-modules. The topologies are the right Ziegler spectrum, Zg_R, of R [17] and the right Gabriel-Zariski spectrum, Zar_R, of R [10], [15]. Ziegler's topology was introduced in his investigation [17] of the model theory of modules: roughly, its closed subsets correspond to complete theories of modules. The Gabriel-Zariski topology on this set was introduced in [10] as the dual of the Ziegler topology: the compact Ziegler-open sets are taken as a basis of closed sets for the Gabriel-Zariski topology (I will recall the relation to the usual Zariski spectrum of a commutative ring below).

In this paper we describe a sheaf of rings (originally introduced in [11]) over Zar_R which generalises the structure sheaf of a commutative noetherian ring. The stalks of this sheaf are rings of definable scalars of certain modules (see [5], [2]) and may be regarded as "localisations" of R.

2 Rings of definable scalars

We recall, rather briefly, some definitions and facts that we will need. We give model-theoretic definitions: for equivalent definitions phrased in terms of the category of functors from mod $-R$, the category of finitely presented R-modules, to **Ab**, the category of abelian groups, see [6] for example. A basis of open sets of Zar_R is given by those sets of the form $[\phi/\psi] = \{N \in \mathrm{Zar}_R : \phi(N) = \psi(N)\}$ where $\phi \geq \psi$ range over pairs of pp formulas in the language of R- modules. The complement of $[\phi/\psi]$ is denoted (ϕ/ψ) and these sets form a basis of open sets of the Ziegler topology. The space Zar_R need be neither T_0 nor compact [10]. To any module M is associated the Ziegler-closed set, $\mathrm{supp}M = \bigcap\{[\phi/\psi] : \phi(M) = \psi(M)\}$ - the **support** of M. To every Ziegler-closed set C (in particular, to every basic Zariski-open set) is associated the corresponding ring of definable scalars R_C which is the ring of functions which are definable (in the language of R-modules) uniformly on all modules with support contained in C (there are more algebraic descriptions - see [2]). The ring R_M of definable scalars of a module M is defined to be $R_{\mathbf{supp}M}$.

3 The presheaf of definable scalars

For background on (pre)sheaves we refer to [16], [4]. In particular, we formally regard a presheaf of objects from the category $\mathcal{C}$, over a given topological space, to be a contravariant functor from the collection of open subsets, regarded as a category (with the inclusion maps as the morphisms), to $\mathcal{C}$.

We construct a presheaf of rings over Zar_R. In fact we define it initially on a basis of open sets but this is enough to yield a presheaf.

Let $\mathcal{U}$ denote the category of open sets in Zar_R. As remarked above there is a basis of open sets of the form $[\sigma/\tau]$ where $\sigma \geq \tau$ are pp formulas: we let $\mathcal{U}_0$ denote this basis (regarded as a full subcategory of $\mathcal{U}$). Define the "presheaf on a basis" $\mathrm{Def}_R : \mathcal{U}_0 \longrightarrow R-\mathrm{Alg}$, where $R-\mathrm{Alg}$ denotes the category of R-algebras, by sending $[\sigma/\tau]$ to $R_{[\sigma/\tau]}$ and by taking the containment map $[\sigma/\tau] \supseteq [\sigma'/\tau']$ to the canonical restriction map from $R_{[\sigma/\tau]}$ to $R_{[\sigma'/\tau']}$. The latter map is defined as follows: let $r' \in R_{[\sigma/\tau]}$ and choose $\rho = \rho(x, y)$ to be any pp formula which defines the function r' on (modules supported on) the set $[\sigma/\tau]$. Then $res_{[\sigma/\tau],[\sigma'/\tau']}(r')$ is the function defined by ρ (on modules supported on $[\sigma'/\tau']$).

We will later construct the associated sheaf, for which construction this data suffices, but we remark that we may immediately extend Def_R to a presheaf (i.e. a functor) from $\mathcal{U}$. The standard way of doing this is to assign to $U \in \mathcal{U}$, the ring $\lim_{\leftarrow}\{R_{[\sigma/\tau]} : [\sigma/\tau] \subseteq U\}$ and to use the natural restriction maps. We prefer here to make a somewhat more conservative extension to arbitrary open sets and to set, for $U \in \mathcal{U}$, $\mathrm{Def}_R(U) = \lim_{\rightarrow}\{R_{[\sigma/\tau]} : U \subseteq [\sigma/\tau]\} = R_U$ say and, when $U \supseteq V$, set the restriction map to be that induced from R_U to R_V by the canonical maps $R_{[\sigma/\tau]} \longrightarrow R_V$ for $[\sigma/\tau] \supseteq U$. Now, $\bigcap\{[\sigma/\tau] : U \subseteq [\sigma/\tau]\}$ is simply the Ziegler-closure $\mathrm{Zg}-\mathrm{cl}(U)$ of U and $\lim_{\rightarrow}\{R_{[\sigma/\tau]} : U \subseteq [\sigma/\tau]\}$ is easily seen to be $R_{\mathrm{Zg-cl}(U)}$ (cf. the computation of the stalks in 3.1 below). So our presheaf Def_R is given by taking a Zariski-open set U to the ring of definable scalars of the Ziegler-closure of U. It is easy to check that this presheaf is a sub-presheaf (proper in general) of that obtained by using the standard definition above (of course, both will have the same sheafification).

We show that the stalk at any point is just the ring of definable scalars of that point.

Proposition 3.1 *The stalk of* Def_R *at the point* $N \in \mathrm{Zar}_R$ *is the ring of definable scalars* R_N *at* N.

Proof. Recall that the stalk at N is defined to be $(\mathrm{Def}_R)_N = \lim_{\rightarrow}\{R_{[\sigma/\tau]} : N \in [\sigma/\tau]\}$. For each basic neighbourhood $[\sigma/\tau]$ of N we have a canonical map from $R_{[\sigma/\tau]}$ to R_N (which takes a function defined by, say, ρ to the function defined by ρ on N) and this system of morphisms is coherent. Therefore we obtain a corresponding map from the direct limit $h : (\mathrm{Def}_R)_N \longrightarrow R_N$. We show that this is an isomorphism.

Let $\rho(x,y)$ define a member r' of R_N. The sentence which says that ρ is total and functional may be written in the form $\sigma \leftrightarrow \tau$ for suitable pp formulas σ,τ. Thus we obtain a basic (Zariski-)open set, $[\sigma/\tau]$, to which N belongs and an element $r'' \in R_{[\sigma/\tau]}$ which is the scalar defined by ρ. Then r' is the image under h of the image $g_{[\sigma/\tau]}(r'')$ of r'' under the canonical morphism $g_{[\sigma/\tau]} : R_{[\sigma/\tau]} \longrightarrow (\mathrm{Def}_R)_N$.

Next suppose $r' \in (\mathrm{Def}_R)_N$ with $hr' = 0$. Choose $[\sigma/\tau]$ and $r'' \in R_{[\sigma/\tau]}$ with $r' = g_{[\sigma/\tau]}(r'')$. Suppose that r'' is defined by the pp formula ρ on $[\sigma/\tau]$. Then on N, ρ defines hr', which is the zero map. Thus $N \in [\exists x\rho(x,y)/y = 0]$. Let $[\sigma'/\tau']$ be a basic open set containing N and contained in both $[\sigma/\tau]$ and $[\exists x\rho(x,y)/y = 0]$: then ρ defines the zero function on $R_{[\sigma'/\tau']}$ and hence $r' = g_{[\sigma'/\tau']}.res_{[\sigma/\tau],[\sigma'/\tau']}(r'') = g_{[\sigma'/\tau']}0 = 0$, as required. □

4 The sheaf of locally definable scalars

We denote by LocDef_R the sheafification of Def_R, the construction of which we now recall. Let LDef_R denote the stalk space of Def_R: the disjoint union $\bigcup\{R_N : N \in \mathrm{Zar}_R\}$ topologised as follows.

Given a basic Zariski-open set $[\sigma/\tau]$ and given $r \in R_{[\sigma/\tau]}$ we define the local section $\hat{r} : [\sigma/\tau] \longrightarrow \mathrm{LDef}_R$ by $\hat{r}(N) = r_N$ where r_N is the image of r under the natural map $R_{[\sigma/\tau]} \longrightarrow R_N$. We take the $\hat{r}[\sigma/\tau]$, as $[\sigma/\tau]$ and r vary, as the basis for the topology on LDef_R. Note that it is a basis since, given $[\sigma/\tau]$, $[\sigma'/\tau']$, $r \in R_{[\sigma/\tau]}$ and $r' \in R_{[\sigma'/\tau']}$ and N such that $r_N = r'_N$ we may take pp formulas ρ_r and $\rho_{r'}$ defining r, respectively r' on $[\sigma/\tau]$, respectively $[\sigma'/\tau']$ and then $\hat{r}$ and $\hat{r}'$ agree on the neighbourhood $[\sigma/\tau] \cap [\sigma'/\tau'] \cap [\exists xyy'(\rho_r(x,y) \wedge \rho_{r'}(x,y') \wedge z = y - y')/z = 0]$ of N. So $\hat{r}[\sigma/\tau] \cap \hat{r}'[\sigma'/\tau']$ is a union of sets of this form.

With this topology on Def_R the projection map π from LDef_R to Zar_R is a local homeomorphism. We then define the sheaf LocDef_R - the **sheaf of locally definable scalars** - to be the functor which takes an open subset U of Zar_R to $\Gamma\mathrm{LDef}_R(U) = \{\sigma : U \longrightarrow \mathrm{LDef}_R : \sigma \text{ is continuous and } \pi\sigma = id_U\}$ - the ring (with pointwise operations) of continuous sections of π over U. Clearly $\mathrm{LocDef}_R = \Gamma\mathrm{LDef}_R$ is a sheaf of R-algebras. Let ν denote the canonical (and obvious: use $r \mapsto \hat{r}$) morphism of sheaves from any presheaf F to its sheafification ΓLF.

The term "locally definable" is explained by the following.

Lemma 4.1 *Given an open subset U of Zar_R and any element r' of $\mathrm{LocDef}_R(U)$ there is a cover $\{U_\lambda\}_\lambda$ of U by (basic) open sets such that, for each λ there is a pp formula ρ_λ such that, for all λ,μ and every point $N \in U_\lambda \cap U_\mu$ the functions ("scalars") defined by ρ_λ and ρ_μ agree on N and such that r' is the section defined by the collection $\{\rho_\lambda\}_\lambda$.*

Proof. This follows directly from the construction and comments above. □

Since basic Zariski-open sets are Ziegler closed we obtain the following description of a (for example) global "locally definable scalar". Let $\{C_\lambda\}_\lambda$ be a set of basic closed subsets of the Ziegler spectrum such that every point of Zg_R lies in at least one set C_λ. For each λ choose a pp formula ρ_λ which defines a function on modules with support contained in C_λ and such that these definable scalars agree on sets of the form $C_\lambda \cap C_\mu$. The result of "patching together" the ρ_λ is a typical element of $\mathrm{LocDef}_R(\mathrm{Zar}_R)$. Such a locally definable scalar will have a well-defined action on any direct sum or product of indecomposable pure-injective R-modules but would not appear to have a natural action on arbitrary R-modules (but also see Section 12 below where we see a natural action on the LocDef_R-module built from any module).

Recall that the sheafification functor has the following universal property.

Proposition 4.2 *(e.g. [16, 4.2]) If F is a presheaf then the sheafification ΓLF is a sheaf. If G is any sheaf and if $F \longrightarrow G$ is a morphism of presheaves then there is a unique factorisation through the canonical map $\nu : F \longrightarrow \Gamma LF$. In particular this applies with $F = \mathrm{Def}_R$ and $\Gamma LF = \mathrm{LocDef}_R$.*

Lemma 4.3 *The canonical map $\nu : \mathrm{Def}_R \longrightarrow \mathrm{LocDef}_R$ is a monomorphism in the category of presheaves over Zar_R.*

Proof. It must be shown (see [16, 3.5]) that for any $[\sigma/\tau]$ and $r, r' \in R_{[\sigma/\tau]}$ if $\nu(r) = \nu(r')$ then $r = r'$, that is, if $r_N = r'_N$ for every $N \in [\sigma/\tau]$ then $r = r'$. So suppose that ρ, respectively ρ', define r, respectively r', on $[\sigma/\tau]$. By assumption, for every $N \in [\sigma/\tau]$ we have $N \models \forall xy(\rho(x,y) \leftrightarrow \rho'(x,y))$. Therefore, on $M = \bigoplus\{N : N \in [\sigma/\tau]\}$ the formulas ρ and ρ' are equivalent. But $\mathrm{supp}M = [\sigma/\tau]$ and so (see [2, p. 189] ρ and ρ' define the same element of $R_{[\sigma/\tau]}$, as required. □

5 Examples

Example 5.1 *Let R be the ring $\mathbf{Z}_4$ of integers modulo 4. Then the Zariski spectrum has just two points, namely $\mathbf{Z}_4$ and $\mathbf{Z}_2$ and it has the discrete topology. On the basic Zariski-open set $\{\mathbf{Z}_4\}$ the ring of definable scalars is the ring $\mathbf{Z}_4$ and similarly the ring of definable scalars at $\mathbf{Z}_2$ is the ring $\mathbf{Z}_2$. The ring of definable scalars on the union of these two open sets is the ring of definable scalars for the largest theory of $\mathbf{Z}_4$-modules and hence [2, 2.4] is $\mathbf{Z}_4$. In any sheaf of rings the ring above the union of two disjoint open sets is the product of the rings above the two sets (indeed, in this case, the stalk space has the discrete topology and so every section is continuous) - which in this case would be $\mathbf{Z}_4 \times \mathbf{Z}_2 \neq \mathbf{Z}_4$. The inclusion $\mathbf{Z}_4 = \mathrm{Def}\{\mathbf{Z}_2, \mathbf{Z}_4\} \longrightarrow \mathrm{LocDef}\{\mathbf{Z}_2, \mathbf{Z}_4\}$ is given by taking an element $r \in \mathbf{Z}_4$ to $(r \bmod 2, r)$.*

So we see that the presheaf of definable scalars need not be a sheaf.

For a ring of finite representation type we can easily identify the ring, $\Gamma\mathrm{LocDef}_R$, of global sections of LocDef_R. Let $\mathrm{Biend}M$ denote the biendomorphism ring of a module M (that is, the ring of endomorphisms of M regarded as a module over its endomorphism ring).

Proposition 5.2 *Suppose that R is a ring of finite representation type. Then $\Gamma\mathrm{LocDef}_R$ is $\prod \mathrm{Biend}(N_i)$ - the product ranging over a set of representatives of isomorphism classes of indecomposable R-modules.*

Proof. Since Zg_R, hence Zar_R is discrete, we have $\Gamma\mathrm{LocDef}_R \simeq \prod R_{N_i}$. But by [2, 3.6], $R_{N_i} \simeq \mathrm{Biend}(N_i)$ since N_i is of finite length over its endomorphism ring. □

Example 5.3 *Take R to be the path algebra, over the field k, of the quiver A_1 (a ring of finite representation type with three indecomposable modules having k-vectorspace dimensions 1, 2, 1 respectively and each with endomorphism* *ring k). Then the ring of locally definable scalars is $k \times M_2(k) \times k$ - the biendomorphism ring of the module which is a direct sum of one copy of each indecomposable (here $M_2(k)$ denotes the ring of 2×2 matrices with entries in k).*

Proposition 5.4 *Let R be a commutative von Neumann regular ring. Then* Def_R *coincides with the Pierce sheaf.*

Proof. The Ziegler and Pierce spectra coincide ([9, Section 4.7 Ex.3]): the points since pure-injective=injective for modules over regular rings and the topologies as a consequence of elimination of quantifiers over these rings. Hence so do the Zariski and Pierce spectra coincide (the Pierce spectrum is self-dual with respect to Hochster's duality [8]). A typical basic open set has the form $[x = x/xe = 0]$ for some idempotent $e \in R$ and the corresponding ring of definable scalars is just R/eR. A general open set has the form $[x = x/xI = 0]$ where I is any ideal of R and so the, already-discussed, process of extending a presheaf on a basis to a presheaf assigns the ring $\lim_{\rightarrow}\{R/eR : e \in I\} = R/I$ to this open set, as does the Pierce sheaf. □

6 The centre: a sheaf of local rings

The stalks of Def_R will not in general be local rings, nor even Morita equivalent to local rings (see [11] which has an example over the path algebra, over $\mathbf{Z}_4$, of the quiver A_1) but we see now that, by a result of Ziegler, their centres will be.

Theorem 6.1 *(based on [17, 5.4]) Let $N \in \mathrm{Zar}_R$. Then $P = P_N = \{r \in C(R) : r$ does not act invertibly on $N\}$ is a prime ideal of the centre $C(R)$ of R and N is naturally a module over the localisation $R_{(P)}$. The natural map $R \longrightarrow R_N$ factorises through the localisation $R \longrightarrow R_{(P)}$ and the image of $R_{(P)}$ in R_N is contained in $C(R_N)$, which is a local ring.*

Proof. Since $\mathrm{End}N$ is local, P_N is an ideal. Also if $r, s \in C(R) \setminus P$ then r, s act invertibly on N and so rs acts invertibly on N, hence $rs \notin P$. Thus P is prime.

The $R_{(P)}$-structure on N is the obvious one.

If an element of R acts invertibly on N then certainly its inverse is a definable scalar on N so we have the factorisation. If $r \in C(R)$ then multiplication by r is an R-endomorphism of N and hence commutes with every element of R_N (since $R_N \subseteq \mathrm{Biend}(N)$): hence also r^{-1}, if it is defined on N, is in $C(R_N)$. That $C(R_N)$ is local follows by applying the first part to N regarded as a module over R_N. □

7 The Zariski structure sheaf

Recall that for any ring R we may define its **Zariski spectrum**, Spec(R), to be the set of two-sided prime ideals endowed with the topology which has as a basis (even in the non-commutative case) of open sets the $D(r) = \{P \in \mathrm{Spec}(R) : r \notin P\}$ (of course, a non-commutative ring may have "few" primes and there are a number of different notions of spectrum for such rings). If R is commutative (or at least if we have a reasonable notion of localisation at sets of primes) we then define the **structure sheaf** $\mathcal{O} = \mathcal{O}_{\mathrm{Spec}(R)}$ on the Zariski spectrum by taking an open set U of Spec(R) to the localisation of R at U.

The Gabriel-Zariski topology Zar_R is named thus because, if one follows [3] in phrasing the above definition in terms of the category Mod $-$ R and if one then applies this definition to the functor category $(R - \mathrm{mod}, \mathbf{Ab})$ then one obtains precisely Zar_R (see [10] for details).

We have seen already that there is a map γ, say, from Zar_R to Spec$C(R)$ given by taking an indecomposable pure-injective N to $P_N = \{r \in C(R) :$ r does not act invertibly on $N\}$ and we saw that we have a canonical morphism of rings $C(R)_{(P_N)} \longrightarrow R_N$. In fact this data gives us a morphism of (pre)sheaves.

Proposition 7.1 *Let R be any ring. Then the above map γ induces a morphism of presheaves* $\mathrm{Def}_R \longrightarrow \mathrm{Spec}C(R)$ *and hence a morphism* $\mathrm{LocDef}_R \longrightarrow \mathrm{Spec}C(R)$ *of sheaves (which we also denote by γ).*

Proof. First we show that, as a map of topological spaces, γ is continuous. So let $r \in C(R)$ and consider $\gamma^{-1}D(r) = \{N \in \mathrm{Zar}_R : r \notin P_N\}$. This is the set of points of Zar_R on which r acts invertibly, that is $[x = x/r|x] \cap [xr = 0/x = 0]$ which is Zariski-open, as required.

Now we must check that we have a morphism from the structure sheaf $\mathcal{O}_{\mathrm{Spec}C(R)}$ to the direct image presheaf $\gamma_*\mathrm{Def}_R$ (see [16] e.g.) where the latter takes an open subset U of Spec$C(R)$ to $\mathrm{Def}_R(\gamma^{-1}U)$. Since the data and condition of being a morphism of sheaves is local, it is enough to check with $U = D(r)$. Now $\mathcal{O}_{\mathrm{Spec}C(R)}(D(r))$ is the ring obtained from $C(R)$ by inverting r, and we have $\mathrm{Def}_R(\gamma^{-1}D(r)) = \mathrm{Def}_R([x = x/r|x] \cap [xr = 0/x = 0])$ - an R-algebra in which r is invertible. So we have a canonical map from the former to the latter. It is immediate from the definitions that these maps commute with the various restriction maps and hence that we do have a morphism of presheaves $\mathrm{Def}_R \longrightarrow \mathrm{Spec}C(R)$ and hence, since Spec$C(R)$ is a sheaf, also, by 4.2, a morphism from the sheafification LocDef_R to Spec$C(R)$. □

8 Representation embeddings

Now suppose that R, S are k-algebras where k is a field. One of the basic notions of the representation theory of finite-dimensional algebras is that of a representation embedding between categories of finitely presented modules. Here I do not wish to restrict to finitely presented modules and so make the following definition, which is an extension of the usual one. By a **representation embedding** from $\mathrm{Mod}-R$ to $\mathrm{Mod}-S$ I mean (as in [12]) a functor $F = - \otimes B : \mathrm{Mod}-R \longrightarrow \mathrm{Mod}-S$, where B is an (R,S)-bimodule which is a finitely generated projective generator as a left R-module, which preserves indecomposability of (at least pure-injective) modules and reflects isomorphism.

It is shown in [12, Thm.7] that any such functor induces a homeomorphic embedding of Zg_R as a Ziegler-closed subset of Zg_S and it follows immediately that F induces a homeomorphic embedding of Zar_R into Zar_S. We use the notation F also for these maps between topological spaces. In general the image of F will just be an intersection of Zariski-open sets but in some cases, in particular if the image of the functor F is a finitely axiomatisable class which is closed under direct summands (and hence a finite intersection of basic Ziegler-closed sets), then it will be Zariski-open.

Now, any representation embedding from Mod-R to Mod-S may be expressed as the composition of the canonical Morita equivalence from Mod-R to Mod-$M_n(R)$ for some n, where $M_n(R)$ denotes the ring of $n \times n$ matrices over R, followed by restriction of scalars from $M_n(R)$ to S (S is regarded as a subring of $M_n(R)$ via its embedding in $\mathrm{End}({}_RB)$ which itself embeds in $\mathrm{End}_R R^n \simeq M_n(R)$ for some n) - see [12]. Let us consider the effect of the Morita equivalence first.

Since the Gabriel-Zariski topology may be defined purely in terms of the category of modules, any Morita equivalence induces a homeomorphism of Zariski spectra. Let G denote the induced homeomorphism from $\mathrm{Zar}_{M_n(R)}$ to Zar_R - which just takes any point $N_{M_n(R)}$ of the former to the unique-to-isomorphism indecomposable direct summand of N_R (that is, N regarded as an R-module via the diagonal map from R to $M_n(R)$). Let $M_n(\mathrm{LocDef}_R)$ denote the ringed space defined by $M_n(\mathrm{LocDef}_R) : U \mapsto M_n(\mathrm{LocDef}_R.(U))$ and let Δ denote the diagonal map from LocDef_R to $M_n(\mathrm{LocDef}_R)$. It is straightforward to check that the direct image sheaf $G_*\mathrm{LocDef}_{M_n(R)}$ is isomorphic to $M_n(\mathrm{LocDef}_R)$ and thus we obtain a morphism of ringed spaces $(G, \Delta) : (\mathrm{Zar}_{M_n(R)}, \mathrm{LocDef}_{M_n(R)}) \longrightarrow (\mathrm{Zar}_R, \mathrm{LocDef}_R)$ (using the notation of [4]). We observe that in order to obtain a morphism in the other direction one should use the presheaf of definable scalars in all sorts which is discussed in Section 13.

In view of the above (rather transparent effect of Morita equivalence) we now confine ourselves to the case where the representation embedding is given simply by restriction of scalars from the ring R to its subring S. In this case it is shown in [12] that there is induced a homeomorphic embedding ρ say - which is, in fact, just restriction of scalars - from Zg_R to Zg_S and hence from Zar_R to Zar_S, with image a Ziegler-closed subset of Zg_S. We identify Zar_R with its image in Zar_S.

Consider the direct image sheaf $\rho_\star\mathrm{LocDef}_R$ which, by definition, takes an open subset U of Zar_S to $\mathrm{LocDef}_R(U \cap \mathrm{Zar}_R)$ - we have to allow the zero ring here in order to cover the case that this intersection is empty. Since S is a subring of R, every scalar definable in the language of S-modules is certainly definable in the language of R-modules and so there is a natural l "restriction of definable scalars" map from LocDef_S to $\rho_\star\mathrm{LocDef}_R$.

Let us suppose from now on that the image of Zar_R in Zar_S is basic Ziegler-closed, hence Zariski-open: this will be enough for what we do later. In particular we can easily compare the restriction (see [16]) of LocDef_S to Zar_R with LocDef_R (strictly, with the direct image of this under the corestriction of ρ to its image). We have already referred to the fact [2] that the ring of definable scalars of a basic Zariski-open set may be obtained as the biendomorphism ring of a sufficiently "large" module with support that set. Now, if M is an R-module then, since $\mathrm{End}M_S$ contains $\mathrm{End}M_R$, we have $\mathrm{Biend}M_S$ contained in $\mathrm{Biend}M_R$. Thus the restriction of LocDef_S to Zar_R is in general just a subsheaf of LocDef_R. If, however, the embedding of S into R is an epimorphism of rings and hence the restriction of scalars functor is full, then $\mathrm{Biend}M_S = \mathrm{Biend}M_R$ and so we conclude that, in this case, the restriction $\mathrm{LocDef}_S \mid \mathrm{Zar}_R = \mathrm{LocDef}_R$. We summarise this discussion as follows.

Proposition 8.1 *Let $F = - \otimes B : \mathrm{Mod}-R \longrightarrow \mathrm{Mod}-S$ be a representation embedding. Suppose that the induced homeomorphic embedding from Zar_R to Zar_S has Zariski-open image. Then the restriction of LocDef_S to (the image of) Zar_R is, up to Morita equivalence, a subsheaf of LocDef_R. If F is a full functor then the restriction of LocDef_S to Zar_R is Morita equivalent to LocDef_R.*

By "Morita equivalence" above (and also below) I simply mean the explicit equivalences which are discussed here. This is a rather ad hoc use of the term. A systematic treatment could be given using the the notion of an "additive space" in the sense of [7] (that is, consider equivalences between the categories of finitely generated projective sheaves over ringed spaces).

9 Wild algebras

Let $k\langle X, Y\rangle$ denote the free associative algebra over the field k in two generators: the archetypal wild algebra. Given any finitely generated k-algebra

R there is a full representation embedding from Mod-R into Mod-$k\langle X, Y\rangle$ the image of which is a finitely axiomatisable subcategory of Mod-$k\langle X, Y\rangle$ [1, (proof of) Thm.3]. Therefore all the conditions of 8.1 are satisfied and we obtain the following.

Corollary 9.1 *Let k be a field and let R be any finitely generated k-algebra. Then there is an open subset of* $\mathrm{Zar}_{k\langle X,Y\rangle}$ *such that the restriction of* $\mathrm{LocDef}_{k\langle X,Y\rangle}$ *to this open subset is Morita equivalent to* LocDef_R.

We say that a finitely generated k-algebra S is **strictly wild** if there is a full representation embedding from Mod-$k\langle X, Y\rangle$ to Mod-S. If the image of this representation embedding is Zariski-open then we may replace $k\langle X, Y\rangle$ by S in 9.1.

Now let R be any commutative finitely generated k-algebra and let $(\mathrm{Spec}R, \mathcal{O})$ be the corresponding affine scheme. The space $\mathrm{Spec}R$ may be identified with the set of indecomposable injective points of Zar_R - a Ziegler-closed subset which is not, however, in general Zariski-open. Furthermore the Zariski structure sheaf may be identified with the restriction of LocDef_R to $\mathrm{Spec}R$ by [2, Thm.2.5]. Therefore we have, in view of the above, that if S is any strictly wild k-algebra then there is a (Ziegler-closed) subset X of Zar_S such that the restriction of LocDef_S to X is Morita equivalent (in the sense we have used earlier) to LocDef_R (although X may not be Zariski-open, it is an intersection of Zariski-open sets and every definable scalar over X extends to a Zariski-open neighbourhood of X: this allows us to calculate the restriction of LocDef_S to X).

This explains our claim in the abstract of the paper.

10 A projective variety

Example 10.1 *As an example of a projective variety within* Zar_R *for some R, let R be the path algebra of the Kronecker quiver $\tilde{A}_1$. Then (see [14])* LocDef_R *consists of a "discrete" part together with a sheaf which may be regarded as a non-separated double cover of the projective line except in that it has a unique generic point.*

11 Interpretations

Many kinds of interpretations of module categories, one in another, induce continuous maps between Ziegler, and hence Zariski, spectra (see [13]) and also induce morphisms between rings of definable scalars. Morita equivalence is one example but (see [13]) there are many more. The natural context in which to consider the induced morphisms of ringed spaces is that of the sheaf of locally definable scalars in all sorts, which is discussed in Section 13 but we do not pursue this here.

12 Modules to sheaves

There is a natural functor which takes any R-module to a module over LocDef_R. Namely, given any R-module M_R define the LocDef_R-module $\tilde{M}$ as follows. First define the presheaf $M \otimes \text{LocDef}_R$ by $(M \otimes \text{LocDef}_R)(U) = M \otimes_R \text{LocDef}_R(U)$ where $U \subseteq \text{Zar}_R$ is open and with the obvious restriction maps. Then let $\tilde{M}$ be the sheafification of $M \otimes \text{LocDef}_R$.

Lemma 12.1 *$\tilde{M}$ is also the sheafification of the presheaf which is defined on basic open sets by taking $[\sigma/\tau]$ to $M \otimes_R R_{[\sigma/\tau]}$.*

Proof. Let F_0 be the presheaf on $\mathcal{U}_0$ described: $F_0[\sigma/\tau] = M \otimes_R R_{[\sigma/\tau]}$ with restriction maps of the form $id_M \otimes res_{[\sigma,\tau],[\sigma',\tau']}$. The extension F of F_0 to a presheaf on $\mathcal{U}$ is given by $F(U) = \lim_{\to}\{M \otimes_R R_{[\sigma/\tau]} : U \subseteq [\sigma/\tau]\} = M \otimes_R \lim_{\to}\{R_{[\sigma/\tau]} : U \subseteq [\sigma/\tau]\} = M \otimes_R \text{Def}_R(U)$ (since $\otimes$ commutes with direct limits). But the sheafification of $M \otimes_R \text{Def}_R$ is isomorphic to $M \otimes_R \text{LocDef}_R$ ($M \otimes_R \text{Def}_R$ and $M \otimes_R \text{LocDef}_R$ have the same stalk space as a set and a basis for the topology on $\text{L}(M \otimes_R \text{Def}_R)$ also serves as a basis for the topology on $\text{L}(M \otimes_R \text{LocDef}_R)$). □

Example 12.2 *Let $R = \mathbf{Z}_4$. Then for any R-module M we have $\tilde{M} = M \otimes \text{LocDef}_R$ with $\tilde{M}(\mathbf{Z}_2\} = M/2M$, $\tilde{M}(\mathbf{Z}_4\} = M$, $\tilde{M}\{\mathbf{Z}_2, \mathbf{Z}_4\} = (M/2M) \times M$ as a $\mathbf{Z}_2 \times \mathbf{Z}_4$-module.*

13 The sheaf of locally definable scalars in all sorts

The sheaf of locally definable scalars is, in fact, a part of a richer structure which we now point out. Model-theorically, we are applying the eq(+)-construction: representation-theoretically, we are replacing the functor $(R_R, -)$ by arbitrary finitely presented functors.

This structure - which we will call the **presheaf of categories** (as opposed to rings) **of definable scalars** or the **presheaf of definable scalars in arbitrary sorts** - is a presheaf of small abelian categories over Zar_R. Namely, to the basic Zariski-open set $[\sigma/\tau]$ we associate the hereditary torsion theory $\mathcal{T} = \mathcal{T}_{[\sigma/\tau]}$ on the functor category $(R-\text{mod}, \mathbf{Ab})$ whose torsion class is generated by the finitely presented functor $F_{[D\tau/D\sigma]}$. The result of localising

$(R - \text{mod}, \mathbf{Ab})$ at this torsion theory of finite type we denote by $(R - \text{mod}, \mathbf{Ab})_{\mathcal{T}}$ or $(R - \text{mod}, \mathbf{Ab})_{[\sigma/\tau]}$. The localised category is again locally coherent and we have (see e.g. [11] , [6], [7]) $(\text{fp} - (R - \text{mod}, \mathbf{Ab}))_{\mathcal{T}} = \text{fp} - ((R - \text{mod}, \mathbf{Ab})_{\mathcal{T}})$, where "fp" denotes the category of finitely presented objects, which we may therefore denote unambiguously by $\text{fp} - (R - \text{mod}, \mathbf{Ab})_{\mathcal{T}}$. We define our presheaf $\mathbf{Def}_R : \mathcal{U}_0 \longrightarrow \mathbf{Cat}$ where the latter is the category of all small categories, by taking $[\sigma/\tau]$ to $\text{fp} - (R - \text{mod}, \mathbf{Ab})_{[\sigma/\tau]}$ and, if $[\sigma/\tau] \supseteq [\sigma'/\tau']$ then we take the restriction map $\text{fp} - (R - \text{mod}, \mathbf{Ab})_{[\sigma/\tau]} \longrightarrow \text{fp} - (R - \text{mod}, \mathbf{Ab})_{[\sigma'/\tau']}$ to be that induced by localising $(R - \text{mod}, \mathbf{Ab})_{[\sigma/\tau]}$ at the "image" under the localisation functor $Q_{[\sigma/\tau]}$ of $\mathcal{T}_{[\sigma'/\tau']}$.

Then the whole of the above discussion may be repeated for this sheaf of categories and we obtain the **sheaf of categories of locally definable scalars** or the **sheaf of locally definable scalars in all sorts, $\mathbf{LocDef}_R$**. Now we note that $(\text{Loc})\text{Def}_R$ is just a small piece of $(\mathbf{Loc})\mathbf{Def}_R$ (a one-point subcategory of the latter in the category of sheaves of categories over Zar_R).

For it is known [5] that for any closed subset C of Zg_R we have (a skeletal version of) $\text{fp} - (R - \text{mod}, \mathbf{Ab})_C$ equivalent to T^{eq+} where T is the theory corresponding to C and where we interpret T^{eq+} as a category whose objects are the pp-sorts ϕ/ψ and whose morphisms are the pp-defined maps between sorts. In particular there is the following result which describes the relation between $(\text{Loc})\text{Def}_R$ and $(\mathbf{Loc})\mathbf{Def}_R$.

Theorem 13.1 *[2] Let C be any closed subset of Zg_R. Then R_C is naturally isomorphic (as an R-algebra) to the endomorphism ring of the object $(R, -)_C$ in $\text{fp} - (R - \text{mod}, \mathbf{Ab})_C$.*

References

[1] S. Brenner, Decomposition properties of some small diagrams of modules, Symp. Math. Ist. Naz. Alta. Mat., 13 (1974), 127-141.

[2] K. Burke and M. Prest, Rings of definable scalars and biendomorphism rings, pp. 188-201 in Model Theory of Groups and Automorphism Groups, London Math. Soc. Lect. Note Ser., Vol. 244, Cambridge University Press, 1997.

[3] P. Gabriel, Des catégories abéliennes, Bull. Soc. Math. France, 90 (1962), 323-448.

[4] R. Hartshorne, Algebraic Geometry, Graduate Texts in Math., Vol. 52, Springer-Verlag, 1977.

[5] I. Herzog, Elementary duality for modules, Trans. Amer. Math. Soc., 340 (1993), 37-69.

[6] I. Herzog, The Ziegler spectrum of a locally coherent Grothendieck category, Proc. London Math. Soc., 74 (1997), 503-558.

[7] H. Krause, The Spectrum of a Module Category, Habilitationsschrift, Universität Bielefeld, 1998.

[8] M. Hochster, Prime ideal structure in commutative rings, Trans. Amer. Math. Soc., 142 (1969), 43-60.

[9] M. Prest, Model Theory and Modules, London Math. Soc. Lect. Note Ser., Vol. 130, Cambridge University Press, 1988.

[10] M. Prest, Remarks on elementary duality, Ann. Pure Applied Logic, 62 (1993), 183-205.

[11] M. Prest, The (pre)sheaf of definable scalars, University of Manchester, preprint, 1995.

[12] M. Prest, Representation embeddings and the Ziegler spectrum, J. Pure Applied Algebra, 113 (1996), 315-323.

[13] M. Prest, Interpreting modules in modules, Ann. Pure Applied Logic, 88 (1997), 193-215.

[14] M. Prest, Sheaves of definable scalars over tame hereditary algebras, University of Manchester, preprint, 1998.

[15] M. Prest, The Zariski spectrum of the category of finitely presented modules, University of Manchester, preprint, 1997.

[16] B. R. Tennison, Sheaf Theory, London Math. Soc. Lect. Note Ser., Vol. 20, Cambridge University Press, 1975.

[17] M. Ziegler, Model theory of modules, Ann. Pure Appl. Logic, 26 (1984), 149–213.

Human Styles of Quantificational Reasoning

Lance J. Rips[1]
Northwestern University

Even middle-school mathematics relies on the use of variables to capture the generality of theorems and proofs. Ordinary nonmathematical language also relies on similar devices -- determiners, pronouns, and sometimes other parts of speech -- to specify information about entities without having to describe or to name them individually. In psychology, variables are a central means of representing, retrieving, and manipulating information in memory, according to many cognitive theories (e.g., Anderson, 1983; Newell, 1990). In these theories, for example, a simple procedure for recognizing a triangle might be spelled out in terms of mental rules such as *IF Closed*(*x*) & *Three-sided*(*x*) & *Two-dimensional*(*x*) *THEN Triangle*(*x*), much as in conventional computer-programming languages.

These psychological theories assume that all people -- even those who have never had math or logic training -- manipulate variables mentally, but until recently there have been no general proposals about reasoning with variables. If people do represent generality in this way, then it is useful to know something about the limits of their ability to deduce information from such representations. It is possible, of course, to capture generality by other means, for example, replacing variables with combinators (e.g., Hindley & Seldin, 1986; Schönfinkel, 1924/1967). The research that I describe here, however, follows the lead of theories such as Anderson's and Newell's in assuming variable-based representations, and it examines people's deductive skills within this framework.

Most earlier theories of human reasoning are of little help in understanding deduction with quantifiers and variables. Although there are many proposals about how people evaluate the validity of arguments, the scope of these proposals is quite limited in this context. Despite the wealth of first-order proof schemes in logic and computer science, psychological investigations of quantifiers (beginning with Gustav Störring in 1908 and continuing to the present) have usually limited themselves to studies of classical syllogisms. In the 1970s and 1980s a few broader proposals

[1]Research reported in this article was supported by National Science Foundation grant SBR-9514491. Thanks are due to Satya Chheda for assistance with the experiment.
Correspondence about this paper should be sent to Lance Rips, Psychology Department, Northwestern University, 2029 Sheridan Road, Evanston IL 60208. (Email: rips@nwu.edu.)

appeared for fragments of first-order logic (Johnson-Laird, Byrne, & Tabossi, 1989; Osherson, 1976). But it is only in the last few years that psychologists have outlined theories that approach the comprehensiveness of full first-order logic (Braine, 1998; Rips, 1994).

My own proposal takes the form of a computer program (called PSYCOP, for Psychology of Proof) that predicts results from laboratory experiments on deduction. Those tests included measures of how quickly and accurately people judge the validity of arguments. The theory assumes that people evaluate arguments by carrying out mental proofs of the conclusion from the premises. According to the theory, they represent the premises and conclusion in memory and then apply mental deduction rules to these internal sentences to see whether the conclusion follows. Their success in finding such a proof for a valid argument depends on the length of the proof, the difficulty of applying individual rules, and other cognitive factors. In what follows, I give a brief description of the theory and then report a new study that provides a further test.

A Theory of Deductive Reasoning

The heart of the proposed theory is a natural deduction system, but one that differs from standard systems in ways that allow it to mimic human styles of reasoning. First, although the system includes versions of most of the introduction and elimination rules for propositional connectives (as in Gentzen, 1935/1969), it also includes other deduction rules as primitives. For example, the system includes a *Disjunctive Modus Ponens* rule that allows reasoners to go from statements of the form *If* P_1 *or* P_2 *then Q* and P_i to *Q*, where $i = 1$ or 2. Such items are unnecessary in standard systems, of course, since they can be derived from the usual introduction and elimination schemas. People appear to treat them as primitive, however, since the ease with which they carry out the corresponding inference is at least as fast and as accurate as that of the component inferences in their derivation.

Second, the system restricts the inference rules in ways that eliminate certain irrelevant applications. Applying a rule like *And Introduction* without restriction -- going from arbitrary statements *P* and *Q* to the single statement *P and Q* -- leads to an infinite number of derived statements (e.g., *P and Q, (P and Q) and Q, ...*), since the rule can apply to its own products. The PSYCOP system never applies this rule unless the system must explicitly derive a conjunction as a step in proving the conclusion. *Backward rules* of this type contrast with *forward rules*, such as Disjunctive Modus Ponens, which the system applies as soon as it encounters the relevant premises in the proof at hand. (Some rules have both forward and backward versions.) As a result, the system's usual proof style is bi-directional, working from the premises of the argument toward the conclusion via the forward rules and working from the conclusion to the premises via the backward rules. The forward rules typically break down premises into components, while the backward rules typically build up conclusions from these components, in a manner similar to natural deduction proofs in normal form (Prawitz, 1965).

More pertinent to the present paper, the system represents first-order statements in Skolem function form (Grandy, 1977, p. 41; Shoenfield, 1967, p. 56), thus eliminating standard quantifiers. Variables that would be universally quantified in the prenex form of a sentence appear as variables in the new notation, whereas variables that would be existentially quantified in prenex form appear as Skolem constants, called *temporary names* here. I will use letters from the end of the alphabet for variables and letters from the beginning for temporary names. These temporary names are subscripted with any variables in whose scope they appear, indicating their dependence on the values of these variables. For example, the statement that would appear in standard notation as $(\forall x)(\exists y)F(x,y)$ appears here as $F(x,a_x)$, whereas $(\exists y)(\forall x)F(x,y)$ corresponds to $F(x,a)$. We also allow proper names in the sentences, symbolizing them with letters from the middle of the alphabet (e.g., m, n).

Because there are no explicit quantifiers in the system's representations, there are no quantifier introduction or elimination rules of the sort found in other natural deduction systems. Instead, most of the bookkeeping required to keep track of variables and temporary names is handled by the rules for propositional connectives (see Rips, 1994, Tables 6.3 and 6.4 for a statement of these rules). For example, to prove a conjunctive conclusion (e.g., *Senator(a) and Nebraskan(a)*), the And Introduction rule must ensure that whatever satisfies the first conjunct also satisfies the second. (If the rule finds that *Senator(a)* is satisfied by *Senator(Kerrey)*, it then tries to determine whether *Nebraskan(Kerrey)*.) The use of Skolem function form also means that distinct English statements, such as *Some teachers don't shout* and *Not all teachers shout*, get mapped into the same representation, *Teacher(a) and ¬Shout(a)*. Thus, the ability to recognize the equivalence of these statements is due to whatever mental operations translate English into the underlying representation rather than to the deduction operations per se. Some such division of labor is necessary, however, in any system that assumes that people carry out inferences on an underlying representation or logical form that is more uniform than natural language itself.

But although the system eliminates quantifiers, it still requires procedures for recognizing that certain sentences are generalizations or instances of others. The need for these procedures arises when a premise and a conclusion are *congruent* -- duplicates, except for their terms. The system must have some way to determine, for instance, that *Kerrey* = *Kerrey* follows from $x = x$, but not the reverse. To handle inferences like this one, the system includes a set of *matching rules* that decide for any pair of congruent sentences whether the first entails the second. Essentially, these rules specify that sentences containing variables ($x = x$) imply those containing temporary or proper names ($a = a$, *Kerrey* = *Kerrey*) and that those containing proper names (*Kerrey* = *Kerrey*) imply those containing temporary names ($a = a$).

The rules must include qualifications, however, to deal properly with scope relations between temporary names and variables.

The appendix to this paper lists the system's four rules for matching (see Rips, 1994, pp. 215-218 for proofs of their soundness). They assume that the premise and conclusion of the argument contain disjoint sets of variables and temporary names, and they operate by successively generalizing the conclusion, one term at a time, in an attempt to mate it to the premise. As an illustration, consider a premise of the form $F(x,y,m,a_{x,y})$ and a conclusion $F(b,n,m,c)$, where x and y are variables, $a_{x,y}$, b, and c are temporary names, and n and m are proper names. (Thus, the premise might represent *Every teacher instructed each student to compare Madonna to some historical figure* and the conclusion *Some teacher instructed Ned (a student) to compare Madonna to some historical figure.*) Figure 1 shows how the rules demonstrate that the first sentence is a generalization of the second. The system notices that the first term of the conclusion, b, is a temporary name and the first term of the premise, x, is a variable. It therefore applies Rule 4, which generalizes temporary names to variables. Before it does so, however, it notes that $a_{x,y}$ in the premise is within the scope of x (has x as subscript). The conclusion also contains a temporary name c in the position corresponding to $a_{x,y}$, so in generalizing b to x it also places x as a subscript to c. In this way, $a_{x,y}$ and c_x both stand in the same scope relation to x. The result of applying Rule 4 is then the formula that appears just above the conclusion in the figure. As a second step, the system generalizes the second term of the new formula, n, to y, using Rule 3 for proper name-to-variable matching. As in the first step, this requires adding y to the subscript of c to maintain the scope relations of the premise. The result of this second rule application is the second formula of Figure 1. Finally, the system requires perfunctory use of Rule 2

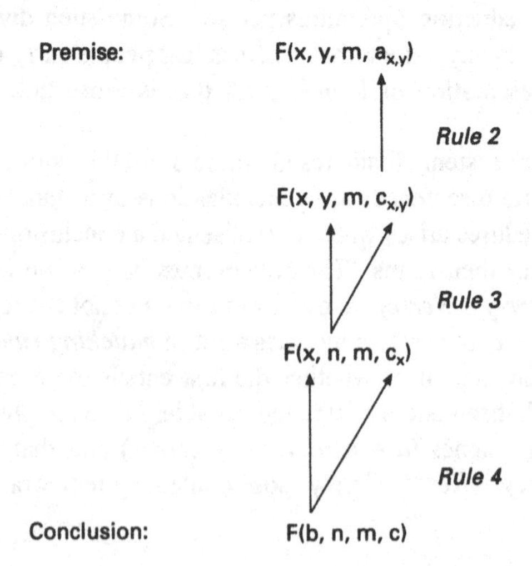

to notice that the second formula is a notational variant of the premise.

An Experiment on Instantiation

Nearly all theories in cognitive psychology predict that people have more difficulty (commit more errors and take longer) in carrying out two independent mental operations than in carrying out the same operation twice. One reason for this is that separate operations make more demands on memory, because people must retrieve two sets of instructions rather than one. A second reason is that executing an operation may facilitate repeating the same operation, a kind of practice or priming effect. Applied to the task of determining whether one sentence entails another, this difference predicts that people should have an easier time recognizing entailments that require applying a single rule twice than recognizing entailments that require two distinct rules. For example, if the rules in the appendix correctly describe human deductive ability, it should be relatively simple to determine that $F(x,y)$ entails $F(a,b)$, since people can accomplish this by applying Rule 4 to each of the two terms of the conclusion. By contrast, the entailment from $F(x,m)$ to $F(n,a)$ should be more difficult, since the (mental) proof requires both Rule 3 for the first term and Rule 2 for the second. Intuitively, this prediction seems correct, since an inference from, say, *Everybody dazzles everyone* to *Somebody dazzles someone* appears more obvious than the one from *Everyone dazzles Ginger* to *Fred dazzles someone*. By measuring the time it takes people to confirm entailments such as these, we can obtain a systematic test of whether PSYCOP correctly divides up the logical work among its rules.

We can call arguments *two-rule problems* if they require two distinct rules for their proof and call them *one-rule problems* if they require one (repeated) rule. To test the prediction that one-rule problems are easier than two-rule problems, I asked participants in one experiment to evaluate arguments similar to the "dazzle" examples (see Rips, 1994, pp. 248-254 for details). The natural-language arguments consisted of a single premise and conclusion, each containing a verb and two terms. Half the arguments were deducible ones, such as the "dazzle" items, and half were not deducible. (The nondeducible arguments were formed from the deducible ones by switching the position of the premise and conclusion). A computer presented the arguments to participants one-at-a-time in random order. The participants read each argument, decided whether the conclusion followed or didn't follow, and pressed one of two keys on a keyboard to indicate their answer. The computer automatically recorded the participants' accuracy and the amount of time they took for their decisions. The participants were undergraduates, none of whom had taken a college course on logic.

The results of this experiment agreed with the theory's predictions. It took participants a mean of 3882 milliseconds (ms) to decide that the two-rule arguments followed (e.g., *Everybody dazzles Ginger; therefore, Fred dazzles someone*), but

3235 ms for the one-rule arguments (e.g., *Everybody dazzles everyone; therefore, somebody dazzles someone*). Likewise, they made errors on 12.0% of the two-rule arguments, but only 2.6% errors on the one-rule arguments. Across problems, the two-rule arguments contained the same number of variables as the one-rule arguments, and the same was true for temporary and proper names. Hence, the frequency of different types of terms cannot account for these differences. Furthermore, none of the sentences contained both a variable and a temporary name, so none of the sentences were potentially ambiguous in their scope relations. (E.g., neither *Everyone dazzles somebody* nor *Somebody dazzles everyone* appeared in any argument.)

It is possible to argue, however, that the differences in the study were due to the fact that all of the one-rule arguments contained either a premise of the form $F(x,y)$ (e.g., *Everyone dazzles everybody*) or a conclusion of the form $F(a,b)$ (e.g., *Someone dazzles somebody*). The former sentence entailed all the others, and the latter sentence was entailed by all the others. Thus, participants who noticed this pattern could respond to the one-rule arguments correctly without having to perform an inference. This could help account for the advantage of one-rule arguments over two-rule items.

To clarify this issue, we've carried out a second experiment in which the difference between one-rule and two-rule arguments is independent of the generality of the terms in the premises and conclusions. To see how this can be done, consider the arguments in Table 1.[2] The first six of these have the same premise form, $F(x,y,n,m)$, but differ in their conclusions. Within this set, the first three arguments require only one rule in their proof, whereas the second triple require two. However, the distribution of types of terms in the conclusions is exactly the same for the one-rule arguments as for the two-rule arguments: Variables occur twice, proper names six times, and temporary names four times across items. Thus, the one-rule and the two-rule arguments maintain the same average level of generality. The same is true for the second set of six arguments. The first three of these are again one-rule arguments, and the second three two-rule arguments. However, all arguments have the same conclusion, $F(a,b,k,l)$, and their premises preserve the same distribution of types of terms for the one-rule items as for the two-rule items. Variables occur four times, proper names six times, and temporary names twice in each argument triple.

Notice also that, according to the theory, Rules 2, 3, and 4 of the appendix appear equally often in the proofs of the one-rule arguments as in the proofs of the

[2] To make Table 1 easier to read, I have sometimes repeated variables and temporary names from the premise in the conclusion. For example, x and y appear in both the premise and conclusion of the second argument in the table. As noted earlier, however, the system assumes that the premise and conclusion contain a disjoint set of variables and temporary names. Hence, instead of $F(x,y,a,b)$ as in the conclusion of the second argument, the system would take $F(z,w,a,b)$, or some equivalent, as input. The appendix rules automatically handle these notational variants, as in the last step of the proof in Figure 1. We can assume that the difficulty associated with these "clean-up" steps is negligible, compared to remaining steps of the proof.

Number of Rule Types	Argument	Rules in Proof	Mean Response Time (ms)	Error Rate (%)
		Constant Premise		
One	F(x,y,n,m) _ F(a,b,n,m)	R4	5271	8.3
	F(x,y,n,m) _ F(x,y,a,b)	R2	6510	6.2
	F(x,y,n,m)_ F(k,l,n,m)	R3	5902	11.4
Two	F(x,y,n,m)_ F(a,y,b,m)	R2, R4	6204	13.5
	F(x,y,n,m)_ F(k,y,a,m)	R2, R3	7343	17.7
	F(x,y,n,m)_ F(a,k,n,m)	R3, R4	6536	18.8
		Constant Conclusion		
One	F(x,y,k,l)_ F(a,b,k,l)	R4	5675	10.4
	F(m,n,k,l)_ F(a,b,k,l)	R2	6475	11.4
	F(a,b,x,y)_ F(a,b,k,l)	R3	6985	14.6
Two	F(x,m,k,l)_ F(a,b,k,l)	R2, R4	6377	10.4
	F(m,b,x,l)_ F(a,b,k,l)	R2, R3	6634	15.6
	F(x,b,y,l)_ F(a,b,k,l)	R3, R4	6577	7.3

two-rule arguments. Rule 2, for example, applies twice in the proof of Argument 2 and twice in the proof of Argument 8 (both one-rule arguments) and once each in the proofs of Arguments 4, 5, 10, and 11 (all two-rule arguments). The difference between the sets of one-rule and two-rule items lies in the way the individual rules are assigned to the arguments, not in the overall frequency with which they apply. This also means that the different types of generalization associated with the rules (i.e., temporary name to permanent name, permanent name to variable, and temporary name to variable) occur just as often in the one-rule as in the two-rule items. Although one of these types of generalization may be more difficult to perform than the others, such variation could not influence the predicted one- versus two-rule difference.

In the experiment, we presented the arguments to the participants using sentences of the form: ______ *reminded* ______ *to compare* ______ *to* ______ (e.g.,

Everyone reminded Martha to compare everybody to Fran). There were 48 valid arguments in all, four versions of each of the 12 basic arguments in Table 1. The versions differed in the ordering of the terms within the premise and conclusion. We can call two terms a *corresponding* pair if they occupy the same argument positions in congruent sentences. For example, in the last argument in the table [i.e., $F(x,b,y,l)$ _ $F(a,b,k,l)$], the first term position is occupied by x (in the premise) and a (in the conclusion), the second position by b and b, the third by y and k, and the fourth by l and l. Hence, the corresponding pairs are $<x,a>$, $<b,b>$, $<y,k>$, and $<l,l>$. Reordering these corresponding pairs within the premise and conclusion will not change the validity of the argument or the number of rules in its proof, so we could equally well have, for example, $F(l,y,b,x)$ _ $F(l,k,b,a)$. To create the four versions, we used a random Latin square to select four re-orderings such that each corresponding pair (e.g., $<x,a>$) appeared once in each term position. As an illustration, one version of the last argument in Table 1 might be: *Everyone reminded somebody to compare everybody to Emily; therefore, someone reminded somebody to compare Lydia to Emily.* A second version of the same argument might be: *Catherine reminded everyone to compare somebody to everybody; therefore, Catherine reminded Liz to compare somebody to someone*. (New female first names appeared as the proper names in each argument.)

The premises and conclusions in this experiment, unlike those of the earlier study, contain possible scope ambiguities. For example, the sentence *Everyone reminded everybody to compare someone to somebody*, which was a possible conclusion for the second argument in Table 1, could have several different readings, depending on whether *someone, somebody*, or both have scope over *everyone* or *everybody*. It is unclear from current research in psycholinguistics (e.g., Kurtzman & MacDonald, 1993), which of the possible readings people extract from such sentences. These potential ambiguities do not affect the validity of the arguments in Table 1, since the premises still entail the conclusions on each of the interpretations. (For convenience, only the wide scope reading for temporary names appears in Table 1.) Similarly, the number of rules in the derivation remains constant over the different readings. Nevertheless, participants may take longer to comprehend ambiguous sentences than unambiguous ones. This means that we will have to examine separately the arguments that contain possible ambiguities (i.e., Arguments 2, 4, 5, 9, 11, and 12 in Table 1) to see whether they behave differently from the remaining unambiguous items.

We shuffled the valid problems among an equal number of invalid arguments, which we obtained from the valid ones by reversing their premise and conclusion. As in the first study, a computer presented the full set of 96 arguments to each participant in random sequence. On each trial, the participant saw the word "ready" on the screen, followed two seconds later by an argument. The participant was to "read both sentences [in the argument] carefully and then decide whether the second sentence MUST be true whenever the first sentence is true." The participant pushed one key on the keyboard to indicate that the conclusion "must be true" on the basis of the premise or a second key to indicate that the conclusion "need not be true." The button press removed the argument from the screen and presented a message informing the participant whether his or her response had been correct or incorrect. The screen then became blank, and the participant could begin the next trial by pressing the space bar. We tested 24 undergraduates, none of whom had college-level logic training and none of whom had participated in the earlier experiment.

The mean response times for the individual arguments in this experiment appear in Table 1. Although there is some overlap between the times for the one-rule and the two-rule arguments, average times are longer for the two-rule items, as predicted. Participants took 6136 ms to recognize the one-rule arguments as valid, whereas they took 6612 ms for the two-rule arguments -- a statistically reliable difference.[3] The main deviation from this trend is the ninth argument in the table, which has longer times than any of the rest of the one-rule arguments. The reason for the difficulty of this argument is unclear, though it is one of the arguments that has potential scope ambiguity. In general, the arguments containing an ambiguous premise or conclusion tend to have longer times than other arguments in their class. The difference between one-rule and two-rule arguments is still present, however, when we remove the ambiguous items. Mean response time is 5831 ms for the unambiguous one-rule problems and 6456 ms for the unambiguous two-rule problems, a slightly larger difference than for the full argument set.

Table 1 also contains the percentage of trials on which participants made an error on each argument, pressing the "need not be true" button when they should have pressed "must be true." In general, these error rates follow the same trend as the response times, though the difference is of fairly small magnitude and is not reliable. Participants made errors on 10.4% of trials for the one-rule problems and on 13.9% for the two-rule problems. The difference is of approximately the same size if we consider only the unambiguous problems: 10.4% errors on the one-rule problems and 14.6% on the two-rule problems. The positive correlation between response time and error rate suggests that participants were not sacrificing accuracy for speed in judging the validity of the arguments.

We can also look at these data from the point of view of the type of rule that appears in the proofs (see Table 1, column 3). For the one-rule arguments, those involving Rule 4 (i.e., Arguments 1 and 7) appear easier than those involving Rules 2 or 3. Mean response time for the former arguments is 5473 ms compared to 6492 ms and 6443 ms for the latter, nearly a 20% increase. The same seems to be true for the two-rule arguments. Arguments that have Rule 4 in their proof (Arguments 4,

[3] The times in Table 1 include only correct ("must be true") responses. A few of the response times in the experiment were extremely long (nearly 30 s), probably a result of participants' inattention. To eliminate outliers, we removed times that were greater than $Q_3 + 1.5(Q_3 - Q_1)$ and those less than $Q_1 - 1.5(Q_3 - Q_1)$, where Q_1 is the first quartile and Q_3 the third quartile for each argument (Mosteller & Hoaglin, 1991). The difference between one-rule and two-rule arguments remains, however, whether or not the outliers are eliminated.

6, 10, and 12) took participants an average of 6424 ms to confirm, those with Rule 2 (Arguments 4, 5, 10, 11) took 6640 ms, and those with Rule 3 (Arguments 5, 6, 11, and 12) took 6772 ms. These differences are not independent of the contrast between ambiguous and unambiguous arguments, discussed earlier. However, the data hint that the differences remain, even when we discard the ambiguous items. Table 1 shows that Argument 1 (whose proof involves only Rule 4) takes less time to verify than Argument 3 (Rule 3 alone) or Argument 8 (Rule 2 alone). None of these arguments contains scope ambiguities. Rule 4 handles cases in which variables are instantiated to temporary names -- from *everyone* to *someone* in the arguments as the participants saw them. Since variables are, in a sense, the most general terms in these arguments and temporary names the least general, the validity of the corresponding arguments may have been more apparent than those in which variables are instantiated to proper names (Rule 3) or proper names to temporary ones (Rule 2). Analogous differences appear in other cognitive domains. It takes people less time, for example, to confirm that 9 is larger than 3 than that 9 is larger than 7 (or that 7 is larger than 3) (Moyer & Landauer, 1967). It is also possible that the difference is due to the constancy of the words *someone* and *everyone* compared to the changing cast list of proper names.

Conclusions

These results support the claim that different forms of generalization (or instantiation) require separate cognitive operations. According to the theory proposed here, people's abilities to generalize from a variable, proper name, and temporary name (i.e., Skolem function or Skolem constant) depend on different rules, since they obey different formal constraints. The rules in the appendix spell out these constraints and allow us to make predictions about people's speed and accuracy in verifying arguments. Arguments that are deducible on the basis of a single rule applied to two terms should be simpler than arguments that require people to switch between rules. The experiment reported here confirms this prediction and shows that the difference isn't due to overall variation in the abstractness of the premises or conclusions.

It is still conceivable, of course, that some uncontrolled difference between one-rule and two-rule arguments could be responsible for the participants' performance. It is unclear at this point, however, what such a factor could be. One possibility, for example, is that sheer repetition of a term (*everyone* or *someone*) in the premise or conclusion could make the task easier, perhaps by reducing the time to read the sentence. A glance at Table 1, however, suggests that this sort of repetition can't account for our main findings. Argument 3 (a one-rule argument) and Arguments 5 and 6 (two-rule arguments) each contain a single repetition of *everyone* in the premise and no repetitions of terms in the conclusion, but mean times are 5271 for the former and 6940 for the latter. It is impossible to rule out all such factors in advance, but there is no simple variable we know of that can predict the obtained pattern of data.

Although the experiment supports the general idea of different cognitive procedures for different types of generalization, this does not necessarily mean that the rules in the appendix give the right description of these procedures. It is certainly possible that other operations for handling terms provide a better account of people's deduction skills. What the results do establish is that any such theory needs to respect the distinctions among the instantiating terms. This constraint is not a trivial one and, in fact, not every psychological theory of deduction satisfies it. A recently published proposal by Martin Braine (1998), for example, has a single rule (his Schema 8i, Table 11.3) that handles instantiating universally quantified terms to both existentially quantified terms and proper names. As such, Braine's proposal can't appeal to a change in rule to explain longer inference times for arguments that require both types of instantiation.

Instantiation or variable-binding is central to most proposals about cognitive architecture. That's because these proposals use variable-binding to hook general information stored in memory to specific cases that fall under them. It would be obviously implausible, for example, to try to remember that every even number is divisible by two by recording this fact with each of the even numbers you happen to encounter; hence, general information about even numbers has to be applied to specific cases. Much the same goes for other generalities that we learn in the course of daily life (how to order a meal in a restaurant) and that we learn in a formal setting (how to prove theorems in complexity theory). It is odd, for this reason, that there have been no attempts to study this mental process directly. The idea behind the present endeavor is to provide a proposal and a technique for filling this gap.

References

Anderson, J. R. (1983). *The architecture of cognition.* Cambridge, MA: Harvard University Press.

Braine, M. D. S. (1998). Steps toward a mental predicate logic. In M. D. S. Braine & D. P. O'Brien (Eds.), *Mental logic* (pp. 273-331). Mahwah, NJ: Erlbaum.

Gentzen, G. (1969). Investigations into logical deduction. In M. E. Szabo (Ed.), *Collected papers of Gerhard Gentzen.* Amsterdam: North-Holland. (Originally published 1935.)

Grandy, R. E. (1977). *Advanced logic for applications.* Dordrecht, Holland: Reidel.

Hindley, J. R., & Seldin, J. P. (1986). *Introduction to combinators and λ-calculus.* Cambridge, England: Cambridge University Press.

Johnson-Laird, P. N., Byrne, R. M. J., & Tabossi, P. (1989). Reasoning by model: The case of multiple quantification. *Psychological Review, 96,* 658-673.

Kurtzman, H. S., & MacDonald, M. C. (1993). Resolution of quantifier scope ambiguities. *Cognition, 48,* 243-279.

Mosteller, F., & Hoaglin, D. C. (1991). Preliminary examination of data. In D. C. Hoaglin, F. Mosteller, & J. W. Tukey (Eds.), *Fundamentals of exploratory analysis of variance* (pp. 40-49). New York: Wiley.

Moyer, R. S., & Landauer, T. K. (1967). Time required for judgements of numerical inequality. *Nature, 215,* 1519-1520.

Newell, A. (1990). *Unified theories of cognition.* Cambridge, MA: Harvard University Press.

Osherson, D. N. (1976). *Logical abilities in children* (vol. 4). Hillsdale, NJ: Erlbaum.

Prawitz, D. (1965). *Natural deduction: A proof-theoretical study.* Stockholm: Almqvist & Wiksell.

Rips, L. J. (1994). *Psychology of proof.* Cambridge, MA: MIT Press.

Schönfinkel, M. (1967). On the building blocks of mathematical logic. In J. van Heijenoort (Ed.), *From Frege to Gödel: A source book in mathematical logic, 1879-1931* (pp. 355-366). Cambridge, MA: Harvard University Press. (Original work published 1924.)

Shoenfield, J. R. (1967). *Mathematical logic.* Reading, MA: Addison-Wesley.

Appendix A: Matching rules for premise and conclusion terms.

Rule Conditions	Rule Output
Rule 1: Generalize variable y in conclusion $P(y)$ to variable x in premise $P(x)$.	
a. Wherever x appears (in nonsubscript position) in $P(x)$, either y or a_y appears in $P(y)$.	a. Substitute x for y in $P(y)$ in all nonsubscript positions.
b. If x and a_y appear in corresponding positions, then other occurrences of a_y also occur in corresponding positions with variables.	b. Substitute x for a_y wherever they appear in corresponding positions.
c. If x appears as a subscript in $P(x)$, y appears as a subscript in corresponding positions.	c. Substitute x for y in each of y's remaining subscript positions.
Rule 2: Generalize temporary name a in conclusion $P(a)$ to temporary or proper name n in premise $P(n)$.	
a. Wherever a appears in $P(a)$ either n or a variable appears in $P(n)$.	Substitute n for all occurrences of a in $P(a)$.
b. Any subscripts of n are included among the subscripts of a.	
c. a does not appear in a premise.	
Rule 3: Generalize proper name m in conclusion $P(m)$ to variable x in premise $P(x)$.	
a. Wherever x appears in (a nonsubscript position) in $P(x)$, m appears in $P(m)$.	a. Substitute x for m at corresponding positions in $P(x)$.
b. If a temporary name a appears in $P(m)$ in corresponding position with a temporary name b_x, then other occurrences of a are also in corresponding positions with b_x.	b. Add x to the subscripts of a wherever a occurs in corresponding positions with b_x.
Rule 4: Generalize temporary name a in conclusion $P(a)$ to variables $x_1,...,x_k$ in premise $P(x_1,...,x_k)$.	
a. Wherever a appears in $P(a)$, x_1 or x_2 or...or x_k appears in (nonsubscript positions) in $P(x_1,...,x_k)$.	a. Substitute x_i for a at corresponding positions in $P(a)$.
b. If temporary name b appears in corresponding position with c_{xi} ($i = 1,...,k$), then other occurrences of b are also in corresponding positions with c_{xi}.	b. Add x_i to the subscripts of b wherever b appears in corresponding positions with c_{xi}.

Recursion Theoretic Memories 1954-1978

Gerald E. Sacks*

No longer then the storytellers,
We become the story.
-Leo Harrington

Not a history lesson. Am not a historian nor was meant to be. A personal memoir, an attempt to recall faint impressions of recursion theoretic events. Recursion theory is a house with many rooms. Today I open the door on classical recursion theory, the science of recursively enumerable sets and degrees, far less so on higher recursion theory, and only from 1954 to 1978. I also turn the light on some personal recursion theoretic struggles during those distant days. Not a survey paper, but a partial reconstruction of what caught my recursion theoretic eye back then. Nature being what it is, I have overlooked a great deal. Here is my list.

1. Incomparable Degrees. There exist incomparable Turing degrees below $0'$ (Kleene and Post 1954). My first year of graduate study was 1958-59. My thesis advisor, J. B. Rosser, conducted a two-hour logic seminar once a week. I was the only student, although one or two Cornell faculty members occasionally attended. Rosser was the principal speaker in the fall. He talked about many-valued logic, set theory, lambda calculus and combinatory logic. At first I thought he preferred formal syntactical arguments, but then he surprised me with slick algebraic proofs of completeness for many-valued systems. Without saying so explicitly, he taught me that logic was just another branch of mathematics to which ideas from other branches could be applied. The lesson, or rather its converse, was reinforced when he gave me a then hard-to-find copy of notes by Tarski on a decision procedure for the elementary theory of the reals. I asked him one day how he found the proofs of obscure identities in combinatory logic. There did not seem to be any intuitive place to start. He said the proofs were found by playing with combinators every day all day for several months. Another valuable lesson. In the spring Rosser

*My thanks to the Logicians of Leeds for their hospitality during the Summer of 1997. A fond memory of recent times.

presented Gödel's proof of the consistency of the continuum hypothesis and then suggested I speak on recursion theory. I presented Kleene's proof that the equation calculus is equivalent to the schemes for partial recursion. Interesting. Then I presented the Kleene-Post paper on degrees and two others discussed below (**4** and **8**). Very interesting. I was struck by the $0'$ bound on the two incomparables, and more so by the construction of a minimal pair of upper bounds for the degrees of the arithmetic sets. The latter result was for me the germ of forcing with perfect conditions (Gandy & Sacks 1967). My short term memory was better in those days than it is now. The Kleene style of argument relied heavily on long formulas. As I wrote them on the blackboard, I deliberately avoided looking at my notes. Rosser took no notice, but was pleased I think.

2. Creative Sets. Any two creative sets are recursively isomorphic (Myhill 1955). During the academic year 1959-1960 Anil Nerode taught a course on recursion theory that broke with Kleene's formal style and followed Post's intuitive lead (1944). Fascinating. Now that I knew recursion theory could be done as informally as any other branch of mathematics, I planned to try my hand at it during the Summer of 1960. Nerode gave an understandable proof of Kleene's fixed point theorem based on a paper of Rogers (1959), and applied it as in Myhill (1955) to show any two creative sets have the same 1-1 degree. I found the fixed point theorem to be mysterious, and Myhill's use of it devoid of the smallest particle of intuition. "How did he ever think of it?" I wondered. A few months later I stopped asking such questions forever. Cf. item **15** below.

3. Hyperarithmetic Sets. Let $X \subseteq \omega$. Then $X \in \Delta^1_1 \longleftrightarrow X \in HYP$ (Kleene 1955). During my first year as an assistant professor at Cornell, 1962-1963, Nerode asked me to give a series of lectures on hyperarithmetic theory. I knew nothing about the subject. Rogers's book was not yet available, so I worked directly from Kleene's papers. This was my first exposure to effective transfinite recursion (ETR), a startling idea, and the beginning of my interest in higher recursion theory. My understanding of ETR was limited until I pondered Rogers 1959, and later conversed with Kreisel. The equivalence of Δ^1_1 and HYP was almost too good to be true. Yet I detected no signs of "rigging" or "setup". The definitions of Δ^1_1 and HYP had been chosen at different times for different reasons. The equivalence was simply a good reason to believe the definitions had been well chosen. In 1969 Kleene told me he worked on this problem for some time and then decided that on a certain day several months off, at 5:30 pm, he would quit. He found the proof at 5:00 pm on the last day. I believe if he had not found it then he would not have thought about it again.

4. Minimal Degrees. There exists a minimal degree below $0''$ (Spector 1956). In the spring of 1959 Rosser asked me to present Spector's paper on minimal degrees. The paper was written in the same style as the Kleene-Post paper. All the recursions were given by explicit formulas. In 1961 Spector told me that he had suppressed an intuitive write-up, because his thesis advisor, Kleene, had insisted on the formalism. In 1960 I remarked to Rosser that Spector's minimal degree construction struck me as being as original as the Friedberg-Muchnik priority argument. He agreed. Spector's construction is now often viewed as an application, somewhat unconscious, of forcing with finite conditions on perfect trees. Spector was an extraordinary problem solver. Kleene asked him if the Turing degrees were dense, and Spector replied with a minimal degree he found during Thanksgiving vacation as a graduate student.

In the Fall of 1959 Nerode drew my attention to several open problems in recursion theory, and in particular to finding a minimal degree below $0'$. He began by asking me to sketch Spector's argument. I had to excuse myself briefly to run downstairs to the mathematics library and review Spector's paper. Nerode didn't seem to mind the delay. We had a long conversation that greatly influenced the course of my thesis begun in earnest during the summer of 1960. I found a minimal degree below $0'$ in January 1961 while waiting in the checkout line of a supermarket outside Princeton, New Jersey. Of course I had thought about it many times before, but I had not thought of combining Spector's approach with the priority method until I leaned against a shopping cart.

5. Post/s Problem. There exist incomparable RE degrees (Muchnik 1956, Friedberg 1957). Nerode explained the argument with a minimum of equations during his 1959-60 course on recursion theory. Fascinating. I very much wanted to continue this line of work. Friedberg's theorem was part of his undergraduate thesis at Harvard. He discovered the proof during a course on recursion theory he took with Hartley Rogers.

6. Inverting the Jump. The Turing jump operator maps onto the cone above $0'$ (Friedberg 1957).

7. Maximal RE Sets. There exists a co-infinite RE set such that every infinitely larger RE set is co-finite (Friedberg 1958). I met Friedberg in New York City in 1962. Over a shared pizza he told me it took several months for him to understand his solution of Post's problem after he found it. I asked him to illuminate the proof of his maximal set result, since I was planning to teach it shortly. He was unable to recover the argument. At that time he was immersed in graduate studies in the Physics department at Columbia.

8. Recursive Functions. The location of the class of total recursive functions in Kleene's arithmetic hierarchy (Schoenfield 1959). Rosser asked me to present this result in his logic seminar in the Spring of 1959. Both of us were impressed by the elegant use of Baire's category theorem in place of the usual diagonal argument.

9. Δ_2^0 Degrees. Splitting $0'$ into two incomparable Δ_2^0 degrees. Inverting the jump on Σ_2^0 sets (Schoenfield 1959). Nerode told me to read this paper soon after it appeared. I was excited by the thinking behind its results. Δ_2^0 functions were characterized as limits of recursive sequences of recursive functions. Finite injury priority arguments were used to construct sets that were not RE. For example, each degree RE in and above $0'$ was the jump of a Δ_2^0 degree. An eye-opener on the power of the priority method. My thesis was largely an effort to continue the work of this rich and varied paper.

10. Recursion in ^{2}**E.** Let $X \subseteq \omega$. Then X is recursive in ^2E iff $X \epsilon HY\dot{P}$ (Kleene 1959). This result, like item **3** above, was almost too good to be true. Two concepts defined far apart in space and time turn out to be equivalent. Yet nothing had been "rigged". The equivalence was an honest mathematical discovery. It convinced me that Kleene's notion of recursion in normal objects of finite type was important, although it was some time before I acted on the conviction (item **25** below).

11. Effective Transfinite Recursion. A clear statement and proof of the Church-Kleene method of effective transfinite recursion (Rogers 1959). My understanding of ETR was formal, hence almost useless, until I read Rogers's intuitive account. Even now it seems remarkable that transfinite recursion can be used to define recursive functions.

12. Hyperarithmetic Quantifiers. Each Π_1^1 predicate $P(n)$ can be put in the form $\exists f_{HYP} A(f,n)$ for some arithmetic A (Spector 1959, Gandy 1960).

13. Compactness for ω **– Logic.** Let Z be a Π_1^1 set of sentences in ω-logic. If every hyperarithmetic subset of Z has a model, then Z has a model (Kreisel 1961, 1965). Item **12** contributed to Kreisel's insight that if Π_1^1 is analogous to RE, then so is hyperarithmetic to finite. His insight led him to item **13,** the beginning of compactness on countable admissible sets beyond $L(\omega)$. He was the first to understand that lifting recursion theory beyond $L(\omega)$ required a lifting not only of "recursively enumerable" but also of "finite" (cf. item **32** below). For me the Spector-Gandy theorem had a non-intuitive aspect that lingers to this day. A number n satisfies the Π_1^1 predicate $P(n)$ thanks to some recursive wellordering far from evident in $\exists f_{HYP} A(f,n)$. The proofs of **12** turn on the nature of non-standard recursive wellorderings.

14. Isols. A general method for extending certain arithmetic functions from integers to the isolic integers (Nerode 1961). An important example of applying mathematical ideas far from logic, in this case difference equations, to problems in recursion theory.

15. Splitting. Each non-recursive RE set is the disjoint union of two RE sets of lower degree (Sacks 1963). During the Summer of 1960, after a year with Rosser (**1** above) followed by a year with Nerode (**2** above), I tried my hand at classical recursion theory. Shoenfield's 1959 paper (**9** above) inspired me to try and split the complete RE set K into two incomplete RE sets, A and B. At first it appeared hopeless. A number n came up in K. It had to be put in A or B. Either choice wreaked havoc on inequalities being preserved in the style of Friedberg (1957). I tried to delay the choice, but to no avail, so I ruled preemptively that the location for n had to be chosen as soon as n came up in K. I tried again and again to find some useful sense in which one choice did less harm than the other. Worst of all I could not see how to generate the inequalities to be preserved. After much head thumping I decided for no reason I can recall to see what would happen if initial segments of equality were preserved. Now priority could be used to choose between A and B for n. There was a sudden realization that inequalities would inadvertently occur otherwise K would be recursive. Succeed by doing the opposite of what you want to do. I was delighted. Recursion theory had given me what life could not.

16. Density of RE Degrees. Between any two distinct comparable RE degrees there lies a third (Sacks 1964). Shoenfield encouraged me to pursue this result, and I began to work on it in the fall of 1961 at the Institute in Princeton. I spent months trying to split an RE degree over a lesser RE degree. My lack of success was the fault of Lachlan's monumental 1975 non-splitting theorem (**29** below), although my student, R. W. Robinson, was able to split over a low RE degree in 1971. Late in the fall I was able to show that each degree above $0'$ and RE in $0'$ was the jump of an RE degree (Sacks 1963). This result opened my eyes to priority schemes in which individual requirements were injured infinitely often. I was led to it by thinking about how to approximate recursively the characteristic function of a Σ^0_2 set. Around this time I gave up on splitting an RE degree over another, and sought density directly. I thought what had been learned from splitting and jump inversion would suffice, but progress was slow. The best I could do as time ran out at the Institute was: if b and c are RE degrees such that $b < c$ and $b' < c'$, then there is an RE degree d such that $b < d < c$ and $b' < d' < c'$. I gave a short talk on this result in Stockholm during the summer of 1962. Shoenfield was present and told me once again to prove the density. Rosser was the chairman of the logic session. He used an alarm clock to prevent speakers from going over time.

In the Fall of 1962 I returned to Cornell. Nerode asked me to give talks on hyperarithmetic theory (**3** above), but I still pondered density. In the spring of 1963 I was waiting for my old friend, S. P., outside of his apartment building on West 96th Street in Manhattan. S. P. was late, so I reviewed once again what I thought of as the principal obstacle to density. Suppose b, an incomplete RE degree, is given. How can one directly construct an RE degree d such that $d \not\leq b$? A typical requirement would be : find a w such that

$$\textit{if } \{e\}^b(w) \textit{ is defined, then } d(w) \neq \{e\}^b(w).$$

The inability of d to wait around while $\{e\}^b$ made up its mind obscured the search for w. After leaning against S. P.'s brick outer walls for forty-five minutes I realized that "planting" an RE degree c in d would lead to the desired w if $c \not\leq b$. Thus an inequality could be found by adding a potential infinity of numbers to a set in conjunction with keeping numbers out to preserve computations. That night at a party in Greenwich Village I told Myhill I could prove the density of the RE degrees. He replied that the Russians might have already proved it. I wrote to

Mostowski in Warsaw who wrote back otherwise.

17. Minimal Pairs. There exist two incomparable RE degrees such that the only degree less than both is 0 (Lachlan 1966, Yates 1966). These papers were full of new ideas. The one that caught my eye was "nesting". A requirement can have several outcomes. Requirement $i+1$has several strategies operating simultaneously, one for each outcome of requirement i. The $(i+1)-th$ successful strategy operates on an infinite subset of the stages where the $i-th$ successful strategy operates. A powerful idea.

18. Anti – Cupping. There exist a non-zero RE degree $d < 0'$ such that $d \cup b < 0'$ for every incomplete RE degree b (Yates 1967, Cooper 1974). To me this result signaled the beginning of a deeper study of the RE degrees that continues to the present.

19. Selection. There exists an effective method of choosing a member from a non-empty subset of ω RE in a normal type 2 object (Gandy 1967). This is one of the results that drew me into recursion in objects of higher type. Another is item **26** below. Gandy gave me an intuitive account of his proof one day after lunch at the Institute in Oberwolfach in 1969. This was my first exposure to an intuitive account of Kleene's theory. I now understood that intuition mattered in recursion in higher types as much as it did in classical recursion theory.

20. Full Approximation. Below each non-zero RE degree is a minimal degree (Yates 1970). The method of this paper was central to further results about minimal degrees below $0'$.

21. Low Splitting. Each RE degree splits over each lesser low RE degree (Robinson 1971). This paper introduced a new application of the fixed point theorem to anticipate the outcome of an RE construction during the construction. The sort of trick that warms the heart of an unregenerate recursion theorist.

22. Finite Initial Segments. Each finite lattice is order isomorphic to an initial segment of degrees below $0'$ (Lerman 1971). A truly difficult theorem. During the years that Lerman worked on this result I gave him a great deal of advice for which he thanked me appropriately (cf. Lerman 1971, page x).

23. Low Basis. Each non-empty Π^1_0 class $\subseteq 2^\omega$ has a low member (Jockusch and Soare 1972).

24. Automorphic Maximality. Any two maximal elements of the lattice of RE sets modulo finite differences are automorphic (Soare 1974). The ground breaking result on the lattice and on connections between inclusion, and degree theoretic, properties of RE sets.

25. k – Sections. Let $0 < k < n$. The k-section of a normal type n object is the k-section of some normal type $k+1$ object (Sacks 1970, 1977).

26. Divergence Witnesses. Co-RE in a normal object F of type ≥ 3 is expressible as Σ_1 over recursive in F (Moschovakis 1967).

27. Reflection. For a normal object of type ≥ 3 the limit of reflection coincides with the location of divergence witnesses (Harrington 1973).

The proof of **25** has two parts: $k = 1$ and $k > 1$. The first part is a forcing argument over an admissible structure. The second is a forcing that succeeds by virtue of a reflection principle for types ≥ 4 that eluded me for several years.

The first part was proved in the spring of 1970, while I was teaching a course on hyperarithmetic theory in Madison, Wisconsin. Grilliot, on leave from Penn State, audited the course, and encouraged my interest in 1-sections. During one of our many Rexall Drug Store lunches, I showed him the proof of **25** for $k = 1$; he responded with a proof of the opposite. His argument consisted of a long sequence of formulas that made no intuitive sense to me but appeared free of error. He suddenly expressed doubt about one of his superscripts and ran home to check it. As I was finishing my jello, he returned and retracted his argument. The superscript was in the wrong place.

After the second part of **25** was proved, I asked Harrington to pursue reflection for types ≥ 3. He succeeded beyond expectation by showing the limit of reflection coincides with the location of Moschovakis's divergence witnesses. A refinement of his result, with initial segments of L in place of objects of finite type, was the key to forcing and priority arguments in E-recursion (Sacks 1990).

Much of my work on recursion in higher types was inspired by the preternatural insight of Moschovakis that divergence witnesses could play a *positive* role.

28. Minimal Jump. Each degree $\geq 0'$ is the jump of a minimal degree. The jump of a minimal degree $< 0'$ is $< 0''$ (Cooper 1973).

29. Non – Splitting. There exist RE degrees b and c such that $b < c$ but c does not split over b (Lachlan 1975). The dawning of a new day in the history of RE degrees. The method is developed masterfully in Soare 1987.

30. Uniform Enumeration Operators. The only extremely uniform enumeration operators are the identity and the jump (Lachlan 1975).

31. Theory of Degrees. The first order theory of the partial ordering of degrees is equivalent to second order number theory (Simpson 1977). Earlier more than one recursion theorist thought the first order theory was of degree $0^{(\omega)}$.

32. α – Density. Let α be Σ_1 admissible. The α-RE degrees are dense (Shore 1976). A delightful combination of fine structure and priority. Both are needed for the proof of: if α is Σ_1 admissible and A is α-RE, amenable and not complete, then $L[\alpha, A]$ is weakly Σ_1 admissible.

33. Measure and Minimality. The set of degrees with minimal predecessors has measure 0 (Paris 1977). Back in 1959 Nerode showed me Spector's 1958 result that the set of pairs of incomparable hyperdegrees has measure 1. This was the first proof of the existence of incomparable hyperdegrees. Intriguing. In 1962 I noticed that the set of minimal degrees has measure 0, but I had no idea how to proceed towards Paris's much deeper result. A year later I wondered about the measure of $\{X \mid \omega_1^X = \omega_1^{CK}\}$. Every senior recursion theorist I queried said 0. The answer was 1.

34. Completeness. Let A be RE. A is complete iff

$$(\exists f \leq_T A)\forall n[W_{f(n)} \neq W_n] \qquad \text{(Arslanov 1977).}$$

A striking extension of the fixed point theorem, the center of classical recursion theory.

35. Δ_2^0 **Hierarchy**. The Δ_2^0 sets constitute a hierarchy analogous to the classical difference hierarchy for $F_\sigma \cap G_\delta$ sets (Ershov 1973).

36. Anti – Post Problem I. Negative solutions to Post's problem for certain inadmissible initial segments of L (S. Friedman 1978).

37. Anti – Post Problem II. There exists a Σ_1 admissible set A such that every A-RE set is either complete or A-recursive (Harrington 1977). A rare and original forcing argument

Harvard University
Massachusetts Institute of Technology
June 30, 1998

References

1944 Post, E. L., Recursively enumerable sets of positive integers and their decision problems, Bull. Amer. Math. Soc. 50, 284-316.

1954 Kleene, S. C. and Post, E. L., The upper semi-lattice of degrees of recursive unsolvability, Ann. of Math. (2) 59, 379-407.

1955 Myhill, J., Creative sets, Z. Math. Logik Grundlag. Math. 1, 97-108.

1955 Kleene, S. C., Hierarchies of number-theoretic predicates, Bull. Amer. Math. Soc. 61, 193-213.

1956 Spector, C., On degrees of recursive unsolvability, Ann. of Math. 64, 581-592.

1956 Muchnik, A. A., On the unsolvability of the problem of reducibility in the theory of algorithms, Dokl. Akad. Nauk SSSR N. S. 108, 194-197.

1957 Friedberg, R., Two recursively enumerable sets of incomparable degrees of unsolvability, Proc. Nat. Acad. Sci. USA 43, 236-238.

1957 Friedberg, R., A criterion for completeness of degrees of unsolvability, J. Symb. Log. 22, 159-160.

1958 Friedberg, R., Three theorems on recursive enumeration, J. Symb. Log. 23, 309-316,

1958 Spector, C., Measure-theoretic construction of incomparable hyperdegrees, J. Symb. Log. 23, 280-288.

1958 Shoenfield, J. R., The class of recursive functions, Proc. Amer. Math. Soc. 9, 690-692.

1959 Shoenfield, J. R., On degrees of unsolvability, Ann. of Math. (2) 69, 644-653.

1959 Kleene, S. C., Recursive functionals and quantifiers of finite type I, Trans. Amer. Math. Soc. 91, 1-52.

1959 Spector C., Hyperarithmetic quantifiers, Fund. Math. 48, 313-320.

1959 Rogers, H., Recursive functions over partial well orderings, Proc. Amer. Math. Soc. 10, 847-853.

1960 Gandy, R., Proof of Mostowski's conjecture, Bull. Polon. Soc. Math. 8, 571-575.

1961 Nerode, A., Extensions to isols, Ann. of Math. (2) 73, 362-403.

1961 Kreisel G., Set theoretic methods suggested by the notion of potential infinity, in Infinitistic Methods, Pergamon, Oxford, 325-369.

1963 Sacks, G., On the degrees less than 0′, Ann. of Math. (2) 77, 211-231.

1963 Sacks, G., Recursive enumerability and the jump operator, Trans. Amer. Math. Soc. 108, 223-239.

1964 Sacks, G., The recursively enumerable degrees are dense, Ann. of Math. (2) 80, 300-312.

1965 Kreisel G., Model-theoretic invariants: applications to recursive and hyperarithmetic operations, in Theory of Models, North-Holland, Amsterdam, 190-205.

1966 Lachlan, A., Lower bound for pairs of recursively enumerable degrees, Proc. London Math. Soc. 16, 537-569.

1966 Yates, C. E. M., A minimal pair of recursively enumerable degrees, J. Symb. Log. 31, 159-168.

1967 Gandy, R. O. & G. Sacks, A minimal hyperdegree, Fund. Math. 61, 215-213.

1967 Gandy, R. O., General recursive functionals of finite type and hierarchies of functionals, Ann. Fac. Sci. Univ. Clermont-Ferrand 35, 202-242.

1967 Moschovakis, Y., Hyperanalytic predicates, Trans. Amer. Math. Soc. 138, 249-282.

1970 Yates, C. E. M., Initial segments of the degrees of unsolvability: part 2: minimal degrees, J. Symb. Log. 35, 243-266.

1971 Robinson, R. W., Interpolation and embedding in the recursively enumerable degrees, Ann. of Math (2) 93, 285-314.

1971 Lerman, M., Initial segments of the degrees of unsolvability, Ann. of Math. (2) 93, 365-389.

1972 Jockusch C. and Soare, R., Degrees of members of Π^0_1 classes, Pacific J. Math. 40, 605-616.

1973 Cooper, S. B., Minimal degrees and the jump operator, J. Symb. Log. 38, 249-271.

1973 Harrington, L., Contributions to recursion theory in higher types, Ph.D. Thesis, Massachusetts Institute of Technology, Cambridge MA.

1974 Cooper, S. B., On a theorem of C. E. M. Yates (handwritten notes).

1974 Soare, R., Automorphisms of the lattice of recursively enumerable sets, part 1: maximal sets, Ann. of Math. (2) 100, 80-120.

1975 Lachlan, A. H., A recursively enumerable degree which will not split over all lesser ones, Ann. Math. Logic 9, 307-365.

1975 Lachlan, A. H., Uniform enumeration operators, J. Symb. Log. 40, 401-409.

1977 Paris, J., Measure and minimality, Ann. Math. Log. 11, 203-216.

1977 Harrington, L., A negative solution to Post's problem, handwritten notes, cf. 1984 Chong, C. T.

1977 Sacks, G., The k-section of a type n-object, Amer. J. Math. 99, 901-917.

1977 Simpson, S. G., First-order theory of the degrees of recursive unsolvability, Ann. of Math. (2) 105, 121-139.

1978 Friedman, S., Negative solutions to Post's Problem I, in Generalized Recursion Theory II, North-Holland, 127-133.

1983 Lerman, M., Degrees of Unsolvability, Springer, Berlin, Heidelberg, New York.

1984 Chong, C. T., Techniques of Admissible Recursion Theory, Springer, Berlin, Heidelberg, New York.

1987 Soare, R., Recursively Enumerable Sets and Degrees, Springer, Berlin, Heidelberg, New York.

1990 Sacks G., Higher Recursion Theory, Springer, Berlin, Heidelberg, New York.

Fields Definable in Simple Groups

Katrin Tent
Mathematisches Institut
Universität Würzburg
97074 Würzburg, Germany
tent@mathematik.uni-wuerzburg.de

Cherlin's Conjecture that an infinite simple group of finite Morely rank is an algebraic group over an algebraically closed field has been around for many years now, and has been the starting point for a considerable amount of research. In this survey paper we describe two approaches towards Cherlin's Conjecture, first without any stability assumption via the theory of algebraic groups and secondly via the theory of Tits buildings in the context of finite Morley rank. While the conjecture is still open, our results cover most classes of classical and algebraic groups and the (twisted) Chevalley groups.

1 Algebraic groups and Cherlin's Conjecture

Restricted to algebraic groups, Cherlin's Conjecture reduces to the following: If the group $G(k)$ of k-rational points of an algebraic group G is a simple group of finite Morley rank, does this imply that k is algebraically closed? In similar form this was asked by Borovik and Nesin in [BN], p.367.

We show that this is (almost) true: whenever G is almost simple (or k-simple) and k-isotropic, the field k (or a finite extension of k) is definable in the pure group structure of $G(k)$; hence if such a group has finite Morley rank, then k has to be either algebraically closed or real closed, thus answering Questions B.46 and B.48 in [BN] for these cases. This result should come as no surprise since in [BT73] it is shown that the group of abstract automorphisms of such a group is essentially the group itself extended by the automorphism group of the field. This suggests from the point of view of model theoretic Galois theory that the field should be 'visible' in the group.

We assume some basic knowledge of linear algebraic groups. Throughout this section, let k be an infinite field, $K \supseteq k$ its algebraic closure, and G a reductive linear algebraic group defined over k. Assume for simplicity that $char(k) \neq 2$, see [KRT] for the general case. Then $G(k)$ denotes the group of k-rational points of G. We usually identify G with $G(K)$. The group G is called k-(almost) simple if it has no proper normal (connected) subgroup defined over k, and it is called absolutely (almost) simple if G is K-(almost) simple. A group may be k-almost simple without being absolutely almost simple. Any reductive k-group has a torus $T \cong K^* \times \ldots \times K^*$ defined over k. If the isomorphism is defined over k, then T is called k-split and G is called k-isotropic.

The Borel subgroups of G are the maximal solvable subgroups of G, they need not be defined over k. Parabolic subgroups are the closed subgroups of G containing a Borel subgroup. If G is k-isotropic, then there is some parabolic subgroup P defined over k, hence $P(k)$ makes sense.

If P is a parabolic subgroup of G defined over k, then the unipotent radical $U = \mathcal{R}_u P$ is defined over k and there exists a complement L defined over k. Then P is a semidirect product $P = L \ltimes U$ and L is called a k-Levi subgroup of P.

1.1 Theorem [KRT] *If G is an isotropic absolutely almost simple k-group, then k is definable in the structure $(G(k), \cdot)$. Hence, if $G(k)$ has finite Morley rank, then it is an algebraic group over an algebraically closed field.*

Proof. Let P be a minimal parabolic k-subgroup of G, with k-Levi decomposition $P = L \ltimes U$, and let $V = Z(U)$ be the center of the unipotent radical U of P. Then V carries the structure of a K-vector space, and essentially by [ABS], it is an irreducible L-module. Hence by Schur's Lemma, the center Z of L acts as K-scalars and non-trivially on V.

First we claim that $L(k)$ and $Z(k)$ are definable in $(G(k), \cdot)$. By [Bo], Prop 8.18, there exists some $x \in G(k)$ such that $L(k)$ is the centralizer of x in $G(k)$. Since L is reductive, the center Z of L is defined over k, and since $L(k)$ is dense in L, it follows that the center of $L(k)$ is exactly $Z(k)$, which shows $Z(k)$ to be definable.

Now let $v \in V(k)$ be a nontrivial element, on which $Z(k)$ acts by conjugation in $G(k)$ as k-scalars. One can show that the orbit M of v under $Z(k)$ contains all elements of the form $t^2 \cdot v$ for $t \in k$. Since any element of k is the difference of two squares, the set $k' = \{a - b;\ a, b \in M\}$ is a 1-dimensional subspace of $V(k)$ and hence carries the additive structure of k. Now define a multiplication on M via $(v^x) \cdot (v^y) = v^{xy}$ for $x, y \in Z(k)$ and linear extension to all of k'. With this multiplication and addition, k' is a field isomorphic to k and we have achieved our claim. □

In the special case where k itself is algebraically closed, a similar result was obtained by Poizat, see e.g. [Po]. In that case, the minimal k-parabolic subgroups are just the Borel subgroups, and their solvability is used to define a field which by the finite Morley rank assumption has to be algebraically closed and isomorphic to k. Related results were also obtained by [PPS] in the o-minimal context (see also Theorem 1.2 and Remark 1.4 below).

Note that it may happen that the structure $(G(k), \cdot)$ has finite Morley rank, where $G(k)$ is a simple group (in the sense of group theory, not in the sense of algebraic groups) and the field k is not algebraically closed: Take the group $\mathbf{PGL}_2(\mathbf{C})$ considered as a matrix group $H(\mathbf{R})$ with the entries being 2×2-matrices over $\mathbf{R}$. Then clearly, also $(H(\mathbf{R}), \cdot)$ is a simple group of finite Morley rank, but obviously, $\mathbf{R}$ cannot be definable in $(H(\mathbf{R}), \cdot)$.

The point is that H is not absolutely simple any more as an algebraic group, hence the assumption of absolute simplicity in Theorem 1.1 which was used to obtain the scalar action of Z on V cannot be weakened to k-simplicity. This example shows that we may have considered the given (abstract) group as an algebraic group over the 'wrong' field.

If the group G is k-simple, and either adjoint or simply connected, then by [Ti66], [BT65] there exists a finite separable field extension l of k and some absolutely almost simple l-group H such that $H(l)$ is isomorphic (as a group) to $G(k)$ via the functor $R_{l/k}$, called *restriction of scalars*. In the above example, the functor $R_{\mathbf{C}/\mathbf{R}}$ takes the group $\mathbf{PGL}_2$ to a group $H = R_{\mathbf{C}/\mathbf{R}}\mathbf{PGL}_2$ with the property that $H(\mathbf{R}) \cong \mathbf{PGL}_2(\mathbf{C})$. So at least for an adjoint or simply connected k-almost simple group G, if we cannot define k in $G(k)$, then it is because we should have considered the abstract group as an algebraic group over a different field.

However, it is not possible to see just from the abstract group $G(k)$ whether G is simply connected or adjoint, e.g. $\mathbf{PGL}_3(\mathbf{R}) \cong \mathbf{SL}_3(\mathbf{R})$ is simple as an abstract group, but $\mathbf{PGL}_3$ is adjoint and $\mathbf{SL}_3$ is simply connected. On the other hand, $\mathbf{SL}_3(\mathbf{Q})$ is still simple while $\mathbf{PGL}_3(\mathbf{Q})$ is far from being simple. Note also that if $char(k) \neq 0$, then a group may be k-simple without being adjoint!

An almost k-simple group sits between an adjoint and a simply connected group and our construction defines the field extension corresponding to these groups:

1.2 Theorem *If G is an almost k-simple k-isotropic group, then a finite extension l of k is definable in $G(k)$. Hence, if $(G(k), \cdot)$ has finite Morley rank, then either k is algebraically closed, or k is a real closed field and $l = k(i)$, where i is a square root of -1.*

Proof. With the notation as above, we let P be a minimal parabolic k-subgroup of G, $P = L \ltimes U$ its Levi decomposition, and $v \in V(k)$. We again want to define the field on the vector space spanned by the orbit of v under $Z(k)$, where Z is the center of L. Note that $Z(k)$ is definable just as before.

If G is simply connected or adjoint, we can find a finite separable field extension l of k and an absolutely simple l-group H such that $G = R_{l/k}H$ is obtained from H by restriction of scalars. Then $G(k)$ is isomorphic via $R_{l/k}$ with $H(l)$. Since we can define l in $H(l)$ by Theorem 1.1 using only the group structure, the same construction yields a definition for l in $G(k)$.

If G is neither simply connected nor adjoint, there might not be any absolutely simple group corresponding to G. In order to see that the same construction defines a field, we have to check (a) that the action of $Z(k)$ on v is by scalars from some field extension l of k, and (b) that the orbit is 'big enough', namely contains at least the squares of l.

To check (a) we take a simply connected k-simple group G' and a central isogeny $\phi : G' \longrightarrow G$. Let H' be the absolutely simple group defined over some finite extension l of k such that $H'(l)$ is isomorphic to $G'(k)$, and let H'' be an adjoint absolutely simple group isogenous with H' via a central isogeny ψ, which corresponds via $R_{l/k}$ to an adjoint group G'' isogenous with G (via some central isogeny ϕ') and such that $H''(l)$ is isomorphic to $G''(k)$. Since the functor $R_{l/k}$ takes central isogenies to central isogenies and central isogenies are essentially unique, we may assume that the following diagram commutes.

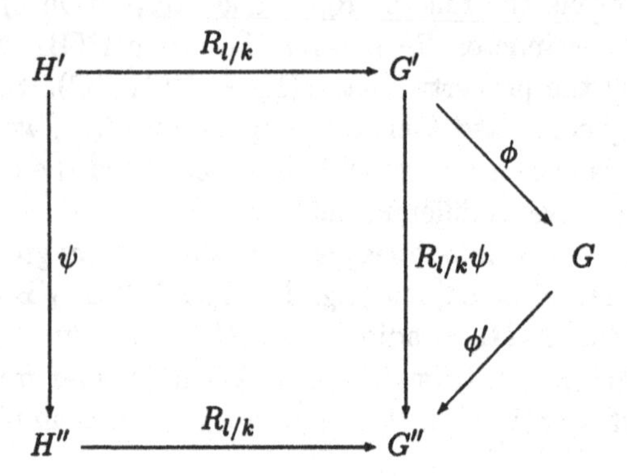

Note that central isogenies preserve the parabolic subgroups and induce isomorphisms on their unipotent radicals (s. [Bo]). This allows us to compare the orbit of v under $Z(k)$ to its image and preimage in G' and G'', respectively. Since we already know that the orbits in G' and G'' define l, we conclude that conditions (a) and (b) above are satisfied (see [KRT] for details).

The last part follows from the Artin-Schreier Theorem since any field whose algebraic closure is a finite extension is real closed (and in particular of characteristic 0). □

1.3 An example. Let q be a quadratic form of Witt index 1 on k^4. The k-group $\mathbf{SO}_4$ corresponding to q is k-almost simple but not absolutely simple. The simply connected group in the k-isogeny class of $\mathbf{SO}_4$ is the spinor group $\mathbf{Spin}_4$, and the adjoint group is the group $\mathbf{PGO}_4$ of projective orthogonal similitudes. Then $\mathbf{SL}_2(l) \cong \mathbf{Spin}_4(k)$ and $\mathbf{PGL}_2(l) \cong \mathbf{PGO}_4(k)$, for some quadratic extension l of k, and we obtain a diagram

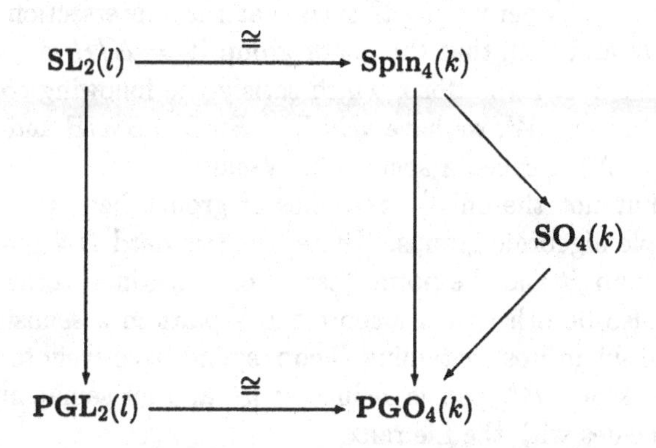

By Theorem 1.2, the field l is definable in the group $\mathbf{SO}_4(k)$. Note that the l-isogeny class of $\mathbf{SL_2}$ does not contain a group which corresponds to the k-group $\mathbf{SO}_4$. In the special case $k = \mathbf{R}$ (the reals) and $q = x_1x_2 + x_3^2 + x_4^2$, $\mathbf{SO}_4(\mathbf{R})$ is the Lorentz group of special relativity, and our construction yields the complex numbers. Since $\mathbf{SO}_4(\mathbf{R})$ has finite Morley rank, the reals cannot be definable in it.

1.4 Remark Assume k perfect and G not necessarily reductive. If $G(k)$ is definably simple, then the connected component G^0 of G is semisimple. If G is connected, then G is k-almost simple.

Proof. Let $G(k)$ be a definably simple k-group. If the connected component of the identity G^0 of G is not semi-simple, then the center of the radical of G is a nontrivial connected abelian normal subgroup A defined over k (see [Bo], p. 219). Since $A(k)$ is dense in A ([Bo], 18.3), we pick $t \in A(k), t \neq 1$. So $N = \bigcap_{g \in G(k)} Z_{G(k)}(t)^g$ is a definable and non-trivial normal subgroup of $G(k)$, contradicting the definable simplicity. Hence G^0 is semisimple.

Now suppose that G is connected and $G = H_1 \cdot H_2$ is an almost direct product where H_1 and H_2 are nontrivial normal k-subgroups of G. Let $t \in H_1(k), t \neq 1$ and consider again $A = \bigcap_{g \in G(k)} Z_{G(k)}(t)^g$. Clearly, $H_2(k) \subseteq A \neq G(k)$ and hence, as above, A is a proper definable normal subgroup of $G(k)$. Thus G must be k-almost simple. □

2 Simple groups and Tits buildings

Now we turn to a geometric approach towards simple groups using the theory of buildings.

Consider a *not necessarily algebraic group* G with a BN-pair; i.e. there are subgroups $B, N \subseteq G$ generating G such that their intersection $H = B \cap N$ is normal in N and such that the factor group $W = N/H$ is generated by a distiguished set S of involutions, which satisfy the following conditions: for all $s \in S$ and all $w \in W$, we have $sBwB \subset BwB \cup BswB$ and $sBs \neq B$. If W is finite, (B, N) is called a *spherical* BN-pair.

Typical (but not the only!) examples of groups having a BN-pair are the semi-simple algebraic groups. There, the standard BN-pair is given by a Borel subgroup B and the normalizer N of a maximal torus $T \subseteq B$ (but there might also be other, non algebraic BN-pairs in a semisimple group). The parabolic subgroups containing B correspond bijectively to subsets of S. The (Tits) rank of a BN-pair is defined as $|S|$ and for semi-simple algebraic groups it coincides with the Lie rank.

It is known [Ti74] that buildings of rank at least 3 satisfy the Moufang condition (see below) and that this property is inherited by their so-called rank 2 residues, i.e. the rank 2 buildings contained in them. Since these residues determine the structure of the building, they are of particular importance. Hence by classifying all rank 2 Moufang buildings of finite Morley rank using the results in [TW], we obtain a classification of all simple groups of finite Morley rank corresponding to these buildings.

2.1 Buildings of rank 2 and polygons

If G has a spherical BN-pair of rank 2, there are two proper parabolic subgroups, P_1 and P_2, say. Then the corresponding building is a coset geometry defined by the incidence structure $(G/P_1; G/P_2; I\ )$ where the incidence relation I is defined by $gP_1\ I\ hP_2$ if and only if $gP_1 \cap hP_2 \neq \emptyset$.

In fact, buildings of rank 2 are nothing but generalized polygons: consider $\mathcal{P} = G/P_1$ as a set of points, $\mathcal{L} = G/P_2$ as a set of lines, and define the distance $d(x, y)$ between $x, y \in \mathcal{L} \cup \mathcal{P}$ as the smallest m such that there is a *chain* $x_0, \ldots, x_m$ with $x_0 = x, x_m = y$ and $x_i\ I\ x_{i-1}$ for $i = 1, \ldots m$. We denote by $D_i(x)$ the set of elements (in either $\mathcal{P}$ or $\mathcal{L}$) at distance i of x. Note that the distance between elements of the same kind is always even.

From the definition of a spherical BN-pair of rank 2 one can check that the incidence structure defined above satifies the following axioms of a generalized n-gon, for $|W| = 2n$.

(i) For all $x, y \in \mathcal{L} \cup \mathcal{P}$, we have $d(x, y) \leq n$.

(ii) If $d(x, y) < n$, then the chain from x to y is unique.

(iii) For any $x \in \mathcal{L} \cup \mathcal{P}$ there are at least three elements incident with x; i.e. $|D_1(x)| \geq 3$.

Since the definition is symmetric in points and lines, interchanging these sets defines again a generalized n-gon, the *dual* of the polygon we started with.

For $n = 2$ the axioms define a complete bipartite graph, and for $n = 3$ the definition yields exactly the projective planes. From now on, we will always assume $n \geq 3$. As examples by Baldwin and DeBonis-Nesin show, there is no hope of classifying all generalized n-gons of finite Morley rank, s. [Ba], [DN]. About all that is known in general is that the set of points always has degree one [Te].

We would like to stress the fact that the rank 2 analysis is crucial for this result. A similar result about buildings of rank at least 3 was claimed without proof in [BN] (Theorem 12.39), based on Fact 12.38, (*loc.cit.*). However, Fact 12.38, which is attributed there to Tits [Ti74] is not correct (and not stated in [Ti74]) - the polar and the metasymplectic spaces yield counterexamples, and these are (exactly for this reason) the difficult cases to deal with.

2.2 Moufang polygons

An automorphism of a polygon $\mathfrak{P} = (\mathcal{P}, \mathcal{L}, I)$ is a permutation of $\mathcal{P} \cup \mathcal{L}$ mapping points to points and lines to lines and preserving the incidence relation. Hence this coincides with the model theoretic notion of an automorphism of the first order structure $\mathfrak{P}$ in the language with predicates for points and lines and the incidence relation. As usual, we will denote the automorphism group of $\mathfrak{P}$ by $Aut(\mathfrak{P})$ and a subgroup of it fixing a set $A \subseteq \mathcal{P} \cup \mathcal{L}$ will be denoted by $Aut(\mathfrak{P}, A)$.

2.1 Definition An n-chain $\alpha = (x_0, \ldots, x_n)$ consisting of $n+1$ distinct elements is called a *root* and $U_\alpha = Aut(\mathcal{P}, D)$ where $D = \bigcup_{i=1}^{n-1} D_1(x_i)$ is the corresponding *root group*.

A generalized n-gon $\mathfrak{P}$ satisfies the *Moufang condition* if for any root $\alpha = (x_0, \ldots, x_n)$ the corresponding root group U_α acts transitively on $D_1(x_n) \setminus \{x_{n-1}\}$.

For an algebraic group of rank 2, these are indeed the root groups in the algebraic sense.

While there are infinite generalized n-gons for all n, the surprising result due to Weiss [W] is that Moufang polygons exist only for $n = 3, 4, 6$ and 8 and a complete classification of all Moufang polygons is currently being written up [TW] (see [vM] for a complete list and a description).

In the context of finite Morley rank we have the following theorem:

2.2 Theorem [KTvM] *If $\mathfrak{P}$ is an infinite Moufang polygon of finite Morley rank, then (up to duality) $\mathfrak{P}$ is (definably) isomorphic to either the projective plane, the symplectic quadrangle, or the split Cayley hexagon over some algebraically closed field K.*

For the proof it is necessary to closely inspect all Moufang polygons, see [KTvM] for details. In most cases, the crucial tool consists in showing that the *projectivity groups* are definable or contain definable subgroups:

2.3 Definition Let $\mathfrak{P}$ be any n-gon. For $x, y \in \mathcal{P} \cup \mathcal{L}$ at distance n, there is a bijection from $D_1(x)$ to $D_1(y)$, denoted $[x; y]$, taking an element in $D_1(x)$ to the uniquely determined element in $D_1(y)$ closest to it. Any composition of such maps is called a *projectivity* and is denoted by $[x_0; \ldots; x_m]$. For any $x \in \mathcal{P} \cup \mathcal{L}$, the set of projectivities $\Pi(x) = \{[x_0 = x; x_1; \ldots; x_m = x] : m \in \mathbf{N}, d(x_i, x_{i-1}) = n, i = 1, \ldots m\}$ is a 2-transitive permutation group on $D_1(x)$.

Note that the isomorphism type of $\Pi(x)$ depends only on whether x is a point or a line. Since the length of the projectivities in $\Pi(x)$ is in general unbounded, these groups are not necessarily definable. This poses the biggest problem in the classification.

$n = 3$ The Moufang projective planes of finite Morley rank have previously been classified by Rose, [Ro]. Their classification is equivalent to the classification of alternative fields of finite Morley rank, which turn out to be exactly the algebraically closed fields.

$n = 4$ The quadrangles contain the most difficult cases of the classification:

- For the orthogonal quadrangles one shows that $\mathbf{PSL}_2(K)$ is a definable subgroup of the projectivity group of a line. Hence K is algebraically closed and the quadrangle is the unique orthogonal quadrangle over K.

- Generalized quadrangles are coodinatized by quadratic quaternary rings, see [vM], similar to the coordinatization of projective planes. Using these coordinates one can define the underlying field (of characteristic 2) in the mixed quadrangles. Again we conclude that the field has to be algebraically closed. Hence the subfield K^2 coincides with K and the quadrangle is the orthogonal quadrangle.

- For the σ-hermitian quadrangles over proper skew fields, we show, using their embedding into projective space and an argument from model theoretic Galois theory, that the skew field is definable in the quadrangle, hence they cannot have finite Morley rank.

- All remaining quadrangles can be shown to contain definable subquadrangles which, by the previous cases, are isomorphic to either an orthogonal or a symplectic quadrangle over an algebraically closed field. But this is impossible.

$n = 6$ To classify the hexagons, we use the fact that in all cases a group of the form $\mathbf{PSL}_2(K)$ for some field K is definable in the hexagon. Hence K has to be algebraically closed. If the hexagon was not the split Cayley hexagon, it would belong either to a field extension of degree 3 over K, to a proper subfield of K containing K^3, where $char(K) = 3$, or to certain simple Jordan algebras over K. Since K is algebraically closed, none of them exist over K (see [Ja] for the Jordan algebras).

$n = 8$ For the octagons, we show that the projectivities of a point have bounded length, and hence the projectivity group is definable. Its commutator subgroup is a definable infinite Suzuki group, which contradicts the finite Morley rank assumption, see [BN] 11.90. Hence there are no Moufang octagons of finite Morley rank.

Tits showed in [Ti74] that all irreducible spherical buildings of rank at least 3 satisfy the (corresponding) Moufang condition, and its rank 2 residues are either Moufang projective planes or Moufang quadrangles. Hence we obtain:

2.4 Theorem [KTvM] *An infinite irreducible spherical building of Tits rank ≥ 3 and of finite Morley rank is the building associated to a simple linear algebraic group G over some algebraically closed field K. Both G and K are definable in the building.*

For the proof one has to check that the rank 2 residues are definable in the building and that all but the buildings mentioned in the theorem have rank 2 residues which are excluded by Theorem 2.2.

This translates into the following statement about groups:

2.5 Theorem [KTvM] *Let G be an infinite group of finite Morley rank with a definable spherical irreducible BN-pair. If the (Tits) rank of G is 2, assume that the BN-pair satisfies the Moufang condition. Then G is isomorphic to $\mathbf{PSL}_3(K)$, $\mathbf{PSp}_4(K) \cong \mathbf{PSO}_5(K)$, or $\mathbf{G}_2(K)$ for some algebraically closed field K. If the rank is at least 3, then G is simple and (definably) isomorphic to a simple algebraic group over an algebraically closed field.*

The condition that the BN-pair be definable is not necessary. It may be replaced by the weaker assumption that the group acts effectively and strongly transitively on an infinite irreducible Moufang building of rank at least 2 and that G contains the subgroup of the automorphism group of the building generated by all the root groups U_α.

These conditions are satisfied in particular whenever the group G is the group of k-rational points of some k-isotropic k-simple algebraic group of k-rank at least 2 for some infinite field k.

References

[ABS] H. Azad, M. Barry, G.Seitz, On the structure of parabolic subgroups, Comm. Algebra 18, 551–562 (1990)

[Ba] J. Baldwin, An almost strongly minimal non-desarguesian projective plane, Trans. Amer. Math. Soc. 342, 695–711, (1994)

[Bo] A. Borel, *Linear algebraic groups, 2nd edition,* Springer (1991)

[BT65] A. Borel, J. Tits, Groupes réductifs, Publ. Math. I.H.E.S 27, 55–150 (1965)

[BT72] A. Borel, J. Tits, Compléments à l' article: Groupes réductifs, Publ. Math. I.H.E.S 41, 253–276 (1972)

[BT73] A. Borel, J. Tits, Homomorphismes "abstraits" de groupes algébriques simples, Ann. Math. 97, 499–571 (1973)

[BN] A. Borovik, A. Nesin, *Groups of finite Morley Rank*, Oxford Science Publication (1994)

[DN] M. J. De Bonis, A. Nesin, There are $2^{\aleph_0}$ many almost strongly minimal generalized polygons that do no interpret an infinite group, to appear in J. Symb. Logic.

[Ja] N. Jacobson, *Lectures on quadratic Jordan algebras,* Tata Institute of Fundamental Research Lecture Notes 45, Bombay (1969)

[KRT] L. Kramer, G. Röhrle, K. Tent, Defining k in $G(k)$, preprint, (1997)

[KTvM] L. Kramer, K. Tent, H. Van Maldeghem, Simple Groups of finite Morley rank and Tits buildings, to appear in Israel J. Math.

[PPS] Y. Peterzil, A. Pillay, S. Starchenko, Simple algebraic and semialgebraic groups over real closed fields, preprint.

[Po] B. Poizat, *Groupes stables,* Nur al-Mantiq wal-Ma'rifah (1987)

[Ro] B.I. Rose, Model theory of alternative rings, Notre Dame J. of Formal Logic 9, 215–243 (1978)

[Te] K. Tent, A note on the model theory of generalized polygons, submitted.

[Ti66] J. Tits, Classification of algebraic semisimple groups, Proc. Symp. Pur. Math. IX, Amer. Math. Soc. 32–62 (1966)

[Ti74] J. Tits, *Buildings of spherical type and finite BN-pairs*, Springer LNM 386, (1974, 1986)

[TW] J. Tits, R. Weiss, *The classification of Moufang polygons*, book in preparation.

[vM] H. Van Maldeghem, *Generalized polygons, a geometric approach*, Birkhäuser (1998)

[W] R. Weiss, The nonexistence of certain Moufang polygons, Inv. Math. 51, 261–266 (1979)

A Combinatory Algebra for Sequential Functionals of Finite Type

Jaap van Oosten*
Department of Mathematics
Utrecht University
The Netherlands
jvoosten@math.ruu.nl

Abstract

It is shown that the type structure of finite-type functionals associated to a combinatory algebra of partial functions from $\mathbb{N}$ to $\mathbb{N}$ (in the same way as the type structure of the countable functionals is associated to the partial combinatory algebra of total functions from $\mathbb{N}$ to $\mathbb{N}$), is isomorphic to the type structure generated by object N (the flat domain on the natural numbers) in Ehrhard's category of "dI-domains with coherence", or his "hypercoherences".

Introduction

PCF, "Gödel's T with unlimited recursion", was defined in Plotkin's paper [17]. It is a simply typed λ-calculus with a type o for integers and constants for basic arithmetical operations, definition by cases and fixed point recursion. More important, there is a special reduction relation attached to it which ensures (by Plotkin's "Activity Lemma") that all *PCF*-definable higher-type functionals have a sequential, i.e. non-parallel evaluation strategy. In view of this, the obvious model of Scott domains is not faithful, since it contains parallel functions. A search began for "fully abstract" domain-theoretic models for *PCF*.

A proliferation of ever more complicated theories of domains saw the light, inducing the father of domain theory, Dana Scott, to lament that "there are *too many* proposed categories of domains and [...] their study has become too arcane" ([18]), a judgement with which it is hard to disagree.

*Research supported by the Dutch National Research Foundation NWO.
AMS Subject Classification: Primary 03D65, 68Q55 Secondary 03B40, 03B70, 03D45, 06B35.

Although most interest in the semantics of *PCF* was shown by computer scientists, it became clear that there is an important overlap with higher-type recursion theory as it was recognized (I believe, initially by Robin Gandy, whose insights were transmitted by Martin Hyland and partially laid down in the paper [9]) that Kleene's late attempts ([11, 12, 13, 14]) to formalize the notion of a recursive functional of higher type, had much in common with the "full abstraction problem" for *PCF*. As far as I am aware however, the *exact* relationship between Kleene's work and the work on *PCF* still remains to be clarified.

An important model of Kleene's axioms is provided by the so-called "continuous (or countable) functionals" (see, e.g., [16]; they were introduced in [10] and [15]). They arise, in a standard way, as the type structure coming from the partial combinatory algebra of "Kleene's function realizability". This is a partial combinatory algebra structure on the set of functions from $\mathbb{N}$ to $\mathbb{N}$.

A surprising result of this paper is, that a natural generalization of function realizability to *partial* functions from $\mathbb{N}$ to $\mathbb{N}$ (yielding a *total* combinatory algebra), gives a type structure of higher-type functionals which coincides with the relevant part of Ehrhard and Bucciarelli's "dI-domains with coherence" ([3, 4, 2]).

This could be interesting for a number of reasons. First, it provides another handle on Ehrhard's work, which is complicated and rather heavily loaded with definitions; however, the fact that dI-domains with coherence have a completely independent generation process (which process is well known in logic), seems to me to enhance their naturalness as a mathematical structure. Of course, the result in this paper calls for comparison with the result in [3], viz. that dI-domains with coherence are the extensional collapse of another domain-theoretic structure, sequential structures and sequential algorithms. My result is essentially different in that it relates the dI-domains with coherence to something which is defined independently of any domain theory. But it might be conjectured that the sequential algorithms, or the part of it that is relevant to *PCF*, can be obtained as a kind of intensional type structure on the combinatory algebra considered here.

Secondly, it shows that Ehrhard's "strongly stable" model of *PCF* lives inside a realizability topos where its domain structure is *intrinsic.* This should be of interest to Synthetic Domain Theory ([7]).

Thirdly it raises the question whether maybe more models of *PCF* (including the fully abstract game models of [9] and [1]) can be induced in this way by combinatory algebras.

In fact the way I see it, the point of this research is to see how, like in this paper, a complicated type structure arises out of a very simple combinatory algebra by the process of Kleene associates. The combinatory algebra then, is some sort of model of 'atomic computations' and the rest is generated by building the finite types.

1 Sequential Functions

We are interested in the following game between partial functions $\alpha, \beta : \mathbb{N} \to \mathbb{N}$. α asks, successively, values of β at given arguments; the game has no outcome if β is undefined at one of these numbers, or if α has no further move; but α may also decide, at some point, that now it has sufficient information about β, and outputs not a question, but an answer.

Formally, we define:

Definition 1.1 *A sequence $u = \langle u_0, \dots, u_{n-1} \rangle$ (coded as a natural number) is called a* dialogue *between α and β if for all i with $0 \leq i < n-1$, writing $u^{<i}$ for $\langle u_0, \dots, u_{i-1} \rangle$, there is j such that*

$$\alpha(u^{<i}) = 2j \text{ and } \beta(j) = u_i$$

We say that the application $\alpha|\beta$ is defined with value n, or $\alpha|\beta = n$, if there is a dialogue u between α and β such that

$$\alpha(u) = 2n + 1$$

Of course, we read $u^{<0}$ as the empty sequence. Note, that dialogues are unique: given α and β, there is a unique (finite or infinite) dialogue between α and β.

Let $\mathcal{B}$ be the set of all partial functions from $\mathbb{N}$ to $\mathbb{N}$; then every $\alpha \in \mathcal{B}$ determines a partial function F_α from $\mathcal{B}$ to $\mathbb{N}$ by

$$F_\alpha(\beta) = n \text{ iff } \alpha|\beta = n$$

Giving $\mathcal{B}$ the topology with as subbase the collection of all

$$\mathcal{U}_p = \{\alpha | p \subseteq \alpha\}$$

for p *finite*, and $N_\perp = \mathbb{N} \cup \{\perp\}$ the topology which is discrete on $\mathbb{N}$ and has $N_\perp$ as only neighborhood of $\perp$, every F_α, considered as total function $\mathcal{B} \to N_\perp$, is continuous; but clearly, not every continuous $F : \mathcal{B} \to N_\perp$ is of the form F_α.

Examples The functions:

i) $F(\alpha) = \begin{cases} 0 & \text{if } \alpha \neq \emptyset \\ \perp & \text{else} \end{cases}$

ii) $F(\alpha) = \begin{cases} 0 & \text{if } \sum_{x \in \mathrm{dom}(\alpha), x \leq 1} \alpha(x) \geq 1 \\ \bot & \text{else} \end{cases}$

iii) $F(\alpha) = \begin{cases} 1 & \text{if } \sum_{x \in \mathrm{dom}(\alpha), x \leq 1} \alpha(x) \geq 1 \\ 0 & \text{if } \alpha(0) = \alpha(1) = 0 \\ \bot & \text{else} \end{cases}$

are all continuous, but not given as F_α.

In order to study the set of functions $\mathcal{B} \to N_\bot$ that are given by some $\alpha \in \mathcal{B}$, it is useful to consider two partial orders on this set: the *pointwise* order is defined by: $F \leq_{\mathrm{pw}} G$ iff for all $\alpha \in \mathcal{B}$ and all $n \in \mathbb{N}$: if $F(\alpha) = n$ then $G(\alpha) = n$.

The *stable* order is defined by: $F \leq_s G$ iff for all $\alpha, \beta \in \mathcal{B}$: if $\alpha \subset \beta$ then $F(\alpha) = n$ if and only if $F(\beta) = G(\alpha) = n$. Clearly, since F and G are continuous, $F \leq_s G$ implies $F \leq_{\mathrm{pw}} G$.

Every continuous $F : \mathcal{B} \to N_\bot$ has a unique *base*, that is a minimal set B of finite functions such that for all α and all n: $F(\alpha) = n$ iff there is a $p \in B$ such that $p \subset \alpha$.

Definition 1.2 *A* sequential tree *is a tree T of finite functions (ordered by $\subseteq$, so the root is the empty function) such that for each $p \in T$ there is $n \in \mathbb{N}$ such that all immediate successors q of p in T have* $\mathrm{dom}(q) = \mathrm{dom}(p) \cup \{n\}$.

Clearly:

Proposition 1.3 *A function $F : \mathcal{B} \to N_\bot$ is of the form F_α for some $\alpha \in \mathcal{B}$, iff its base is the set of leaves of a sequential tree.*

With any F_α therefore we can associate a sequential tree with set of leaves B, and a function $v : B \to \mathbb{N}$. Conversely every such pair (B, v) defines a function F which is given as F_α for some (non-unique) α. We call the pair (B, v) the *trace* of F. This is in harmony with usage in the literature of this term, cf. [5].

When is a base B of a continuous function the set of leaves of a sequential tree? Answer:

Proposition 1.4 *B is the set of leaves of a sequential tree if and only if for each nonempty finite subset B' of B we have: if $p \subseteq \bigcap B'$ then either $B' = \{p\}$ or $\bigcap_{q \in B'} \mathrm{dom}(q \setminus p) \neq \emptyset$.*

This proposition doesn't seem very informative, but yields at once:

Corollary 1.5 *Let $(B_i | i \in I)$ be a directed system of sets of finite functions such that each B_i is the set of leaves of a sequential tree. Then $\bigcup_{i \in I} B_i$ is the set of leaves of a sequential tree.*

From now on, we shall call functions $F : \mathcal{B} \to N_\perp$ which are given as F_α, *sequential* functions.

Let F, G be two sequential functions, with traces (B_F, v_F) and (B_G, v_G).

Proposition 1.6 *i)* $F \leq_{\text{pw}} G$ *iff for every* $p \in B_F$ *there is* $q \in B_G$ *with* $q \subseteq p$ *and* $v_G(q) = v_F(p)$;

ii) $F \leq_s G$ *iff* $B_F \subseteq B_G$ *and* v_F *and* v_G *coincide on* B_F;

iii) $\alpha \subseteq \beta$ *implies* $F_\alpha \leq_s F_\beta$

Definition 1.7 *In a partially ordered set* $(D, \leq)$ *we say that* d *is the* least upper bound *(or* lub, *or* join, *or* supremum*) of* $A \subseteq D$ *if* d *is least such that* $\forall a \in A . a \leq d$. *Write* $d = \bigvee A$.

We say A *is* bounded *if it has an upper bound in* D.

We say that $d \in D$ *is* compact *if for every directed* $I \subseteq D$, *if* $d \leq \bigvee I$ *then* $\exists i \in I . d \leq i$.

D *is called* ω-algebraic *if the set of compact elements of* D *is countable and for each* $d \in D$, *the set* $\{k \in D | \, k \text{ compact} \wedge k \leq d\}$ *is directed and has* d *as least upper bound.*

D has the I-property *if for every compact* $k \in D$, $\{d \in D | \, d \leq k\}$ *is finite.*

D *is* distributive *if* $(x \vee x') \wedge y = (x \wedge y) \vee (x' \wedge y)$ *whenever these lubs exist* ($x \vee x' = \bigvee\{x, x'\}$ *and* $x \wedge x' = \bigvee\{y | \, y \leq x$ *and* $y \leq x'\}$).

D *is a* dI-domain *if it has least upper bounds of directed subsets and of bounded subsets, is distributive and* ω*-algebraic, and has the I-property.*

The conclusion is:

Proposition 1.8 *The set of sequential functions* $\mathcal{B} \to N_\perp$, *with the stable order, is a dI-domain.*

Moreover, the set of sequential functions $\mathcal{B} \to N_\perp$ is *atomic*, which means that every element is the supremum of the atoms below it (an atom is a non-bottom element which has no non-bottom elements stricly below it). Atomic dI-domains are known in the literature as *qualitative domains* ([5]).

2 A type structure of sequential functionals

In this section we restrict ourselves to the following types: o is a type; if σ is a type, then $\sigma \to o$ is a type.

To every such type we assign a set $\mathcal{O}_\sigma$ of *sequential functionals* of type σ, and to any $f \in \mathcal{O}_\sigma$ a nonempty set $\text{Ass}(f) \subset \mathcal{B}$, the set of *associates* of f. The definition is:

$\mathcal{O}_o = N_\perp$; $\mathrm{Ass}(n) = \{\alpha | \alpha(0) = n\}$ and $\mathrm{Ass}(\perp) = \{\alpha | 0 \notin \mathrm{dom}(\alpha)\}$. $\mathcal{O}_{\sigma\to o}$ consists of those functions $f : \mathcal{O} \to N_\perp$ such that there is a $\beta \in \mathcal{B}$ such that for all $x \in \mathcal{O}_\sigma$ and all $\alpha \in \mathrm{Ass}(x)$, $\beta|\alpha = n$ if and only if $f(x) = n$, and $\mathrm{Ass}(f)$ is the set of β satisfying this condition. We write $\mathrm{Ass}(\sigma)$ for the set of associates of elements of $\mathcal{O}_\sigma$.

By an easy induction on σ, if $\alpha \in \mathrm{Ass}(\sigma)$ there is a unique $f \in \mathrm{O}_\sigma$ with $\alpha \in \mathrm{Ass}(f)$; this f is denoted by $[\alpha]$ (strictly speaking, $[\alpha]$ depends on the type, but there will never be ambiguity). We write $\alpha \sim \alpha'$ if $[\alpha] = [\alpha']$, for $\alpha, \alpha' \in \mathrm{Ass}(\sigma)$.

Again, on $\mathcal{O}_\sigma$ one can define the pointwise and stable orders: on $N_\perp$, we have $\perp \leq n$ for all n, and that is all for both orders; on $\mathcal{O}_{\sigma\to o}$, $f \leq_{\mathrm{pw}} g$ iff for all $x \in \mathcal{O}_\sigma$, $fx = n$ implies $gx = n$; and $f \leq_s g$ iff for all $x \leq_s y \in \mathcal{O}_\sigma$, $fx = n \Leftrightarrow gx = fy = n$.

We extend the notation $\leq_s$ to associates and say: $\alpha \leq_s \beta$ if $[\alpha] \leq_s [\beta]$. In general on associates, $\leq_s$ can at most be a preorder. Whether $\leq_s$ is reflexive or not is equivalent to whether elements of $\mathcal{O}_{\sigma\to o}$ are monotone w.r.t. $\leq_s$. Associates are, of course, also ordered by inclusion, so the question arises what the relation between these orders is. These matters will be resolved by the following theorem.

Some more terminology: elements x, y of $\mathcal{O}_\sigma$ are called *compatible* if they have a common upper bound w.r.t. $\leq_s$. Similarly for associates.

Theorem 2.1 *i) If $\sigma = \tau\to o$ and $f \in \mathcal{O}_\sigma$ then f is monotone w.r.t. the stable order on $\mathcal{O}_\tau$. In particular, $\leq_s$ is a partial order on $\mathcal{O}_\sigma$;*

ii) If $\gamma \in \mathrm{Ass}(\sigma)$ and $q \subset \gamma$ is finite, there is an element $\mu_\sigma(q)$ of $\mathrm{Ass}(\sigma)$ with the properties:

a) $q \subset \mu_\sigma(q)$;

b) $\mu_\sigma(q) \leq_s \gamma'$ whenever $\gamma' \in \mathrm{Ass}(\sigma)$ and $q \subset \gamma'$;

c) if $\mu_\sigma(q) \leq_s \gamma'$ for $\gamma' \in \mathrm{Ass}(\sigma)$, there is a finite $q' \subset \gamma'$ such that $\mu_\sigma(q) \leq_s \mu_\sigma(q')$;

d) $[\mu_\sigma(q)]$ is a compact element of $\mathcal{O}_\sigma$.

iii) For $\alpha, \beta \in \mathrm{Ass}(\sigma)$, $\alpha \subseteq \beta$ implies $\alpha \leq_s \beta$, and $f \leq_s g$ in $\mathcal{O}_\sigma$ implies that for every $\beta \in \mathrm{Ass}(g)$ there is $\alpha \in \mathrm{Ass}(f)$ with $\alpha \subseteq \beta$;

iv) If $x, y \in \mathcal{O}_\sigma$ are compatible then their meet $x \wedge y$ exists;

v) If $\sigma = \tau\to o$, $f \in \mathcal{O}_\sigma$, then f preserves meets of compatible elements: for $x, y \in \mathcal{O}_\tau$ compatible, $f(x \wedge y) = n$ iff $fx = fy = n$.

Proof. The proof is somewhat involved; it is a simultaneous induction on the type σ.

For $\sigma = o$, the first part of i) is vacuous and it's clear that $\leq_s$ is a partial order on $\mathcal{O}_o$; ii) take $\mu_o(q) = q$; the rest is left to the reader; iii) and iv) are obvious, and v) is vacuous.

Now let $\sigma = \tau \to o$.

i). If $f \in \mathcal{O}_\sigma$ and $x \leq_s y$ in $\mathcal{O}_\tau$ then since f has an associate γ and, by the induction hypothesis of iii), every associate β of y contains an associate α of x, $fx = n \Leftrightarrow \gamma|\alpha = n \Rightarrow \gamma|\beta = n \Leftrightarrow fy = n$. So, f is monotone and $f \leq_s f$; clearly then, $\leq_s$ is a partial order on $\mathcal{O}_\sigma$.

ii). Let q be finite such that $q \subset \gamma$ for some $\gamma \in \mathrm{Ass}(\sigma)$. There is a finite set E of finite functions p such that $p \subset \beta$ for some $\beta \in \mathrm{Ass}(\tau)$, and $q|p$ is defined. If $E = \emptyset$, we simply put $\mu_\sigma(q) = q$; in that case it's clear that q itself is an associate of the function $\lambda x.\bot$. Assume now that $E \neq \emptyset$; let $E' = \{p_1, \ldots, p_n\} \subseteq E$ be such that for each $p \in E$ there is a unique $p_i \in E'$ with $\mu_\tau(p_i) \leq_s \mu_\tau(p)$. For a finite function r we put

$$\Delta_r = \{\delta \in \mathrm{Ass}(\tau) | r \subset \delta \text{ and for some } i, \mu_\tau(p_i) \leq_s \delta\}$$

Define $\Gamma = \mu_\sigma(q)$ as follows:

$$\Gamma(u) = \begin{cases} q(u) & \text{if } u \in \mathrm{dom}(q) \\ \text{undefined} & \text{if } u \text{ is not a dialogue between } \Gamma \text{ and} \\ & \text{some finite function } r, \text{ or if } \Delta_r = \emptyset \\ 2k+1 & \text{if there is a } \gamma \in \mathrm{Ass}(\sigma) \text{ with } q \subset \gamma,\ \gamma|r = k \text{ and} \\ & \text{there is } r' \supseteq r \text{ and } i \text{ with } \mu_\tau(p_i) \leq_s \mu_\tau(r') \\ 2l & \text{for } l = \min(\bigcap_{\delta \in \Delta_r} \mathrm{dom}(\delta \setminus q)) \text{ else, if} \\ & \bigcap_{\delta \in \Delta_r} \mathrm{dom}(\delta \setminus q) \neq \emptyset \\ \text{undefined} & \text{else} \end{cases}$$

First, let us remark that the case $\Delta_r \neq \emptyset, \bigcap_{\delta \in \Delta_r} \mathrm{dom}(\delta \setminus r) = \emptyset$ can only apply if $\Delta_r = \{r\}$ since if $q \subset \gamma \in \mathrm{Ass}(\sigma)$, $\gamma|\delta$ is defined for all $\delta \in \Delta_r$ (by induction hypothesis i), since $\gamma|p_i$ is defined, hence $\gamma|\mu_\tau(p_i)$ is defined).

A second remark is, that if $\gamma|r = k$ for $\gamma \in \mathrm{Ass}(\sigma)$ with $q \subset \gamma$, and $r' \supseteq r$ is such that $\mu_\tau(p_i) \leq_s \mu_\tau(r')$, we must already have that $\mu_\tau(p_i) \leq_s \mu_\tau(r)$; because in that case, by induction hypothesis ii), $\mu_\tau(p_i)$ and $\mu_\tau(r)$ are compatible and $\gamma|\mu_\tau(p_i) = \gamma|\mu_\tau(r) = k$; by induction hypothesis there is an associate ε of their meet, with $\varepsilon \subseteq \mu_\tau(p_i)$, and $\gamma|\varepsilon = k$; but then, $p_i \subseteq \varepsilon$ (because $q|p_i = k$, $q \subseteq \gamma$ and $p_i \subseteq \mu_\tau(p_i)$); hence $\mu_\tau(p_i) \leq_s \varepsilon$ and so, $\mu_\tau(p_i) \leq_s \mu_\tau(r)$.

Therefore, $\Gamma|\delta = k$ if and only if there is $r \subseteq \delta$, r finite, and i with $q|p_i = k$ and $\mu_\tau(p_i) \leq_s \mu_\tau(r)$ (note, that in a dialogue u between γ and r, if $\gamma(u) = 2l$

and $l \notin \text{dom}(r)$, $l \in \bigcap_{\delta \in \Delta_r} \text{dom}(\delta \setminus r)$ will always hold, so in a continuing dialogue between Γ and δ, these questions will ultimately be posed). Now if $\Gamma|\delta = k$ and $\delta \sim \delta'$, there is $r \subseteq \delta$ with $\mu_\tau(p_i) \leq_s \mu_\tau(r) \leq_s \delta'$ hence by induction hypothesis there is $r' \subseteq \delta'$ with $\mu_\tau(p_i) \leq_s (r')$; so $\Gamma|\delta' = k$. This proves that $\Gamma \in \text{Ass}(\sigma)$.

Suppose $q \subseteq \gamma$, $\gamma \in \text{Ass}(\sigma)$. Let $x \leq_s y \in \mathcal{O}_\tau$. By induction hypothesis iii) there are associates $\theta \subseteq \zeta$ of x, y respectively. If $\Gamma|\zeta = \gamma|\theta = k$ let $q_1 \subseteq \zeta$, $q_2 \subseteq \theta$ finite with $\Gamma|q_1 = \gamma|q_2 = k$. Since $\mu_\tau(p_i) \leq_s \mu_\tau(q_1 \cup q_2)$ for some i, we have that $\Gamma|\theta = k$; so, $\Gamma \leq_s \gamma$.

We still have to prove that if $\Gamma \leq_s \gamma \in \text{Ass}(\sigma)$, there is a finite part r of γ such that $\Gamma \leq_s \mu_\sigma(r)$. For this, it is sufficient to note that for $\gamma \in \text{Ass}(\sigma)$ the following two conditions are equivalent:

a) $\Gamma \leq_s \gamma$

b) for all i, $\gamma|\mu_\tau(p_i) = q|p_i$, and if p is such that $\mu_\tau(p)$ is compatible with $\mu_\tau(p_i)$ and $\gamma|p = \gamma|p_i$, then $\mu_\tau(p_i) \leq_s \mu_\tau(p)$.

For a)$\Rightarrow$b), that $\gamma|\mu_\tau(p_i) = q|p_i$ is clear, and if $\mu_\tau(p)$ is compatible with $\mu_\tau(p_i)$, and x is their meet in $\mathcal{O}_\tau$, and $\mu_\tau(p_i) \not\leq_s \mu_\tau(p)$, then x is strictly below $[\mu_\tau(p_i)]$, and $[\Gamma](x)$ is undefined but $[\gamma](x)$ is defined (induction hypothesis!); contradiction with $\Gamma \leq_s \gamma$.

For b)$\Rightarrow$a), if (using induction hypothesis) $\theta \subseteq \zeta$ are in $\text{Ass}(\tau)$ and $\Gamma|\zeta = \gamma|\theta = k$, there are finite $r_1 \subseteq \zeta$, $r_2 \subseteq \theta$ with $\Gamma|r_1 = \gamma|r_2$. Then there is i with $\mu_\tau(p_i) \leq_s \mu_\tau(r_1)$, and $\mu_\tau(r_1)$ and $\mu_\tau(r_2)$ are compatible since $\theta \subseteq \zeta$. By b), $\mu_\tau(p_i) \leq_s \mu_\tau(r_2)$, so $\Gamma|\theta = k$. The other implication is left to the reader.

Now if $\Gamma \leq_s \gamma \in \text{Ass}(\sigma)$ there is a finite $r \subseteq \gamma$ such that for all i, $r|\mu_\tau(p_i) = q|p_i$. Since the second condition of b) clearly remains true if we replace γ by something $\leq_s \gamma$, we have $\Gamma \leq_s \mu_\sigma(r)$.

By a similar argument, left to the reader, $[\mu_\sigma(q)]$ is compact in $\mathcal{O}_\sigma$.

iii). If $\alpha, \beta \in \text{Ass}(\sigma)$ with $\alpha \subseteq \beta$, and $x \leq_s y \in \mathcal{O}_\tau$ then by induction hypothesis every associate ζ of y contains an associate θ of x, hence, $[\alpha](x) = n$ iff $\alpha|\theta = n$ iff $\alpha|\zeta = \beta|\theta = n$ iff $[\alpha](y) = [\beta](x) = n$; so $\alpha \leq_s \beta$.

Now let $f \leq_s g \in \mathcal{O}_\sigma$ and $\beta \in \text{Ass}(g)$. We define $\alpha \subseteq \beta$ by stipulating that $u \in \text{dom}(\alpha)$ iff the following hold:

a) u is a dialogue between β and some finite function q;

b) there is $\gamma \in \text{Ass}(\tau)$ with $q \subseteq \gamma$ and $f([\gamma]) \neq \bot$;

c) if $\beta(u) = 2k+1$ then we must have: for all $\gamma \in \text{Ass}(\tau)$, if $q \subseteq \gamma$ then $f([\gamma]) = k$.

Then for $\gamma \in \text{Ass}(\tau)$, the implication $\alpha|\gamma = k \Rightarrow f([\gamma]) = k$ clearly follows. For the converse, if $f([\gamma]) = k$ then certainly $\beta|\gamma = k$ so $\beta|q = k$ for finite

$q \subseteq \gamma$. The only way that $\alpha|\gamma = k$ can fail to hold is that there is another $\gamma' \in \mathrm{Ass}(\tau)$ with $q \subseteq \gamma'$, and $f([\gamma']) \neq k$. Then $f([\gamma']) = \bot$ since $f \leq_s g$. For $\mu_\tau(q)$ from induction hypothesis ii) however, we have $\beta|\mu_\tau(q) = k$ because $q \subseteq \mu_\tau(q)$, and $f([\mu_\tau(q)]) = \bot$ by i), since $\mu_\tau(q) \leq_s \gamma'$. But also $\mu_\tau(q) \leq_s \gamma$, and we obtain a contradiction with $f \leq_s g$.

iv) and **v)** are now easy: if x, y are compatible with upper bound z, let $\gamma \in \mathrm{Ass}(z)$ and pick associates α, β for x, y with $\alpha, \beta \subseteq \gamma$. Then $\alpha \cap \beta$ is an associate of $x \wedge y$, and such meets are clearly respected by any $f \in \mathcal{O}_{\sigma\to o}$. ∎

From this theorem we shall obtain a series of corollaries, which culminate in the theorem that every $\mathcal{O}_\sigma$ is a qualitative domain, and that every $f \in \mathcal{O}_{\sigma\to o}$ is a so-called *stable* function (Theorem 2.8).

Corollary 2.2 *Let $\sigma = \tau \to o$. Then to any $f \in \mathcal{O}_\sigma$ an associate β_f can be assigned in such a way, that $f \leq_s g$ if and only if $F_{\beta_f} \leq_s F_{\beta_g}$ as sequential functions: $\mathcal{B} \to N_\bot$.*

Proof. Define β_f by:

$$\beta_f(u) = \begin{cases} \text{undefined} & \text{if } u \text{ is not a dialogue between } \beta_f \text{ and some} \\ & \text{finite function } q, \text{ or if there is no } \gamma \in \mathrm{Ass}(\tau) \cap \mathcal{U}_q \\ & \text{for which } f([\gamma]) \neq \bot; \\ 2k+1 & \text{if } f([\gamma]) = k \text{ for all } \gamma \in \mathrm{Ass}(\tau) \cap \mathcal{U}_q; \\ 2k & \text{for } k = \min(\bigcap\{\mathrm{dom}(\gamma \setminus q) | \gamma \in \mathrm{Ass}(\tau) \cap \mathcal{U}_q, \\ & f([\gamma]) \neq \bot\}), \text{ else} \end{cases}$$

If $\delta \in \mathrm{Ass}(f)$, u a dialogue between δ and q, and p a finite part of some $\gamma \in \mathrm{Ass}(\tau) \cap \mathcal{U}_q$ such that $f([\gamma]) \neq \bot$, then if $\delta(u) = 2l$, either $l \in \mathrm{dom}(p)$ or

$$l \in \bigcap\{\mathrm{dom}(\gamma' \setminus p) | \gamma' \in \mathrm{Ass}(\tau) \cap \mathcal{U}_p, f([\gamma']) \neq \bot\}$$

Therefore, if $f([\gamma]) = k$ then $\beta_f|\gamma = k$. The converse is obvious, so $\beta_f \in \mathrm{Ass}(f)$.

Now suppose $f \leq_s g$ in $\mathcal{O}_\sigma$ and $\theta \subseteq \zeta$ in $\mathcal{B}$. Definitely if $\beta_f|\theta = k$ then $\beta_g|\theta = \beta_f|\zeta = k$; conversely if $\beta_g|\theta = \beta_f|\zeta = k$ there are q_1, q_2 finite, $q_1 \subseteq \zeta$, $q_2 \subseteq \theta$ such that f is constant k on $\mathrm{Ass}(\tau) \cap \mathcal{U}_{q_1}$ and g is constant k on $\mathrm{Ass}(\tau) \cap \mathcal{U}_{q_2}$ (and the sets $\mathrm{Ass}(\tau) \cap \mathcal{U}_{q_i}$ are nonempty).

Then $f([\mu_\tau(q_1 \cup q_2)]) = g([\mu_\tau(q_2)]) = k$ hence by $f \leq_s g$, $f([\mu_\tau(q_2)]) = k$ hence f is constant on $\mathrm{Ass}(\tau) \cap \mathcal{U}_{q_2}$, hence $\beta_f|\theta = k$.

This shows that $f \leq_s g$ implies $F_{\beta_f} \leq_s F_{\beta_g}$ as sequential functions; conversely, since for $x \leq_s y$ in $\mathcal{O}_\tau$ there are associates $\theta \subseteq \zeta$ of x, y respectively, we have

$$fy = gx = k \Leftrightarrow F_{\beta_f}|\zeta = F_{\beta_g}|\theta = k \Leftrightarrow F_{\beta_f}|\theta = k \Leftrightarrow fx = k.$$

∎

Corollary 2.3 *$\mathcal{O}_\sigma$ has directed joins for all σ, and they are preserved by any $f \in \mathcal{O}_{\sigma\to o}$.*

Proof. The first statement is a straightforward combination of the previous corollary and the theorem that the set of sequential functions has directed joins. To see that they are respected by $f \in \mathcal{O}_{\sigma\to o}$, suppose $I \subseteq \mathcal{O}_\sigma$ directed, $\gamma \in \mathrm{Ass}(\bigvee I)$, $f(\bigvee I) = k$. There is finite $q \subseteq \gamma$ such that $f([\mu_\sigma(q)]) = k$; since $[\mu_\sigma(q)]$ is compact there is $i \in I$ with $f(i) = k$. ∎

Corollary 2.4 *$\mathcal{O}_\sigma$ has joins of bounded subsets.*

Proof. Let $A \subseteq \mathcal{O}_\sigma$ have upper bound z with associate γ; pick for each $a \in A$ an associate $\gamma_a \subseteq \gamma$. We may assume that each γ_a is only defined on dialogues. Then $\bigcup(\{\gamma_a | \, a \in A\}) \in \mathrm{Ass}(\sigma)$ and is an associate of the pointwise join of A. ∎

Corollary 2.5 *If $A \subseteq \mathcal{O}_\sigma$ is nonempty and bounded, then $\bigwedge A$ exists and is preserved by any $f \in \mathcal{O}_{\sigma\to o}$.*

Proof. That $\bigwedge A$ exists follows from the previous corollary. But, in the notation of that proof, $\bigcap\{\gamma_a | \, a \in A\}$ is an associate of $\bigwedge A$. If $\beta | \gamma_a = k$ for all a, then $\beta | \bigcap_{a\in A} \gamma_a = k$. So $f(\bigwedge A) = k$ iff for all $a \in A$, $f(a) = k$. ∎

Corollary 2.6 *$\mathcal{O}_\sigma$ is distributive.*

Proof. To show that $(x \vee x') \wedge y \leq (x \wedge y) \vee (x' \wedge y)$ (the other inequality always holds), we may assume that $x \vee x'$ and y are compatible (otherwise we replace y by $(x \vee x') \wedge y$); then the statement follows from ordinary distributivity of $\cap$ over $\cup$. ∎

Corollary 2.7 *$\mathcal{O}_\sigma$ has the I-property.*

Proof. For $\sigma = o$ this is trivial, and for $\sigma = \tau \to o$, first since for every $x \in \mathcal{O}_\sigma$, $x = \bigvee\{[\mu_\sigma(q)] | \, q \subseteq \gamma\}$ for any $\gamma \in \mathrm{Ass}(x)$ and this join is directed, every compact element c of $\mathcal{O}_\sigma$ is less than some $\mu_\sigma(q)$. For $\mu_\sigma(q)$ there is a finite set $\{p_1, \ldots p_n\}$ such that if $\mu_\sigma(q)|\delta = k$ then $\mu_\tau(p_i) \leq_s \delta$ for some i. So if $c([\delta]) = k$ then by $c \leq_s [\mu_\sigma(q)]$ we have $c([\mu_\tau(p_i)]) = k$. So c determines a subset of $\{p_1, \ldots, p_n\}$ on which it is defined. Hence $c = \mu_\sigma(q')$ for some q', and there are only finitely many elements of $\mathcal{O}_\sigma$ below $[\mu_\sigma(q)]$. ∎

Let us summarize:

Theorem 2.8 *Every $\mathcal{O}_\sigma$ is a dI-domain, and every $f \in \mathcal{O}_{\sigma\to o}$ is a* stable *function, meaning that it preserves directed joins and meets of nonempty bounded subsets.*

Moreover, from the proof of Corollary 2.7 it follows that every $\mathcal{O}_\sigma$ is atomic, hence a qualitative domain. Note, that this gives another proof of Corollary 2.5, since a stable function between qualitative domains automatically preserves meets of nonempty bounded subsets.

Now we want to characterize the structure $\{\mathcal{O}_\sigma | \sigma \text{ a type}\}$ as subcategory of the category of qualitative domains and stable functions. Clearly, not every stable function from $\mathcal{O}_\sigma$ to $N_\perp$ is an element of $\mathcal{O}_{\sigma\to o}$.

Example. Let f_1, f_2, f_3 be the partial functions:

$$\begin{array}{lll} f_1(0)=0 & f_1(1)=0 & \\ & f_2(1)=1 & f_2(2)=0 \\ f_3(0)=1 & & f_3(2)=1 \end{array}$$

Note that f_1, f_2, f_3 are pairwise incompatible but $\{f_1, f_2, f_3\}$ is not the set of leaves of a sequential tree. Considering f_1, f_2, f_3 as elements of $\mathcal{O}_{o\to o}$, there is a stable function $\phi : \mathcal{O}_{o\to o} \to \mathcal{O}_o$ with basis $\{f_1, f_2, f_3\}$, but ϕ is not an element of $\mathcal{O}_{(o\to o)\to o}$.

In the next section we shall see that the sequential functionals as defined here, are part of a structure known in the literature, namely Ehrhard's *strongly stable model*([3, 4]).

3 Sequential Functionals and Strong Stability

The definitions of dI-domains with coherence, strongly stable functions etc. below, are all due to Thomas Ehrhard ([3]).

Definition 3.1 *A* dI-domain with coherence *is a pair* $\mathcal{D} = (D, C(D))$ *where* D *is a dI-domain and* $C(D)$ *a set of finite, nonempty subsets of* D *(called the* coherent *subsets), satisfying:*

i) *for every* $d \in D$, $\{d\} \in C(D)$;

ii) *if* $A \in C(D)$, B *finite with* $\forall b \in B \exists a \in A\, b \leq a$ *and* $\forall a \in A \exists b \in B\, b \leq a$, *then* $B \in C(D)$;

iii) *if* $E_1 \subseteq D, \ldots, E_n \subseteq D$ *are directed and such that for all* $x_1 \in E_1, \ldots, x_n \in E_n$, $\{x_1, \ldots, x_n\} \in C(D)$ *then* $\{\bigvee E_1, \ldots, \bigvee E_n\} \in C(D)$.

From this definition it follows immediately that the set $C(D)$ is determined by the set of all coherent sets of compact elements. Note also that every finite set that has an upper bound in D, is coherent.

Definition 3.2 *Let $\mathcal{D} = (D, C(D))$ and $\mathcal{E} = (E, C(E))$ be two dI-domains with coherence. A continuous function $f : D \to E$ is called* strongly stable *if for every $A \in C(D)$, $f[A] \in C(E)$ and $f(\bigwedge A) = \bigwedge f[A]$.*

Note, that every strongly stable function is stable. Evidently there is a category **dIC** of dI-domains with coherence and strongly stable functions, and it is a subcategory of the category of dI-domains and stable functions. Ehrhard shows that **dIC** is cartesian closed: the product $\mathcal{D} \times \mathcal{E}$ is $(D \times E), C(D \times E))$ where $D \times E$ is the product of dI-domains, and $A \subseteq D \times E$ is coherent iff both its projections are coherent. The function space $\mathcal{D} \Rightarrow \mathcal{E}$ is $(D \Rightarrow E, C(D \Rightarrow C))$ where $D \Rightarrow E$ is the set of strongly stable functions from D to E (which is a dI-domain, with the stable order), and $\{f_1, \ldots, f_n\}$ is coherent iff for every coherent $\{d_1, \ldots d_m\} \subseteq D$ and every $K \subseteq \{1, \ldots, n\} \times \{1, \ldots, m\}$ such that K projects surjectively onto $\{1, \ldots, n\}$ and $\{1, \ldots, m\}$, one has that $\{f_i(d_j) \mid (i,j) \in K\}$ is coherent in E and

$$\bigwedge \{f_i(d_j) \mid (i,j) \in K\} = (\bigwedge_{i=1}^{n} f_i)(\bigwedge_{j=1}^{m} d_j)$$

In **dIC** we have the object $N = (N_\perp, C(N))$ where $A \subseteq N_\perp$ is coherent if either $\perp \in A$ or $A = \{n\}$ for some $n \in \mathbb{N}$. Using this object N and the cartesian closedness of **dIC** we have an obvious interpretation of the types of section 2. We shall show that this interpretation yields exactly the type structure of sequential functionals from section 2.

To begin with, we have noticed in the previous section that each $\mathcal{O}_\sigma$ is a qualitative domain, so a word about continuous functions between qualitative domains is in order; here I restrict to functions $\mathcal{O}_\sigma \to N_\perp$. Every such function f has (and is conversely determined by) its *trace* (B, v) where $B \subseteq \mathcal{O}_\sigma$ a set of compact elements b which are minimal w.r.t. the property that $f(b) \neq \perp$, and $v : B \to \mathbb{N}$ a function. We call B the *base* of f. In order to check stability-like properties for f it is the base B that matters (v being redundant), e.g.:

- f is stable if and only if $b \neq b'$ implies that b and b' are incompatible, for $b, b' \in B$;
- f is strongly stable iff every coherent subset of B is a singleton;
- $f : \mathcal{O}_\sigma \to N_\perp$ is sequential iff the set

$$\{p \text{ finite} \mid [\mu_\sigma(p)] \in B\}$$

is the set of leaves of a sequential tree.

I leave the verification of these facts to the reader.

We now turn to the dI-domains $\mathcal{O}_\sigma$. Even without knowing that the elements of $\mathcal{O}_{\sigma\to o}$ are strongly stable functions we can still define a coherence on them as if they were, i.e.:

- $A \subset N_\perp$ is coherent iff $\perp \in A$ or $A = \{n\}$ for some $n \in \mathbb{N}$;
- $A \subset \mathcal{O}_{\sigma\to o}$ is coherent iff for each coherent $B \subset \mathcal{O}_\sigma$ and each $E \subseteq A \times B$ such that $\pi_1(E) = A$ and $\pi_2(E) = B$, we have that $\{f(b) \mid (f, b) \in E\}$ is coherent, and, for each $n \in \mathbb{N}$, if for all $(f, b) \in E$, $f(b) = n$ then $(\bigwedge A)(\bigwedge B) = n$.

Note that this definition is equivalent to Ehrhard's for function spaces. We shall prove that in fact, $\mathcal{O}_{\sigma\to o}$ is the dIC of strongly stable functions from $\mathcal{O}$ to $N_\perp$.

First some simple remarks about associates of type $(\tau\to o)\to o$.

Lemma 3.3 *Let $\gamma \in \mathrm{Ass}((\tau\to o)\to o)$. Then there is $\gamma' \sim \gamma$ with the property that for every $\beta \in \mathrm{Ass}(\tau\to o)$ such that $\gamma'|\beta$ is defined and every dialogue u between γ' and β:*

> *$\gamma'(u^{<i})$ is of the form $2v$ where $v = \langle v_0, \ldots, v_{n-1}\rangle$ is a dialogue between β and some $\delta \in \mathrm{Ass}(\tau)$ with $\beta|\delta$ defined, and for every $m < n$ there is $j < i$ with $\gamma'(u^{<j}) = 2v^{<m}$*

Proof. Since we may assume that β is only defined on dialogues between β and some $\delta \in \mathrm{Ass}(\tau)$ with $\beta|\delta$ defined, $\gamma(\langle\rangle)$ must be $2v$ where v is such a dialogue; let γ' first question β on all subdialogues of v, etc. ∎

So basically, what $\gamma \in \mathrm{Ass}((\tau\to o)\to o)$ can do when confronted with a hypothetical $\beta \in \mathrm{Ass}(\tau\to o)$ is: feed it some δ, and see. But since γ a priori knows nothing about β except for the arguments at which β wants to know δ, the following lemma (which formalizes this idea) should be clear:

Lemma 3.4 *Let $\gamma \in \mathrm{Ass}(f)$, $f \in \mathcal{O}_{(\tau\to o)\to o}$ and suppose $\{c_1, \ldots, c_n\} \subset \mathrm{Base}(f)$. Then one of the following three possibilities occurs:*

i) $n = 1$;

ii) *there is an $x \in \mathcal{O}_\tau$ such that $c_i(x) \neq \perp$ for all $i \leq n$, but there are $i, j \leq n$ with $c_i(x) \neq c_j(x)$;*

iii) *there are finite functions $p_1, \ldots, p_n$, $p_i \subseteq \delta_i \in \mathrm{Ass}(\tau)$, and associates β_i of c_i, with $\beta_i|p_i$ defined but no β_i defined on a proper subfunction of p_i, and a $q \subseteq \bigcap_{i=1}^n p_i$ such that $\bigcap\{\mathrm{dom}(p_i \setminus q) \mid i \leq n\} = \emptyset$.*

Proof. Suppose $n > 1$ and assume γ satisfies Lemma 3.3. Take any $\beta_1 \in \mathrm{Ass}(c_1)$ and let u be the dialogue between γ and β_1. There must be a least index i such that for some $j \neq 1$, for no $\beta_1' \sim \beta_1$ and $\beta_j \in \mathrm{Ass}(c_j)$, $\langle u_0, \ldots, u_i \rangle$ is both a dialogue between γ and β_1', and γ and β_j. Now pick for each $j > 1$ an associate β_j such that the dialogue between γ and β_j starts with $u^{<i}$. $u^{<i}$ may contain already several finished dialogues between the β's and some finite functions p, but at point i we have $\gamma(u^{<i}) = 2v$ where v is a dialogue between some p and *all* β_j's. Pick for each j, now $1 \le j \le n$, a p_j such that $\beta_j|p_j$ is defined, $p_j \subseteq \delta_j \in \mathrm{Ass}(\tau)$, and v is a dialogue between β_j and p_j. Then $p \subseteq \bigcap_{j=1}^n p_j$.

If $\beta_1(v) = 2l$ then for some j, l cannot be in $\mathrm{dom}(p_j)$ since otherwise there would be an associate $\beta_j' \sim \beta_j$ which also asks l at this point. So then (iii) holds. If $\beta_1(v) = 2l + 1$ and (iii) does not hold, then for all j, $\beta_j|p$ is defined but the values must be different hence (ii) holds. ∎

Theorem 3.5 *For every type σ we have:*

i) A set $\{c_1, \ldots, c_n\}$ of compact elements of $\mathcal{O}_\sigma$ is coherent $\iff n = 1$ or there are $p_1, \ldots, p_n$ finite with $c_i = [\mu_\sigma(p_i)]$, and for some $q \subseteq \bigcap_{i=1}^n p_i$, $\bigcap\{\mathrm{dom}(p_i \setminus q) \mid 1 \le i \le n\} = \emptyset$;

ii) For a continuous function $f : \mathcal{O}_\sigma \to N_\perp$ we have: $f \in \mathcal{O}_{\sigma\to o} \iff f$ is strongly stable.

Proof. Induction on σ. For $\sigma = o$ the facts are obvious; so let $\sigma = \tau\to o$. We prove i)$\Leftarrow$, ii)$\Leftarrow$, ii)$\Rightarrow$, i)$\Rightarrow$.

i)$\Leftarrow$. If $n = 1$ then $\{c_1, \ldots, c_n\}$ is coherent by the first axiom of coherence. If $n > 1$, $p_1, \ldots, p_n$ finite with $c_i = [\mu_\sigma(p_i)]$ and $q \subseteq \bigcap_{i=1}^n p_i$ with $\bigcap\{\mathrm{dom}(p_i \setminus q) \mid 1 \le i \le n\} = \emptyset$, let $\{x_1, \ldots, x_m\}$ be a coherent set of compact elements of $\mathcal{O}_\tau$ and $K \subseteq \{1, \ldots, n\} \times \{1, \ldots, m\}$ which projects surjectively onto $\{1, \ldots, n\}$ and $\{1, \ldots, m\}$. By induction hypothesis ii) we may assume that all elements of $\mathcal{O}_\sigma$ are strongly stable functions from $\mathcal{O}_\tau$ to $N_\perp$; we apply induction hypothesis i) to $\{x_1, \ldots, x_m\}$. If $m = 1$ then we must have that either some $c_i(x_1) = \perp$ or $\{c_1(x_1), \ldots, c_n(x_1)\} = \{[\mu_\sigma(q)](x_1)\}$ by the assumption on $\{c_1, \ldots, c_n\}$. So this is coherent but also if $r_1, \ldots, r_m$ finite with $x_i = [\mu_\tau(r_i)]$, and $s \subseteq \bigcap_{j=1}^m r_j$ with $\bigcap\{\mathrm{dom}(r_j \setminus s) \mid 1 \le j \le m\} = \emptyset$ and $p_i|r_j$ is defined for $(i, j) \in K$, we must have that $q|s$ is defined. This proves that $\{c_1, \ldots, c_n\}$ is coherent.

ii)$\Leftarrow$ follows from i)$\Leftarrow$ just proved: if $f : \mathcal{O}_\sigma \to N_\perp$ is strongly stable then no nonempty, finite, non-singleton subset of $\mathrm{Base}(f)$ can be coherent hence $\{p \text{ finite} \mid [\mu_\sigma(p)] \in \mathrm{Base}(f)\}$ is the set of leaves of a sequential tree, so $f \in \mathcal{O}_{\sigma\to o}$.

ii)$\Rightarrow$ Let $f \in \mathcal{O}_{\sigma\to o}$ and $\{c_1,\ldots,c_n\} \subseteq \text{Base}(f)$. Suppose $n > 1$. Apply Lemma 3.4: if case ii) holds then clearly $\{c_1,\ldots,c_n\}$ cannot be coherent. So suppose case iii) holds, i.e. there are finite functions $p_1,\ldots,p_n$, $p_i \subseteq \delta_i \in \text{Ass}(\tau)$ and $\beta_i \in \text{Ass}(c_i)$ with $\beta_i|p_i$ defined, and $q \subseteq \bigcap_{i=1}^n p_i$ with $\bigcap\{\text{dom}(p_i \setminus q) \,|\, 1 \le i \le n\} = \emptyset$. By induction hypothesis i) we have that the set $\{[\mu_\tau(p_i)] \,|\, 1 \le i \le n\}$ is coherent; hence if $\{c_1,\ldots,c_n\}$ were coherent we would already have that $\beta_i|q$ defined (verify that $[\mu_\tau(q)] = \bigwedge\{[\mu_\tau(p_i)] \,|\, 1 \le i \le n\}$), which contradicts the choice of the p_i. So $\{c_1,\ldots,c_n\}$ is not coherent, hence f is strongly stable.

i)$\Rightarrow$ follows from ii)$\Rightarrow$ just proved: if the conclusion of i)$\Rightarrow$ does not hold for $\{c_1,\ldots,c_n\}$ then $\{c_1,\ldots,c_n\} \subseteq \text{Base}(f)$ for some $f \in \mathcal{O}_{\sigma\to o}$ which by ii)$\Rightarrow$ contradicts coherence of $\{c_1,\ldots,c_n\}$. ∎

There is a full sub-ccc of the ccc **dIC** on objects which are qualitative domains and whose coherence is generated by coherence on atoms. Ehrhard (l.c.) gives a presentation of this category in the style of Girard's qualitative domains. He calls the objects *hypercoherences.* Since $N_\perp$ is a hypercoherence, it turns out that in fact our whole type structure lands in the category of hypercoherences.

4 $\mathcal{B}$ as a combinatory algebra

For $\alpha \in \mathcal{B}$ and $x \in \mathbb{N}$ let α_x denote the partial function which sends $y \in \mathbb{N}$ to $\alpha(\langle x, y\rangle)$ (if this is defined), now $\langle\cdot,\cdot\rangle$ referring to some (recursive) bijection $\mathbb{N} \times \mathbb{N} \to \mathbb{N}$.

Definition 4.1 *Given $\alpha, \beta \in \mathcal{B}$ let $\alpha \bullet \beta$ denote the partial function*

$$\lambda x.\alpha_x|\beta$$

Theorem 4.2 *With $(\cdot)\bullet(\cdot)$ as defined in 4.1, $\mathcal{B}$ is a combinatory algebra.*

I record this fact without proof. My own proof was a laborious calculation of the combinators **k** and **s**, which is not very illuminating. Another proof could consist in showing that every recursive operator $\{e\}^{F_1,\ldots,F_n}$ in n partial oracles is in fact of the form:

$$\{e\}^{F_1,\ldots,F_n}(x) = y \Leftrightarrow (\cdots(\alpha \bullet F_1)\cdots \bullet F_n)(x) = y$$

for some α. A third approach would establish a characterization of those functions $F : \mathcal{B}^n \to \mathcal{B}$ which are of form $F(\beta_1,\ldots,\beta_n) = (\cdots(\alpha\bullet\beta_1)\cdots\bullet\beta_n)$ for some α. This involves some combinatorics with sequential trees.

Let me just make clear in what way the type structure $\{\mathcal{O}_\sigma \,|\, \sigma \text{ type}\}$ of section 2, and hence the corresponding part of Ehrhard's Hypercoherences, fits into the realizability topos generated by the combinatory algebra $\mathcal{B}$ (for realizability toposes consult [8, 6]). Let us call it $\mathcal{E}ff(\mathcal{B})$. An important subcategory (the subcategory of $\neg\neg$-separated objects) of $\mathcal{E}ff(\mathcal{B})$ can be described as follows:

Let $\mathcal{B}$-Set be the category with objects pairs (X, E_X) where X is a set and $E_X : X \to \mathcal{P}(\mathcal{B})$ a function. A function $f : X \to Y$ is a morphism from (X, E_X) to (Y, E_Y) if for some $\alpha \in \mathcal{B}$:

$$\forall x \in X \forall \beta \in E_X(x)\, \alpha \bullet \beta \in E_Y(f(x))$$

α is said to *track* f.

The category $\mathcal{B}$-Set is cartesian closed: the function space $(Y, E_Y)^{(X, E_X)}$ may be rendered as $(Y^X, E_{X \Rightarrow Y})$ where $\alpha \in E_{X \Rightarrow Y}(f)$ iff α tracks f.

For each type σ, $(\mathcal{O}_\sigma, \text{Ass})$ is an object of $\mathcal{B}$-Set and it is an easy exercise to verify that $(\mathcal{O}_{\sigma \to o}, \text{Ass})$ is isomorphic in $\mathcal{B}$-Set to $(N_\perp, \text{Ass})^{(\mathcal{O}_\sigma, \text{Ass})}$.

References

[1] S. Abramsky, R. Jagadeesan & P. Malacaria, *Full abstraction for PCF* (extended abstract), in: Theoretical Aspects of Computer Software: TACS '94. Springer (1994), LNCS 789

[2] A. Bucciarelli & T. Ehrhard, *Sequentiality and strong stability*, Proc. Logic Comput. Sci. Amsterdam (1991), pp. 138-145

[3] T. Ehrhard, *Projecting sequential algorithms on strongly stable functions*, Annals of Pure and Applied Logic **77**(1996), pp. 201-244

[4] T. Ehrhard, *A relative PCF-definability result*, manuscript, May 1996

[5] J.-Y. Girard, *The System F of Variable Types, Fifteen Years Later*, Theoretical Computer Science **45**(1986), pp. 159-192

[6] J. M. E. Hyland, *The Effective Topos*, in: Troelstra (ed.), The L. E. J. Brouwer Centenary Symposium, North-Holland 1982

[7] J. M. E. Hyland, *First Steps in Synthetic Domain Theory*, in: Carboni et al (eds.), Category Theory, Springer (1991), LNM 1488

[8] J. M. E. Hyland, P. T. Johnstone & A. M. Pitts, *Tripos Theory*, Math. Proc. Camb. Phil. Soc. **88**(1980), pp. 205-232

[9] J. M. E. Hyland & L. Ong, *On Full Abstraction for PCF* (1994), typeset manuscript, ftp-able at `theory.doc.ic.ac.uk` in directory `papers/Ong`

[10] S. C. Kleene, *Countable Functionals*, in: A. Heyting (ed.), Constructivity in Mathematics, North Holland (1959), pp. 81-100

[11] S. C. Kleene, *Recursive Functionals and Quantifiers of Finite Type I*, Trans. AMS **91**(1959), pp. 1-52

[12] S. C. Kleene, *Recursive Functionals and Quantifiers of Finite Type II*, Trans. AMS **108**(1963), pp. 106-142

[13] S. C. Kleene, *Recursive Functionals and Quantifiers of Finite Type Revisited I*, in: Fenstad et al (eds), Generalized Recursion Theory II, North-Holland (1978), pp. 185-221

[14] S. C. Kleene, *Recursive Functionals and Quantifiers of Finite Type Revisited II*, in: Barwise et al (eds), The Kleene Symposium, North-Holland (1980), pp. 1-29

[15] G. Kreisel, *Interpretation of Analysis by means of Constructive Functionals of Finite Types*, in: A. Heyting (ed.), Constructivity in Mathematics, North Holland (1959), pp. 101-128

[16] D. Norman, Recursion on the Countable Functionals, Springer (1980), LNM 811

[17] G. Plotkin, *LCF considered as a programming language*, Theoretical Computer Science **5**(1977), pp. 227-255

[18] D. S. Scott, *A New Category?*, manuscript, December 1996

Model Theory of Analytic and Smooth Functions

Alex J. Wilkie
Mathematical Institute
Oxford

The title of this short survey is, I now realise, ambiguous. I did not intend to discuss spaces of functions forming *domains* of structures in some suitable first-order language (e.g. the language of vector spaces or rings with, possibly, symbols for differential or integral operators). I did, and still do, intend to consider particular (interesting?, naturally occurring?,...) functions, or classes of functions, as part of the *language* for structures whose domain is the set of real or complex numbers, usually the reals. However, since I believe that the future might lie more in considerations of the former type I shall conclude with a short discussion of that area. For the main theme, my starting point is the ordered field of real numbers $\bar{\mathbb{R}} := \langle \mathbb{R}; +, \cdot, 0, 1, < \rangle$, its theory $\mathrm{Th}(\bar{\mathbb{R}})$, and the celebrated quantifier elimination theorem of Tarski ([29]): given any formula $\phi(\bar{x})$ of $L(\bar{\mathbb{R}})$ one can effectively find a quantifier free formula $\psi(\bar{x})$ of $L(\bar{\mathbb{R}})$ such that $\mathrm{Th}(\bar{\mathbb{R}}) \models \forall \bar{x}(\phi(\bar{x}) \leftrightarrow \psi(\bar{x}))$. Since one can easily decide the truth or falsity (in $\bar{\mathbb{R}}$) of quantifier free *sentences* of $L(\bar{\mathbb{R}})$, the theorem immediately implies the decidability of $\mathrm{Th}(\bar{\mathbb{R}})$. This was Tarski's main concern in his paper but applications of quantifier elimination for $\mathrm{Th}(\bar{\mathbb{R}})$ have been many and diverse. (For a recent account of an old example in differential equations see [20].) In a subsequent paper [28],[1] Tarski himself infers that every $\bar{\mathbb{R}}$-definable subset of $\mathbb{R}$ (even allowing parameters) is a finite union of open intervals and points. This must have seemed an observation hardly worth emphasising at the time, but turned out to be a remarkable insight. For in the 1980's van den Dries showed that many of the finiteness properties enjoyed by the semi-algebraic sets (i.e. the $\bar{\mathbb{R}}$-definable subsets of $\mathbb{R}^n$, for arbitrary n) follow for subsets of $\mathbb{R}^n$ definable in any structure with domain $\mathbb{R}$ provided that the ordering is in the language and that the definable subsets of $\mathbb{R}$ have this simple form. Such structures were called *o-minimal* (order-minimal) by Knight, Pillay and Steinhorn who generalized and investigated

[1]Subsequent, that is, to his original proof which did not appear for over twenty years.

the notion for arbitrary totally ordered domains (see [23],[26]). In this paper, however, I shall not be concerned with general properties of o-minimal structures (even those with domain $\mathbb{R}$) as several surveys already exist (see [10],[14]) and, indeed, van den Dries' excellent book on ths subject ([11]) has finally appeared. Rather, I shall be trying to give some idea of the methods that have been used in order to study the definable sets in structures where we suspect there to be no wild behaviour (cf the title of van den Dries' book). I shall also restrict myself to expansions of the real ordered field $\bar{\mathbb{R}}$ by functions with various smoothness assumptions, which I now explain.

Let U be an open subset of $\mathbb{R}^n$ and $f : U \to \mathbb{R}$ a function. Then f is said to be C^m (on U) if f is m-times continuously differentiable on U. If f is C^m (on U) for all m then f is said to be C^∞ (on U). If f is C^∞ on U and $p \in U$ then one may form the formal Taylor series of f at p. If this series converges to $f(x)$ for all x in some neighbourhood of p (lying within U) then f is said to be *analytic* at p. If f is analytic at p for all $p \in U$ then f is said to be analytic on U.

For the moment we shall only be concerned with expansions of $\bar{\mathbb{R}}$ by some set of analytic functions. Unfortunately no such expansion has elimination of quantifiers (unless, of course, all the new functions are already $\bar{\mathbb{R}}$-definable). This was proved by van den Dries in [8] by modifying an argument of Osgood who treated the case $\langle \bar{\mathbb{R}}, \exp \rangle$ (where $\exp(x) = e^x$ for $x \in \mathbb{R}$). Notice that I have not been precise about the domains of the new functions in our expanded structures, but this turns out to be unimportant for the van den Dries result: he showed that we do not have quantifier elimination even locally. Let us examine a simple (but, in fact, fairly typical) case to see what goes wrong even if we allow ourselves arbitrary, locally defined analytic functions as terms.

Consider a function f analytic on an open neighbourhood U of the origin in $\mathbb{R}^{n+1}$. We attempt to eliminate the quantifier from the formula[2]

$$\exists x_{n+1}(\langle \bar{x}, x_{n+1} \rangle \in V \wedge f(\bar{x}, x_{n+1}) = 0) \tag{1}$$

on some sufficiently small open neighbourhood V of $\langle \bar{0}, 0 \rangle$ with $V \subseteq U$. (I am using the vector notation $\bar{x}$ for an n-typle $\langle x_1, ..., x_n \rangle$; thus $\bar{0}$ denotes the origin in $\mathbb{R}^n$.)

If $f(\bar{0}, 0) \neq 0$ then, by the continuity of f, we can choose V so that (1) defines the empty set.

If $f(\bar{0}, 0) = 0$, but $\frac{\partial f}{\partial x_{n+1}}(\bar{0}, 0) \neq 0$ then the Implicit Function Theorem tells us that for sufficiently small $\varepsilon > 0$ there is a function (in fact an analytic one) $\phi : (-\varepsilon, \varepsilon)^n \to \mathbb{R}$ whose graph is precisely the zero set of f in some neighbourhood of $\langle \bar{0}, 0 \rangle$. It follows that (1) defines a set of the (quantifier free

[2] As usual in this subject, and especially when only one structure is under consideration, I do not distinguish between functions and function symbols or sets and predicate symbols.

definable) form $(-\varepsilon,\varepsilon)^n$ for suitably chosen V. Now there is a generalization of the Implicit Function Theorem which is crucial in this kind of work. It is the Weierstrass Preparation Theorem and it states that if, for some m, $f(\bar{0},0) = \frac{\partial f}{\partial x_{n+1}}(\bar{0},0) = \cdots = \frac{\partial^m f}{\partial x_{n+1}^m}(\bar{0},0) = 0 \neq \frac{\partial^{m+1} f}{\partial x_{n+1}^{m+1}}(\bar{0},0)$, then there exists $\varepsilon > 0$ such that $(-\varepsilon,\varepsilon)^{n+1} \subseteq U$ and, on $(-\varepsilon,\varepsilon)^{n+1}$,

$$(2) \qquad f(\bar{x},x_{n+1}) = \psi(\bar{x},x_{n+1}) \cdot (x_{n+1}^{m+1} + \phi_m(\bar{x}) \cdot x_{n+1}^m + \cdots + \phi_0(\bar{x}))$$

for some (unique) analytic functions $\phi_0, \ldots, \phi_m : (-\varepsilon,\varepsilon)^n \to \mathbb{R}$ (which take the value 0 at $\bar{0}$) and $\psi : (-\varepsilon,\varepsilon)^{n+1} \to \mathbb{R}$, with ψ nonvanishing.

Taking $V = (-\varepsilon,\varepsilon)^{n+1}$, (2) implies that (1) is equivalent to

$$(3) \exists x_{n+1}(\langle \bar{x}, x_{n+1}\rangle \in (-\varepsilon,\varepsilon)^{n+1} \wedge x_{n+1}^{m+1} + \phi_m(\bar{x}) \cdot x_{n+1}^m + \cdots + \phi_0(\bar{x}) = 0).$$

If we replace $\phi_j(\bar{x})$ by a new variable y_j in (3) (for $j = 0, \ldots, m$) we obtain a formula of $L(\bar{\mathbb{R}})$ which, by Tarski's theorem, is equivalent to a quantifier free formula, $A(\bar{x}, y_0, \ldots, y_m)$ say, of $L(\bar{\mathbb{R}})$. But then $A(\bar{x}, \phi_0(\bar{x}), \ldots, \phi_m(\bar{x}))$ is equivalent to (3), and hence to (1), and is a quantifier free formula in the language that we are allowing ourselves.

The difficulty comes, therefore, when $\frac{\partial^m f}{\partial x_{n+1}^m}(\bar{0},0) = 0$ for all m which, since f is analytic, is equivalent to the function $x_{n+1} \mapsto f(\bar{0}, x_{n+1})$ being identically zero. This situation is analysed in the important paper [7] and the idea is as follows.

For each $p \geq 1$ let F_p denote the ring of (real) formal power series in p variables and let O_p denote the subring consisting of those series that converge (so define analytic functions) on some open neighbourhood of the origin in $\mathbb{R}^p$. It is known that both F_p and O_p are Noetherian rings. Now write $f(\bar{x},x_{n+1}) = \sum_{m=0}^{\infty} g_m(\bar{x}) \cdot x_{n+1}^m$ where $g_m(\bar{x}) = \frac{1}{m!}\frac{\partial^m f}{\partial x_{n+1}^m}(\bar{x},0) \in O_n$, the equality being valid on some open neighbourhood of $\langle \bar{0}, 0\rangle$. Choose M large enough so that $g_0(\bar{x}), \ldots, g_M(\bar{x})$ generate the same ideal as $\{g_m(\bar{x}) : m \geq 0\}$ in O_n. Say $g_{m+j}(\bar{x}) = \sum_{m \leq M} a_{j,m}(\bar{x}) \cdot g_m(\bar{x})$ for $j \geq 1$ where the $a_{j,m}$'s lie in O_n. Then

$f(\bar{x},x_{n+1}) = \sum_{m \leq M} g_m(\bar{x}) x_{n+1}^m + \sum_{j=1}^{\infty}\left(\sum_{m \leq M} a_{j,m}(\bar{x}) \cdot g_m(\bar{x})\right) x_{n+1}^{M+j}$ which, after re-arrangement, becomes

$$(4) \qquad f(\bar{x},x_{n+1}) = \sum_{m \leq M} g_m(\bar{x}) \cdot x_{n+1}^m \cdot u_m(\bar{x},x_{n+1})$$

where, for $m \leq M$, $u_m(\bar{x},x_{n+1}) = 1 + x_{n+1}^{M+1-m} \cdot \sum_{j=1}^{\infty} a_{j,m}(\bar{x}) \cdot x_{n+1}^{j-1}$ so, in particular,

$$(5) \qquad u_m(\bar{0},0) = 1 \quad \text{for } m \leq M.$$

Actually, the above argument is only correct formally (i.e. in the ring F_{n+1}). But it can be shown (by general algebraic arguments based on the fact that O_{n+1} is *flat* in F_{n+1}) that if (4) has a formal solution for the u_m's satisfying (5) (as we have shown it has) then it has a solution in O_{n+1}. Hence we have a representation for f given by (4) with the g_m's and u_m's analytic on some neighbourhood of $\bar{0}$ and $\langle \bar{0}, 0 \rangle$ respectively where, because of (5), we may assume the u_m's are nonvanishing.

The point of (4) is that it makes the singularity of f at $\bar{x} = 0$ more amenable. For let $p \leq M$ and denote by R_p the (quantifier free definable) region $\{\bar{x} \in \mathbb{R}^n : g_p(\bar{x}) \neq 0 \text{ and } |g_p(\bar{x})| \geq |g_m(\bar{x})| \text{ for all } m \leq M\}$. We shall analyse (1) on each R_p, which is sufficient since (1) is clearly true elsewhere. To do this let $\tilde{y} = y_0, \ldots, y_{p-1}, y_{p+1}, \ldots, y_M$ be new variables and, for $\tilde{c} \in [-1,1]^M$ (similarly indexed), consider the function, denoted $f_p^*(\bar{x}, x_{n+1}, \tilde{y})$, obtained by dividing the right hand side of (4) by $g_p(\bar{x})$ and replacing $g_m(\bar{x})/g_p(\bar{x})$ by $y_m + c_m$ (for each $m \leq M$ with $m \neq p$). Now because of (5), the coefficient of x_{n+1}^q in the Taylor series for $f_p^*(\bar{0}, x_{n+1}, \tilde{0})$ is nonzero, where q is the least element of the (nonempty!) set $\{p\} \cup \{m \leq M : m \neq p \text{ and } c_m \neq 0\}$. Hence we may use the Preparation Theorem method described above to obtain a quantifier free formula equivalent to the formula $\exists x_{n+1}(\langle \bar{x}, x_{n+1}, \tilde{y} \rangle \in W \wedge f_p^*(\bar{x}, x_{n+1}, \tilde{y}) = 0)$, for some open neighbourhood W of $\langle \bar{0}, 0, \tilde{0} \rangle$ in $\mathbb{R}^{n+1+M}$. If we now substitute back the (*nonanalytic*) expressions $(g_m(\bar{x})/g_p(\bar{x})) - c_m$ for y_m we obtain a "formula" equivalent to (1) for those $\bar{x} \in R_p$ with the ratio $g_m(\bar{x})/g_p(\bar{x})$ sufficiently close to c_m for each $m \leq M$ with $m \neq p$. The compactness of $[-1,1]^M$ guarantees that it suffices to consider only finitely many $\tilde{c}$'s, and so we obtain a "formula" equivalent to (1) for all $\bar{x} \in R_p$. This "formula" is quantifier free apart from the use of *division*, and it turns out (via an elaboration of the argument just sketched) that this is the only obstruction to analytic quantifier elimination. This is the main result of [7] and the precise statement is as follows:

Theorem 1 (Denef–van den Dries, 1988)
Let $\mathbb{R}_{an}$ be the expansion of $\bar{\mathbb{R}}$ obtained by adding, for each $n \geq 1$ and each analytic function $f : U \to \mathbb{R}$, where U is some open neighbourhood of $[-1,1]^n$ in $\mathbb{R}^n$, the function $f^* : \mathbb{R}^n \to \mathbb{R}$ defined by

$$f^*(\bar{x}) = \begin{cases} f(\bar{x}) & \text{if } \bar{x} \in [-1,1]^n, \\ 0 & \text{otherwise.} \end{cases}$$

Then the theory of the structure $\langle \mathbb{R}_{an}, D \rangle$ has elimination of quantifiers, where $D : \mathbb{R}^2 \to \mathbb{R}$ is the function defined by

$$D(x,y) = \begin{cases} x/y & \text{if } y \neq 0 \text{ and } |x| \leq |y|; \\ 0 & \text{otherwise.} \end{cases}$$

As I have already mentioned, the theory of the structure $\mathbb{R}_{an}$ does not eliminate quantifiers. But clearly the theorem above implies that it is *model complete*, i.e. every formula of $L(\mathbb{R}_{an})$ is equivalent to an *existential* formula of $L(\mathbb{R}_{an})$. This latter result, or rather its reformulation in the language of analytic geometry in which it states that the complement of a subanalytic set is subanalytic, was proved much earlier by Gabrielov (see [17]).

His proof used a direct induction on n to show that the complement of every existentially definable subset of $\mathbb{R}^n$ is existentially definable, which is clearly sufficient to establish model completeness. The argument involves intricate use of differential and analytic geometric techniques rather than power series methods but suffers from the disadvantage that it does not seem capable of obtaining the refined result of Denef and van den Dries. This becomes apparent in one of its applications! For in the paper [18] Gabrielov adapts (and, in fact, considerably simplifies) his method to show that in any reduct of $\mathbb{R}_{an}$ (expanding $\bar{\mathbb{R}}$) we only need the class of terms to be closed under differentiation for the theory of that reduct to be model complete. But, as Gabrielov himself shows in [19], there is *no* such quantifier elimination result even with division (see below).

Theorem 2 (Gabrielov, 1996)

Let F be any collection of real valued analytic functions with domains open neighbourhoods of $[-1,1]^n$ in $\mathbb{R}^n$ (for various n's) and suppose that F is closed under partial differentiation. Then the theory of the structure $\mathbb{R}_F := \langle \bar{\mathbb{R}}, \{f^* : f \in F\}\rangle$ is model complete (where the f^* are defined as in Theorem 1).

Thus, for example, the theory of the structure $\langle \bar{\mathbb{R}}, \exp^* \rangle$ is model complete but, as is shown in [19], that of the structure $\langle \bar{R}, \exp^*, D\rangle$ does not have quantifier elimination (cf. the remark above) and it is an interesting open problem to find natural auxiliary functions turning Theorem 2 into a quantifier elimination result.

Of course, it is desirable here that the auxiliary functions be definable in $\mathbb{R}_F$, but minimal conditions on F guaranteeing that the theory of $\langle \mathbb{R}_F, D\rangle$ eliminates quantifiers would also be interesting. One can axiomatise the proof method of Theorem 1 in order to obtain such a result and this was done by van den Dries. The condition he imposes on F is that the ring of functions (in arbitrarily many variables) it generates be closed under *Weierstrass division*, which is a process ensuring that this ring (a) is closed under differentiation, (b) is flat in the formal power series ring and (c) satisfies the Weierstrass Preparation Theorem (see [9] for details). Unfortunately, it is now known that Weierstrass division does not hold definably: an example of Bianconi shows that *if* the collection of those functions definable in the structure $\langle \bar{\mathbb{R}}, \exp^* \rangle$ and analytic on an open neighbourhood of the origin is closed under Weierstrass division *then* it would have to contain the sine function (restricted to $(-\varepsilon, \varepsilon)$ for some $\varepsilon > 0$) which, he showed several years later in [4], is impossible.

The point is, I suppose, that the classical Weierstrass Preparation Theorem (and the Division Theorem) is genuinely a result of *complex* analysis — it does hold *definably* in the analogous structures $\mathbb{C}_F$ (for F closed under differentiation) — and only gives information for real functions after taking real and imaginary parts of complex analytic functions. Hence the intrusion of the sine function when investigating the (real) exponential function.

The two methods I have discussed so far have both been *local*. This should be clear from my sketch above in the case of the Denef–van den Dries method. As for the Gabrielov method, suffice it to say that it relies heavily on (variations and generalizations of) the fact that the boundary (in $\mathbb{R}^n$) of a set of the form $\{\bar{x} \in \mathbb{R}^n : \exists x_{n+1} f(\bar{x}, x_{n+1}) = 0\}$ is contained in the set $\{\bar{x} \in \mathbb{R}^n : \exists x_{n+1}(f(\bar{x}, x_{n+1}) = 0 \wedge \frac{\partial f}{\partial x_{n+1}}(\bar{x}, \bar{x}_{n+1}) = 0)\}$ for functions f under consideration. (The latter set ought to have "lower dimension" than the former and this idea forms the basis of Gabrielov's induction.) This is true, by the Implicit Function Theorem, provided that the quantified variable x_{n+1} cannot become unbounded on a bounded region of $\bar{x}$'s. This is why the structures considered above involve the *restricted* functions f^* and not the functions f themselves, which may exhibit unruly behaviour on *open* sets.

In [32] I overcame this difficulty in one particular case: I showed that the theory of the structure $\langle \bar{\mathbb{R}}, \exp \rangle$ (with exp unrestricted) is model complete. The proof uses Robinson's Test for model completeness together with the simple, but extremely useful, fact that any existential formula (in free variables $\bar{x} = x_1, \dots, x_n$ say) of the language of a structure expanding $\bar{\mathbb{R}}$ *by functions* is equivalent to one of the form $\exists x_{n+1}, \dots, x_m\ F(\bar{x}, x_{n+1}, \dots, x_m) = 0$ where F is a polynomial in the new functions (and the variables indicated). Thus, I only had to show that if k, K are models of $\mathrm{Th}(\langle \bar{\mathbb{R}}, \exp \rangle)$ with $k \subseteq K$ and $p(x_1, \dots, x_n, y_1, \dots, y_n)$ is a polynomial with coefficients in k, then the equation $p(x_1, \dots, x_n, e^{x_1}, \dots, e^{x_n}) = 0$ has a solution in k provided that it has one in K.

To do this one first shows that such equations always have solutions in k provided that they have them in K *with coordinates bounded* between elements of k, and I had already done this in the paper [31]. The proof used definable versions of the Intermediate Value Theorem and Rolle's Theorem (hence the need for the bounds) in much the same way as some proofs of Tarski's theorem do. The argument here is rather more complicated, however, because there seems to be no way of reducing to the case of functions in one variable (and, presumably, if there were then some simple expansions of $\langle \bar{\mathbb{R}}, \exp \rangle$ would have a theory with quantifier elimination which, by results of Macintyre and Marker in [24], seems unlikely). My methods in [31] (and the first part of

[32]) were, in fact, fairly ad hoc, but I have recently come across a many-variable version of the Intermediate Value Theorem which might simplify the argument. I have not been able to do so, but I mention it in case the reader has a use for it. (The proof is an easy exercise using the Brouwer Fixed Point Theorem.) It states that if $\bar{f} : B \to \mathbb{R}^n$ is a continuous function, where B is a closed (euclidean) ball in $\mathbb{R}^n$ centred at $\bar{0}$, and $\bar{f}(\bar{x}) \cdot \bar{x} > 0$ (scalar product) for all $\bar{x} \in \partial B$ (the boundary of B), then $\bar{f}$ has a zero in B. Notice that this statement has the logical form $\forall \to \exists$, which is promising, but for applications one also needs some "definability-friendly" form of the reverse implication (which is false as it stands).

To return to the proof of the model completeness of $\langle \bar{\mathbb{R}}, \exp\rangle$ it remains, then, to establish the boundedness property for the solutions of exponential-polynomial equations. For this I made use of the natural valuation associated with an ordered field, in this case the *ordered field* K, the point being that K has the extra structure of a logarithm defined on it which behaves very much like the valuation. It is the interplay between the two that clinches the result and it is precisely here that one uses the special properties of the exponential function — the method does not seem adaptable to any other (unrestricted) expansion. It has been shown, however, that one *can* generalize the ground structure $\bar{\mathbb{R}}$.

This was first done by van den Dries and Miller in [13] who modified my argument to show that the theory of the structure $\langle \mathbb{R}_{an}, \exp\rangle$ is model complete. This was improved upon by van den Dries, Macintyre and Marker in [12] who elaborated on ideas of Ressayre to show that this theory has quantifier elimination if a symbol for log is added to the language, and that it has a very natural axiomatization (over one for $\mathrm{Th}(\mathbb{R}_{an})$). These latter results, which my original method seems incapable of achieving (and, indeed, it would be very surprising if $\mathrm{Th}(\langle \bar{\mathbb{R}}, \exp, \log\rangle)$ had quantifier elimination, but this is still unsolved — see [27]), require a much more sophisticated approach to the valuation theory mentioned above. One in fact produces, en route, embeddings of nonstandard models of $\mathrm{Th}(\mathbb{R}_{an})$ into explicitly constructed power series models. Most recently, van den Dries and Speissegger managed to establish the essential valuation-theoretic facts in situations where these embeddings are not known to exist and thereby obtained the most general result in this direction concerning exponential expansions:-

Theorem 3 (van den Dries–Speissegger, 1997, [16])

Let $\tilde{\mathbb{R}}$ be any expansion of $\bar{\mathbb{R}}$ in which the restricted exponential function $\exp^*$ is definable. Suppose further that $\tilde{\mathbb{R}}$ is *o*-minimal and polynomially bounded (i.e. every $\tilde{\mathbb{R}}$-definable function $f : \mathbb{R} \to \mathbb{R}$ is eventually bounded by a polynomial). Then if we add all $\tilde{\mathbb{R}}$-definable functions as terms to the language, the theory of the structure $\langle \tilde{\mathbb{R}}, \exp, \log\rangle$ admits quantifier elimination.

Further, this theory is axiomatized over Th($\tilde{\mathbb{R}}$) by the natural (universal) axioms for exp and log.

I now turn to expansions of $\bar{\mathbb{R}}$ by C^∞, but not necessarily analytic, functions. The first remark is that although the Preparation Theorem still holds (just replace "analytic" everywhere by "C^∞" in the formulations above, but omit the uniqueness statement, to get the *Malgrange–Mather Preparation Theorem* (see [30] for a discussion)) one cannot immediately follow the first method discussed above because the awkward case becomes intractible: the function f under consideration may have zeroes of infinite order with respect to x_{n+1} (for certain values of $\bar{x}$) without being identically zero. Further, even if we exclude such functions from our initial considerations we have no guarantee that they will not crop up as a result of applying the Preparation Theorem. This difficulty was partially overcome by van den Dries and Speissegger ([16]) who discovered a large class of C^∞-functions (defined around the origin) with good closure properties (including an analogue of Weierstrass Division) and yet each function in the class is uniquely determined by its Taylor series at the origin. So this class behaves sufficiently like the class of analytic functions except that the Taylor series may be divergent.

I shall not attempt here to give the rather complicated statement of their results (see also [15]) but suffice it to mention an example. In conjunction with Theorem 3 they show that there exists a structure expanding $\bar{\mathbb{R}}$ which has a model complete and o-minimal theory and in which the classical gamma function (on $(0, \infty)$) is definable. (The gamma function is, of course, analytic on $(0, \infty)$ but, as was shown in [12], it is not definable in $\langle \mathbb{R}_{an}, \exp \rangle$. Hence, the application of Theorem 3 here really does have to be a structure $\tilde{\mathbb{R}}$ in which genuinely non-analytic functions are definable.)

Another approach to non-analyticity has been taken by Maxwell ([25]) and myself ([32]) based on work of Charbonnel in [6]. Charbonnel's paper claims to prove the o-minimality of $\langle \bar{\mathbb{R}}, \exp \rangle$ without going via model completeness, but many of the arguments do seem, at least to my mind, incomplete. However, the main idea is fundamentally sound and rather elegant.

One begins with the observation that the quantifier-free definable subsets of $\mathbb{R}^n$ in structures of interest to us are well-behaved in the sense that they do not contain infinitely many connected components. Indeed, we would hardly be embarking on a model-theoretic investigation of the structure had this not been already established. In fact, since we would want this property to be preserved under elementary equivalence (for some definable version of connectedness) we would expect there to be a finite upper bound on the number of connected components independent of any parameters in the quantifier-free formula defining the set, a property I shall here call *tameness*.

Now tameness is clearly preserved under the operations of taking projections (from $\mathbb{R}^n$ to $\mathbb{R}^m$ for $n \geq m \geq 1$), finite unions and products, and topological closure (in the ambient $\mathbb{R}^n$). Charbonnel showed that by setting up things carefully (and cleverly) one could also preserve tameness under taking finite intersections. (The reason for including topological closure, which is not one of the usual logical operations, will become clear shortly.) The only (first-order) operation now missing is that of taking complements. It turns out that this is automatic under various smoothness assumptions on the expansion of $\bar{\mathbb{R}}$ under consideration. However, no unnatural restriction of functions is required because, and this was Charbonnel's crucial idea, limiting behaviour of a set at say, $x_{n+1} = \infty$ can be treated by inverting this variable and then taking the intersection with the hyperplane $x_{n+1} = 0$ of the closure (in $\mathbb{R}^{n+1}$) of the resulting inverted set.

Maxwell's result is as follows:-

Theorem 4 (Maxwell, 1998, [25])

Let $\tilde{\mathbb{R}}$ be any expansion of $\bar{\mathbb{R}}$ by C^1-functions with domains $\mathbb{R}^n$ for various n) and suppose that quantifier-free $\tilde{\mathbb{R}}$-definable sets are tame. Suppose further that the closure of any existentially $\tilde{\mathbb{R}}$-definable set is existentially $\tilde{\mathbb{R}}$-definable. Then the theory of $\tilde{\mathbb{R}}$ is model complete (and o-minimal).

Although the second hypothesis on $\tilde{\mathbb{R}}$ here might seem artificial, Theorem 4 does isolate the use of analyticity and restriction in Theorem 2. For one of Gabrielov's first lemmas in his proof is to the effect that this hypothesis holds for the structures $\mathbb{R}_F$. So it is fair to say that Theorem 4 is a generalization of Theorem 2, albeit one that is difficult to apply. It turns out that if we strengthen the smoothness assumption then we can explain Theorem 4 as follows:

Theorem 5 (Wilkie, 1997, [33])

Let $\tilde{\mathbb{R}}$ be any expansion of $\bar{\mathbb{R}}$ by C^∞-functions (with domains $\mathbb{R}^n$ for various n) and suppose that quantifier-free $\tilde{\mathbb{R}}$-definable sets are tame. Then any $\tilde{\mathbb{R}}$-definable set can be obtained from them by finitely many applications of projection, union, intersection and taking topological closure. In particular (by the remarks above) $\tilde{\mathbb{R}}$ is o-minimal.

It follows from Theorem 5 that any structure of the form $\langle \bar{\mathbb{R}}, f_1, ..., f_r \rangle$, where $f_1, ..., f_r$ is a *Pfaffian chain* on $\mathbb{R}^n$ (i.e. each $\frac{\partial f_i}{\partial x_j}$ can be expressed as a polynomial in $\bar{x}, f_1, ..., f_i$), is o-minimal. (In fact, Bianconi had already obtained a partial result in this direction in [3] and his methods, as well as Charbonnel's, feature strongly in the proof of Theorem 5.) The tameness assumption in this case follows from work of Khovanskii ([22]). Indeed, it is a pleasure to acknowledge here the motivation and impetus that Khovanskii's ideas have given to model-theoretic research into analytic and smooth functions.

I now conlude this survey, as promised, with some results on rings and spaces of functions.

Let U be a connected, open subset of the complex plane $\mathbb{C}$ and let $H(U)$ denote the set of all holomorphic functions on U (with values in $\mathbb{C}$). Then $H(U)$ is naturally the domain of a ring (with pointwise operations). Notice that if $\phi : U_1 \to U_2$ is a *conformal equivalence* between U_1 and U_2 (i.e. a bijective holomorphic map), then ϕ induces a ring isomorphism $f \mapsto f \circ \phi$ from $H(U_2)$ to $H(U_1)$. A theorem of Bers [2] asserts that the converse holds: if $H(U_1)$ and $H(U_2)$ are isomorphic as rings via an isomorphism that maps the constant function $\sqrt{-1}$ (on U_1) to $\sqrt{-1}$ (on U_2) (respectively, $\sqrt{-1}$ to $-\sqrt{-1}$), then U_1 and U_2 are conformally (respectively, anti-conformally) equivalent.

A model-theorist might now ask what, if anything, happens if we weaken isomorphism here to elementary equivalence. This question was "answered" by Becker, Henson and Rubel in their authoratitive paper [1]: the statement that every connected, open U is determined up to conformal or anti-conformal equivalence by the first-order theory of the ring $H(U)$ is independent of ZFC! This paper also makes a detailed study of the theory of the ring $H(\mathbb{C})$ and proves that it is highly undecidable, in fact it interprets second-order arithmetic. (See also Rubel's book [27] where the author makes the light-hearted, but valid, remark that it follows that all of complex analysis is contained within just the ring structure of the class of entire functions.) Actually, the same is true for $H(U)$, for any connected, open subset U of $\mathbb{C}$, but this turned out to be more difficult. In [1] the problem was reduced to that of giving a first-order definition of the set of constant functions (which is easy for $H(\mathbb{C})$ since, by Picard's Little Theorem, an entire function f is constant if and only if it is either 0,1 or else both f and $f-1$ are invertible in the ring $H(\mathbb{C})$) and this was finally achieved in 1994 by Huuskonen [21].

It seems to me an interesting project to investigate invariants of connected, open subsets U of $\mathbb{C}$ coming from theories of structures on $H(U)$ in languages less expressive than the ring language. A contribution in this direction has been made by my student H.T.F. Braun who considered $H(U)$ as a $\mathbb{C}$-vector space with a unary function symbol for differentiation. Notice that differentiation is surjective on $H(\mathbb{C})$ but not on $H(\mathbb{C}-\{0\})$, so not all such structures are elementarily equivalent (as they are when considered merely as $\mathbb{C}$-vector spaces). But it turns out, as Braun showed in [5], that this is all that can happen: if U is simply connected then the differential, $\mathbb{C}$-vector spaces $H(U)$ and $H(\mathbb{C})$ are elementarily equivalent and, in fact, their theory eliminates quantifiers, whereas if U is not simply connected (however badly) then $H(U)$ is elementarily equivalent to $H(\mathbb{C}\backslash\{0\})$. In both cases the theory, when scalar multiplication is restricted to algebraic numbers say, is decidable.

Unfortunately, even the gentlest of extensions of the language considered by Braun introduce major mathematical difficulties, but nevertheless seem worth pursuing.

References

[1] J. Becker, W. Henson and L.A. Rubel, First-order conformal invariants, Ann. of Maths. (2nd series) 112 (1980) 123–178.

[2] L. Bers, On rings of analytic functions, Bull. Amer. Math. Soc. 54 (1948), 311–315.

[3] R. Bianconi, On sets ∀-definable from Pfaffian functions, J. Symb. Logic 57, No.2, (1992), 688–697.

[4] ———, Nondefinability results for expansions of the field of real numbers by the exponential function and by the restricted sine function, J. Symb. Logic 64, No.4, (1997), 1173–1178.

[5] H.T.F. Braun, Some rings and modules of analytic functions, dissertation, Oxford, 1997.

[6] J.-Y. Charbonnel, Sur certains sous-ensembles de l'espace euclidean, Ann. Inst. Fourier. Grenoble 41, 3 (1991), 679–717.

[7] J. Denef and L. van den Dries, p-adic and real subanalytic sets, Ann. of Maths. (2nd series) 128 (1988), 80–138.

[8] L. van den Dries, Remarks on Tarski's problem concerning $\langle R, +, \cdot, \exp\rangle$, Logic Colloquium 1982, North Holland, 1984, 97–121.

[9] ———, On the elementary theory of restricted elementary functions, J. Symbolic Logic 53 (1988), 796–808.

[10] ———, *o*-minimal structures, in Logic Colloquium '93 (edited by W. Hodges, J.M.E. Hyland, C. Steinhorn and J. Truss), OUP 1997, 137–185.

[11] ———, Tame topology and *o*-minimal structures, LMS Lecture Note Series 248, CUP, 1998.

[12] ———, A. Macintyre and D. Marker, The elementary theory of restricted analytic fields with exponentiation, Ann. of Math. 140 (1994), 183–205.

[13] ——— and C. Miller, On the real exponential field with restricted analytic functions, Israel J. of Maths., 85 (1994), 19–56.

[14] ——— and ———, Geometric categories and *o*-minimal structures, Duke J. 84 (1996), 497–540.

[15] ——— and P. Speissegger, The real field with convergent generalized power series is model complete and o-minimal, to appear in Trans. Amer. Math. Soc.

[16] ——— and ———, The field of reals with multisummable series and the exponential function, to appear.

[17] A. Gabrielov, Projections of semianalytic sets, Funct. Anal. Appl., 2 (1968), 282–291.

[18] ———, Complements of subanalytic sets and existential formulas for analytic functions, Inv. Math. 125 (1996), 1–12.

[19] ———, Counterexamples to quantifier elimination for fewnomial and exponential expressions, Preprint, Purdue University, 1996.

[20] L. Garding, Some points of analysis and their history, University lecture series 11, AMS, 1997.

[21] T. Huuskonen, Constants are definable in rings of analytic functions, Proc. Amer. Math. Soc. 122 (1994), 697–702.

[22] A.G. Khovanskii, On a class of systems of transcendental equations, Soviet Math. Dokl. 22 (1980), 762–765.

[23] J.F. Knight, A. Pillay and C. Steinhorn, Definable sets in ordered structures, II, Trans. Amer. Math. Soc. 295 (1986), 593–605.

[24] A. Macintyre and D. Marker, A failure of quantifier elimination, Revista Mat, 10 (1997), 209–216.

[25] S. Maxwell, A general model completeness result for expansions of the real ordered field, Annals of Pure and Applied Logic 95 (1998), 185–227.

[26] A. Pillay and C. Steinhorn, Definable sets and ordered structures I, Trans. Amer. Math. Soc. 295 (1986), 565–592.

[27] L.A. Rubel, Entire and meromorphic functions, Springer, 1995.

[28] A. Tarski, Sur les ensembles definissables de nombres reels, Fund. Math. 17 (1931), 210–239.

[29] ———, A decision method for elementary algebra and geometry, second edition revised, Rand Corporation, Berkeley and Los Angeles, 1951.

[30] C.T.C. Wall (ed.), Proceedings of Liverpool singularities — symposium I, LNM 192, Springer-Verlag (1971).

[31] A.J. Wilkie, On the theory of the real exponential field, Illinois J. Math. 33 (1989), 384–408.

[32] ———, Model completeness results for expansions of the ordered field of real numbers by restricted Pfaffian functins and the exponential function, J. Amer. Math. Soc. 9, No.4, 1996, 1051–1094.

[33] ———, A general theorem of the complement and some new *o*-minimal structures, submitted.

Printed in the United States
By Bookmasters